T0138331

METRICAL STRESS THEORY

METRICAL S•T•R•E•S•S THEORY

Principles and Case Studies

BRUCE HAYES

THE UNIVERSITY OF CHICAGO PRESS
Chicago and London

Bruce Hayes is professor of linguistics at the University of California, Los Angeles.

The University of Chicago Press, Chicago 60637
The University of Chicago Press, Ltd., London
© 1995 by The University of Chicago
All rights reserved. Published 1995
Printed in the United States of America
03 02 01 00 99 98 97 96 95 94 12345
ISBN: 0-226-32103-7 (cloth)
0-226-32104-5 (paper)

Library of Congress Cataloging-in-Publication Data

Hayes, Bruce, 1955–
Metrical stress theory : principles and case studies / Bruce Hayes.
p. cm.
Includes bibliographical references and indexes.
1. Accents and accentuation. 2. Metrical phonology.
I. Title.
P231.H38 1995
414′.6—dc20 93-34063
CIP

∞ The paper used in this publication meets the minimum requirements of the American National Standard for Information Sciences—Permanence of Paper for Printed Library Materials, ANSI Z39.48-1984.

This book is dedicated to
MORRIS HALLE

One of the things I discovered when I studied phonology with Morris was his unparalleled ability to get students excited about, and actively involved in, new topics of research. Not only did Morris's teaching give me something to work on that I found very exciting, but his encouragement made it easier for me to go out on a limb from time to time, something which hardly came naturally to me.

This book is my second effort on the problem—parametric metrical theory—that I took on in my dissertation. This time I am guided by a much more substantial literature, of which an important part is Morris's own work. Writing the book has at many times reminded me of how much I have enjoyed and benefited from arguing the issues with Morris, both during and after my days as a graduate student.

CONTENTS

Abbreviations *xiii*
Acknowledgments *xv*

1. INTRODUCTION 1
   1.1 Theoretical Content 2
   1.2 Data 3

2. DIAGNOSING STRESS PATTERNS 5
   2.1 The Phonetic Correlates of Stress 5
   2.2 A Definition of Stress 8
   2.3 An Empirical Base for the Phonology of Stress 9
      2.3.1 Diagnostic I: Attraction of Nuclear Intonational Tones 10
      2.3.2 Diagnostic II: Vowel Quality and Segmental Rules 12
      2.3.3 Diagnostic III: Non-Nuclear Intonational Tones 16
      2.3.4 Diagnostic IV: The Rhythm Rule 18
   2.4 Conclusions 21

3. BACKGROUND 24
   3.1 Typological Properties of Stress 24
   3.2 Metrical Theory: Stress as Rhythm 26
      3.2.1 Rhythmic Structure 26
      3.2.2 Stress as Rhythm 28
      3.2.3 The Limits of Rhythm in Language 30
   3.3 The Typology of Stress Rules 31
   3.4 Formal Properties of Stress Rules 33
      3.4.1 Locality 33
      3.4.2 The Continuous Column Constraint and Its Consequences 34
   3.5 Grouping in Metrical Structure 37
   3.6 The Metrical Foot 40
   3.7 The Faithfulness Condition 41

3.8 Arguments for Bracketing Structure in Metrical Theory 41
3.8.1 Deletion of Stressed Vowels 41
3.8.2 Constituency in Phrasal Phonology 43
3.8.3 "Natural Classes" in Stress Rules 45
3.8.4 Prosodic Morphology 47
3.8.5 Word Minima 47
3.8.6 Summary 48
3.9 Syllables and Syllable Quantity 48
3.9.1 The Syllable as Stress-bearing Unit 49
3.9.2 Syllable Weight 50
3.10 Parametric Metrical Theory 54
3.11 Extrametricality 56
3.12 Labeling Rules 61
3.13 Conclusion 61

4. FOOT INVENTORY 62
4.1 Some Examples of Bounded Systems 62
4.1.1 Syllabic Trochees in Pintupi 62
4.1.2 Iambs in Seminole/Creek 64
4.1.3 Moraic Trochees in Cairene Arabic 67
4.2 Justifying the Foot Inventory 71
4.3 Alternatives 75
4.3.1 Hayes 1981 75
4.3.2 HV 76
4.4 Further Arguments 77
4.5 The Rhythmic Basis of the Foot Inventory 79
4.5.1 The Iambic/Trochaic Law 79
4.5.2 The Iambic/Trochaic Law and Foot Inventories 81
4.5.3 The Iambic/Trochaic Law and Segmental Rules 82
4.6 Summary 85

5. FURTHER ELEMENTS OF THE THEORY 86
5.1 Degenerate Feet 86
5.1.1 Strong and Weak Prohibitions on Degenerate Feet 87
5.1.2 Word Minima 87
5.1.3 Degenerate Feet and Word Layer Labeling 89
5.1.4 Latin and Similar Cases 91
5.1.5 Malayalam and Similar Cases 92
5.1.6 Priority in Foot Construction 93
5.1.7 Repair of Degenerate Feet 95
5.1.8 Assessing the Case for Degenerate Feet in Weak Position 98
5.1.9 Defining "Degenerate" for Syllabic Trochee Systems 101

5.2 More on Extrametricality 105
5.2.1 Extrametricality on Multiple Levels 105
5.2.2 Extrametricality in Clash 108
5.2.3 Stray Adjunction? 108
5.3 The Unstressable Word Syndrome 110
5.4 Modes of Parsing 113
5.4.1 Persistent versus Non-persistent Footing 114
5.4.2 More on the Free Element Condition 115
5.4.3 Top-Down versus Bottom-Up Stressing 116
5.4.4 Unstressable Words from Top-Down Stressing 117
5.5 The Number of Grid Layers 119
5.5.1 Cola 119
5.5.2 Conflation 119
5.6 Further Issues in Syllable Theory 120
5.6.1 The Quantity of CVC Syllables 120
5.6.2 Assessing the Evidence for Syllable-splitting Foot Construction 121
5.7 Summary 124

6. CASE STUDIES 125
6.1 Moraic Trochees 125
6.1.1 Palestinian Arabic 125
6.1.2 Egyptian Radio Arabic 130
6.1.3 Cahuilla 132
6.1.4 Wargamay 140
6.1.5 Trochaic Shortening in Fijian and Other Languages 142
6.1.6 Maithili 149
6.1.7 Hindi 162
6.1.8 Lenakel 167
6.1.9 Other Moraic Trochee Systems 178
6.2 Syllabic Trochees 182
6.2.1 Auca 182
6.2.2 Icelandic 188
6.2.3 Other Syllabic Trochee Systems 198
6.3 Iambs 205
6.3.1 Hixkaryana and Other Cariban Languages 205
6.3.2 Rhythmic Lengthening in Choctaw and Chickasaw 209
6.3.3 Unami and Munsee Delaware 211
6.3.4 Further Algonquian Languages: Malecite-Passamaquoddy, Eastern Ojibwa, Menomini, Potawatomi 215
6.3.5 Cayuga and Other Lake Iroquoian Languages 222
6.3.6 Negev Bedouin Arabic 226

6.3.7 Syncope and Stress in Cyrenaican Bedouin Arabic 228
6.3.8 Stress and Syllabification in the Yupik Languages 239
6.3.9 Other Iambic Systems 260
6.3.10 Right-to-Left Iambs? 262
6.3.11 Even Iambs? 266
6.3.12 Conclusion 269

7. SYLLABLE WEIGHT 270
7.1 Quantity versus Prominence 270
7.1.1 How Rules Refer to Prominence 273
7.1.2 Hindi Dialects 276
7.1.3 Stress Based on Tone 278
7.1.4 Klamath 279
7.1.5 Mam 281
7.1.6 Negev Bedouin Arabic (continued) 283
7.1.7 Pirahã 285
7.1.8 Asheninca 288
7.2 Unbounded Stress Systems 296
7.3 Dual Criteria of Weight in Moraic Theory 299
7.3.1 Laryngeal Metathesis in Cayuga 301
7.3.2 Yupik Quantity 302
7.3.3 Summary and Comparison 304
7.4 Miscellanea 305

8. TERNARY ALTERNATION AND WEAK LOCAL PARSING 307
8.1 Outline of the Theory 307
8.2 Cayuvava 309
8.2.1 Analysis 310
8.2.2 Earlier Accounts of Cayuvava 312
8.3 An Alternative Version of Weak Local Parsing 314
8.4 Summary of the Arguments 315
8.5 Estonian 316
8.5.1 Data 316
8.5.2 Prince's Account 319
8.5.3 A Revised Analysis 322
8.5.4 General Issues Related to Estonian Stress 328
8.6 Other Finno-Ugric Languages 329
8.7 Sentani 330
8.8 Pacific Yupik 333
8.8.1 Words with All Light Syllables 334
8.8.2 Heavy Syllables 335
8.8.3 Phonetic Fortition 341
8.8.4 Fortition, Foot Boundaries, and Persistent Stressing 343

8.8.5 Further Phenomena in Pacific Yupik 343
8.8.6 Other Approaches; Summary 345
8.9 Winnebago 346
8.9.1 Basic Data 346
8.9.2 Dorsey's Law 351
8.9.3 The Proposals of Hale/White Eagle and HV 352
8.9.4 Miner's Analysis 355
8.9.5 Extending Miner's Account 358
8.9.6 Conclusions 364
8.10 Other Cases of Ternary Alternation 365

9. PHRASAL STRESS 367
9.1 Background 367
9.2 Phrasal Stress Operations 368
9.2.1 The End Rule 368
9.2.2 Move X 370
9.2.3 Destressing 371
9.2.4 Beat Addition 371
9.2.5 Eurhythmy 372
9.3 The Need for Beat Addition 373
9.4 The End Rule at Phrasal Levels 376
9.4.1 Making the Shorter Taller 377
9.4.2 Making the Taller Taller 379
9.5 Move X and Its Restrictions 382
9.5.1 Maximality 383
9.5.2 The Strong Domain Principle 386
9.5.3 Summary 391
9.6 More on Beat Addition 391
9.6.1 Application of Beat Addition to Deeply Embedded Constituents 391
9.6.2 Directional Effects in Beat Addition 394
9.7 Conclusion 398

10. THEORETICAL SYNOPSIS AND CONCLUSION 400
10.1 Synopsis of Formal Proposals 400
10.2 Conclusion 402

References 405
Index of Names 433
Index of Languages and Language Families 439
Index of Subjects 443

ABBREVIATIONS

abl.	ablative
acc.	accusative
aor.	aorist
caus.	causative
conj.	conjunctive
declar.	declarative
dir.	direct
excl.	exclusive
f.	feminine
fut.	future
gen.	genitive
hon.	honorific
HV	Halle and Vergnaud 1987b
immed.	immediate
imperf.	imperfect
imv.	imperative
incl.	inclusive
indic.	indicative
interr.	interrogative
intr.	intransitive
m.	masculine
n.	noun
nom.	nominative
obj.	object
obl.	oblique
part.	participle
pl.	plural
refl.	reflexive
sg.	singular
sing.	singulative
SPE	*The Sound Pattern of English* (Chomsky and Halle 1968)
subj.	subject
trans.	transitive
v.	verb
μ	mora
σ	syllable
ᴗ	light syllable
–	heavy syllable
=	superheavy syllable
<α>	designates that α is extrametrical
§	cross reference within this book
	Sources of examples are cited by the initial(s) of the author(s) of a work listed closely above.
ex.	number of an example cited from another work
p.c.	personal communication

ACKNOWLEDGMENTS

Over the many years during which this book has been in preparation, it has been a great pleasure to interact with a great number of linguists who have provided useful ideas, inspiration, data, advice on languages, and comments on individual sections. These include Filippo Beghelli, Gösta Bruce, Henry Churchyard, Abigail Cohn, Megan Crowhurst, Stuart Davis, Isidore Dyen, Caroline Fery, Paul Frank, Ives Goddard, Chris Golston, Carlos Gussenhoven, Kenneth Hale, Michael Hammond, Morris Halle, José Hualde, Ellen Kaisse, Patricia Keating, Michael Kenstowicz, Paul Kiparsky, Jeff Leer, Fred Lerdahl, John Lynch, Jack Martin, John McCarthy, Armin Mester, Karin Michelson, Kenneth Miner, K. P. Mohanan, Pamela Munro, Pramod K. Pandey, Alan Prince, Iggy Roca, Geoffrey Russom, Hansjakob Seiler, Lisa Selkirk, Sigriður Sigurjonsdottir, Norval Smith, Donca Steriade, Irene Vogel, Don Weeda, and Anthony Woodbury. Many thanks to them all, and apologies to anyone whom I may have forgotten. Special thanks to René Kager and to another, anonymous, reviewer for the University of Chicago Press, both of whose comments were extremely helpful. Thanks as well to Geoffrey Huck, Karen Peterson, and Peter T. Daniels of the University of Chicago Press, and to Allyson Carter, Maria Theresa Gelvoria, and Heather Morrison, who served as research assistants.

I would also like gratefully to acknowledge the support of the National Science Foundation (grant BNS-9007403), and several grants from the Committee on Research of the UCLA Academic Senate.

To my wife Patricia Keating go special thanks for her encouragement, moral support, and everything else.

An earlier version of portions of this manuscript appeared as Hayes 1992.

• 1 •

Introduction

Metrical stress theory is a branch of the theory of generative phonology that deals with stress patterns. The central claim of the theory, in my view, is that stress is the linguistic manifestation of rhythmic structure, and that the special phonological properties of stress can be explicated on this basis. Expressed pretheoretically, the question addressed in this book is: What is the role of rhythm in determining linguistic stress patterns? The proposed answer consists of a set of specific formal principles, which collectively define a particular version of metrical theory.

I present the theory in logical order, starting from first principles. This has two purposes: to make the book useful to readers who have no prior acquaintance with metrical theory, and to permit the reader to evaluate the theory in a systematic and rigorous way, without relying on any tacit assumptions carried over from earlier work. At the same time, I hope to give an idea of the range of current opinion in the field; thus at various points the views being argued for are compared with alternatives from the literature.

As the subtitle suggests, the book alternates theory sections with applications of the theory to the analysis of particular languages. This occasionally raises expository difficulties: all the evidence comes from the individual analyses, and I have attempted to do justice to each language in its own right. The theoretical points that a language makes may be deeply embedded in other material. For this reason the exposition contains extensive cross references. Chapters 3 through 5 present the core of the theory, giving the arguments in outline (but with few full analyses of individual languages) and cross references to the later analyses that exemplify the theoretical ideas. Chapter 6 consists of analyses, with summaries of what they argue for and cross references to the theoretical development in earlier chapters. The same procedure is followed within chapter 7 (which covers syllable quantity) and chapter 8 (ternary stress). The reader is encouraged to turn to the cross-referenced analyses if, in reading the theory sections, more illustrations are wanted. For convenience, and to permit the reader to evaluate the theory as a whole, a theoretical synopsis is provided in chapter 10.

Two other chapters are also included in the book: chapter 2 is a discussion of the phonetics of stress and how stress patterns can be inferred from the phonetic record, and chapter 9 is a discussion of phrasal stress.

Readers wishing to tailor their reading to their own background may note the following. The book does not assume any prior knowledge of metrical stress theory but does assume elementary concepts of generative phonology, as presented for example in Wolfram and Johnson 1982 or at a more advanced level in Kenstowicz and Kisseberth 1979. Occasionally reference is made to more recent theoretical developments in lexical and autosegmental phonology, covered for instance in Goldsmith 1990 and Kenstowicz 1994. The latter texts are also recommended for additional background and different perspectives on metrical theory. Readers wishing a brief overview of the present book may want to focus on chapters 2–4, chapter 5 through § 5.1.6, and representative analyses of chapter 6 (e.g. § 6.1.1, § 6.1.3, § 6.2.2, § 6.3.1, § 6.3.3). Readers with expertise in metrical stress theory may wish to skim chapter 3 and § 4.1.

To readers interested in testing out a new hypothesis in metrical theory, I modestly suggest reading the book from cover to cover, if for no other reason than to assimilate the data. One purpose of writing a book of this length is to provide an easily accessible means of testing and developing new proposals.

1.1 THEORETICAL CONTENT

To preview the main theoretical content of the book:

(a) I adopt as the representation for metrical structure the **bracketed grid,** which is a hierarchy of rhythmic beats, grouped into a hierarchy of constituents. Bracketed grids obey two fundamental principles: (i) a requirement that grid columns be continuous; (ii) a requirement that grid marks and the constituents they head be in one-to-one correspondence. A broad range of phenomena follows from these two assumptions.

(b) The smallest constituent in metrical structure is the foot. I present three commonly encountered foot types: the **moraic trochee,** the **syllabic trochee,** and the **iamb,** and argue that this small inventory suffices as a complete inventory of bounded feet, accounting for widespread asymmetries in the typology of bounded stress assignment.

(c) The basis of the foot inventory is a principle to be called the **Iambic/Trochaic Law,** which forms part of the theory of rhythm, not of language proper. The Iambic/Trochaic Law determines the set of possible feet, as well as motivating a large number of segmental rules that adjust metrical structure.

(d) Metrical structure creation is **non-exhaustive;** that is, it need not exhaust the string of syllables. This assumption simplifies the stress rules of a number of languages, and crucially allows us to describe ternary systems without expansion of the basic inventory of metrical feet.

(e) Many stress languages impose a **ban on "degenerate" feet,** that is, feet that consist of a single mora in languages that respect quantity, feet of one syllable in languages that do not. A number of consequences follow from such a ban.

(f) Syllable weight is not a unitary phenomenon; instead, languages distinguish syllable **quantity** and syllable **prominence.** Quantity is represented by mora count, and the criteria defining it vary only slightly across languages. Prominence may be based on many other properties of the syllable, and is formally represented with grid columns of varying height. Only quantity may be referred to by rules of foot construction, whereas prominence may be referred to by other metrical rule types, for example end rules and destressing.

The proposals of this book are partly my own but also reflect a great deal of work by others. To give some important examples, the idea of stress as rhythm, and its formalization in metrical stress theory, is due to Liberman 1975 and Liberman and Prince 1977; the idea of a typological theory based on foot templates originated with Prince 1976a; the form of representations used here is due to Hammond 1984a and Halle and Vergnaud 1987b; the attempt to relate foot types to word minima and prosodic morphology is based on the proposals of McCarthy and Prince 1986; and my views on degenerate feet were inspired by Kager 1989.

Since Halle and Vergnaud 1987b is (justly) a dominant work in current literature and is referred to very frequently here, it is abbreviated as HV.

1.2 DATA

While I have been able to hear taped or live data for some of the languages described here, most data are from secondary sources. To the authors of these works I owe a very large debt, and I have tried to show respect for their work by reporting their data and findings as accurately as possible. Where the family membership of a language is listed, this has been taken from Comrie 1981, Foley 1986, Grimes 1992, and the individual language sources.

For clarity, I have usually converted an author's practical orthography into a standardized phonetic notation. The version of phonetic transcription used here is the 1989 revision of the International Phonetic Alphabet (published in Ladefoged 1990), adapted to the special needs of this book. In particular, (a) [č, ǰ], not [tʃ, dʒ], are used for the palato-alveolar affricates, to avoid confusion with consonant clusters; (b) [š, ž], not [ʃ, ʒ], are used for the palato-alveolar fricatives, to parallel [č, ǰ]; (c) [y] is used for the high front unrounded glide, to avoid confusing [ǰ] with IPA [j]; (d) hence [ü, ö], not [y, ø], are used for front rounded vowels; and (e) [ʹ], [ˋ] are usually used to mark primary and secondary stress, since these are more compact and legible than IPA [ˈ], [ˌ].

One final remark on data: readers conducting their own research in metrical theory should be warned that as a reporter of other people's data, I am fallible, and they are therefore urged to consult the original secondary sources (obtainable in research libraries and by interlibrary loan), or if possible to find a native speaker consultant. Future researchers who cite this book as a data source without bothering to consult and cite the original references will hear from my lawyer.[1]

1. Actually, there is more involved here than just accuracy or giving credit where it is due: in researching this book, I often found that it was precisely by moving beyond theory-centered writings to the original sources on which they were based that the data could be found to support a sharply different analysis. It is only natural that theorists, pressed for space, will focus on the data most relevant to their own analyses. The same holds true, of course, for this book.

• 2 •

Diagnosing Stress Patterns

This chapter addresses a question that seems logically prior to any other in developing a theory of stress, namely: When we talk about stress, what is the physical phenomenon we are talking about? The definition of stress is one of the perennially debated and unsolved problems of phonetics. A body of careful experimental work has established that no one physical correlate can serve as a direct reflection of linguistic stress levels; see Lehiste 1970; Lea 1977; Berinstein 1979; Ladd 1980; Beckman 1986; and references cited there.

The absence of a clear physical definition of stress means that any theory of stress is in an indirect relation with the facts that support it. In most areas of phonology, it is not too difficult to ascertain when the observed facts confirm or falsify a hypothesis. This is because the distinctive features which form the core of phonological representations have relatively clear acoustic or articulatory correlates.

In the case of stress, however, even the facts are often in doubt. For example, mutually inconsistent systems exist for recording English stress contours (compare, for instance, Jones 1956 with Kenyon and Knott 1944). Our inquiry would obviously be served by the discovery of a clear and unambiguous phonetic correlate of stress. But the phonetic literature suggests that no such thing exists.

In this chapter, I first review the evidence against an invariant phonetic correlate for stress. Second, I try to show that the absence of such a correlate does not exclude the possibility of defining and investigating stress in a rigorous way.

2.1 THE PHONETIC CORRELATES OF STRESS

The nicest available definition of stress, if only it were true, would be the proposal made by Stetson 1928. Stetson believed that the prosodic organization of speech is reflected directly in the actions of the muscles that control expiration. Roughly, Stetson claimed that every syllable in an utterance is accompanied by a **breath pulse:** a contraction of the muscles of the chest wall, which gives the syllable peak its increased sonority over the segments of the syllable margin.

Further, Stetson believed that *stressed* syllables carry an extra-strong breath pulse, executed by abdominal rather than chest-wall muscles.

The work of Ladefoged (1967, and work cited there) refuted many of Stetson's claims. In particular, individual syllables do not have separate breath pulses, and the division of labor among the expiratory muscles is not governed phonologically as Stetson thought. However, Ladefoged did endorse Stetson's claim that stress can always be diagnosed by examining expiratory effort.

Other work, however (Lieberman et al. 1967; Lieberman 1968; van Katwijk 1974; J. Ohala 1977), has strongly argued against the Stetson/Ladefoged view of stress. In the observations made by these researchers, speakers produced "breath pulses" for a *subset* of stressed syllables (especially syllables receiving emphatic stress), but, crucially, not for all of them. That is, one often finds syllables that by other criteria must clearly be counted as stressed which nevertheless are not accompanied by a breath pulse.

Peterson 1958, 402–4, describes a paralyzed patient who had normal control of the larynx but who breathed entirely under the control of a respirator. Although this patient was "unable to produce strongly stressed speech," she could nevertheless speak in a way that "closely resembles normal conversational speech in all aspects." It appears that breath pulses (which this patient could not produce) are not obligatorily present in normal speech.

The research above can be summed up as follows: if a breath pulse is present, it is probable that stress is present, but there are many stressed syllables produced without a breath pulse.

Work in the perceptual domain has further advanced our understanding of the phonetic correlates for stress. Fry 1955, 1958 developed an experimental paradigm that used synthetic speech stimuli. The idea was to vary certain physical parameters systematically, keeping segmental content the same, and let listeners judge if the stress contour was affected. Schematic stimuli are as in (1):

(1) a. **Loudness** *PERmit* vs. *perMIT*

b. **Duration** [pr̩ːmɪt] vs. [pr̩mɪːt]

c. **Pitch**

per		*m*
mit	vs.	*per* *i* *t*

d. **Listener's judgment** *pérmit* or *permít*?

Fry found that loudness had the least effect on stress perception, despite its intuitive status as the most natural correlate of stress. Duration changes had a greater effect, with longer syllables more likely to be perceived as stressed. The strongest effects on stress perception were achieved by altering the pitch contours, as shown. Thus pitch and duration, rather than loudness, seem to be the principal perceptual cues for stress. Later work, such as Bolinger 1958; Morton and Jassem 1965; and Nakatani and Aston 1978, supported Fry's findings. Loudness does have some effect, since if one integrates loudness over the duration of the syllable, this yields a measure that correlates with stress judgments slightly better than pitch does (see Beckman and Pierrehumbert 1986, 272, and references cited there).

The multiple phonetic cues for stress, and the subordinate role of loudness, are particularly interesting when one considers that languages use duration and pitch in their phonological systems for entirely different purposes. Duration is the phonetic cue for vowel length, which is phonemic in many languages. Duration is also widely used to mark phonological phrasing: the right edges of major phrases typically receive extra duration (Klatt 1975; Wightman et al. 1992). Further, pitch is the phonetic cue for tone, in languages with phonemic tone systems, and also is the phonetic basis of intonation. The basic point is this: aside from the marginal role of loudness, stress is **parasitic,** in the sense that it invokes phonetic resources that serve other phonological ends.

Partly as a consequence of this, stress is phonetically realized on a language-specific basis. For example, languages with phonemic vowel length contrasts have been shown to avoid using duration as a correlate for stress; see Berinstein 1979. This makes sense, since using duration to mark stress in these languages would obscure the phonemic vowel length contrast.

A more dramatic example of this type can be found in Finnish. Carlson 1978 points out that Finnish emphatic stress can involve lengthening of unstressed, rather than stressed, syllables. This occurs when the emphasized word begins with a short-voweled open stressed syllable. For example, an emphatic utterance of the word *vítut* would involve a lengthened (but unstressed) second syllable: [vítuːt]. For discussion of the motivation for this pattern, see Carlson 1978 and Prince 1980.

The use of pitch to realize stress is likewise dependent on the character of a language's intonational system. For example, Chung 1983, 38, notes that in Chamorro, the standard declarative intonational contour places the lowest rather than the highest pitch on what is perceived to be the stressed syllable of a word. In English, it is also clear that high pitch does not invariantly signal stress, since in the intonational pattern for yes/no questions the stressed syllable receives the lowest pitch:

(2) *Pennsy l vá n i a ?*

It will not do either to say that stress is signaled by a "pitch excursion" or some similar notion. English has an intonational contour called "scooped" by Ladd 1980, in which the main pitch excursion comes one or two syllables after the stressed syllable:

(3) a. *I don't know*

b. *But Mánitowoc has a library!*

In fact, it is possible to construct examples from which pitch information is excluded, with the stress pattern emerging intact. Huss 1975 compared contrasting forms like *tórment* ~ *tormént,* placed after the main stress of a declarative utterance, so that the pitch contours were neutralized to a uniform low.

(4) a. *We FIRST tormént, he said.* b. *His FIRST tórment, he said.*

The stress distinction in pairs like *tórment* ~ *tormént* was systematically reflected in relatively greater duration assigned to the syllable bearing the word stress. Listeners showed only a marginal ability to recover the intended stress contour out of context, but the fact that the phonetic outputs were different shows that the speaker's own internalized representation does include a stress distinction here.

To sum up the point of these examples: it is certainly true that pitch and duration are both intimately linked with stress. However, because the relation between stress and pitch/duration is both indirect and language-specific, it is impossible to "read off" stress contours from the phonetic record.

At this point, the situation may seem rather unpleasant: How can we investigate a phenomenon that has no consistent physical correlates? There is a reasonable answer to this, which has tacitly or overtly governed the practice of many phonologists. I present this answer in two parts, providing first a definition of stress, then discussing how it can be investigated rigorously.

2.2 A DEFINITION OF STRESS

The central claim of metrical stress theory, argued in Liberman 1975 and Liberman and Prince 1977, is that stress is the linguistic manifestation of rhythmic structure. That is, in stress languages, every utterance has a rhythmic structure which serves as an *organizing framework* for that utterance's phonological and phonetic realization. One reason for supposing that stress is linguistic rhythm is that stress patterns exhibit substantial formal parallels with extra-linguistic rhythmic structures, such as those found in music and verse; see § 3.1–2, § 4.5.

If the equation of stress and rhythmic structure is valid, then we automatically account for why there is no invariant physical realization for stress. The reason is that rhythm in general is not tied to any particular physical realization;

one can detect and recognize the same rhythm irrespective of whether it is realized by (for example) drumbeats, musical notes, or speech. Because of this independence, we are not bound to the prediction that any particular phonetic correlate will invariably realize stress in any particular language.

Naturally, certain phonetic correlates serve more readily as cues for stress than others: it is only natural that strong rhythmic beats should coincide with breath pulses, with greater duration, and with raising of pitch. But these are only tendencies, and since phonology serves many ends other than rhythmic ones, they sometimes override the natural correlation between strong rhythmic beats and particular phonetic phenomena. The prediction corresponds to what we observe: there are general tendencies in how stress is manifested across languages, but nothing is iron-clad.

This conception gives rise to a practical difficulty: given that rhythm is an abstract notion that cannot be directly observed, how can we give our observations about stress a solid empirical basis? Chomsky and Halle (1968, hereafter *SPE,* 24–26) were aware of this problem, and took an approach fairly similar to that taken in syntax: if a large group of native speakers who understand the task can agree on a linguistic observation (such as a grammaticality judgment, or the presence of a particular ambiguity, or a stress contour), then we are justified in taking that agreement as a datum. This seems reasonable, and my own judgments about English stress are in broad agreement with the stress contours claimed in *SPE.* However, not everyone's judgments are, and *SPE* was criticized in the literature (e.g. Lehiste 1970; Stockwell 1972; Vanderslice and Ladefoged 1972) for its view of the data.

Part of the difficulty is that prosodic phenomena are among the least accessible to consciousness. For example, teachers of beginning phonetics often encounter students who, although native speakers of English, simply cannot hear where the main stress of an English word falls.

2.3 AN EMPIRICAL BASE FOR THE PHONOLOGY OF STRESS

The difficulties that arise from using intuitive judgments of stress are to some extent avoidable. By carefully examining and comparing various phonological diagnostics for stress, it is possible to study the stress system, at least in English, with little recourse to intuition. The basis for this is the fact that the stress phonology of English forms a tightly organized system: segmental rules, intonational patterns, and phonotactic constraints agree with each other in diagnosing a particular stress pattern. The underlying stress pattern is most clearly demonstrable in that it unifies in a coherent way a broad set of phenomena. The following discussion is intended to support this claim.

Before proceeding, a note on the data: in a number of areas to be discussed, there is variation across dialects of English. We will be considering patterns that are common in American English. Dialectal differences will affect individual arguments, but the basic points could probably be made in other dialects as well.

2.3.1. Diagnostic I: Attraction of Nuclear Intonational Tones

Most generative work on English stress has maintained that every intonational phrase (roughly speaking, every phrase of approximately sentence length; Selkirk 1984; Nespor and Vogel 1986) has one and only one primary stress. The best evidence for this comes from the English intonational system. Research on intonation, e.g. Liberman 1975 and Pierrehumbert 1980, indicates that the English intonational system can be analyzed in part as an inventory of **tunes.** These are abstract tonal sequences, each paired with a somewhat elusive meaning, which may be assigned to any given text. I present an encapsulated and simplified version of four of these tunes; see Pierrehumbert 1980 for a more serious account.

The **declarative** tune consists of the tonal sequence M H* L (Mid–High–Low), where H* is linked to a specified syllable and M and L are linked to the initial and final edges. Aligning this declarative tune with the word *assimilation* gives (5) (for association of tones to edges, see Liberman 1975; Pierrehumbert and Beckman 1988; Hayes and Lahiri 1991a):

(5)
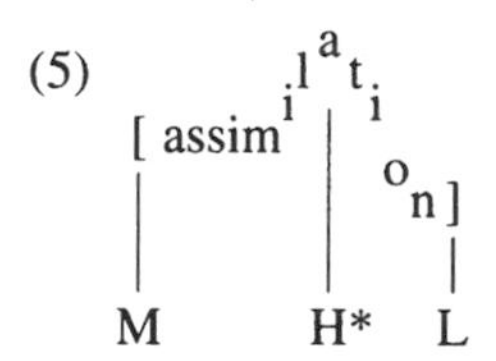

Note that between the phonological tones, pitch is in continuous movement up or down. This is analyzed by Pierrehumbert as interpolation, carried out within the phonetic component; phonologically, such sequences are toneless.

The same text as in (5) can be aligned with different tunes. For example, the **question** tune (6a) has the tonal sequence M L* H; it is used for yes–no questions and other purposes. The **downstepping** tune (6b) is H M* L; it implies that what is being said is in some sense predictable from context (Pierrehumbert and Hirschberg 1990).

(6)
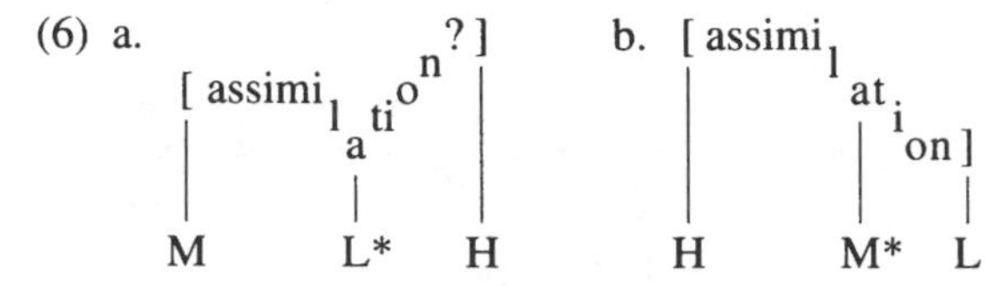

The **delayed rise** or **scooped** tune begins low, places a L*+H (rising) sequence starting on a particular syllable, and ends low. (For its meaning, see Ward and Hirschberg 1985; Pierrehumbert and Hirschberg 1990.)

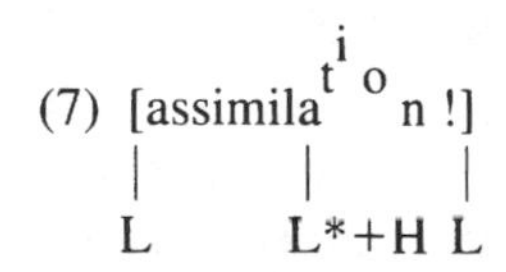

What is of interest here is the way in which the various tunes line up against different texts. Consider the three examples in (8), each lined up with all four tunes:

(8)

a. [assimilation]			b. [preliminary]			c. [pontoon]		
M	H*	L	M	H*	L	M	H*	L
M	L*	H	M	L*	H	M	L*	H
H	M*	L	H	M*	L	H	M*	L
L	L*+H	L	L	L*+H	L	L	L*+H	L

The point is that the starred tone of each tune consistently docks to the same syllable in each word, namely the one marked in dictionaries (and agreed upon by most native speakers) as the main stressed syllable: *assimi**lá**tion, pre**lí**minary, pon**tóon**.* Thus English intonation can be used as a diagnostic for the position of main stress.

These examples illustrate the view adopted here concerning pitch and stress (cf. Ladd 1980, chaps. 1–2): pitch is directly determined by the intonational system, but the rules linking tones to texts refer to the position of stress. As a result, pitch can serve as a powerful phonetic cue for stress location. However, in locations where the intonational system places no tones, pitch cannot serve as a cue for stress, and other cues such as duration take over; cf. (4) above, as well as Nakatani and Aston 1978.

The diagnostic of intonation can be used to bring evidence to bear on stress patterns that are not so clear as main word stress. For example, there is disagreement in the literature about whether simple two-word phrases, such as *tall trees* or *divine right,* bear a weak stress followed by a strong one, or two equal stresses. The intonational evidence argues for the first interpretation, because in a neutral reading the starred tone of an intonation contour must fall on the second word, as in (9a).

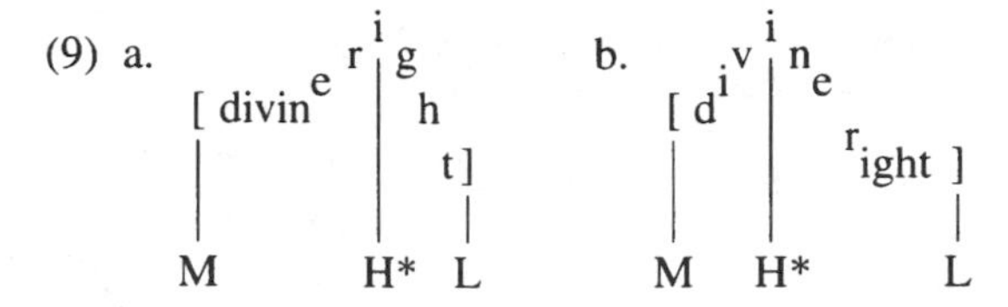

The alignment in (9b) would be well-formed only in a context in which it was being emphasized just what sort of right was at issue; that is, in a context with contrastive stress on *divine.*

2.3.2 Diagnostic II: Vowel Quality and Segmental Rules

It is a fairly uncontroversial assumption that a syllable of English is completely stressless if its vowel is schwa; examples are the boldfaced vowels in *about* [**ə**báwt], *comet* [kám**ə**t], *medicine* [mɛ́d**ə**s**ə**n], *connect* [k**ə**nɛ́kt], and *August* [ɔ́g**ə**st]. By schwa is meant not the mid back unrounded vowel of *cup* [kʌ́p], but rather a reduced vowel, which is shorter, higher, and perceptually less distinct than [ʌ]. Some dialects have two reduced vowel phonemes, /ɨ/ and /ə/; as in *American* [əmɛ́rɨkən]. I assume that both of these vowels are stressless.

If schwa is always stressless, we would expect that it would never bear main stress. The evidence of Diagnostic I from the previous section bears this out: a schwa is never assigned the starred tone of an intonational contour. If one constructs examples in which it is, they sound odd because of the phonological contradiction:

(10) a. **medícine* [mɛd ə sən]
M H* L
M L* H
etc.

b. **comét* [kam ət]
M H* L
M L* H
etc.

Thus there is reason to believe that all schwas are stressless vowels. However, it can be argued that not all stressless vowels are schwa. Stress and vowel quality in English are in a fairly intricate relationship, carefully analyzed in *SPE* and earlier work. I will not go into the *SPE* analysis but will only try to establish the existence of stressless vowels other than schwa on a rigorous basis.

The best way to do this is to use the segmental phonological rules of English as diagnostics for stress. Numerous rules of English refer to the distinction between completely stressless vowels and other vowels and thus can serve as diagnostics. I will describe four such rules here, drawing on Kahn 1976 and others. The rules are formulated in purely descriptive terms; more insightful versions would rely on syllable structure (Kahn 1976; Selkirk 1982) or on foot structure (Kiparsky 1979; Hayes 1982a).

(a) The phonemes /t, d/ may be realized as flap [ɾ] in certain environments. Word-internally, they become flap only when preceded by a vowel or glide and followed by a completely stressless vowel. I use /t/ examples here, since flapping is easier to hear for them.

(11) a. **Flapping** *t, d* → *ɾ* / [−consonantal] ____ $\begin{bmatrix} \text{V} \\ -\text{stress} \end{bmatrix}$

b. *data* [déyɾə] vs. *attain* [ətéyn]

(b) In many dialects the sequence /ns/ optionally receives a brief transitional epenthetic stop when a stressless vowel follows. I refer to this rule as **/t/ Insertion,** without intending to claim that a full /t/ is actually inserted (see Hayes 1986b; Clements 1987 for autosegmental analyses).

(12) a. **/t/ Insertion** $\emptyset \rightarrow \check{t}\ /\ n \_\_\_ s \begin{bmatrix} V \\ -\text{stress} \end{bmatrix}$

b. *Mensa* [mɛ́**nt̆s**ə] vs. *insane* [ɪ**ns**éyn]

(c) The phoneme /l/ optionally becomes voiceless through most of its duration when it occurs between /s/ and a stressless vowel (Kiparsky 1979):

(13) a. **/l/ Devoicing** $l \rightarrow [-\text{voice}]\ /\ s \_\_\_ \begin{bmatrix} V \\ -\text{stress} \end{bmatrix}$

b. *Iceland* [áys**l̥**ənd] vs. *Icelandic* [àys**l**ǽndɪk]

There is some devoicing in other contexts as well, but it is not as severe as in the context of (13).

(d) Word-medial voiceless stops are aspirated provided they are in the onset of a stressed syllable and are not preceded by a strident:

(14) a. **Medial Aspiration**

$$\begin{bmatrix} -\text{son} \\ -\text{cont} \\ -\text{voice} \end{bmatrix} \rightarrow \begin{bmatrix} +\text{spread} \\ \text{glottis} \end{bmatrix} /\ [-\text{strid}] \_\_\_ ([+\text{son}]) \begin{bmatrix} V \\ +\text{stress} \end{bmatrix}$$

b. *append* [ə**pʰ**ɛ́nd] vs. *campus* [kʰǽm**p**əs]

accost [ə**kʰ**ɔ́st] vs. *chicken* [čʰɪ́**k**ən]

Other rules could be added to this list. In general, the consonant allophones found before schwa in English differ from those found before stressed vowels.

Consider now the vowels that are not schwa, but do not bear main stress either. The chart in (15) illustrates the behavior of some of these vowels with respect to the four rules above.

(15)	/ey/	/ɛ/	/ay/	/a/
Flapping	*imitate* [ɪ́mə**tʰ**eyt], *[ɪ́mə**ɾ**eyt]	*protest*$_{\text{Noun}}$ [prówt**ʰ**ɛst], *[prów**ɾ**ɛst]	*maritime* [mǽrə**tʰ**aym], *[mǽrə**ɾ**aym]	*proton* [prów**tʰ**an], *[prów**ɾ**an]
/t/ Insertion	*compensate* [kʰámpə**ns**eyt], *[kʰámpə**nt̆s**eyt]	*incest* [ɪ́**ns**ɛst], *[ɪ́**nt̆s**ɛst]	*insight* [ɪ́**ns**ayt], *[ɪ́**nt̆s**ayt]	*consolidation* [kə**ns**alədéyšən], *[kə**nt̆s**alədéyšən]
/l/ Devoicing	*legislate* [lɛ́ǰə**sl**eyt], *[lɛ́ǰə**sl̥**eyt]	(no cases found)	(no cases found)	*Islam* [ɪ́**sl**am], *[ɪ́**sl̥**am]
Medial Aspiration	*octane* [ák**tʰ**eyn], *[ák**t**eyn]	*appendicitis* [ə**pʰ**ɛndəsáyɾəs], *[ə**p**ɛndəsáyɾəs]	*Malachi* [mǽlə**kʰ**ay], *[mǽlə**k**ay]	*coupon* [kʰú**pʰ**an], *[kʰú**p**an]

In these data, /ey/, /ɛ/, /ay/, and /a/ that lack main stress behave just like main-stressed vowels. For example, the word *imitate* shows that a /ey/ that lacks main stress blocks Flapping of a preceding /t/, just like the main-stressed /ey/ of *attáin.*

These observations lead to the following tentative conclusion: English has at least three degrees of stress. Complete stresslessness is diagnosed by the presence of schwa or by segmental rules including (11)–(14). Main stress is diagnosed by its attraction of the starred tone of the intonational contours in (8). Finally, secondary stress is diagnosed by having none of these characteristics. Henceforth secondary stress is marked with / ̀/, as in *ímitàte* or *appèndicítis.* Note that by "degrees of stress" I do not mean three absolute levels, but simply that syllable X has more stress than syllable Y, which in turn has more than syllable Z. I continue this use of the term "degree" below.

An alternative to positing three degrees of stress is to suppose that the segmental rules of (11)–(14) are triggered by the actual vowel [ə], rather than by stress. For example, Flapping would look like (16):

(16) **Flapping** *t, d* → *ɾ* / [−consonantal] ____ ə

This position is untenable, however, because schwa is not the only stressless vowel of English. A subset of the English vowels are able to appear in both main-stressed and completely stressless positions, as we have diagnosed them with intonational and segmental evidence. Although there is dialect variation, the vowels in (17) typically show this behavior:

(17) **Vowel**		**Main Stressed**		**Stressless**		
/r̩/		*burr*	[bŕ̩]	(a)	*butter*	[bʌ́ɾr̩]
		refer	[rəfŕ̩]	(b)	*cancer*	[kʰǽ**nt͡s**r̩]
				(c)	*wrestler*	[rɛ́**sl̥**r̩]
				(d)	*upper*	[ʌ́**p**r̩]
/i/	in word-final position	*bee*	[bí]	(a)	*pity*	[pɪ́ɾi]
				(b)	*fancy*	[fǽ**nt͡s**i]
				(c)	*Leslie*	[lɛ́**sl̥**i]
				(d)	*hockey*	[há**k**i]
	before a vowel	*Ian*	[íən]	(a)	*Whittier*	[hwɪ́ɾir̩]
		museum	[myuzíəm]	(b)	*fanciest*	[fǽ**nt͡s**iəst]
				(c)	*Wes-leyan*	[wɛ́**sl̥**iən]
				(d)	*Gropius*	[gró**wp**iəs]
/ow/	in word-final position	*go*	[gów]	(a)	*motto*	[máɾow]
		Trudeau	[trudów]	(b)	*Rinso*	[rɪ́**nt͡s**ow]
				(c)	*Oslo*	[á**sl̥**ow]
				(d)	*Harpo*	[hár**p**ow]

	before a vowel	*boa*	[bówə]	(a) *Ottawa* [áɾowə] (b) — (c) — (d) —
/ɪ/	before /ŋ/	*ring*	[ríŋ]	(a) *Keating* [kʰíɾɪŋ] (b) *tensing* [tʰɛ́**nt͡s**ɪŋ] (c) *whistling* [hwí**sl̥**ɪŋ] (d) *hoping* [hó**wp**ɪŋ]

In the right column of (17) are examples showing that these vowels can behave just like schwa in triggering the four phonological rules noted above under (11)–(14). For example, syllabic [ɹ̩] behaves just like schwa in triggering Flapping of a preceding /t/, as shown by *butter* [bʌ́ɾɹ̩]. Using such data, one can show that in a full taxonomy of English vowels we must distinguish three categories: always stressed (e.g., (15)); variably stressed or stressless (17), and never stressed (schwa). Since some vowels can be either stressed or stressless, and it is the stress that determines the outcome, phonological rules like (11)–(14) cannot refer to schwa vowel quality rather than stress.

To complete the picture, here is the full taxonomy of vowels as found in my own speech. Naturally, other dialects will differ (sometimes greatly) in their vowel inventories and the ways in which vowel quality can be used to diagnose stress.

(18) a. **Never Stressed:** [ə, n̩, m̩]

b. **Variable:** [ɹ̩]
[l̩] (*pull* [pʰĺ̩] vs. *apple* [ǽ**p**l̩])
[i, ow] / ____ #, / ____ V, and in prefixes
(*comprehend* = [kʰàm**pr**ihɛ́nd, kʰàm**pr**əhɛ́nd])
[ɪ] / ____ ŋ
[yu] ~ [yə] (*occupy* = [á**ky**upʰày], [á**ky**əpʰày])

c. **Always Stressed:** [ey, ɛ, æ, a, ɔ, ʌ, ʊ, u], and [i, ow, ɪ] when not in the contexts of (b)

I conclude from this that a distinction of at least three degrees of stress in English is well motivated. Note that the most important aspect of the evidence is its highly systematic nature; a large number of independent criteria agree with each other. For example, the fact that the agentive suffix *-er,* with syllabic [ɹ̩], fails to trigger Medial Aspiration in *hiker* [háy**k**ɹ̩] shows that it is stressless, which in turn explains why the same suffix triggers Flapping in *fighter* [fáyɾɹ̩] and /l/ Devoicing in *wrestler* [rɛ́**sl̥**ɹ̩]. The fact that Flapping cannot apply before stressed syllables in the same word, together with the restriction of starred tones in intonational contours, means that a main intonation peak can never occur on a medial syllable beginning in a flap, as in the hypothetical form (19).

(19) *petola*

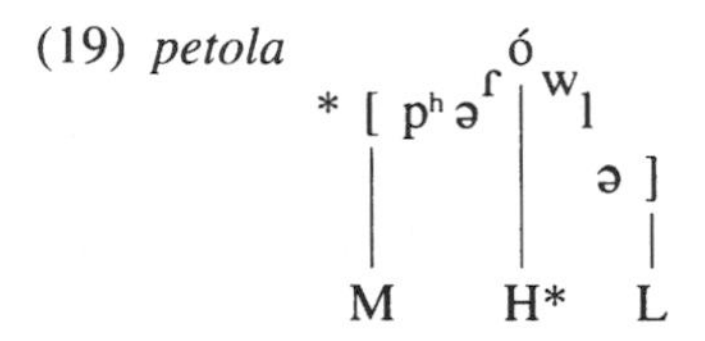

With patience, one could make a very large number of empirical connections of this sort. It is these connections that demonstrate most clearly that the stress system constitutes a unifying framework for English phonology.

2.3.3 Diagnostic III: Non-Nuclear Intonational Tones

The discussion so far has motivated three degrees of stress in English: stressless, secondary stress, and main stress. In fact, it can be argued that additional intermediate degrees exist. To do this, it will be necessary to examine more diagnostics.

All the English intonational contours diagnose the place of main stress. In addition, some of them diagnose weaker stresses as well. As an example, consider the "surprise-redundancy" contour (Sag and Liberman 1975; Liberman 1975; Pierrehumbert 1980), illustrated in (20):

(20)

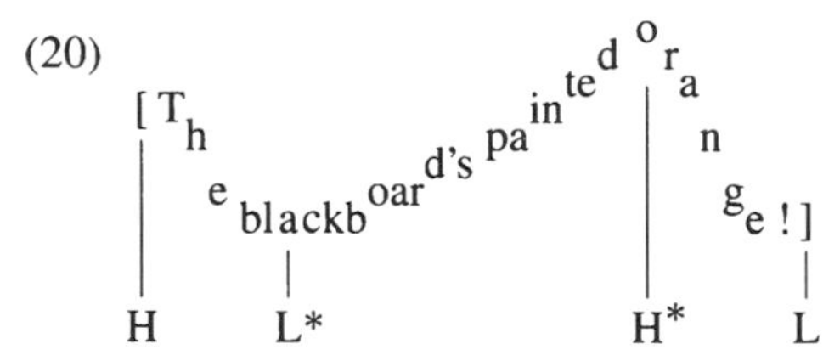

Roughly, this contour is used to express surprise or the view that one's interlocutor should already know what one is saying. (Readers not experienced in performing out-of-context intonation might try placing the examples in the context "Well, I'll be damned," ____, as a reviewer has aptly suggested.)

Note that in texts with initial stress, there is no room available for the initial H tone, and it goes unrealized:

(21)

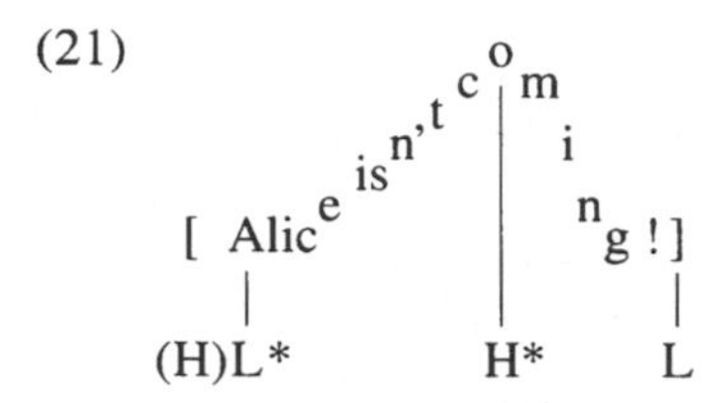

For present purposes, the crucial question is how the tune is aligned with a given text. It is straightforward to establish that the H* tone is aligned with the strongest stress. More interesting is how the L* is aligned: it is associated with the strongest stress to the left of the main stress. This can be argued for in a number of ways. First, consider utterances that contain a single long word, in which only one syllable to the left of the main stress has a non-schwa vowel:

(22) a. *collaboration!* b. *classification!*

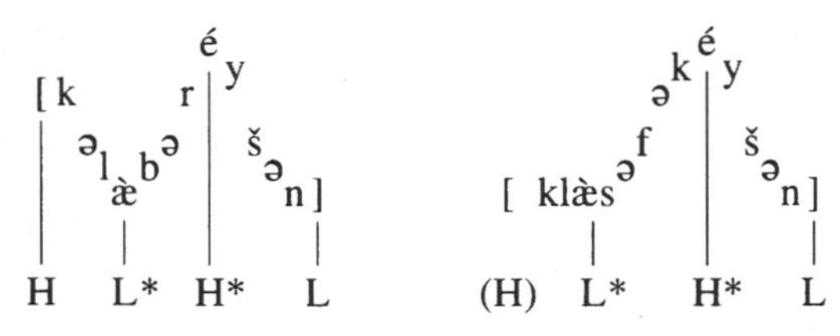

Apparently, the L* seeks out the vowel with secondary stress in preference to stressless vowels. Each word of (22) sounds odd if pronounced with the contour of the other.

Another argument that the L* tone seeks out the strongest stress to the left of the main stress is as follows. Note that the compound *blackboard* bears initial stress and that the phrase *black board* bears final stress. These statements can be justified straightforwardly by Diagnostic I (§ 2.3.1). If we place these words in an utterance to the left of the main stress, then we get a corresponding difference in the placement of the L* tone. Compare (20) with (23):

(23) 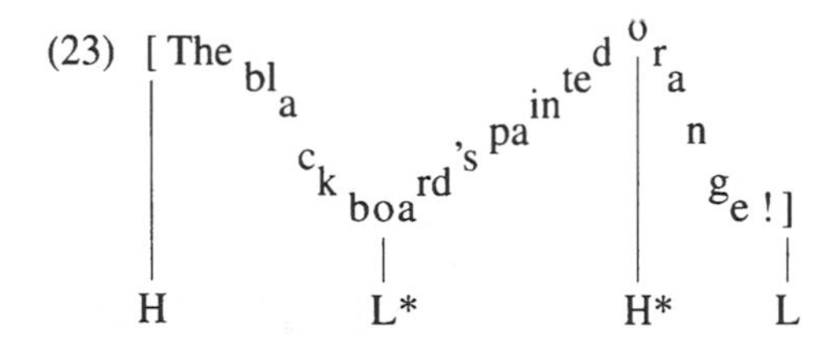

Under the plausible assumption that the stress contrast between *bláckboard* and *black bóard* persists even when neither one bears the main stress, the intonational difference follows, since the L* tone seeks out the strongest stress available to it.

At this point I have at least indirectly motivated four degrees of stress. In (23), the main stress is the first syllable of *orange;* it is stronger than the stress on *board* (Diagnostic I). *Board* bears stronger stress than *black,* since it attracts the L* tone (Diagnostic III). Finally, *black* must bear a stronger stress than the preceding word *the,* since by Diagnostic II syllables containing /æ/ have a higher degree of stress than syllables containing /ə/.

At least in some dialects, four degrees of stress can be observed in a single word. Consider the contrast in (24):

(24) a. *sensationality!* b. *Constantinople!*

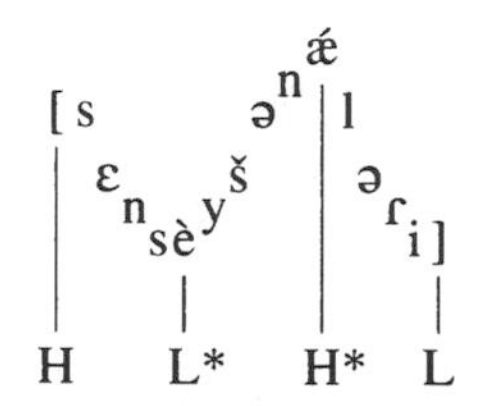

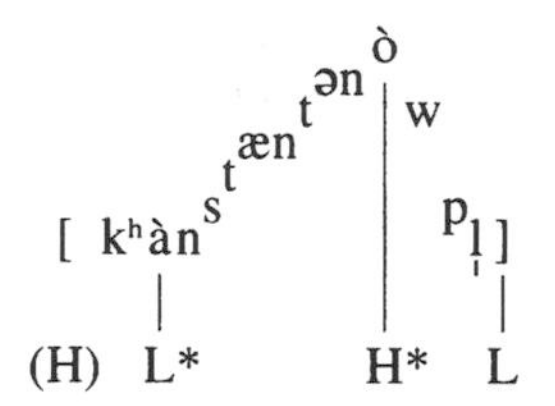

These two words begin with two weak stresses (the vowels /ɛ, ey, a, æ/ always bear some stress), followed by a stressless syllable and the main stress. Yet for many speakers the alignment with the surprise-redundancy contour differs, with the L* falling on the first syllable in *sensationality* but on the second syllable in *Constantinople* (for why this is so, see Kiparsky 1979). By the same reasoning as above, one can argue that these words show four levels of stress.

Not all dialects of English have this contrast; I have found speakers who level out the distinction in the direction of (24b). However, even these speakers arguably have at least four levels in multi-word utterances.

Another English intonation contour that can diagnose secondary stresses is the so-called "chanted vocative," which differs from unchanted contours in having sequences of level pitch, assigned by tonal spreading. Here, a H* tone falls on the main stress, and a M* tone on the strongest stress that follows the H*. This results in a different alignment when the position of the secondary stress varies, as (25a, b) show. Where there is no stress after the main stress, M* docks onto the final syllable by default, as in (25c).

(25) a. Poindexter! b. Annabel! c. Pamela!

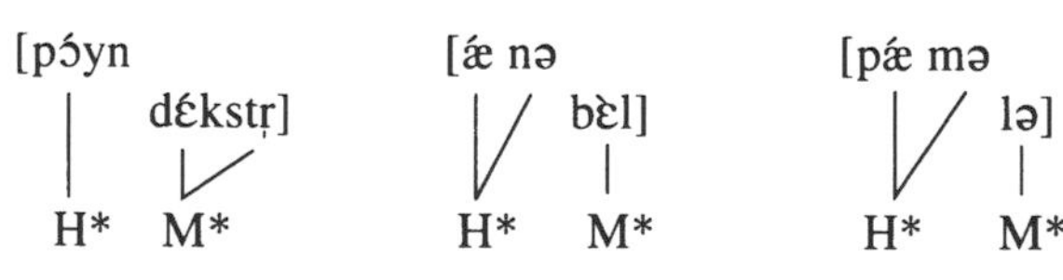

For discussion and analysis, see Liberman 1975; Ladd 1978; and Hayes and Lahiri 1991b. A similar Dutch case is analyzed in Gussenhoven 1993.

2.3.4 Diagnostic IV: The Rhythm Rule

Consider now one final diagnostic for stress in English. Like many languages, English has a rule that shifts a stress leftward when a stronger stress follows, creating alternations like *thirtéen* ~ *thìrteen mén* (rather than *thirtèen mén*). Informally, the rule can be stated as in (26):

(26) **Rhythm Rule** 3 . . . 2 . . . 1 → 2 . . . 3 . . . 1

where 1, 2, and 3 indicate the relative rankings of the stresses of a phrase. For references and full discussion of this rule, see § 3.4.2, § 9.5.

The existence of the Rhythm Rule is easily motivated by the diagnostics already presented. The various stress contrasts can be justified as follows. The word *Mississíppi* has penultimate main stress, as can be shown by Diagnostic I. From the evidence of vowel quality (Diagnostic II), the first syllable of this word bears secondary stress: [mɪ̀səsɪ́pi]. Suppose we embed *Mississippi* in the phrase *Mississippi múd.* By Diagnostic I, the main stress falls on *mud.* We can now inquire where the second strongest stress falls, using Diagnostic III, the surprise-redundancy contour.

Interestingly, there are two possible outcomes, shown in (27):

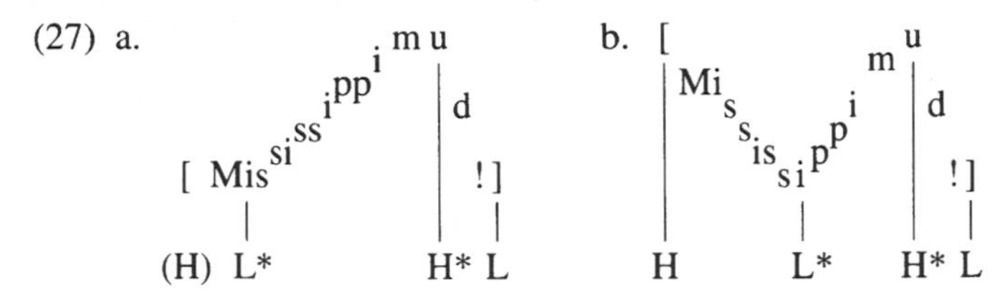

Case (27a), the more normal reading, is the result of applying the Rhythm Rule: the stress levels of the first and third syllables of *Mìssissíppi* have reversed, so that the L* of the surprise-redundancy contour seeks out the first syllable.

```
(28)  3    2    1          2      3     1
     Mississippi mud →  [ Mississippi  mud ]
                              |          |  |
                         (H) L*          H* L
```

Case (27b) shows that the Rhythm Rule is optional, at least in certain contexts: the L* falls on *-sip-*, showing that it has retained its prominence over *Miss-*.

It would be incorrect to interpret these data as simply free assignment of the L* to any stressed syllable. To see why, consider the alignments in (29):

```
(29) a.  2    3    1              b.   2    3    1
        [alligator alley! ]          * [ alligator alley! ]
         |         |     |              |    |     |     |
         L*        H*    L              H    L*    H*    L
```

The word *álligàtor* has a falling stress contour, and must retain this contour in all contexts, since there is no "Anti-Rhythm Rule" that takes 2 . . . 3 . . . 1 to 3 . . . 2 . . . 1. Accordingly, there is only one option for docking the L* tone of the surprise-redundancy contour. This shows that alignment of this tone is not free in general. The varying alignment of the surprise-redundancy contour with *Mississippi mud* must be due to varying stress.

The Rhythm Rule can be used to shed more light on the English stress system. First, it is a diagnostic for stresslessness. The reason is that the rule is unable to retract stress onto a completely stressless syllable. The inventory of names in (30) illustrates this:

(30) a.
3 2 1 → 2 3 1
Christine Smith → *Christine Smith* ([krìstín])

3 2 1 → 2 3 1
Dawnelle Allen → *Dawnelle Allen* ([dɔ̀nɛ́l])

3 2 1 → 2 3 1
Maureen Reagan → *Maureen Reagan* ([mɔ̀rín])

b.
2 1
Lamont Cranston → no change ([ləmánt])

2 1
Collette Craig → no change ([kəlɛ́t])

2 1
Patrice French → no change ([pətrís])

Again, it is possible to show that the relevant constraint applies to stressless vowels in general, not just to schwa:

(31)
2 1
Jerome Smith → no change ([ǰr̩ówm])

2 1
Pierre Jones → no change ([piɛ́r])

The Rhythm Rule can also serve as a confirming diagnostic for other stress levels. The crucial point is that the retracted secondary stress seeks out the most prominent "landing site" on its left to move to. Consider the following example. In *Sùnset Párk,* familiar diagnostics show that the main stress falls on *Park* and the second strongest stress on *Sun-*. However, since the vowel of *-set* is [ɛ], it too must bear a weak degree of stress.

Observe now that when the Rhythm Rule applies to *Sunset Park,* the stress on *Park* is retracted all the way to *Sun-*, and cannot land on *-set.* This can be shown by Diagnostic III, the surprise-redundancy contour.

(32)
3 4 2 1 → 2 4 3 1 , not 3 2 4 1
Sunset Park Zoo → [Sunset Park Zoo], not * [Sunset Park Zoo]
(H) L* H* L — H L* H* L

That is, the fact that *Sun-* has stronger stress than *-set* makes it the more qualified landing site when the stress on *Sunset Park* is retracted. I take this as evidence that the stress contrast between the two syllables of *Sunset* is preserved even when the word is embedded in a more complex expression. Notice that all four syllables in the input to (32) bear some degree of stress; if we add in a stressless syllable (e.g. *Sunset Park Gardens*), we have evidence for five levels of stress in English.

The "landing site" property of the Rhythm Rule confirms some conclusions arrived at earlier. For example, I argued that the difference between schwa and most other vowels of English involves stress as well as vowel quality. This is confirmed by the fact that secondarily stressed vowels take precedence over schwa in receiving a retracted Rhythm Rule stress:

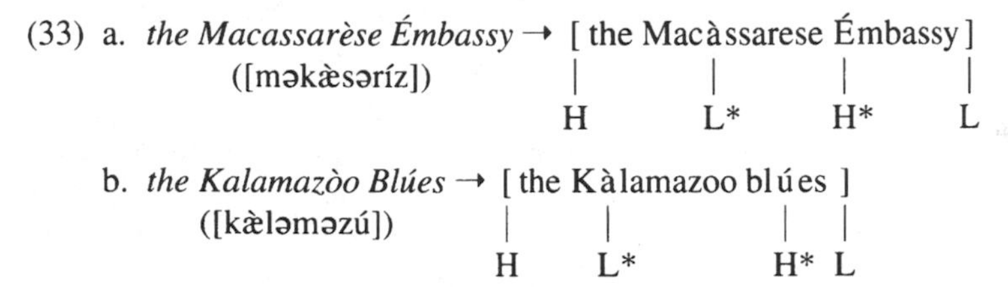

Further, the Rhythm Rule can be used to verify the contrast of secondary versus tertiary stress within the same word. Those speakers who get a stress contrast between *totàlitárian* and *Cònstantinóple* can, if pushed, apply the Rhythm Rule to these words. The result comes out as expected:

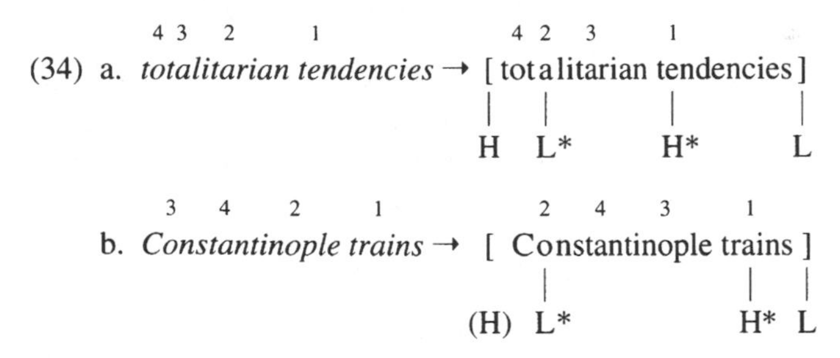

This is probably pushing the method as far as it can go, as the stress contours of (34) are somewhat strained. This is arguably due to the difficulty of applying the Rhythm Rule to sequences of this length; the most natural outcome here is not to apply the Rhythm Rule at all. However, to the extent that it is possible to apply the Rhythm Rule, the landing sites seem to be the predicted ones.

2.4 CONCLUSIONS

I have now covered several phonological diagnostics for stress in English. The overall picture is summed up as follows.

(a) The diagnostics invariably agree with each other. The evidence is strong that underlying all of these diverse phenomena is a single organizing system, namely, the stress system.

(b) Although the exposition above was devoted to the gradual discovery of more stress levels, I would not want to claim that there is a specified number of levels of stress available in English. The maximum found was five, but we looked only at utterances that are syntactically fairly simple. Slightly longer utterances may include more levels; for example (35a) includes six (the five numbered ones, plus stressless), and in (35b) it seems possible to intuit seven.

2 5 4 3 1
(35) a. *non-totalitarian tendencies*

3 6 5 4 2 1
b. *Non-totalitarian tendencies appall him.*

With utterances of increasing complexity, it gets more difficult to judge the number of stress levels; they become "bleached out" in a complex wash of pitch, duration, and rhythm.

It is not clear how such utterances should be categorized. My own view (borrowed from *SPE*) is to let the number of stress levels be determined by the theory, that is, by the most general and explanatory rules that one can devise on the basis of the simple and clear cases. As we will see, rule systems of this sort do generate a large number of levels when applied to complex utterances. Some of these levels may be essentially undetectable in the phonetic record; after all, although the phonetic resources of English for revealing stress contours are substantial, they are not infinite.

An alternative view is to limit the number of stress levels in the phonological system to the number that (it is claimed) can be clearly realized in the phonetics. The postulated number varies from author to author; for instance three (Stockwell 1972), four (Vanderslice and Ladefoged 1972), or five (Gussenhoven 1991). In such a view, the phonological rules simply stop generating more levels once the maximum is reached.

A reason to be skeptical of this view is that there is no clearly defined cutoff point where, as these accounts would predict, the phonology stops creating new levels. The data rather suggest that the phonology can create an unbounded number of levels, and that these gradually blur out as they increase in number and the phonetic cues signal them with progressively less clarity.

(c) The kind of evidence examined here does not require instruments to gather. The relevant facts of segmental phonetics and pitch contours are clear to anyone with a reasonably good ear and a little practice. Since the auditory data involve discrete patterns rather than physical quantities, they are more reliable and easier to interpret.

This is not to say that experimental work on the phonetics of stress is uninteresting; quite to the contrary. But such work is more likely to lead to insights concerning the nature of phonetic rules and speech perception, rather than to any kind of automatic diagnostic for stress.

(d) I believe a similar approach would prove fruitful in the first-hand study of stress in other languages; that is, many of the diagnostics for stress in English carry over directly, especially those involving attraction of intonational tones.

The diagnostic of schwa vowel quality can be stated in more general terms. What seems crucial concerning English schwa is not so much its actual quality (though centralization often is a characteristic of stresslessness), but the fact that the number of phonemic vowel contrasts in English is greatly reduced in stressless position. It is the reduction of contrasts that generalizes most readily across languages. For example, Italian (Vincent 1987) contrasts seven vowels in stressed position, but only five in stressless; Russian reduces its five vowel phonemes to three in stressless position (Jones and Ward 1969); and Catalan (Mascaró 1975) reduces seven vowels to three (or five in borrowed words). In Mandarin, stresslessness is marked by loss of all tonemic contrasts, in syllables bearing the "neutral tone" (Yip 1980).

(e) Lastly, one must consider how to analyze stress in other languages when one is relying on evidence from secondary sources. Such data are not necessarily unreliable. In particular, linguists describing stress in other languages often provide descriptions of its phonetic correlates and of phonological rules that diagnose stress. Where the facts reflect purely auditory intuition, we are on shakier ground. Still, it seems that most people have fairly good intuitions about stress. The studies of Lea 1977 and Thompson 1980 showed good inter-speaker reliability concerning where stress was judged to fall, even though the subjects had no phonetic training. Further, people's intuitions about stress agreed fairly well with what emerges from phonological argumentation.

The situation with regard to the data thus seems short of optimal, but still good enough to justify at least preliminary cross-linguistic investigations of the sort conducted here. The crucial point made in this chapter is that by intensive study of the intonational and segmental phonology of a language, it is possible to make the investigation of stress patterns more rigorous. It is to be hoped that further study of this nature will be carried out in the future, strengthening the empirical basis for theories of stress.

• 3 •

Background

This chapter begins with the typology of stress and stress rules, then treats the basic ideas of metrical theory, and concludes with specific theoretical proposals that will be important in what follows.

3.1 TYPOLOGICAL PROPERTIES OF STRESS

The preceding chapter reviewed evidence that stress has distinct phonetic characteristics that motivate assigning it a rather different phonological representation. The phonological patterning of stress leads to the same conclusion, as is argued in this section, based on earlier literature. The typological patterns discussed here provide a number of criteria of adequacy for proposed representations of stress.

(A) CULMINATIVITY. One distinctive phonological characteristic of stress is that it is normally **culminative,** in the sense that each word or phrase has a single strongest syllable bearing the main stress (Liberman and Prince 1977, 262).

The requirement of culminativity typically exempts grammatical words; consider English *the,* normally realized as stressless [ðə]. Arguably, this is because grammatical words are phonologically cliticized to neighboring content words (e.g. as in Hayes 1989a); that is, the requirement of culminativity applies to phonological, not grammatical words. Note that when function words are pronounced alone, they are typically stressed. For example, the isolation form of *the* is [ðʌ́ː] or [ðíː].

The domain of culminativity may differ from language to language. For example, in English, stress is culminative at the word level (every content word has a single strongest stress), at the level of the intonational phrase, and possibly at other levels as well, such as the phonological phrase (Nespor and Vogel 1986). In French (Dell 1984) and Italian (Nespor 1988, 225–26), stress is culminative at the phrasal level, but not necessarily at the word level, since rules of destressing may eliminate word stresses on the surface.

Possible exceptions to culminativity are noted in the literature. For instance,

Dixon 1977 claims that in Yidiɲ (§ 6.3.9), the multiple stresses that occur in a long word are equal in prominence. The same is said for Central Alaskan Yupik by Woodbury 1987 and for Tübatulabal by Voegelin 1935. Further cases are noted by Hyman 1977b, 38–39. Such cases often lead to disagreement among linguists. For instance, Kenneth Hale (p.c.) notes that in his fieldwork with Yidiɲ speakers, he heard the first of the several stresses in a long word as stronger than the others. Miyaoka 1985 argues that the rightmost stress of a Central Alaskan Yupik word is the strongest; and there is some evidence (§ 6.3.10) that in Tübatulabal, the rightmost stress in a word is stronger than the others. In any event, in the version of metrical theory adopted below, these languages still respect culminativity at least in a modest sense: they all involve parsing each word into metrical feet (§ 3.6) and display culminativity at the foot level. The overall picture, then, is that culminativity may be a universal of stress systems, which is subject to parametric variation for the level at which it holds.

Some languages have been claimed to lack culminativity at all levels; that is, there can be completely stressless utterances. Bickmore 1989, 1992 argues that certain words of Kinyambo are completely stressless. However, since under Bickmore's analysis Kinyambo prosody involves a late conversion from stress to tone, Kinyambo does not violate culminativity in surface forms, but only at the abstract level in which its phonology acts as a stress system. Other languages claimed to have atonic content words include Cayuga and Seneca (§ 6.3.5), Sierra Miwok (§ 6.3.9), and Yupik Eskimo (§ 6.3.8.8). In all of these, I will suggest a tonal account of the facts that preserves the principle of culminativity.

(B) RHYTHMIC DISTRIBUTION. Stress is **rhythmically distributed** (Selkirk 1984), in the sense that syllables bearing equal levels of stress tend to occur spaced at roughly equal distances, falling into alternating patterns. Thus in many languages, six-syllable words are regularly assigned the stress pattern σ́ σ σ́ σ σ́ σ (/σ/ = syllable); but there appear to be no languages in which six-syllable words regularly receive the pattern σ σ σ σ́ σ́ σ́.

Rhythmic distribution of phonological properties other than stress, at least in most cases, involves phenomena that are themselves stress-conditioned. Thus in long English words such as *Apalachicola* [æ̀pəlæ̀čəkólə], the vowel [ə] occurs spaced every other syllable; but this is obviously due to the restriction of schwa to stressless syllables in English, and is therefore parasitic on the even spacing of stress.

(C) STRESS HIERARCHIES. Stress is **hierarchical** (Liberman and Prince 1977, 262), in the sense that most stress languages have multiple degrees of stress: primary, secondary, tertiary, and so on. Such degrees of stress can appear within the phonology, rather than being the result of late phonetic rules. In contrast, ordinary features have a limited, predetermined number of contrasting phonological values, held by some scholars to be just two. An example of multiple "deep" phonological values for stress was presented in the preceding chapter: *Constantinople* (23010) contrasts with *sensationality* (320100), arguably as a result of the cyclic derivation of the latter from *sènsátion.*

(D) LACK OF ASSIMILATION. To my knowledge, it is an exceptionless phonological universal that stress does not assimilate. That is, a stressed syllable does not induce stress on the immediately preceding or following syllable. In this respect stress differs from most substantive features: assimilation of [round] or [back] or [nasal] and so on is a characteristic phonological process. Note that this is not a corollary of (B) above: in exceptional cases, languages do tolerate stresses on adjacent syllables, but the lack of stress assimilation appears to be absolute.

3.2 METRICAL THEORY: STRESS AS RHYTHM

Metrical stress theory posits that the phonetic and phonological differences between stress and ordinary features can be best accounted for if one abandons the assumption that stress is a feature. Instead, the theory represents stress as a hierarchically organized rhythmic structure (Liberman 1975; Liberman and Prince 1977). The discussion of rhythm that follows is based largely on Liberman 1975 for language and Lerdahl and Jackendoff 1983 for music.

3.2.1 Rhythmic Structure

Consider the rhythm of a simple nursery tune:

(1)

```
×                                       ×
×                   ×                   ×                   ×
×         ×         ×         ×         ×         ×         ×         ×
×    ×    ×    ×    ×    ×    ×    ×    ×    ×    ×    ×    ×    ×    ×    ×
This      old       man,                he        played    one,

×                                       ×
×                   ×                   ×                   ×
×         ×         ×         ×         ×         ×         ×         ×
×    ×    ×    ×    ×    ×    ×    ×    ×    ×    ×    ×    ×    ×    ×    ×
He        played    knick-    knack     on        my        thumb,    with a

×                                       ×
×                   ×                   ×                   ×
×         ×         ×         ×         ×         ×         ×         ×
×    ×    ×    ×    ×    ×    ×    ×    ×    ×    ×    ×    ×    ×    ×    ×
Knick-    knack,    pad - dy  wack,     give your dog  a    bone,
```

```
×                               ×
×               ×               ×               ×
×       ×       ×       ×       ×       ×       ×       ×
×   ×   ×   ×   ×   ×   ×   ×   ×   ×   ×   ×   ×   ×   ×   ×
This    old     man     came    rol-    ling    home
```

Above the words is placed a **metrical grid,** which is intended to depict the temporal structure of the tune. What we see is a sequence of beats (**grid columns**), equally spaced in time, which vary in strength according to their height. Beats alternate in prominence, and syllables and notes tend to begin on stronger beats.

A crucial aspect of a grid is that the **rows** are just as relevant as the columns. For every row of the grid above, the reader will find that it is possible to tap in a natural way, one tap per grid mark, in time to the music. (Tapping to the lowest row may require slowing down the tempo.) Going down through the columns, each level of tapping proceeds twice as fast as the level above. Such behavior is evidence that the various levels of rhythm are intuitively present for the listener, even when not every beat is signaled by a note onset.

Following Halle and Vergnaud (1987b, hereafter HV), I refer to each row of the grid as a **layer.**

Grids such as (1) are useful in characterizing various rhythmic phenomena; see in particular Lerdahl and Jackendoff 1983. To give a simple case, I suggest that syncopation can be roughly defined as the placement of a note onset on a relatively weak early beat when a later stronger beat is available. For example, Pete Seeger (1954) sings the nursery rhyme above with a syncopation on *this old man:*

```
(2) ×
    ×               ×
    ×       ×       ×       ×
    ×   ×   ×   ×   ×   ×   ×   ×
    This old        man,
```

A number of general characteristics of rhythm can be seen in (1). First, rhythmic structure is **hierarchical,** with sequences of beats having multiple levels of strength. Second, rhythmic structure obeys a tendency to **even spacing** at all intervals of repetition. Third, rhythm obeys a law of **downward implication:** any beat on a high (i.e. broadly spaced) layer must also serve as a beat on all lower layers. Thus (3) is a logically conceivable rhythmic structure, but appears to be unattested.[1]

```
(3) ×        ×        ×        ×        ×        ×        ×
    ×     ×      ×     ×     ×      ×     ×     ×      ×     × . . .
```

1. "Two against three" structures can occur, but only when two musical voices are present; arguably, each voice has a distinct rhythm.

The three crucial characteristics of rhythm just given are aptly summarized by Liberman and Prince's phrase "hierarchy of intersecting periodicities" (1977, 333): the word "intersecting" crucially conveys the notion of downward implication. These characteristics are invoked for musical rhythm by Lerdahl and Jackendoff 1983, 19–20, and appear to characterize much verse rhythm as well (Piera 1980; Hayes 1988).

A caveat concerning the conception of rhythm just presented is that while it holds for many musical traditions, it is not universal. For discussion of differing, non-hierarchical rhythmic systems, see Liberman 1975, 179–82, and Lerdahl and Jackendoff 1983, 18. The extent to which alternative conceptions of rhythm are linguistically relevant remains unexplored.

3.2.2 Stress as Rhythm

Consider now a grid employed as a representation of stress:

(4)

```
                   x
x        x         x
x   x    x   x     x   x
x x x x  x x x  x x x  x x
twenty-seven Mississippi legislators
```

At first glance, grid formalism may appear to be a notational variant of a multivalued stress feature: for example, we could roughly translate a grid into an *SPE* stress representation by assigning the syllable with the highest column the value [1 stress], the syllable(s) with the second highest column(s) [2 stress], and so on. But when we consider the grid as a rhythmic structure, its essentially different nature becomes clear.

First, it is not just the columns of the grid that are phonologically relevant, but the rows as well. For example, the grid in (4) indicates that the sets of syllables {*twen-*, *sev-*, *Mis-*, *-sip-*, *leg-*, *-lat-*} and {*twen-*, *Mis-*, *leg-*} are phonologically relevant. The cross-linguistic patterning of stress suggests that such sets (i.e. sets defined as containing all the syllables with a given degree of stress or greater) are indeed phonologically relevant. Rules of stress assignment, of intonational pitch accent association, and of rhythmic adjustment (chap. 9) all characteristically refer to notions such as "rightmost syllable with at least *n* degrees of stress" or "consecutive syllables bearing at least *n* degrees of stress." Such notions have a straightforward expression when they are formalized on a grid layer but seem complex and arbitrary when stated on a sequence of [stress] feature values.

Note further that rules referring to predicates such as "consecutive syllables bearing *n* stress OR LESS" or "rightmost syllable bearing AT MOST *n* stress" cannot be straightforwardly formulated in grid notation. Rules of this sort in fact appear to be missing from natural languages.

The relevance of grid rows is a notable characteristic of rhythmic structure in general. In stress patterns, such sequences of beats seem to have a perceptual reality even for speakers who cannot hear stress, as I have found in teaching beginning phonetics. For example, if a speaker of English is given the word *reconciliation,* which has the stress pattern shown in (5), (s)he will find it natural to tap once, twice, three times, or six times, as indicated; but the task of tapping four or five times produces confusion.

```
(5)
                                        ×
                         ×              ×
                         ×         ×    ×
                         ×   ×     ×  × ×    ×
                         r e c o n c i l i a t i o n
        1 tap:                          T
        2 taps:          T              T
        3 taps:          T         T    T
        4 taps:                   ???
        5 taps:                   ???
        6 taps:          T   T     T  T T    T
```

The reason should be clear: there are grid layers corresponding to one, two, three, and six beats, but none for four or five.

A second way in which grid notation differs from a [stress] feature is that the absolute height of a grid column has no intrinsic significance. A three-mark grid column simply indicates a syllable that is rhythmically more prominent than a two-mark column and less prominent than a four-mark column in the same representation. In contrast, the values of the stress feature proposed in *SPE* are intended to be phonologically relevant; and *SPE* indeed contains rules that refer to [3 stress] and other values.

A slight caveat should be mentioned here: phonological rules may refer to notions such as "strongest stress of a word" or "strongest stress of a phrase." However, it is clear that such notions are distinct from actual levels of stress.

A metrical representation of stress allows the special properties of stress reviewed above to be stated perspicuously, as follows.

(A) CULMINATIVITY. Culminativity in stress systems can be deduced from a metrical representation, given certain postulates (Liberman and Prince 1977, 263). The crucial assumption is that parsing of phonological material into hierarchical structure is obligatory at all levels. Examples of this principle in other domains include syllabification (see for example Ito 1986), or the theory of phonological phrasing (i.e. the Prosodic Hierarchy: Selkirk 1980a; Nespor and Vogel 1986; Hayes 1989a).

To bring this principle to bear on metrical phonology, I assume that metrical grids are not simple sequences of columns, but in fact relational structures. Given two grid marks $\times_1$ and $\times_2$ on the same layer, one of the following must hold: either one mark is subordinated to the other, or both marks are subordinated to a third.

(6) a. $\times$ b. $\times$ c. $\times$

$\times_1$ $\times_2$ $\times_1$ $\times_2$ $\times_1$ $\times_2$ $\times_3$

In other words, the forms in (6) depict not simply a sequence of stacks of $\times$'s, but a relative prominence relation between the grid marks on the subscripted layer. This notion of stress as structure will become clearer below in § 3.5, which discusses bracketing structure in metrical theory.

If one assumes that stress involves structure, and that structure is created exhaustively, it is easy to see that stress will be culminative within the domain of the stress rules: since prominence relations are obligatorily defined on all layers, then no matter how many layers there are, there will be a topmost layer with just one grid mark.

In contrast, genuine phonological features such as [round] or [nasal] are not distributed culminatively because they do not form hierarchical domains across the phonological string.

(B) Rhythmic Distribution. As we have seen, even spacing of beats is characteristic of rhythm in other domains. In addition, note that equal spacing of stresses tends to occur at multiple levels, as can be seen in (4); see Hayes 1984 for further discussion.

(C) Multiple Levels. The existence of multiple levels of stress reflects the hierarchical nature of rhythmic structure (Liberman and Prince 1977, 263).

(D) Lack of Assimilation. The absence of stress assimilation follows from the absence of a feature [stress] to assimilate. In principle, we might expect grid marks to be associated to more than one syllable, but this would go against the nature of rhythmic structure: a rhythmic beat, which is what a grid mark represents, forms a point in time rather than a sequence (Lerdahl and Jackendoff 1983, 18).

The general picture is that a metrical representation is to be preferred to a stress feature for its ability to capture the typological properties of stress.

3.2.3 The Limits of Rhythm in Language

Before going on, a caveat may be useful: while stress appears to be generally rhythmic in character, it should not be imagined that all natural language stress patterns will sound like musical sequences, with perfectly regular intervals. (Speech deliberately produced in this way in fact sounds quite odd.) Dauer 1983, surveying the literature, notes that there is at best only weak evidence to support a tendency toward physically even spacing of stresses. Rather, the even-spacing effect of rhythm makes itself felt primarily in two other areas: (a) the phonological rules of languages tend to space stresses evenly; (b) in the

PERCEPTUAL domain, listeners hear stresses as more evenly spaced than they really are (Lehiste 1977; Donovan and Darwin 1979; Darwin and Donovan 1980); that is, they impose a regular rhythm on the incoming physical signal.

Functionally, there is good reason for stresses not to be produced with perfect rhythm. Stress serves multiple purposes: it creates phonemic contrasts, marks morphological and syntactic structure, signals the distribution of focus, and so on. The timing or duration of speech sounds is also multifunctional, as noted in § 2.1. Thus many factors compete with rhythm in determining the location and timing of stresses, and the outcome can be thought of as a compromise between differing goals. The basic claim being made here is that rhythm plays an important role in stress placement, but it is by no means the only factor.

3.3 THE TYPOLOGY OF STRESS RULES

In this section, I return again to typology, this time focusing not on surface distribution of stress, but rather on the rules that assign it.

In many languages, stress is predictable and therefore can be accounted for by phonological rules. In Polish, for example, words of more than one syllable have penultimate stress; and other languages have different rules, sometimes far more complex. Stress relations among words at the phrasal level are also rule-governed: for example, in English, syntactic phrases typically have a rising stress contour: *tàll trées.* This section focuses primarily on the basic typology of word stress rules; for the typology of stress rules at the phrasal level, see § 9.1–2.

(A) FREE VERSUS FIXED STRESS. The oldest notion in stress rule typology is that of **fixed** versus **free** stress languages. These terms simply refer to the phonemic status of stress in a language: fixed stress is predictable in its location, and may be derived by rule, while free stress is unpredictable and must be lexically listed. I assume that this is typically carried out by placing metrical structure in lexical entries, though as we will see there are other possibilities as well.

Like most such typological distinctions, the fixed/free stress opposition is a blurry one. For example, many languages have phonemic stress, but the possible locations of stress are limited to only certain positions within a word, for instance in Spanish to the last three syllables. In languages with phonemic primary stress, secondary stress often is predictable. Polish (§ 6.2.3) has predictable penultimate stress in most of its vocabulary, yet tolerates a set of mostly borrowed words with antepenultimate stress.

(B) RHYTHMIC VERSUS MORPHOLOGICAL STRESS. Independent of the free/fixed division is a division of stress systems into what I call **rhythmic** and **morphological** varieties. In a rhythmic stress system, stress is based on purely phonological factors, such as syllable weight (§ 3.9.2) or limitations on the distance between stresses and between stress and word boundaries. In a morphological system, stress serves to elucidate the morphological structure of a word. Often, a particular syllable of the root bears the main stress, and affixes are stressless or bear weak stress. The productive morphology of English (i.e. the Level II morphology; Kiparsky 1982a) is a good example of a morphological stress system: main stress falls on whatever syllable of the stem is assigned main stress at the stem level (i.e. Level I), and most affixes are subordinated to this main stress. Thus the fact that *un#bóund#ed#ness* has antepenultimate stress has nothing to do with any rhythmic principles, but merely reflects that fact that the stem syllable is in antepenultimate position in this word.

In another type of morphological system, surface stress is the result of a complex interplay of stem type (accented vs. unaccented) and diacritic properties of affixes: affixes can be inherently stressed, inherently stressless, can remove stresses from the domain to which they are attached, assign a stress to the preceding syllable, and so on. Such systems often have a rhythmically determined default pattern, which is found where none of the morphemes of the word asserts its own accentual preferences. A system of this sort has been described for Indo-European and various of its daughter languages in the work of Halle and others (Halle and Kiparsky 1977, 1981; Kiparsky 1982b; HV). Other systems of this sort are found in Modern Hebrew (Bat-El 1992), Pashto (Bečka 1969), and Cupeño (Hill and Hill 1968).

Naturally, the notions of "morphological" and "rhythmic" stress system are not usually manifested in pure form; most stress systems are a mix of the two. For example, English has a rhythmic stress system in stems (Level I) but mostly morphological stress for productive affixes (Level II).

Morphological stress systems, particularly of the stem-stress type, are fairly common but in my opinion have played a less significant role in the development of parametric metrical theory, since the greatest strength of the metrical approach has been in describing rhythmic influences on stress. I will not treat morphological systems here in detail; see the references just cited and Halle and Vergnaud 1987a for interesting proposals in this area.

(c) Bounded and Unbounded Stress. Consider now the typology of rhythmic systems. These can be roughly divided into **bounded** and **unbounded** types. In a bounded stress system, the stresses fall within a particular distance of a boundary or another stress; stem stress in English (i.e. Level I stress) is an example. In an unbounded system, stress can fall an unlimited distance from a boundary or another stress, provided the appropriate conditions are met. An example of an unbounded system would be the following: stress the rightmost heavy syllable in a word; if there is no heavy syllable, stress the initial syllable. (For "heavy", see § 3.9.2.) A survey of unbounded systems appears in § 7.2.

For a time, unbounded systems were of great interest because they appeared to follow a universal pattern: all languages with "rightmost heavy" conditions appeared to have "leftmost" as their default clause for words with no heavy syllable; and languages with "leftmost heavy" clauses had "rightmost" as their default case. However, this universal has turned out to be an accident of the order in which theorists located the relevant examples in the literature; to my knowledge, there are about as many cases in which the default stress position for all-light words is exactly the opposite, namely, leftmost in leftmost-heavy systems and rightmost for rightmost-heavy systems (see § 7.2). In other words, all four logical possibilities derivable from the choices (right/left)most heavy, (right/left)most default, actually occur.

This completely symmetrical pattern makes it difficult to argue for any really explicit mechanism to derive unbounded stress. Since the facts in this area are quite simple and fill out the logical possibilities, it is hard to develop a theory that goes much beyond just describing the facts. I suggest a particular mechanism, modeled on the proposal in Prince 1983a, in § 7.2.

The cases covered intensively in this book involve only a subset of the typology just outlined: I focus on BOUNDED, RHYTHMIC stress systems, or in the case of mixed systems, the parts of the system that are bounded and rhythmic. It is these systems where rhythmic structure seems to have the most pervasive influence on stress; thus these are the systems that arguably are of the greatest interest from a metrical perspective.

3.4 FORMAL PROPERTIES OF STRESS RULES

I now turn from the descriptive generalizations of stress rule typology to the formal properties of stress rules as expressed in metrical theory.

3.4.1 Locality

Rules of stress assignment differ formally from rules that manipulate features, in several ways. While most featural rules obey constraints of locality, applying between adjacent segments or syllables, stress rules are often non-local. For instance, the rule assigning phrasal stress in English applies roughly as follows: it locates the main stress of the rightmost word in a phrase and promotes it to the strongest of the phrase. In order to locate its target, this rule must often scan several syllables, as in the phrase *hypothètical ímitators,* where the target syllable *im* is four syllables from the end. Other rules of stress assignment are comparably non-local in their application.

The non-locality of stress rules can be explicated on the basis of the hierarchical nature of metrical representation (Liberman and Prince 1977, 262–63), because in a metrical grid, the higher layers are phonologically relevant. The English phrasal stress rule amplifies the highest grid mark that is string-adjacent to the right phrase boundary (Prince 1983a, 27).

(7) a.
```
        ×    ×
 ×      ×    ×   ×
 × ×  ××× ×  ×××
hypothetical imitators
```
→ b.
```
             ×
        ×    ×
 ×      ×    ×   ×
 × ×  ××× ×  ×××
hypothetical imitators
```

For purposes of locality, the rule need only "see" the highest layer of the input grid. On this representation, its application is local:

(8) a.
```
[    ]
[× × ]
```
→ b.
```
[   ×]
[× × ]
```

The ability of metrical representations to capture superficially non-local rule application with local mechanisms is quite crucial in light of other developments in modern phonological theory. In particular, it appears that long-distance SEGMENTAL phenomena (e.g. assimilation and dissimilation) are likewise best treated as only apparent; they, too, constitute local phenomena when treated within a hierarchical model of segmental representation (see, for example, Clements 1985 and Steriade 1987). At this stage, many investigators believe that locality is a property that holds for all areas of phonology, and that the formal devices proposed in *SPE* that achieve non-local rule application on a brute force basis (e.g. parentheses, variables, ⟨ ⟩, ( )*) should be dispensed with. Metrical theory forms part of a general research program to define the ways in which phonological rules may apply non-locally by characterizing such rules as local with respect to a particular representation.

3.4.2 The Continuous Column Constraint and Its Consequences

Another property of stress rules is a tendency to exaggerate pre-existing contrasts, by making strong syllables stronger and weak ones weaker. Part of the research program in metrical phonology, initiated by Prince 1983a, is to deduce this pattern from basic theoretical postulates.

The central idea here is to impose an inviolable constraint on metrical representations, to the effect that a grid column may never have a gap. This idea, due to Prince 1983a, 33, can be expressed as in (9):

(9) **Continuous Column Constraint**
A grid containing a column with a mark on layer $n + 1$ and no mark on layer n is ill-formed. Phonological rules are blocked when they would create such a configuration.

Intuitively, the Continuous Column Constraint imposes the following requirement: if a syllable forms a rhythmic beat on a given layer, it must also form a rhythmic beat on all lower layers. As we have seen (§ 3.2.1), this principle of downward implication seems to be a basic characteristic of rhythmic structure, and thus arguably has extralinguistic motivation. In what follows, I discuss a number of areas in which the Continuous Column Constraint predicts the pattern of making the strong stronger and the weak weaker.

(A) LANDING SITES FOR STRESS SHIFT. One such area involves cases of stress shift, as in the Rhythm Rule of English (§ 2.3.4). This rule accounts for the appearance of leftward-shifted stress in certain words and phrases when a stronger stress follows, as in *thirtéen,* but *thìrteen mén;* see § 9.5 for full discussion. Prince 1983a, 33, suggests that such rules of stress shift take the form **Move X.** This rule schema can be defined as in (10):

(10) **Move X**
Move one grid mark at a time along its layer. Where Move X resolves a stress clash, movement must take place along the row where the clash occurs.

As Prince points out, the Move X schema and the Continuous Column Constraint together make the following prediction: if there is more than one possible "landing site" for a retracted stress, the site onto which stress actually retracts must be that which already bears the strongest pre-existing stress (Prince 1983a, 33). Consider for example the phrase *Sunset Park Zoo,* from § 2.3.4. In this form, the retracted stress must land on *Sun-* rather than *-set:*

(11) a.
```
                  ×                   ×
              ×   ×            ×      ×
     ×        ×   ×    →       ×    × ×
     ×   ×    ×   ×            × ×  × ×
     Sunset Park Zoo           Sùnset Park Zóo
```

b.
```
                  ×                   ×
              ×   ×              ×    ×
     ×        ×   ×    →       ×    × ×
     ×   ×    ×   ×            × ×  × ×
     Sunset Park Zoo          *Sunsèt Park Zóo
```

As can be seen in (11b), if the moved /×/ were to lodge on the syllable *-set,* it would create a discontinuous grid column. Thus the Continuous Column Constraint predicts the correct landing site, *Sun-*. Note that this is an instance of "making the strong stronger."

The pattern of seeking out the highest available landing site for retracted stresses is found widely across languages. Move X and the Continuous Column Constraint derive the pattern from general principles, which is clearly an improvement over stating it as a set of language-particular idiosyncrasies.

(B) IMMOBILITY OF STRONG BEATS. Another universal predicted by Move X and the Continuous Column Constraint is that when stress clashes are resolved, it is always by movement of the weaker, rather than the stronger stress. Since Move X must occur within the layer of the clash (i.e. where neighboring × 's occur), movement of the stronger stress cannot take place without creating a discontinuous column (Prince 1983a, 33). Thus, for example, movement is impossible in (12):

```
(12)         ×                       ×                             ×
             × ×                           × ×              ×          ×
       ×     × ×  ×       →        ×       × ×  ×           ×      × ×  ×
       ×   × × × ×××               ×     × × × ×××          ×    × × × ×××
    kangaroo imitators          *kangaroo imitators,     *kangaroo imitators
```

In this example, neither of the two highest × 's on the syllable *-roo* is movable. Assuming with Prince that only one × may move at a time, the impossible output **kángaroo ìmitators* is avoided; and similarly in other cases.

(C) MAIN STRESS PLACEMENT. Another pattern of the "strong gets stronger" type is as follows: stress rule A selects some subset of the syllables of the word to bear stress; stress rule B then selects the left- or rightmost of these stressed syllables as the main stress. This is shown schematically in (13):

```
(13)                                                                ×
                              →     ×   × ×   ×        →     ×   × ×   ×
      σ σ σ σ σ σ σ σ             σ σ σ σ σ σ σ σ          σ σ σ σ σ σ σ σ
```

In certain cases, it can be argued that this "bottom up" stressing is the only plausible analysis, since assigning the main stress directly would lead to an extremely complex rule. Languages with this property include Cairene Arabic (§ 4.1.3), Seminole/Creek (§ 4.1.2), Wargamay (§ 6.1.4), Munsee/Unami (§ 6.3.3), Eastern Ojibwa (§ 6.3.4), and Malecite-Passamaquoddy (§ 6.3.4). The crucial point is that the Continuous Column Constraint guarantees that the higher grid mark may only be assigned to a syllable that already bears stress; otherwise, representations like (14) are derived:

```
(14) *                ×
       ×   × ×   ×
     σ σ σ σ σ σ σ σ
```

(D) DESTRESSING. The Continuous Column Constraint implies "make the weak weaker" when it governs destressing rules. As Prince 1983b and Hammond 1984a point out, such rules appear always to involve the removal of one stress on a syllable adjacent to another stress; that is, they resolve "stress clashes." Another universal property is that if the two stresses are unequal in strength, it is always the weaker stress that is removed (Hammond 1984a). Suppose that we constrain destressing in the same way that Move X is constrained above: it must apply within the layer where the clash occurs. Under such a proposal, the general schemata for Destressing in Clash are as in (15):

(15) **Destressing in Clash** a. × → ∅ / ___ ×
b. × → ∅ / × ___

It is easy to see that when combined with the Continuous Column Constraint, these schemata predict that only the weaker of two adjacent stresses may delete, because application to a stronger stress would create a discontinuous column. In (16a), the cyclically assigned stress on the stem *párent* is deleted under clash with the following stress. Deleting the stronger stress (16b) is impossible: deletion on the top layer is impossible because there is no clash there, and deletion on the middle layer is blocked by the Continuous Column Constraint.

```
(16) a.    ×                ×
          × ×               ×
          × ×  ×    →     × ×  ×
         parental        parental

     b.    ×                              ×
          × ×             × ×            ×
          × ×  ×    →     × ×  ×         × ×  ×
         parental       *parental,    *parental
```

In this book, I posit destressing rules in a number of languages. In several instances I invoke the general schemata of (15), relying on the Continuous Column Constraint to ensure that the rule only removes weaker stresses adjacent to stronger, and not vice versa; see Spanish (§ 5.1.7), Egyptian Radio Arabic (§ 6.1.2), Cahuilla (§ 6.1.3), Wargamay (§ 6.1.4), Lenakel (§ 6.1.8.3), and Icelandic (§ 6.2.2.4). The same strategy has earlier been applied to Italian by Nespor and Vogel (1989).[2]

To summarize, the Continuous Column Constraint crucially constrains the application of three rule schemata: Move X, Destressing in Clash, and main stress assignment (formalized in § 3.12 as the End Rule). The interaction of these principles has as a result the general pattern noted earlier, whereby the strong get stronger, the weak weaker.

3.5. GROUPING IN METRICAL STRUCTURE

Traditional notions of rhythmic structure often attribute more to rhythm than just a hierarchy of beats: rhythm is held to involve grouping of consecutive beats into phrases as well. This is true, for instance, of the theories of musical rhythm developed in Cooper and Meyer 1960 and Lerdahl and Jackendoff

2. Looking ahead slightly, note that in a theory with bracketed grids (§ 3.5), the Destressing schemata given above would always violate the Faithfulness Condition (24). For this reason, I will assume that by convention, Destressing removes the brackets associated with the deleted grid mark.

1983. In both theories, grouping structure forms a hierarchy, corresponding closely to the hierarchy of strong and weak beats.

The existence of grouping in metrical stress theory has been the subject of a debate between "tree theory" (e.g. Liberman and Prince 1977; Hayes 1981), which includes bracketing, and "grid theory" (e.g. Prince 1983a; Selkirk 1984), which excludes it. Below, I review the evidence in favor of bracketing. First, however, I will discuss how to formalize it.

Earlier work in metrical theory (e.g. Liberman 1975; Liberman and Prince 1977; Hayes 1984) adopted a tree formalism (as in (17)) as the basic linguistic representation of stress.

(17)

The nodes of the tree were labeled s(trong) and w(eak) to mark relative prominence; grids were essentially read off the tree, serving as a kind of extraphonological rhythmic interpretation. Subsequently, work in pure-grid theories (Prince 1983a; Selkirk 1984) made clear that grids cannot be extraphonological, since there are language-particular phonological rules that are best formulated to refer to the grid. This led to efforts to create hybrid representations, in which bracketing (i.e. tree-like) information was incorporated into the structure of the grid.

Here, I adopt representations of this type, called **bracketed grids,** as developed by Hammond 1984a and HV. Bracketed grids are similar to the pure-grid representations of Prince and Selkirk but include brackets at all layers of the grid to indicate the constituency that would appear in a metrical tree. Examples (18a, b) depict a bracketed grid for an English phrase, in the formats used by Hammond and by HV. These may be compared to the pure tree representation in (17) and the pure grid representation in (18c).

(18) a. **Bracketed Grid** (Hammond 1984a) b. **Bracketed Grid** (HV) c. **Pure Grid** (Prince 1983a, Selkirk 1984)

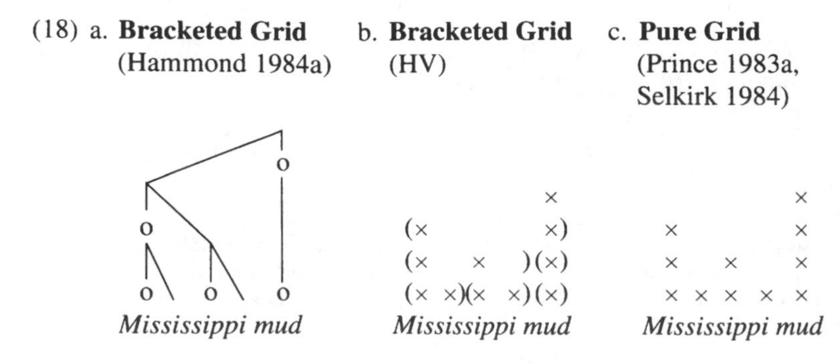

I adopt HV's representations below, altered for brevity as follows: the bottom row of /×/'s (which serve only as place markers) is eliminated, and /./ is used to mark the location of completely stressless syllables. In the grids that result, all pairs of brackets enclose just one grid mark:

```
(19) (             ×)
     (×         )(×)
     (× .)(×  .)(×)
     Mississippi mud
```

Both the HV representation and the abbreviated notation proposed in (19) are only graphic conveniences, which leave unsettled in a number of crucial cases just what is intended. I briefly describe a more precise notation here; see HV chapters 4–5 for a formalized interpretation. Taking the text *Mìssissíppi*, we first incorporate phonemic transcription and syllable nodes:

```
(20) a. (      ×    ) = b. (         ×    )
        (× .)(× .)         (×   .)(×   .)
        Mississippi          σ  σ  σ  σ
                             /\ /\ /\ /\
                            mɪ sə sɪ pi
```

More elaborate syllable structure could be added here, depending on the theory adopted.

The next version shows the crucial step: like HV (p. 41), I regard the brackets of metrical representation as actually enclosing elements of the terminal string (with the difference that for me, the terminal string consists of actual syllables, rather than a sequence of terminal grid marks).

```
(21)               ×3
         ×1        ×2
         σ  σ      σ  σ
         /\ /\     /\ /\
     ((m ɪ s ə)1(s ɪ p i)2)3
```

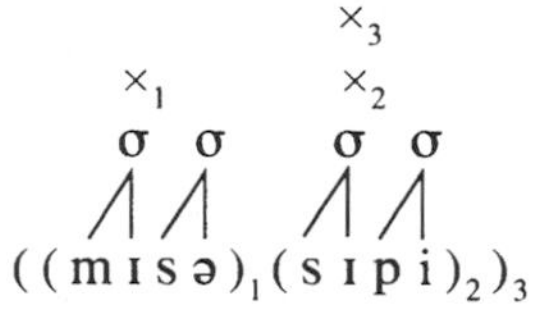

The form in (21) can be read: (mɪsə) is a metrical constituent (for clarity, indexed 1), whose "head" is a rhythmic beat on the lowest layer of the grid, namely the × indexed 1. (sɪpi) is a metrical constituent, whose head is $\times_2$; and (mɪsəsɪpi) is a metrical constituent, whose head is $\times_3$. Vertical alignment of ×'s and σ's has formal significance, indicating alignment of syllables and beats in time.

We will see below that in certain cases, more precise predictions can be made from the formalism if we examine "official" representations like (21), with bracketing on the terminal string. However, for clarity I normally use representations like (20a).

3.6 THE METRICAL FOOT

How are metrical bracketings determined? At the phrasal level, this is fairly straightforward: metrical bracketing is based on syntactic bracketing; or alternatively, on the phonological bracketings motivated for phrasal phonology by Selkirk 1980a; Nespor and Vogel 1986; and others. In simple morphological stress systems (§ 3.3), metrical bracketing within the word tends to be based on morphological bracketing.

The interesting cases are the systems of rhythmic word stress, where the metrical bracketing is not based on any pre-existing structure. The key to determining bracketing in such cases is a close examination of the rules that assign stress.

One of the seminal ideas in metrical stress theory is this: the best way to express stress rules might not actually be the most direct one, that is, to place stress on a particular syllable. The alternative is to state the possible structures for metrical constituents and construe stress placement as the parsing of a word into such constituents. These constituents, the minimal bracketed units of metrical theory, are called **feet.**

Let us consider a simple case: in many languages, the appropriate foot for assigning rhythmic word stress is disyllabic, with prominence on the initial syllable:

(22) (× .)
σ σ

If we parse the syllables of a word into a sequence of such feet, placing the feet adjacent to one another, we derive alternating stress: on odd syllables if the direction of parsing is from left to right (23a), and on the (2nd, 4th, 6th, . . .) from the end if parsing is from right to left (23b).

(23) a. (× .)(× .)(× .)(× .)(× .)
σ́σ σ́σ σ́σ σ́σ σ́σ . . .
⟶

b. (× .)(× .)(× .)(× .)(× .)
. . . σ́σ σ́σ σ́σ σ́σ σ́σ#
⟵

Example (23a) would correspond, for example, to the stress pattern of Pintupi (§ 4.1.1), (23b) to Nengone (§ 6.2.3). I argue in § 3.8.3 and § 4.5 that there are explanatory advantages to assigning stress by foot-parsing, rather than directly.

The notion of assigning stress by parsing into accentual units has apparently been invented a number of times independently; e.g. by Scott 1948; Lehiste 1965; Miyaoka 1971; Allen 1973; and Dixon 1977. As a formal concept within generative phonology, the foot appears in Prince 1976a; Halle and Vergnaud 1978; McCarthy 1979a, 1979b; Selkirk 1980b; and other early work. The issue of what kinds of feet exist, and what their role is in the theory, is the central one of this book.

Summing up so far, the version of metrical theory adopted here is a bracketed grid theory. It involves (a) a hierarchy of rhythmic beats, expressed as a grid; and (b) a grouping structure: feet serve as the lowest metrical constituent, and they are grouped into higher level units, which are themselves grouped into still higher units, and so on.

3.7 THE FAITHFULNESS CONDITION

What is the relation between grid structure and bracketing structure? Earlier work in bracketed grid theory (Hammond 1984a; HV) postulates that there is a one-to-one correspondence between the two: in terms of informal representations like (19), every domain encloses a single grid mark, and every grid mark is enclosed by a single domain. This requirement is termed the **Faithfulness Condition** by HV (pp. 15–16); I state it as in (24):

(24) **Faithfulness Condition**
Grid marks must be in one-to-one correspondence with the domains that contain them.

A more precise version of the Faithfulness Condition is proposed in § 9.4.2. My use of the condition basically follows HV, except that I intend to impose it as a well-formedness condition on all stages of the derivation, rather than just on the output. Thus, the condition is capable of blocking the application of rules that would create violations of it.

On occasion I will refer to the grid mark that is paired with a particular domain as the "head" of that domain, following HV.

The Faithfulness Condition makes it unnecessary to include indices (cf. (21)) in representations of metrical structure, because in a grid that obeys it, it is always possible to tell which domain is paired with which grid mark (see § 9.4.2 (30)).

3.8 ARGUMENTS FOR BRACKETING STRUCTURE IN METRICAL THEORY

Bracketing structure is harder to support empirically than grid structure, because it has no directly observable correlates. Any arguments for bracketing will necessarily be indirect; that is, by positing bracketing we can attain a deeper account of phenomena that would otherwise seem arbitrary. In this section, I review general arguments that support the existence of bracketing.

3.8.1 Deletion of Stressed Vowels

Until recently, most phonologists would not have imagined that rules of vowel deletion could apply to stressed vowels. However, recently the existence of a fair number of cases of this sort has become clear. The cases found so far behave contrary to naive expectation: the stress borne by the deleted vowel does not disappear, but migrates to the right or left.

This preservation of stress under vowel deletion is a plausible consequence of grid theory. As HV note (pp. 28–30), grid marks may be interpreted as having an existence independent of the segmental string, and thus may survive the deletion of the segments supporting them. In essence, the theory views stress as partly autosegmental (HV 5), so that one finds "stress stability" as an analogue to the well-known phenomenon of tonal stability (Goldsmith 1976). For arguments from speech errors anticipating this result, see Fromkin 1977.

What is especially interesting about these cases is that one can apparently predict whether the dislodged /×/ will reassociate with the syllable to its left or to its right. In languages whose stress pattern must be analyzed with initially stressed feet, migration is rightward; whereas in languages with finally stressed feet, migration is leftward, as shown in (25):

(25) a. (× .) (× .) → (× .) (×)
CVCV C**V**CV → CVCV CØCV

b. (. ×)(. ×) → (×)(. ×)
CVC**V** CVCV → CVCØ CVCV

As HV 28–30 observe, the direction of migration is a direct consequence of bracketed grid theory, in that movement in the opposite direction is excluded by the Faithfulness Condition. For example, if the stranded /×/ in (25a) moved leftward instead of rightward, the constituent on its left would have two heads and the constituent it abandoned would have none. Note that the foot bracketing is in many cases independently establishable, in that only one bracketing can account adequately for the stress pattern, so that a genuine prediction is being made here.

A number of instances of bounded stress migration have been presented in the literature and interpreted metrically: Tiberian Hebrew (Prince 1975; Churchyard 1990), Bedouin Hijazi Arabic (Al-Mozainy 1981; Al-Mozainy, Bley-Vroman, and McCarthy 1985), Bani-Hassan Bedouin Arabic (Kenstowicz 1983; Irshied and Kenstowicz 1984; § 8.10), Tripoli Arabic (§ 6.3.7), and Seminole/Creek (Tyhurst 1987; § 4.1.2).[3] Below, I discuss some additional cases: Unami (§ 6.3.3), Central Alaskan Yupik (§ 6.3.8.6), Pacific Yupik (§ 8.8.5), Asheninca (§ 7.1.8.2), and Cyrenaican Bedouin Arabic (§ 6.3.7). The Bedouin Arabic dialects are particularly instructive, in that they share one of the segmental rules that deletes stressed vowels. In Bani-Hassan and Hijazi, the

3. In addition, Halle and Vergnaud note examples of stress migration under deletion in Russian, Sanskrit, and Lithuanian. These cases carry less force because the constituency involved is unbounded (1987a, 53). As § 3.3 argues, constituency for unbounded systems is difficult to prove. In addition, the Sanskrit and Lithuanian examples involve pitch accent rather than stress, and could have an alternative tonal analysis (see discussion in Al-Mozainy, Bley-Vroman, and McCarthy 1985 and § 8.9.6).

feet can independently be shown to be stress-initial, and stress migration is to the right. In Cyrenaican, the feet are stress-final, and stress migration is to the left. This difference is predicted by the Faithfulness Condition.

A class of cases that should probably be kept distinct is those involving hiatus resolution; that is, deletion or gliding of a vowel next to another vowel. Examples of this occur in Modern Greek (/ɣríɣor**a é**rxome/ → [ɣríɣor**á**rxome] 'quickly I-come'; Kaisse 1982, 66), and in Chicano Spanish (/kom**í u**bítas/ → [kom**yú**βítas] 'I-eat grapes'; Hutchinson 1974, 186). As Kaisse 1982 and Clements and Keyser 1983 suggest, such examples arguably involve a special mechanism: the merger of two adjacent syllables into one, with inheritance of the prosody of the input syllables. Under such an account, these examples would not necessarily be problematic for the Faithfulness Condition.

There is independent evidence from tonal phonology that hiatus resolution is carried out by syllable merger. When tones are set adrift by hiatus resolution, they do not dock at random; rather, they land on the vowel that "triggered" the deletion (Clements and Ford 1979, 207). Such an outcome follows straightforwardly if hiatus resolution is syllable merger, but would otherwise require a global constraint, so that the floating tone could "remember" the vowel that triggered its stranding.

Summing up, the migration of stress under vowel deletion appears to be predictable in a large set of cases. Where vowel deletion resolves hiatus, the mechanism is syllable merger with inheritance of prosody. In other cases of vowel deletion, stress migrates to an adjacent syllable, obeying the Faithfulness Condition.

3.8.2 Constituency in Phrasal Phonology

Bracketed grids can also explain an observation made in Hayes 1984 concerning an odd property of the English Rhythm Rule. Examples like (26) show that Move X applies to position a stress in the middle of a four-syllable interval, dividing it into two equal sequences. In this derivation, the even division of a quadrisyllabic interval takes place with the second application of Move X, going from (26b) to (26c).

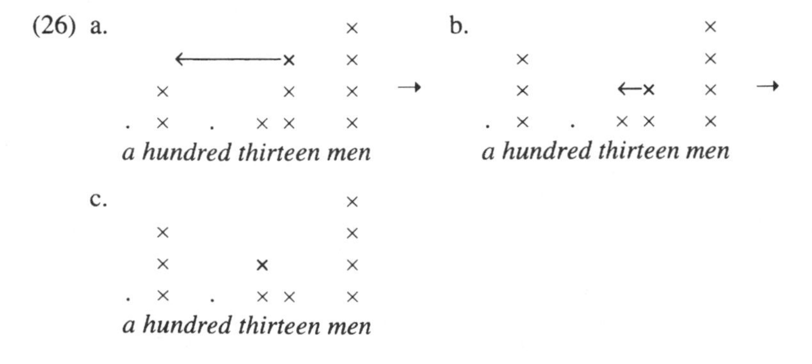

Given that rhythmic pressure for equal spacing induces the second application of Move X in this and similar examples, it is surprising that Move X should not apply in (27):

```
(27) a.               ×           b.                    ×
                 ×    ×                 ←—————————×     ×
      ←—×        ×    ×    →        ×             ×     ×    →
      × .  ×     ×    ×             × .  ×        ×     ×
      overdone steak blues          overdone steak blues

     c.               ×           d.                    ×
      ×      ?        ×             ×                   ×
      ×     ←—×       ×    →        ×    ×              ×
      × .  ×     ×    ×             × .  ×        ×     ×
      overdone steak blues         *òverdòne steak blúes
```

The transition from (27c) to (27d) is essentially the same as that from (26b) to (26c), yet (27d) and similar examples are ill-formed. The output representation must remain (27c), despite the uneven division of the quadrisyllabic interval. Arguably, there must be some overriding factor that blocks Move X in (27d). This factor must have something to do with the different phrasal bracketing of (26) and (27), as there is otherwise no potentially relevant difference between the two.

I claim that the relevant factor is the Faithfulness Condition. In (28) and (29), the examples of (26) and (27) are assigned bracketed grids based on their phrasal structure. In (28), where Move X is not blocked, both applications of the rule are confined within the constituent of which the moved element is the head, so the Faithfulness Condition is not violated:

```
(28) a.  (                  × )        b.  (           )(    × )
         (←———————— × )(    × )            (×           )(   × )
         (×       ) (  × )( × )   →        (×     ) ( ←× )(  × )   →
       . (×     . ) (×)(× )( × )         . (×    .) ( ×)(× )( × )
       a hundred thirteen men            a hundred thirteen men

     c.  (                  × )
         (×              )( × )
         (×       )(×    )( × )
       . (×     . )( ×)(× )( × )
       a hundred thirteen men
```

But the application of Move X taking (29c) to (29d) violates the Faithfulness Condition, since the constituent *overdone* has two heads and the second-layer constituent *steak* has none. Thus the surface representation must remain as in (29c).

```
(29) a. (                ×  )      b.  (                 ×  )
        (            ×  )( ×  )        ( ←——————×  )( ×  )
        ( ←—×    )( ×  )( ×  )  →      (×           )( ×  )( ×  )  →
        (× .)( ×  )( ×  )( ×  )        (× .)( ×  )( ×  )( ×  )
        overdone steak blues           overdone steak blues

     c. (                ×  )      d. *(                 ×  )
        (×         ?  )( ×  )          (×                )( ×  )
        (×      ←→)(×  )( ×  )  →      (×     ×  )(     )( ×  )
        (× .)( ×  )( ×  )( ×  )        (× .)( ×  )( ×  )( ×  )
        overdone steak blues           òverdòne steak blúes
```

This argument is made in greater detail in § 9.5.1, where the structure of the crucial input representations is defended as part of a more general theory of phrasal stress assignment.

Summarizing, the Faithfulness Condition constrains the migration of stress both (a) within the foot, when stressed vowels are deleted; and (b) within phrasal phonology, where movement is specified by versions of the rule schema Move X.

3.8.3 "Natural Classes" in Stress Rules

Another argument for constituency can be based on the typology of word stress patterns. For expository purposes I adopt here the notion of **predicted natural class.** Just as a particular theory of features predicts that certain combinations of segments will form natural classes, so does a particular parametric metrical theory predict this of certain combinations of stress patterns. In feature theory, we can test whether a predicted natural class is valid by examining the sets of segments that pattern together in phonological rules. In this section, I describe the natural classes predicted by different versions of metrical stress theory, and discuss which predicted natural classes are empirically confirmed.

Consider the stress patterns of (23), repeated below as (30): (30a) is stressed on odd-numbered syllables, counting from the left, while (30b) is stressed on the second, fourth, sixth, etc. syllables from the end.

```
(30) a.    (× .)(× .)(× .)(× .)(× .)         b.      (× .)(× .)(× .)(× .)(× .)
         #  σ́σ   σ́σ   σ́σ   σ́σ   σ́σ . . .          . . .  σ́σ   σ́σ   σ́σ   σ́σ   σ́σ #
            ————————→                                          ←————————
```

In a theory where stress is assigned by foot parsing, these two patterns form a predicted natural class, since both involve an initially stressed disyllabic foot template:

```
(31) (× .)
      σσ
```

Another predicted natural class consists of the pair of stress patterns in (32): even-numbered syllables counting from the left (32a), and the first, third, fifth, etc. from the end (32b). Both would involve parsing with the final-stressed template in (32c).

(32) a. (. ×)(. ×)(. ×)(. ×)(. ×)
σσ́ σσ́ σσ́ σσ́ σσ́ . . .
⟶

b. (. ×)(. ×)(. ×)(. ×)(. ×)
. . . σσ́ σσ́ σσ́ σσ́ σσ́ #
⟵

c. (. ×)
σσ

Compare now a theory that predicts a different set of natural classes. In Prince 1983a, grids lack any kind of bracketing and are constructed directly by rule. Here, the predicted natural classes follow from Prince's parameter **peak first versus trough first,** meaning 'start the alternation with stress' versus 'start the alternation with stresslessness'. The peak first versus trough first system predicts the natural classes of stress patterns in (33):

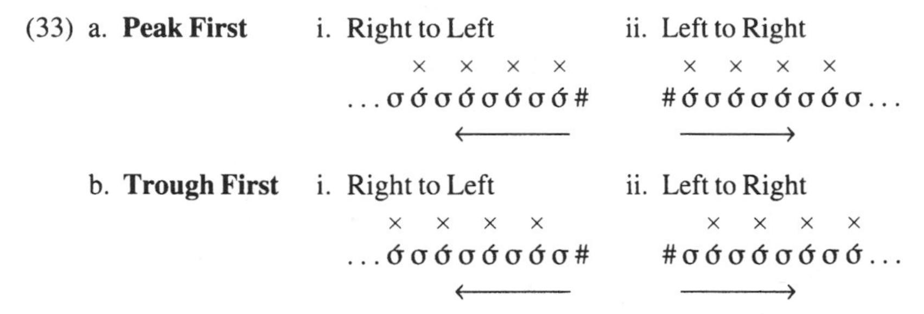

What facts might we use to assess different predicted natural classes? First, consider cases where a language has two stress rules, assigning stress in different contexts. If the foot structure of a language forms a basic organizing principle of its phonology (see § 4.5.3, § 5.4.1), then we would expect that the two rules should typically use the same foot type, and the stress patterns defined in (30) and (32) will be empirically confirmed as natural classes.

Lenakel (§ 6.1.8) is a case that bears on this issue. In Lenakel, the pattern of (30a) is found in verbs and adjectives, whereas the pattern of (30b) is found in nouns. Other languages that have both (30a) and (30b) in different contexts include Modern Greek (§ 6.2.3), Polish (§ 6.2.3), Auca (§ 6.2.1), and Southwest Tanna (§ 6.1.9).

This asymmetry has an immediate formal interpretation in the foot theory: the foot type is (31), and it may be assigned in both directions. In a theory without constituency, there is no plausible account of the co-occurrences. In particular, we must suppose that for some reason the peak first/trough first parameter switches its value whenever direction is reversed. There appear to be no languages that show the predicted natural classes of the peak first/trough first theory.

A second way to test the predicted natural classes of a stress theory is through typology. It turns out that stress patterns that can be described by (31) occur widely, in either direction (§ 6.2), whereas stress patterns derived by (32c) are quite rare (§ 6.3.11). The foot-based theory can state that the unmarked foot type is (31), thus capturing the generalization about markedness in a single statement. In contrast, a theory based on the notion of peak first versus trough first must suppose that (for some reason) peak first alternation is unmarked from left to right, but trough first from right to left.

3.8.4 Prosodic Morphology

Another source of evidence for grouping in metrical structure is the role it plays in the theory of **Prosodic Morphology** (McCarthy and Prince 1986, 1990). The central notion of this theory is that the categories that are relevant for templatic morphology (i.e. reduplication, Semitic-type word formation, truncation, etc.) are precisely the categories made available by prosodic theory in general, including mora, syllable, and prosodic word.

Crucially for present purposes, the set of categories also includes a purely metrical constituent, the foot. As McCarthy and Prince show, many templatic morphological processes have the foot as their prosodic target; in fact, the set of feet they suggest as the set of targets coincides precisely with the feet that I argue for here (chap. 4) on the basis of stress.

Typically (though not universally), the kind of foot required by a language's morphological system is the same as that required by its stress system; thus in many languages where stress is derivable using the foot of (31), reduplication is carried out by copying the first two syllables of a word, that is, by parsing out a foot. Such connections could not be made under a theory of stress where stress is simply assigned to the appropriate syllables.

3.8.5 Word Minima

The following argument is due to McCarthy and Prince 1986. In many languages, there is a minimum placed on the size of a word. For example, in Mohawk (Michelson 1988), every content word must contain at least two syllables. In Fijian (§ 6.1.5), every word must contain at least two moras; that is, it must consist of at least one heavy syllable or two lights (for heavy and light syllables, see § 3.9.2). In the version of metrical theory developed here, these requirements can be stated simply as the requirement that every word contain at least one foot. The connection between the two can be made straightforwardly if the stress assignment rules work by parsing syllables into feet (i.e., every word must undergo parsing); but is arbitrary under a theory that has no bracketing, and whose rules simply place stress in the right locations.

This argument has a potential hole in it: we might say as an alternative that in languages with word minima, EVERY WORD MUST BE ABLE TO UNDERGO THE STRESS RULE. For example, in Mohawk the word minimum is two syllables, and the stress rule, stated without bracketing, is as in (34):

$$\text{(34)}\quad \upsilon \rightarrow \overset{\times}{\upsilon} \;/\; \underline{\quad\quad}\; \sigma\,]_{word}$$

The principle that every word must be stressable would also establish a connection between word length and stress pattern.

To see why this alternative will not work, consider cases like Bidyara/Gungabula, Pitta-Pitta, and Wangkumara, all from § 6.2.3. Here, the word minimum is again two syllables, but stress is different, falling on the first, third, fifth, etc. syllables. In parsing terms, we would say that the feet are disyllabic and initially stressed:

(35)
```
  (× .)(× .)(× .)(× .)(× .)
# σσ  σσ  σσ  σσ  σσ ...
  ——————→
```

The word minimum matches the foot that is needed for stress assignment. Now, in a theory that lacks bracketing, the stress rules would be something like this:

(36) a. **Main Stress**

$$\sigma \rightarrow \overset{\overset{\times}{\times}}{\sigma} \;/\; [_{word}\; \underline{\quad\quad}$$

b. **Secondary Stress**

$$\sigma \rightarrow \overset{\times}{\sigma} \;/\; \overset{\times}{\sigma}\, \sigma\; \underline{\quad\quad} \quad \text{(iterative)}$$

Such a theory fails to exclude monosyllables, since they can perfectly well undergo the main stress rule (words do not have to undergo the secondary stress rule, as is shown by the absence of secondary stress in disyllables). The upshot is that only a bracketing-based theory of stress can capture connections between stress patterns and word minima.

3.8.6 Summary

The arguments above—stress shift under deletion, phrasal stress shift, foot typology, prosodic morphology, and word minima—are what I take to be the main general reasons for adopting bracketing structure in metrical theory. There are also interesting language-specific arguments in the literature, summarized below: see Everett 1988 on Pirahã (§ 7.1.7) and Leer 1989 and Rice 1990 on Pacific Yupik (§ 8.8).

3.9 SYLLABLES AND SYLLABLE QUANTITY

The metrical theory of stress is tightly bound up with the theories of syllable structure, particularly those aspects of syllable theory concerned with weight. I review relevant parts of such theories below.

The general relevance of syllables to rules of stress assignment has long been clear (see for instance McCawley 1974; Kahn 1976). A theory in which stress rules refer only to the structural properties of syllables is far more constrained than a theory (such as that of *SPE*) in which stress rules refer directly to segments. In addition, the attempt to write stress rules using segments alone often leads one to recapitulate the syllable structure of the language in virtually every stress rule one writes. An example of this loss of generality is found in *SPE*, where the recurring expression for a "weak cluster" introduces the principles of English syllable division on an ad hoc basis into several of the English stress rules.

3.9.1 The Syllable as Stress-bearing Unit

Following earlier work (e.g. Jakobson 1931), I adopt the view that in stress languages, the stress-bearing unit is the syllable. This means, for instance, that in disyllabic words there are only two possibilities for stress placement, irrespective of how many segments the word contains. Formally, I assume that syllables are the units which are grouped together in metrical structure and to which grid marks are associated.

Halle and Vergnaud's view on stress-bearing units is quite different: they argue that stress rules may include statements of the form "the stress-bearing elements are X," where X can be for example vowels (HV, p. 49), phonemes in the rhyme (p. 61), or lexically designated segments (p. 193). I suggest that this proposal states as part of language-particular grammars what arguably should be part of universal grammar. It seems unlikely that we would ever find a language in which the stress-bearing units are consonants, or nasals, or coronals, and so on.[4] The phenomena that lead HV to suppose that segments can be stress-bearing can be better accounted for by adopting an explicit theory of syllable structure and syllable weight.

There are in fact languages in which actual contrasts of accentual pattern within the syllable may be found; that is, a heavy syllable may have rising or falling prominence. Examples include Ancient Greek, Lithuanian, Hopi, and some dialects of Serbo-Croatian. Superficially, such languages might be taken as evidence that units smaller than the syllable may bear stress. However, these cases have traditionally been analyzed as **pitch accent** languages (see, for example, Jakobson 1931). In generative phonology, they can be treated as involving tonal representations within the word phonology, either in addition to or instead of metrical representations. An example is Lithuanian, analyzed by HV (pp. 190–203) as involving segment-level stress, but by Halle and Kiparsky 1981 and Blevins 1991 as involving tone.

From the evidence of pure-tone languages, it is uncontroversial that tones may associate to elements smaller than the syllable. Because pitch accent languages are tonal in character, they do not counterexemplify the claim that the STRESS-bearing unit is universally the syllable.

The invocation of pitch accent is not some kind of terminological escape hatch. Pitch accent languages must satisfy the criterion of having **invariant**

4. Cohn 1989, 174–75, suggests that the schwa vowel of Indonesian is non–stress-bearing, which would force us to add to the grammar of this language a statement that only syllables with full vowels may bear stress. However, Cohn notes the possibility that schwa in Indonesian is epenthetic, which would also account for its stresslessness.

tonal contours on accented syllables, since tone is a lexical property. This is not so for pure stress languages, where the tonal contours of stressed syllables can vary freely, being determined postlexically by the intonational system (cf. chap. 2, (8)). In § 8.9, I discuss a case where an improved metrical analysis is made possible by dealing explicitly with the tonal component of a pitch accent system.

Besides disallowing stress-bearing units smaller than the syllable, I also follow earlier work (e.g. Prince 1976a) in assuming that rules of foot construction may not split syllables; for example, we cannot allow the first part of a heavy syllable to belong to one foot and the second part to belong to the next. For further discussion, see § 3.11, § 5.6.2, and § 6.1.3.

3.9.2 Syllable Weight

Among the more interesting stress rules are those that refer to a distinction between **heavy** and **light** syllables. By this it is meant that all syllables may be grouped into two such classes, and it is the class membership (heavy or light) of a syllable, rather than its segmental content, that determines the syllable's influence on stress. Heavy syllables characteristically attract stress, whereas light syllables receive stress only in the absence of an eligible heavy syllable.

A well-known example is the stress rule of Latin. In Latin, a syllable is heavy if it contains a long vowel or if it is closed; otherwise it is light. Words with a heavy penultimate syllable receive penultimate stress, words with a light penult receive antepenultimate stress, and in all cases where a word is too short to obey these laws, stress falls as far as possible to the left. The examples below are from Mester 1992 and Jacobs 1989, 7.

(37) a. **Heavy Penultimate Syllable, Penultimate Stress** (CVː)

a.míː.kus	'friend, kind'
gu.ber.náː.bunt	'they will reign'

b. **Heavy Penultimate Syllable, Penultimate Stress** (CVC)

or.na.mén.tum	'equipment'
sa.pi.én.teːs	'wise (nom.pl.)'

c. **Light Penultimate Syllable, Antepenultimate Stress**

i.ni.miː.kí.ti.a	'hostility'
sí.mu.laː	'simulate (2 sg.imp.)'
do.més.ti.kus	'belonging to the house'

d. **Initial Stress in Disyllables**

mán.daː	'entrust (2 sg.imp.)'
ká.nis	'dog'
hé.ri	'yesterday'

Note that for purposes of weight computation, the segment count of a syllable is quite irrelevant, in that a light syllable can sometimes have more segments than a heavy one; thus /tri/ is light but /iː/ and /it/ are heavy.

The Latin weight system, where closed syllables are grouped together with long-voweled syllables in the heavy class, is quite common across languages. The other system that is commonly found is a division between long-voweled and short-voweled syllables, irrespective of whether a syllable is closed. An example is St. Lawrence Island Yupik (§ 6.3.8.1). Numerous examples of both types appear in this book.

What is common to both systems is the principle that prevocalic segments in the syllable (i.e. onset segments) are **prosodically inert:** that VC is prosodically equivalent to CVC and CCVC, Vː to CVː and CCVː, and so on. While this claim is not fully valid at the observational level (§ 7.1, § 7.4), it is so well supported across languages that it serves as the central observation for formal theories of syllable weight.

Two approaches dominate such formal theorizing. In the theory of syllabic constituency (Pike and Pike 1947; McCarthy 1979a; Levin 1985a), the syllable is assigned a particular internal constituent structure, and it is stipulated that only certain constituents may be prosodically active. If constituents like those in (38) are assumed, then the heavy/light distinction exemplified by Latin may be characterized as **branching versus non-branching rhyme.** The weight criterion that counts vowel length only may be characterized as **branching versus non-branching nucleus.** The representations below use an X-tier (Prince 1984; Levin 1985a) as a representation for the segmental level; a long segment is interpreted as a single feature complex linked to two X-slots (O = Onset, R = Rhyme, N = Nucleus).

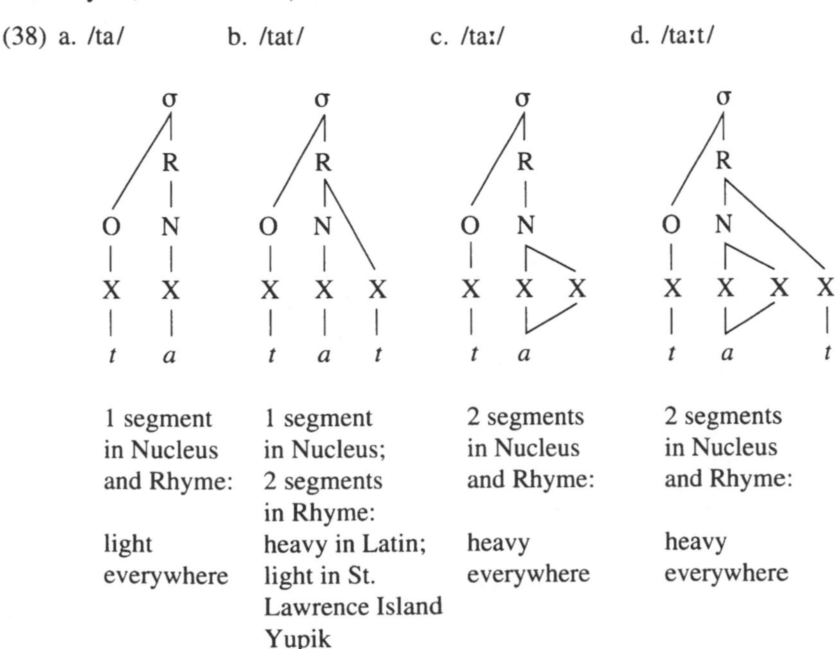

Another approach is to posit explicit units of syllable weight in the representation, namely, moras (McCawley 1968; Prince 1976a, 1983a; van der Hulst 1984; Hyman 1985; McCarthy and Prince 1986; Hayes 1989b; J. Ito 1989; Zec 1988). In this approach, the segments that are prosodically active in a particular language are marked as such by assigning them a mora (or two, for long vowels). The moras are the units to which metrical structure may refer. Cross-linguistic differences in the criterion of syllable weight are expressed by language-specific conditions on mora assignment: languages like Latin allow a postvocalic consonant within the syllable to bear a mora (symbolized /μ/), whereas languages like St. Lawrence Island Yupik do not. Thus weight distinctions can be made on the basis of the mora count of a syllable:

(39) a. **CV Light; CVC, CVV, CVVC Heavy**

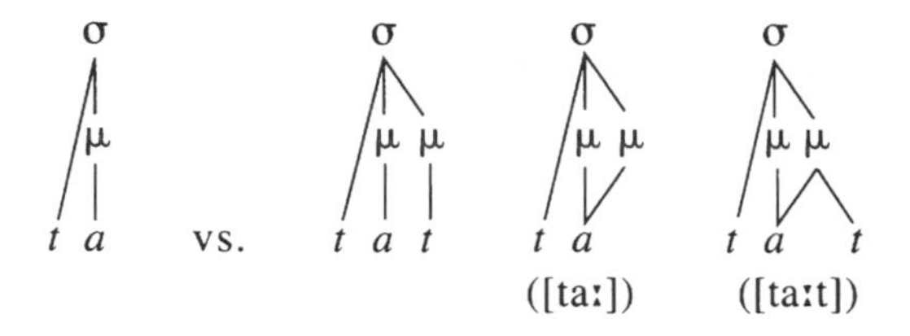

b. **CV, CVC Light; CVV, CVVC Heavy**

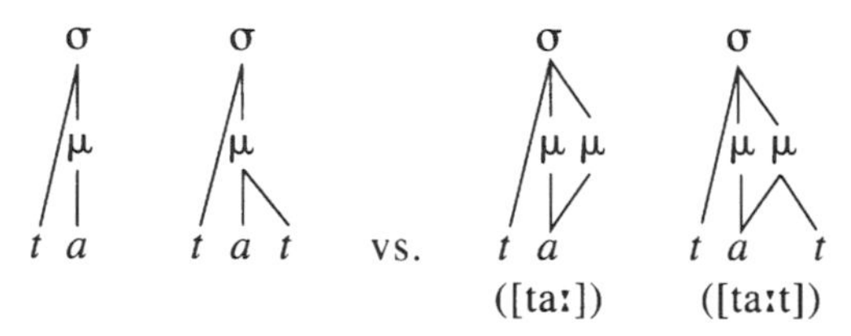

Hayes (1989b) uses the term **Weight by Position** to refer to the rule or principle of syllabification that assigns a mora to a postvocalic consonant within the syllable; thus (39a) illustrates a language that has Weight by Position; (39b), a language that lacks it. Crucially, in no language is a mora licensed by an onset consonant, which accounts for the universal absence of syllable quantity distinctions based on onsets (for discussion of apparent exceptions, see § 7.1, § 7.4).

The debate concerning how syllable weight is to be represented (constituency theory vs. moraic theory) is bound up with a related debate concerning the representation of contrastive segment length. The syllabic constituency theory is characteristically coupled with an "X-theory" account of segment length, where the length of a segment is depicted by the number of X's it is associated with. In moraic theory, the syllabic terminals are usually fewer in number than the segments, and long segments are defined simply as multiply linked. Note that a moraic theory could in principle invoke syllabic constituents, too. But since the main purpose of constituency is to depict weight, and this can be done by the moras alone, moraic theories usually dispense with constituency.

There are two central arguments in favor of the moraic theory.

First, there is the argument of **counting,** from McCarthy and Prince 1986. Phonological processes that count moras are commonplace: stress assignment, weight-based segmental rules, minimal word requirements, quantitative meter. But it appears that there are no phonological processes that count segments, in the sense defined by X-theory. Such a process would treat, for example, the strings /aː/, /ta/, and /at/ as equivalent, since they all bear two X's. Moraic theory predicts this absence, since its representations do not depict segment count.

Second, there is the argument from **moraic conservation,** stated in its most general form in Hayes 1989b. I define a compensatory process as one in which one segment is shortened or deleted, with another becoming simultaneously longer. The world's languages include a very wide array of such processes. Crucially, in every case, the number of moras in the input string is equal to the number of moras in the output. Moraic theory derives this result straightforwardly, under the assumption that compensatory processes are expressed as the rearrangement of segmental material with respect to an invariant moraic frame. In contrast, compensatory processes do not conserve the number of X's in the representation, so the law of moraic conservation is essentially an accident under the X-theory.

For these reasons, I use moraic theory here as the account of syllable weight. However, recent work has made it clear that moraic theory in the pure form described above is insufficiently powerful as a theory of syllable weight; in particular, it cannot account for languages in which both criteria for heavy syllable status are found for different phonological processes (Crowhurst 1991a; Steriade 1991). Two options, then, are to try to constrain X-theory appropriately, or to try to extend moraic theory to account for multiple criteria of syllable weight. Since the former strategy seems unworkable, I pursue the latter here; see § 7.3.

Moraic theory comes in more than one version. In the representations of (39), onset consonants are attached to the syllable node, following McCarthy and Prince 1986 and Hayes 1989b. Another possibility (Hyman 1985; Zec 1988; J. Ito 1989; Katada 1990) is to attach onset consonants to the initial mora, as in (40), a revised version of (39a):

(40)
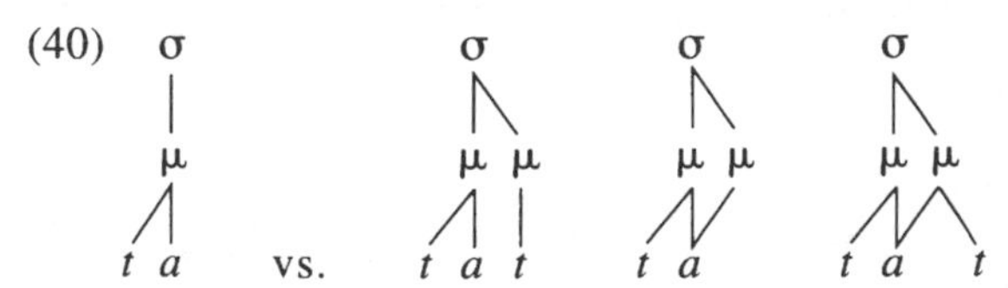

The arguments in favor of associating onsets to the syllable nodes (see Steriade 1988a and Hayes 1989b, 298–99, for review) hinge on the need to depict formally the notion **weight-bearing segment.** That is, we want the representations in (40) to indicate that it is the vowel /a/ that is weight-bearing, not the preceding consonant. Possibly some other means could be adopted to do this (e.g. by labeling the segmental daughters of a mora for relative sonority, as in Kiparsky 1979). Assuming this is the case, we can adopt representations like (40), where onsets belong to the first mora. These seem conceptually simpler for purposes of stress assignment, since to determine weight one can simply count the daughters of the syllable node, without any kind of stipulation that onsets are ignored.

Although this book uses moraic representations, I emphasize that many of the issues to be addressed are independent of any particular theory of syllable quantity (i.e. any theory that encodes the distinction between light and heavy syllables). To underscore this, in the representations of metrical structure I will usually just use the traditional symbols macron /–/ and breve /⏑/ to designate heavy and light syllables respectively.

3.10 PARAMETRIC METRICAL THEORY

A notion that has been crucial to metrical studies of word stress is the idea of **parameterizing** the theory. In a parametric theory, a rule system is regarded as a particular choice from a limited list of options, or parameters. To give an idea of the kind of parameters that have been proposed, here is a sample list, incorporating proposals from throughout the literature.

(41) a. **Choice of foot type**

i. SIZE	Maximally unary/binary/ternary/ unbounded
ii. QUANTITY SENSITIVITY	Heavy syllables (may/may not) occur in weak position of a foot
iii. LABELING	Feet have (initial/final) prominence
iv. OBLIGATORY BRANCHING	The head of a foot (must/need not) be a heavy syllable

b. **Direction of parsing** Left to right/right to left
c. **Iterativity** Foot construction is (iterative/once only)
d. **Location** (Creates new metrical layer/applies on existing layer)

By setting all the relevant parameters, one derives a stress rule. For example, the alternating stress rule found in one variety of Hungarian (§ 8.6) sets the parameters as in (42):

(42) a. i. Feet are maximally binary.
ii. Heavy syllables may occur in weak position.

iii. Feet have initial prominence.
iv. The head of a foot need not be a heavy syllable.
b. Parsing is left to right.
c. Foot construction is iterative.
d. Creates new layer.

This set of parameter settings places stress on odd-numbered syllables, going from left to right:

(43) (× .)(× .)(× .)(× .)(× .)
σσ σσ σσ σσ σσ . . .
⟶

A parametric theory can be contrasted with a theory such as that of *SPE* or the syntactic theory of Chomsky 1965: such theories posit a set of primitives with which rules may be expressed, and all rules are allowed which are well-formed combinations of those primitives. A parametric theory is usually more constrained and capable of stronger predictions.

An interesting problem within parametric metrical theory is to what extent the parameters characterize rules versus grammars. Here, we will conservatively assume that parameters characterize rules. However, the possibility that they have more general scope, as suggested by HV, is an appealing one: for example, it predicts that when more than one rule creates feet, the feet created should be the same. This in general appears to be true, though a difficult case is found in Onondaga, § 6.3.11.

A proposed parametric theory of stress is successful to the extent that it is well defined, is maximally restrictive, and is capable of describing all the stress systems of the world's languages. The latter criterion is the most difficult, as the stress systems of many languages are remarkably complex. A theory that reduces such complexity to a small set of general principles is a substantial result. Moreover, such a theory may help account for the ease with which children acquire complex stress systems, in that a system can be learned by setting the parameters one by one (Dresher and Kaye 1990).

Important early work in parametric metrical stress theory includes Prince 1976a, 1976b, McCarthy 1979b, and perhaps most influentially, Halle and Vergnaud 1978. Halle and Vergnaud laid out a number of parameters, including choice of foot shape, labeling conventions, parameters for the representation of quantity, and the **Obligatory Branching** parameter (41a.iv).

Hayes 1981 attempted to extend the parametric research program along two fronts: to broaden the database, and to constrain the theory by cutting back on the range of parametric choices. Most crucially, Hayes argued that the inventory of foot templates can be limited to **binary** and **unbounded,** with ternary templates excluded. Since the number of logically conceivable ternary templates is quite large, the exclusion of ternary units has a correspondingly large constraining effect on the theory. A brief review of the Hayes 1981 theory, comparing it to the present proposal, appears in § 4.3.1.

The central content of this book is a parametric theory, presented in the next two chapters. On the basis of the typology of rhythmically based bounded stress rules, I attempt to develop a theory that matches the data closely, at the same time being as constrained as possible. The theory shares some of the foot types proposed in earlier theories but is quite different in various other respects.

3.11 EXTRAMETRICALITY

The proposal I make here inherits from Hayes 1981 the restriction that feet come in only two maximal sizes, binary and unbounded. Such a restriction appears to be tenable only if the theory includes a subtheory of **extrametricality.** I briefly review this subtheory here.

To begin with an example, Estonian (§ 8.5) has a complex stress system that refers to the distinction between light (CV) and heavy (CVC, CVː, and longer) syllables. I will not review the Estonian rules here, but note only the following crucial fact: in word-final position, a special definition of heavy versus light is used: final CVC counts as light rather than heavy.

To illustrate this fully would require complete discussion of the Estonian stress pattern. Instead, I will use the diagnostics in (44), which are justified and refined in § 8.5.

(44) a. **Nonfinal Syllables** The third syllable of a word may be stressless only if it is light.

b. **Final Syllables** If the third syllable is also the final syllable, it is stressless if light and stressed if heavy.

These diagnostics seem arbitrary in isolation but fall into place in a complete analysis. The examples in (45) illustrate the basic pattern (page references from Hint 1973).

(45) a. **CV Is Light in All Contexts**

Nonfinal:	*pímes**ta**vàle* (or *pímestàvale*)	H 161
Final:	*ósa**va***	H 157

b. **CVC Is Heavy in Nonfinal Syllables, Light Finally**

Nonfinal:	*válu**sàt**tele* (only)	H 161
Final:	*pála**val***	H 157

c. **CVV Is Heavy in All Contexts**

Nonfinal:	*vára**sèi**mattèle* (only)	H 163
Final:	*lúːlet**tài***	H 157

d. **CVCC Is Heavy in All Contexts**
Nonfinal: (no examples found)
Final: *só:yemàks* H 157

The crucial case is (45b), where CVC is shown to be heavy nonfinally but light finally. Parallel examples where final CVC is exceptionally light occur in English (Hayes 1982b), Arabic dialects (McCarthy 1979a; § 4.1.3; § 6.1.1; § 6.1.2), a dialect of Hindi (Hayes 1981, 79–81, citing Mohanan 1979), Spanish (Harris 1983, 1992), Romanian (Steriade 1984), Ancient Greek (Steriade 1988b), and Menomini (§ 6.3.4).

A unitary treatment of syllable weight in final versus nonfinal position is possible if we are allowed to stipulate that the relevant rules IGNORE a consonant in final position. In (46) I give some abstract forms for final Estonian syllables. The notation ⟨ ⟩ surrounding a consonant (adopted from HV) means that the consonant is ignored:

(46) a. **Final Position**		b. **Nonfinal Position**	
CV	=	CV	(light)
CV⟨C⟩	=	CV	(light)
CVC⟨C⟩	=	CVC	(heavy)
CV:	=	CV:	(heavy)
CV:⟨C⟩	=	CV:	(heavy)

A formal basis for this procedure is provided by the theory of **extrametricality rules.** Extrametricality as a notion of metrical theory was put forth by Liberman and Prince 1977, and the idea of general rules of extrametricality was proposed in Hayes 1979. Subsequent work in this area includes Hayes 1981; Harris 1983; Archangeli 1984; Poser 1984, 1986; Franks 1985, 1989; Pulleyblank 1986b; Ito 1986; HV; Sauzet 1989; Inkelas 1989; Barker 1989; Buckley 1991; and Roca 1992. An extrametricality rule designates a particular prosodic constituent as invisible for purposes of rule application: the rules analyze the form as if the extrametrical entity were not there. Thus in Estonian, final consonants are extrametrical.

To keep this notion as constrained as possible, Hayes 1981 proposed the restrictions on extrametricality in (47):

(47) a. **Constituency** — Only constituents (segment, syllable, foot, phonological word, affix) may be marked as extrametrical.

b. **Peripherality** — A constituent may be extrametrical only if it is at a designated edge (left or right) of its domain.

c. **Edge Markedness** — The unmarked edge for extrametricality is the right edge.

d. **Nonexhaustivity** An extrametricality rule is blocked if it would render the entire domain of the stress rules extrametrical.

Constraint (47b) was first expressed as a well-formedness condition (rather than as a constraint on the format of extrametricality rules) by Harris (1983), who referred to it as the **Peripherality Condition;** I follow Harris's conception and terminology here. Provision (47c) is supported by a very strong skewing toward right-edge extrametricality in the attested examples. Provision (47d) is included to allow stressing of monosyllables in languages with syllable extrametricality.

Somewhat tentatively, I exclude "mora" from the list in (47a), based on the absence of plausible cases. Unambiguous cases of mora extrametricality can be excluded by rigorous enforcement of the principle that foot boundaries cannot occur syllable-internally (§ 3.9.1). In this view, extrametricality theory comprises two domains: there is segmental extrametricality, which exempts segments from mora assignment, and higher level extrametricality, which exempts syllables and feet from rules creating metrical structure (cf. Roca 1992). For discussion of a possible counterexample, see § 4.1.3.

Here is an example of syllable extrametricality: in Macedonian and other languages listed in § 6.2.3, stress normally falls on the antepenultimate syllable of a word (and on the initial syllable of shorter words). This can be derived by marking final syllables as extrametrical, then forming a binary, left-strong foot:

(48) $(\times\ .)$
$\ldots\ \sigma\sigma\ \langle\sigma\rangle\ \#$

The notation used here for extrametricality rules is given in (49):

(49) $X \rightarrow \langle X \rangle\ /\ \_\_\_\ ]_D$

where X is some phonological constituent and $]_D$ is the edge of the domain (usually the word) of the stress rules. This notation is used only in the interest of clarity, since it includes several redundancies: (a) the identity of D is given by the stress rules in general; (b) adjacency to $]_D$ is invariantly required by the Peripherality Condition; (c) given the markedness of left-edge extrametricality, the fact that we have $/\_\_\_\ ]_D$ instead of $/\ [_D\ \_\_\_$ can be assumed as a default specification, in the sense of Archangeli and Pulleyblank, in press.

Extrametricality strikes many people as intuitively non-obvious. Thus special thoroughness in marshaling formal arguments in favor of it is required.

(a) Extrametricality permits a sharp reduction in the class of possible foot templates. For example, it is the principal mechanism that makes it possible to eliminate basic ternary templates from metrical theory (for the residual cases, see chap. 8). A simple example is the case of antepenultimate stress just mentioned, where the extrametricality account permits us to dispense with the ternary template (50a). Syllable extrametricality also allows us to dispense with the ternary template (50b) for Latin and similar cases (§ 5.1.4):

(50) a. (× . .)   b. (× . .)
σ σ σ   σ ◡ σ

The crucial point is that what looks like stress assigned by the templates of (50) is in fact found only where the templates would occur at the right edge of a word. A better strategy is to analyze the word-final cases with extrametricality. This avoids foot templates that, when used other than word-finally, generate totally unattested stress systems. For example, if we parse words iteratively from left to right with (50b), we derive a stress system that is nowhere attested.

(b) In measuring syllable weight, word-final position is likewise often a special case: a word-final syllable must have more consonants to be counted as heavy, since word-final consonants are often extrametrical.

(c) Extrametricality is crucial to the theory of syllabification in Ito 1986, which posits that the convention of **Stray Erasure** (Harris 1983) applies to any segment that is neither syllabified nor extrametrical. This allows Ito to derive cases of consonant deletion previously thought to require idiosyncratic deletion rules.

(d) In various languages different lexical classes (e.g. nouns vs. verbs, regular vs. exceptional words) have distinct but related stress patterns. Extrametricality permits us to capture the unifying principles of such systems with the foot template and parsing algorithms, while characterizing the distinct aspects of stress in different lexical classes with extrametricality. Instances of this may be found in English (Hayes 1982b), Spanish (Harris 1983, 1992; den Os and Kager 1986), Romanian (Steriade 1984), Onondaga (Chafe 1977, 175; § 6.3.11), Yawelmani (Archangeli 1984), Djingili (§ 6.2.3), central Macedonian dialects (Franks 1987, 141), Chamorro (Chung 1983), Lenakel (§ 6.1.8), Polish (§ 6.2.3), Paamese (§ 6.1.9), Pirahã (§ 7.1.7), and Cayuvava (§ 8.2). In a dialect of Hindi (§ 6.1.7), optional extrametricality creates free variation in stress.

(e) The Peripherality Condition accounts for cases in which idiosyncratic stressing of a stem is replaced by regular stressing when a suffix is added, as in Spanish, Polish, Yawelmani, and Chamorro. Once suffixation has taken place, an idiosyncratically extrametrical stem-final syllable is no longer peripheral and thus loses its extrametricality. The normal stress rules then apply.

(f) Extrametricality permits a simple account of stress rules that include **avoidance clauses** (Hayes 1982b). In such languages, stress is assigned from the left edge of the word, with an overriding proviso that it not fall on final syllables. Such cases can be analyzed with a normal stress rule computing stress from the left, but with final syllables extrametrical.

(g) As Prince 1983a showed, extrametricality permits a simplification in the theory of **labeling rules,** the rules that determine prominence relations within a single domain. In early metrical theory, it was thought that there must be four basic algorithms:

(51) **Labeling Rules** (Halle and Vergnaud 1978)
  a. Rightmost elements are strong.
  b. Leftmost elements are strong.
  c. Rightmost elements are strong if and only if they branch.
  d. Leftmost elements are strong if and only if they branch.

(An element is said to branch if it contains more than one constituent.)

An example of a rule that was believed to require reference to branching is that responsible for labeling the word layer in English nouns: here, the right node is normally labeled strong if it branches, as in (52a), but if the right node does not branch, prominence lands on a foot to its left, as in (52b):

```
(52) a. Branching  (     ×  )        b. Non-branching  (×        )
                   (× .)(× .)                          (× .)(×   )
                   Ìsidóra                             Ísidòre
```

Now, on independent grounds it can be shown that the last syllable of English nouns is extrametrical: nouns often display antepenultimate stress, which Hayes 1982b derives in the manner of (48). Assuming that extrametrical elements are not accessible to word layer labeling, the correct outcome is obtained simply by placing the /×/ of the word layer on the rightmost VISIBLE subordinate /×/ at the foot layer, invoking the simpler "right strong" labeling procedure:

```
(53) a. (     ×)          b. (×  )
        (× .)(×)             (× .)(×  )
        Ìsidó⟨ra⟩            Ísi⟨dòre⟩
```

As Hayes 1982b notes, the morphological classes that allow antepenultimate stress are approximately the same as those which (in the earlier account) require the "right strong if branching" algorithm (the others simply use "right strong"). This correlation follows directly from the extrametricality account.

All the other cases that earlier appeared to require labeling based on branching can likewise be reanalyzed using extrametricality or other means (see for example Prince 1983a for English compounds, § 4.1.2 for Seminole/Creek, § 4.1.3 for Cairene Arabic, § 7.1.8.4 for Asheninca, and Barker 1989 and § 6.3.10 for Turkish). Thus extrametricality permits labeling theory to be simplified, limiting it to the two fundamental cases "right strong" and "left strong."

It can be seen that the range of phenomena for which extrametricality can provide a formal account is fairly broad. For more on extrametricality, see § 5.2.

3.12 LABELING RULES

As just noted, extrametricality theory makes possible a very simple account of how prominence relations among feet are established. Adapting the terminology of Prince 1983a, I use the term **End Rule (Left/Right)** to describe a metrical rule with the effects listed in (54):

(54) **End Rule (Left/Right)**
  a. Create a new metrical constituent of maximal size at the top of the existing structure.
  b. Place the grid mark forming the head of this constituent in the (leftmost/rightmost) available position.

The English word layer rule can be seen as an instance of End Rule Right, moderated by syllable extrametricality. The notion **available position** in (54b) means a position where a grid mark may be placed without violating the Continuous Column Constraint (9). We will see reasons for revising the End Rule later on (§ 9.4), but for the moment (54) will suffice.

A large number of languages have simple initial or final stress; see the listings in Hyman 1977b. Where there is no secondary stress (e.g. as in Bengali; Hayes and Lahiri 1991a), stress can be assigned exclusively by the appropriate End Rule.

3.13 CONCLUSION

This review of the typology and theory of stress has covered: (a) parallels between stress and rhythmic structure, and how they are characterized in grid notation; (b) universal patterns of stress behavior and how they are characterized by grid theory, particularly by the Continuous Column Constraint; (c) the evidence for including grouping structure in metrical representations (i.e. bracketed grids); (d) the notion of a parametric metrical theory, with an inventory of foot templates as its central content; (e) theoretical notions crucial to parametric metrical theory: syllable weight and its formal representation, extrametricality rules, labeling rules. In the next two chapters, I present a proposal for a parametric metrical theory of word stress rules.

• 4 •

Foot Inventory

The central question addressed in any parametric metrical theory concerns the basic foot shapes it allows. This chapter begins with three case studies, given as examples of the three basic foot types to be assumed here. I then argue that these three types suffice as a complete set of bounded feet—that the set is both highly restrictive and empirically sufficient. The last part of the chapter argues that the inventory is grounded in a basic law of rhythmic structure, to be called the **Iambic/Trochaic Law.**

The three-member inventory presented here was first proposed independently by McCarthy and Prince 1986 and Hayes 1987; both proposals were based on observations about foot typology made in Hayes 1985.

4.1 SOME EXAMPLES OF BOUNDED SYSTEMS

4.1.1 Syllabic Trochees in Pintupi

Pintupi is a Pama-Nyungan language of Australia, described by Hansen and Hansen 1969, 1978; for a related dialect see Douglas 1958. The metrical analysis given here follows that of Hammond 1986 in most respects. This presentation covers most of the Pintupi stress system; a further detail is presented in § 5.1.9. The simplifications assumed here do not crucially affect the status of Pintupi as an illustration of a particular foot type, namely the **syllabic trochee.**

Main stress in Pintupi falls on the initial syllable of a word. Secondary stress falls on every other syllable thereafter, but not on final syllables. Content words in Pintupi usually have at least two syllables (§ 5.1.9 discusses the exceptions), and they may have up to nine syllables in the data presented. The examples below are from Hansen and Hansen 1969, 163.

(1)	a. σ́ σ	*páɳa*	'earth'
	b. σ́ σ σ	*tʲúʈaya*	'many'
	c. σ́ σ σ̀ σ	*máɭawàna*	'through from behind'
	d. σ́ σ σ̀ σ σ	*púɭiŋkàlatʲu*	'we (sat) on the hill'

e. σ́ σ σ̀ σ σ̀ σ	*tʲámulìmpatʲùŋku*	'our relation'
f. σ́ σ σ̀ σ σ̀ σ σ	*tíɭirìŋulàmpatʲu*	'the fire for our benefit flared up'
g. σ́ σ σ̀ σ σ̀ σ σ̀ σ	*kúranʲùlulìmpatʲùɻa*	'the first one (who is) our relation'
h. σ́ σ σ̀ σ σ̀ σ σ̀ σ σ	*yúmaɻìŋkamàratʲùɻaka*	'because of mother-in-law'

In support of their stress description, Hansen and Hansen note that certain vowels have distinct allophones in stressless syllables; cf. [tʲárʌnʲtʲàrʌnʲpʌ] (phonemic /tʲaranʲtʲaranʲpa/) 'ant (species)' (1969, 157), where stressless /a/ is realized phonetically as [ʌ]. Word-final vowels that are also phrase-final may be devoiced, again supporting the view that they are stressless.

The analysis of Pintupi stress is straightforward. The calculation of stress clearly must go from left to right, since the alternating pattern is "left-justified." Since stresses occur every two syllables, and odd syllables get stressed, the foot structure is determined: feet must be disyllabic, with prominence on the initial syllable of the foot. This structure will be referred to throughout this book as the "syllabic trochee":

(2) **Syllabic Trochee**
```
(×  .)
 σ  σ
```

Syllabic means that the foot template simply counts syllables, ignoring their internal structure; **trochee** is borrowed from classical metrics and means 'disyllabic foot with initial prominence'.

The difference between primary and secondary stress in Pintupi can be derived by constructing an additional layer of metrical structure (the **word layer**) atop the feet, using End Rule Left (§ 3.12). Thus the full analysis is as in (3).

(3) **Foot Construction** Parse words into syllabic trochees, going from left to right.

Word Layer Construction End Rule Left

```
a. (×  .)(×  .)(×  .)       b. (× .)(× .)(×  .)          Foot
    σ  σ σ   σ σ  σ             σ σ σ σ σ  σ σ           Construction
   tʲamulimpatʲuŋku             tiɭiriŋulampatʲu

   (×               )           (×               )       Word Layer
   (×  .)(×  .)(×  .)           (× .)(× .)(×  .)         Construction
    σ  σ σ   σ σ  σ             σ σ σ σ σ  σ σ
   tʲámulìmpatʲùŋku             tíɭirìŋulàmpatʲu
```

Note that the final syllable in (3b) is a constituent of the word layer but belongs to no foot, since the template does not allow monosyllabic feet. Unfooted syllables are in general assumed to be stressless. It can be seen that the theory adopted here explicitly rejects the view that all syllables must be footed. For further discussion, see § 5.1, § 5.2.3, and chapter 8.

Further examples of syllabic trochees are given in § 6.2.

4.1.2 Iambs in Seminole/Creek

Seminole and Creek are the names used for two dialects of a single Muskogean language. Both are spoken in Oklahoma, Seminole in Florida as well. Seminole/Creek has played an influential role in the history of metrical theory. Earlier metrical analyses include Halle and Vergnaud 1978; Hayes 1981; Prince 1983a; HV; Jackson 1987; Tyhurst 1987; and Blevins 1990. For all but Jackson and Tyhurst's analyses, the data were taken from the work of Haas (1977), who discovered the basic factual generalizations. Below, I have relied in addition on data and comments from Jack Martin and Pamela Munro (p.c.).

Seminole/Creek syllables fall into two classes: light (CV) and heavy (CVC, CVː, and longer). I will first describe the patterning of accent in bare stems, in some cases accompanied by closely cohering affixes. The relevant forms are according to Jackson 1987 those derived at Level I of the morphology; I will refer to them as **simplex words.**

The accentual pattern of simplex words is as follows. If a word consists of only light syllables, accent falls either on the penult or on the final syllable, whichever is preceded by an odd number of syllables. In the data below, /c/ represents a phoneme variously realized as [č] or [t͡s].

(4) a.	/⏑ ⏑́/	*cokó*	'house'	Tyhurst 162
		ifá	'dog'	T 163
b.	/⏑ ⏑́ ⏑/	*am-ífa*	'my dog'	T 163
		osána	'otter'	T 163
c.	/⏑ ⏑ ⏑ ⏑́/	*pom-osaná*	'our otter'	T 163
		apataká	'pancake'	T 163
d.	/⏑ ⏑ ⏑ ⏑́ ⏑/	*am-apatáka*	'my pancake'	T 163
		anokicíta	'to love'	J.M. p.c.
e.	/⏑ ⏑ ⏑ ⏑ ⏑ ⏑́/	*isimahicitá*	'one to sight at one'	Haas 203
		am-anokicitá	'to love mine'	J.M. p.c.
f.	/⏑ ⏑ ⏑ ⏑ ⏑ ⏑́ ⏑/	*itiwanayipíta*	'to tie each other'	H 203
		amanokic-ak-íta	'to love mine (pl.subj.)'	J.M. p.c.

The presence of heavy syllables affects stress: simplex words ending in /–/ receive final accent (5), whereas simplex words ending in /– ⏑/ receive penultimate accent (6).

(5) a.	/–́/	*fóː*	'bee'	T 162
b.	/⏑ –́/	*niháː*	'lard'	T 162
c.	/– –́/	*hoktíː*	'woman'	Jackson 91

d.	/⏑ ⏑ –́/	*hitot-íː*	'snow'	T 163
e.	/⏑ – –́/	*haɬiːssíː*	'moon'	J.M. p.c.
f.	/– ⏑ –́/	*akhasíː*	'lake'	T 165
g.	/– – –́/	*tiːniːtkíː*	'thunder'	T 162

(6)	a.	/–́ ⏑/	*ícki*	'mother'	T 165
	b.	/⏑ –́ ⏑/	*kofócka*	'mint'	T 162
	c.	/– –́ ⏑/	*akcáwhka*	'stork'	J.M. p.c.

If the heavy syllable occurs earlier in the word, stress falls either on the penult or on the final syllable, whichever is separated by an odd number of light syllables from the preceding heavy. This resembles the pattern found in words with all light syllables (4).

(7)	a.	/– ⏑ ⏑́/	*taːskitá*	'to jump (sg. subj.)'	J 82
		/⏑ ⏑ – ⏑ ⏑́/	*atiloːyitá*	'to gather (pl. obj.)'	J.M. p.c.
		/ – ⏑ ⏑ – ⏑ ⏑́/	*nafkitikaːyitá*	'to hit (pl. obj.)'	J 82
	b.	/– ⏑ ⏑́ ⏑/	*taːshokíta*	'to jump (dual subj.)'	J 82
		/⏑ – ⏑ ⏑́ ⏑/	*tokoɬhokíta*	'to run (dual subj.)'	J 82
	c.	/– ⏑ ⏑ ⏑ ⏑́/	*iŋkosapitá*	'one to implore'	H 204

The analysis given here for these facts is an adaptation of earlier foot-based accounts, cited above. The foot template allows at most two syllables and is right-strong; and any disyllabic foot must have a light syllable as its left member. I will call this foot template the **iamb,** again following the terminology of classical metrics.

(8) **Iamb** Form (. ×) ⏑ σ if possible; otherwise form (×) –.

This formulation of the iamb differs slightly from earlier accounts in that it does not allow a foot to consist of a single light syllable. For discussion of such feet, see § 5.1. Anticipating this later discussion, one could say that an iamb is basically /⏑ –́/ or anything shorter, with a separate provision (not specific to iambs) that excludes /⏑́/.

Halle and Vergnaud 1978 first observed that the odd/even count respected in Seminole/Creek can be derived by means of parsing into maximally binary feet. The particular rule adopted here is given in (9):

(9) **Foot Construction** Form iambs from left to right.

a. (. ×)(. ×)	b. (. ×)(. ×)	c. (×) (. ×)
⏑ ⏑ ⏑ ⏑	⏑ ⏑ ⏑ ⏑ ⏑	– ⏑ ⏑ ⏑
apataka	*amapataka*	*taːshokita*
d. (. ×) (. ×)	e. (×) (×)	f. (×) (×) (×)
⏑ – ⏑ ⏑ ⏑	– – ⏑	– – –
tokoɬhokita	*akcawhka*	*tiːniːtkiː*

The three legitimate iamb shapes, /⏑ –́/, /⏑ ⏑́/, and /–́/, can be seen in these forms. When a single light syllable is left over at the end of the parse, as in (9b–e), no foot is formed, since /⏑́/ is excluded as a foot.

Once feet are in place, the derivation of the position of accent is straightforward: we create a word layer with End Rule Right (§ 3.12):

(10) **Word Layer Construction** End Rule Right

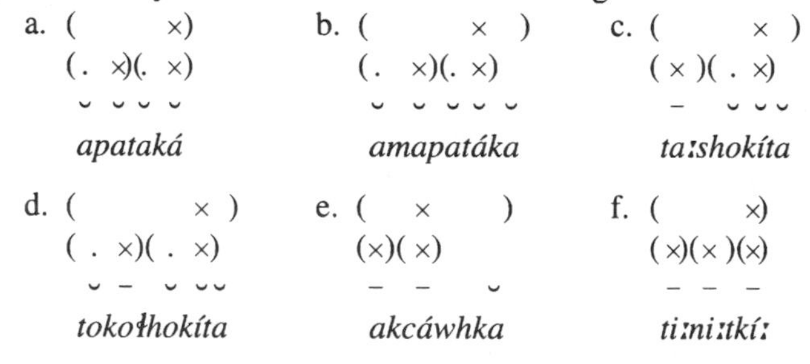

In general, this procedure will derive the accent patterns outlined above. Notice the crucial mechanism in avoiding accent on the final syllables of (10b–e): the final syllable cannot support a foot, since feet of the form /⏑́/ are disallowed. Thus the final syllable provides no docking site for the word layer grid mark, due to the Continuous Column Constraint (§ 3.4.2):

(11) *(×)
(×) (. ×)
– ⏑ ⏑ ⏑
ta:shokitá

If /⏑́/ feet are disallowed, then in principle Seminole/Creek could not have words consisting of a single light syllable, assuming (see § 3.8.5, § 5.1) that every word must be footed. This is indeed the case; content words of the form /⏑/ are missing (Jack Martin, p.c.), and (as Martin points out) the phonology of the language suggests that this absence is not accidental. In particular, an optional rule of initial short vowel deletion fails to apply if it would create a /⏑́/ word: compare /ifa+wo:hka/ → *ifawó:hka, fawó:hka* 'dog-barker' = 'hound' with /ifa/ → *ifá* (only) 'dog'. CV function words do occur: *ma* 'that, there'; *ya* 'this, here'. According to Martin, such words usually do not occur as isolation forms; they either are amplified by a case marker (nom. *ma-t, ya-t,* acc. *ma-n, ya-n*) or occur unaccented in the presence of an adjacent noun, where they might be analyzed as having been phonologically cliticized. When mentioned, as in "the word *ma,*" they are lengthened, but retain the typical schwa vowel quality of short /a/.

The metrical structures of (10) in principle predict the existence of secondary stresses. The extent to which these are actually realized in phonetic forms is an unsettled issue (Haas 1977; Blevins 1990; Martin p.c.). The difficulties arise from the fact that Seminole/Creek lacks the usual phonetic correlates of metrical structure, because its accentual system is essentially tonal (Haas 1977). Typically, a High tone is placed on the metrically strongest syllable, and other tones may appear as well, as part of a verbal grade system. As Martin 1992 shows, these additional tones sometimes dock onto the strongest syllables of feet other than the rightmost one in the word. This suggests that full metrical structure is indeed present in Seminole/Creek surface representations, although its phonetic manifestation is often incomplete. For further discussion, see § 5.5.2.

The crucial aspect of Seminole/Creek for our purposes is that its foot structure forms a clear illustration of the iamb, one of the basic foot templates adopted here. Further illustrations appear in § 6.3.

4.1.3 Moraic Trochees in Cairene Arabic

The stress pattern of Arabic as spoken in Cairo has also played a central role in the development of metrical theory. The first formal metrical account of Cairene stress was McCarthy 1978, 1979a, and numerous variants have been proposed in subsequent literature, including Halle and Vergnaud 1978; Hayes 1981; Prince 1983a; Selkirk 1984; and HV. The analysis below is based on these accounts, as well as the earlier quasi-metrical analysis of Allen 1973, 165.

The facts treated include data from "Cairene Classical Arabic," which is the pronunciation of Classical Arabic employed in universities of Cairo. Data of this sort are from Mitchell 1960 and Kenstowicz 1980, 40. With a possible minor exception, such words are pronounced with the same stressing that is assigned to colloquial Cairene forms. The Classical forms are of interest since they show a wider array of possible syllabic shapes and thus provide a stiffer test for any proposed account. The claimed difference between stress in Cairene Classical Arabic and in colloquial is that words with final long vowels are said to receive nonfinal stress in Cairene Classical but final stress in the colloquial (where they are rare).

Syllables in Cairene fall into three categories: light CV (/⌣/), heavy CVC and CVː (/–/), and superheavy CVCC and CVːC, symbolized /=/. Superheavy syllables are largely restricted to final position. The basic stress pattern, which was originally discovered by Mitchell 1960, is stated in (12). In the data, Cairene Classical forms are marked *Cl.* Most of the glosses are as provided by McCarthy 1979a; and page numbers for forms from Mitchell 1960 are from the 1975 reprinting of his article.

(12) a. Stress the final syllable if it is superheavy or colloquial /CVː/:

/⌣ =́/	*katábt*	'I wrote'	Harrell 1957, 15
/– =́/	*haǰǰáːt*	'pilgrimages' *Cl.*	Mitchell 1975, 77
/⌣ –́/	*gatóː*	'cake'	Mi 81

b. Otherwise, stress the penult provided it is heavy:

/–́ –/	*béːtak*	'your (m.sg.) house'	H 15
/⏑ –́ ⏑/	*katábta*	'you (m.sg.) wrote' *Cl.*	Mi 78
/⏑ –́ –/	*mudárris*	'teacher'	McCarthy 1979a, 446
/– –́ ⏑/	*haːðáːni*	'these (m.dual)' *Cl.*	Mi 77

c. Otherwise, stress the penult or the antepenult, whichever is separated by an even number of syllables from the closest preceding heavy syllable (A), or (if there is no such syllable) from the beginning of the word (B):

i. Penultimate Stress

A.	/– ⏑́ ⏑/	*qattála*	'he killed' *Cl.*	Mi 77
	/⏑ – ⏑́ –/	*mudarrísit*	'teacher (f. construct)'	McC 446
	/– ⏑ ⏑ ⏑́ ⏑/	*ʔadwiyatúhu*	'his drugs (nom.)' *Cl.*	Mi 79
B.	/⏑́ –/	*fíhim*	'he understood'	Kenstowicz 1980, 42
	/⏑ ⏑ ⏑́ –/	*šaǰarátun*	'tree (nom.)' *Cl.*	Mi 78
	/⏑ ⏑ ⏑́ ⏑/	*katabítu*	'she wrote it (m.)'	H 15
	/⏑ ⏑ ⏑ ⏑ ⏑́ –/	*šaǰaratuhúmaː*	'their (dual) tree (nom.)' *Cl.*	Mi 79

ii. Antepenultimate Stress

A.	/– ⏑́ ⏑ ⏑/	*ʔinkásara*	'it got broken' *Cl.*	Mi 77
	/– ⏑ ⏑ ⏑́ ⏑ ⏑/	*ʔadwiyatúhumaː*	'their (dual) drugs' *Cl.*	Mi 79
B.	/⏑́ ⏑ ⏑/	*kátaba*	'he wrote' *Cl.*	Mi 77
	/⏑ ⏑ ⏑́ ⏑ ⏑/	*šaǰarátuhu*	'his tree (nom.)' *Cl.*	Mi 80

The complexity of this pattern is striking. McCarthy's analysis constituted one of the strong early arguments for the metrical approach, since it reduced this complexity to the combination of a small number of elements, each in itself simple. I suggest below that theoretical ideas developed since McCarthy's account can produce a slight further simplification.

In analyzing this system, the first issue that must be dealt with is syllable quantity. In nonfinal position, the weight distinction is ordinary, opposing light CV syllables to all heavier syllables. In final position, the opposition is again binary, but on a different basis: the syllable types CVːC, CVCC, and CVː attract stress, while CVC and CV do not (for CVː in Classical forms, see below).

I analyze this as a case of consonant extrametricality (§ 3.11): this demotes heavy CVC to light CV⟨C⟩ in final position, while retaining final CVː, CVC⟨C⟩, and CVː⟨C⟩ as heavy. The rule is given below in (14a).

In Classical words, final CVː is counted as light, not heavy. We can add an additional rule for this, marking the second mora of a long vowel as extrametrical in Classical words (14b). Comments by Harrell (1960, 25) suggest that the status of such syllables is doubtful in any event: he claims that vowel length distinctions in final position normally go unrealized in the pronunciation of Classical forms; see also Harms 1981, 434. Thus Mitchell's data, in which final unstressed /CVː/ is claimed to be pronounced, may be somewhat artificial in any event. This bears on the suggestion made in § 3.11 that mora extrametricality does not exist.

A striking aspect of the Cairene data is the distinction between odd and even sequences of light syllables (12c), which determines whether the penult or antepenult will be stressed. Allen 1973 and McCarthy saw that this can be analyzed as a tacit alternating pattern, with light syllables grouped into pairs going from left to right. My analysis follows Allen in assuming that single heavy syllables form a metrical unit as well. Thus feet are defined as in (13):

(13) **Moraic Trochee** (× .) or (×)
⏑ ⏑ or –

The term **moraic trochee** is adopted since the foot consists of two moras, of which the first is stronger. This is obvious for a disyllabic moraic trochee. The trochaic character of a heavy syllable is argued for by Prince (1983a), who notes the characteristic sonority profile, with the first mora of the heavy syllable more sonorous than the second. Even in long vowels, it is arguable that the first mora of /–/ is in an intuitive sense stronger: it normally serves as the docking site for High tone, in languages that limit H tones to one mora per syllable, such as Winnebago (§ 8.9), Kinyambo (Bickmore 1989, 1992), and the deeper phonology of Serbo-Croatian (Inkelas and Zec 1988).

The moraic trochee as just defined provides a simple account of the Cairene pattern, for which I state the first three rules in (14). Mora Extrametricality is placed in parentheses because of the doubts that have been expressed about its descriptive validity.

(14) a. **Consonant Extrametricality** $C \rightarrow \langle C \rangle$ / ___ $]_{word}$

b. **(Mora Extrametricality)** $\mu \rightarrow \langle \mu \rangle$ / μ ___ $]_{word}$ in Classical words
(μ μ linked to α)

c. **Foot Construction** Parse the word from left to right into moraic trochees.

In representative examples, these rules create the structures in (15). Note that syllable weight is adjusted to correspond to the application of the extrametricality rules; mora extrametricality on a long vowel is indicated as /V⟨ː⟩/.

(15) a. **Final Superheavies, Final CVː in Colloquial Words**

(×) (×)
◡ – ◡ –
katab⟨t⟩ gatoː

b. **Heavy Penults**

(×) (×)(×)
◡ – ◡ – – ◡
mudarri⟨s⟩ haːðaːni

c. **Light Penult, Even Parity**

(× .) (× .)(× .) (× .)(× .)(× .) (×)(× .)(× .) (×)(× .)
◡ ◡ ◡ ◡ ◡ ◡ ◡ ◡ ◡ ◡ ◡ ◡ – ◡ ◡ ◡ ◡ ◡ – ◡ ◡
fihi⟨m⟩ katabitu šaǰaratuhuma⟨ː⟩ ʔadwiyatuhu mudarrisi⟨t⟩

d. **Light Penult, Odd Parity**

(× .) (× .)(× .) (×)(× .) (×)(× .)(× .)
◡ ◡ ◡ ◡ ◡ ◡ ◡ ◡ – ◡ ◡ ◡ – ◡ ◡ ◡ ◡ ◡
kataba šaǰaratuhu ʔinkasara ʔadwiyatuhuma⟨ː⟩

It can be seen that the alternating count of light syllables is carried out by the bimoraic foot structure. The count is restarted after heavy syllables, since heavy syllables are necessarily followed by foot boundaries.

The main stress can then be located in the correct position by creating a higher metrical layer, as in (16):

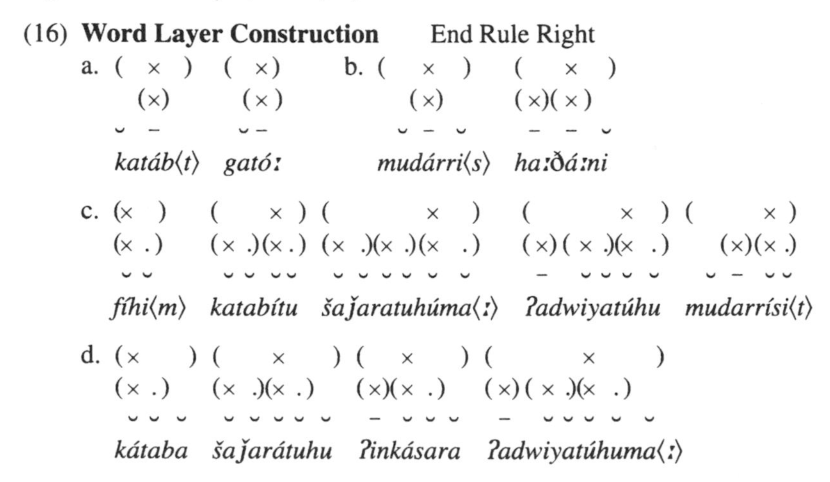

An important aspect of the analysis is that a single light syllable (including syllables counted as light due to extrametricality) cannot form a foot. This is in fact independently supported by the absence of CV(C) content words in Cairene (Broselow 1988; for the minimal word in Arabic in general see McCarthy and Prince 1990, 251–60). Because /◡́/ feet are disallowed, in the cases of (16d) the final syllable is not footed. End Rule Right cannot promote this syllable, since to do so would violate the Continuous Column Constraint (cf. (11) for Seminole/Creek):

```
(17) *(      ×)
      ( × .)
       ˘ ˘ ˘
      kataba
```

The same facts are accounted for in McCarthy's analysis by invoking a labeling rule based on branching. My assumption is that with appropriate devices this kind of labeling rule can be dispensed with (§ 3.12).

The metrical structures that generate primary stress also predict secondary stresses on the heads of non-primary stressed feet. Kenstowicz 1980 notes that at least for certain word shapes, this prediction is independently supported, since only vowels in weak position within the foot undergo a rule of phrasal Syncope. Secondary stress is also supported by the observations of Harrell 1960 on a different speech style (§ 6.1.2). However, the analysis cannot derive the secondary stress pattern described by Welden 1980; if Welden's observations are correct, then we must posit additional rules that would go beyond the scope of the theory proposed here. Welden's observations are in any event controversial; see HV, p. 60.

Further examples of stress systems based on the moraic trochee are presented in § 6.1.

4.2 JUSTIFYING THE FOOT INVENTORY

The three analyses just presented were chosen as representative cases of the three basic bounded foot types adopted in this book. They are restated in (18):

```
(18) a. Syllabic Trochee   (× .)
                            σ σ

     b. Moraic Trochee     (× .)       (×)
                            ˘ ˘    or   –

     c. Iamb               (. ×)       (×)
                            ˘ σ    or   –
```

I attempt to justify this inventory in two ways: by showing that it best fits the data, and by showing that (initial appearances to the contrary) it forms a natural and expected set. The first task is a very large one, and in this chapter I give only the form of the arguments, filling in data in the later chapters. The second task is undertaken in § 4.5.

One way of determining what feet are needed is to consider the typology of stress patterns from a different point of view. In particular, the framework of Prince 1983a provides a way of classifying the logical possibilities straightforwardly, so we can check what is empirically attested. As noted in § 3.8.3, Prince's theory assigns stress with direct placement of grid marks instead of foot parsing, using two basic parameters:

(19) a. Is the alternation of stresses computed left to right or right to left?
b. Is the first syllable encountered by the rule made stressed ("peak first") or stressless ("trough first")?

In a quantity-insensitive language, that is, one that treats all syllables the same, the four possibilities obtained by combining (19a) and (19b) would come out as in (20):

(20)	i. Even-Syllabled Words	ii. Odd-Syllabled Words
a. Left to Right, Peak First	σ́ σ σ́ σ σ́ σ σ́ σ ⟶	σ́ σ σ́ σ σ́ σ σ́ ⟶
b. Left to Right, Trough First	σ σ́ σ σ́ σ σ́ σ σ́ ⟶	σ σ́ σ σ́ σ σ́ σ ⟶
c. Right to Left, Peak First	σ σ́ σ σ́ σ σ́ σ σ́ ⟵	σ́ σ σ́ σ σ́ σ σ́ ⟵
d. Right to Left, Trough First	σ́ σ σ́ σ σ́ σ σ́ σ ⟵	σ σ́ σ σ́ σ σ́ σ ⟵

When we add to this classification the possibility of contrasting syllable quantity, more patterns arise. Normally, heavy syllables receive stress, unless overriding factors such as destressing rules intervene. Moreover, it appears that heavy syllables invariably interrupt any alternating count of light syllables; for example, the odd/even relations among light syllables seen in Cairene Arabic and Seminole/Creek are recalculated whenever the alternation hits a heavy syllable and starts in on a new sequence of lights. Thus in quantity-based systems, we must add a third question to (19a, b):

(21) When a heavy syllable is found, does the alternating count resume peak first or trough first?

For example, Cairene Arabic resumes the count peak first, Seminole/Creek trough first. Adding this option to (19a, b), we obtain eight logical possibilities.

The striking fact is that of the twelve logical possibilities outlined above (four quantity-insensitive, eight quantity-sensitive), many are completely unattested, or attested only marginally.

For example, there appear to be no right-to-left quantity-sensitive systems that are peak first both departing from the word edge and after hitting a heavy syllable. Such a system would result from right-to-left construction of a foot consisting of the mirror image of the moraic trochee:

(22) ×)(. ×)(. ×)(×)(×)(×)(×)(. ×)(. ×)(×)(×)(×)(. ×)(. ×)
. . . ˘ ˘ ˘ ˘ ˘ – – – – ˘ ˘ ˘ ˘ – – – ˘ ˘ ˘ ˘
←

Since the present system lacks such a foot, it will not generate the pattern of (22).

Some of the gaps are not absolute but involve cross-linguistic asymmetries. For example, in the great majority of cases (in this book, sixty-one of sixty-five), quantity-insensitive alternation is peak first going from left to right, but trough first going from right to left. This asymmetrical pattern would hardly be expected if peak first versus trough first were a fundamental parameter of the system, but it makes sense under the theory proposed here (§ 3.8.3): these patterns are the result of parsing a word into syllable trochees, going in either direction.

My specific proposal is this: the syllabic trochee is the basic mechanism available for quantity-insensitive alternation, whereas its mirror image is excluded from the inventory of basic foot types. This generates peak-first left-to-right and trough-first right-to-left patterns, and not their mirror images.

The question remains of what is to be done with the rare cases that go the other way. One option, suggested by Jacobs 1990, is simply to adopt another foot type, the mirror image of the syllabic trochee, and stipulate that it is marked, owing to the Iambic/Trochaic Law (discussed below in § 4.5.1). I believe, however, that it is feasible to maintain a stronger theory, analyzing the systems that appear to require such a mirror image foot using other formal devices (see § 6.3.11), and retaining the inventory of (18) as the complete set of bounded feet.

In cases of quantity-sensitive alternation, there are also asymmetries. The overall picture is given in (23), which presupposes that an alternating stress rule can be described by selecting a foot type and a direction of parsing. In addition, one must also consider cases where a syllable is made extrametrical at the edge from which the parse originates, creating in some cases a **double trough** at word edge.

(23) Dir.	At Word Edge	After / – /	Generated	Attested
a. L–R	peak	peak	yes, moraic trochees	yes (§ 6.1)
b. L–R	peak	trough	no	no
c. L–R	trough	peak	marginally; moraic trochees + extrametricality	no
d. L–R	trough	trough	yes, iambs	yes (§ 6.3)

e. L–R	double trough	peak	no	no
f. L–R	double trough	trough	marginally, iambs + extrametricality	marginally
g. R–L	peak	peak	no	no
h. R–L	peak	trough	yes, iambs	maybe
i. R–L	trough	peak	no	no
j. R–L	trough	trough	yes, moraic trochees	yes (§ 6.1)
k. R–L	double trough	peak	no	no
l. R–L	double trough	trough	yes, moraic trochees + extrametricality	yes (§ 6.1)

Some comments follow. For a possible case of (23b), Bani-Hassan Arabic, and a counteranalysis, see § 8.10. Cases (23c, f) are generated only as marginal options under the theory, since they would require left-edge extrametricality. Since left-edge extrametricality appears to be unusual under any circumstances (§ 3.11), we would expect such cases to be rare or unattested. The one case of which I am aware is of type (23f): Kashaya, discussed in Buckley 1991. The question of whether cases of right-to-left iambs (23h) are found is discussed in § 6.3.10.

Cases (23b, g, i), of which I have found no instances, can be generated under some versions of metrical theory, but not under the theory proposed here. In particular, (23b) would require the use of feet that looked like the mirror image of the iamb, while (23g, i) would require a kind of iamb that looked like the mirror image of the moraic trochee.

The general picture that emerges from the chart (and more crucially, from the analyses in chapter 6 on which the chart is based) is that there is a fairly close match between the predictions of the foot inventory proposed here and what is observed typologically.

The overall claim is that the stress rules of the world's languages are **skewed:** there are substantial differences in the frequency of a given pattern and its mirror image, for example (20a) versus (20c), or (23a) versus (23g). I suggest that accounting for such asymmetries, which would hardly be expected a priori, should be taken as one of the central tasks of parametric metrical theory. The core of my account is the basic foot inventory: syllabic trochee, moraic trochee, iamb. This inventory can account for the typological asymmetries because it is itself asymmetrical.

4.3 ALTERNATIVES

In arguing further for the inventory, it will be helpful to consider alternatives explicitly. I discuss here the theories of Hayes 1981 and HV.

4.3.1 Hayes 1981

Hayes 1981 proposed a set of bounded foot templates, all maximally disyllabic, which are defined by four parameters. The two parameters that play the most important role in the theory are as in (24):

(24) a. **Quantity-Sensitive** yes/no

b. **Dominance** left strong/right strong

For purposes of comparison, it is useful to unpack this two-by-two arrangement into its component foot construction algorithms:

(25) a. **Quantity-Insensitive Left Dominant** Form $\begin{array}{cc}(\times & .)\\ \sigma & \sigma\end{array}$ [otherwise form $\begin{array}{c}(\times)\\ \sigma\end{array}$]

b. **Quantity-Insensitive Right Dominant** Form $\begin{array}{cc}(. & \times)\\ \sigma & \sigma\end{array}$ [otherwise form $\begin{array}{c}(\times)\\ \sigma\end{array}$]

c. **Quantity-Sensitive Left Dominant** Form $\begin{array}{cc}(\times & .)\\ \sigma & \smile\end{array}$ [otherwise form $\begin{array}{c}(\times)\\ \sigma\end{array}$]

d. **Quantity-Sensitive Right Dominant** Form $\begin{array}{cc}(. & \times)\\ \smile & \sigma\end{array}$ [otherwise form $\begin{array}{c}(\times)\\ \sigma\end{array}$]

Comparison of theory (25) with the present proposal (18) is made more complex by the distinct ways in which the two theories treat cases where the available syllable string does not permit a maximal foot to be constructed; see § 5.1. For now, I will compare only the maximal foot shapes posited in the theories, since what one does when a maximal foot cannot be constructed is an independent issue.

The following equivalences can be seen:

(26) a. Iamb = Quantity-Sensitive Right Dominant = $\begin{array}{cc}(. & \times)\\ \smile & \sigma\end{array}$

b. Syllabic Trochee = Quantity-Insensitive Left Dominant = $\begin{array}{cc}(\times & .)\\ \sigma & \sigma\end{array}$

Elsewhere the systems are distinct. The proposal here does without quantity-insensitive right-dominant feet (25b) and uses moraic trochees for some of the stress systems that the earlier theory described with quantity-sensitive left-dominant feet (§ 5.1.4). The comparison of the moraic trochee with the quantity-sensitive left-dominant foot will be especially important here. For brevity, I will refer to the latter as the **uneven trochee,** uneven because the two sides of the foot may be unequal in the instantiation /– ⏑/. The competing proposals for a quantity based trochaic foot are repeated in (27):

(27) a. **Moraic Trochee** b. **Uneven Trochee**

a.		b.		
(× .)	(×)	(× .)		(×)
⏑ ⏑	–	σ ⏑	otherwise	σ

Analogously, the other foot we will do without, the right-dominant quantity-insensitive foot /σ σ́/, will be called the **even iamb.**

Two other parameters are posited in the Hayes 1981 proposal. One is a parameter of labeling, adopted from Halle and Vergnaud 1978: while in the unmarked case the "dominant" node of a foot is made strong, in the marked case foot labeling is carried out by the rule "dominant nodes are strong if and only if they branch," where under the theory of syllable weight employed by Hayes 1981, a heavy syllable is treated as branching. In effect, this provides two additional marked templates:

(28) a. **Quantity-Sensitive, Right Nodes Dominant, Dominant Strong iff Branching**

(. ×)		(× .)		(×)
⏑ –	or	⏑ ⏑	otherwise	σ

b. **Quantity-Sensitive, Left Nodes Dominant, Dominant Strong iff Branching**

(× .)		(. ×)		(×)
– ⏑	or	⏑ ⏑	otherwise	σ

The present proposal does without these feet.

Hayes 1981 also adopts from Halle and Vergnaud 1978 the **Obligatory Branching Parameter,** which requires that the head of a foot be a heavy syllable. This parameter adds two additional feet (left- and right-headed) to the system. Hammond 1986 revises the Obligatory Branching Parameter in a way that significantly expands its scope. Here, I propose to eliminate Obligatory Branching entirely. For alternative analyses of languages that have been argued to support the parameter, see § 5.1.5 (Malayalam and similar cases), § 6.3.10 (Turkish), § 6.1.8 (Lenakel), § 7.1.4 (Klamath), and § 7.2 (Khalkha Mongolian and similar cases).

4.3.2 HV

HV's theory factors out the parameter "quantity-sensitive" into a separate phonological rule: assign a grid mark to all heavy syllables. The distinction between quantity-sensitive and quantity-insensitive feet is thus not specified in the foot construction rules per se; rather, foot parsing gives the same results because a pre-existing grid mark must serve as the head of a foot. Coupled with a formal means of creating binary feet, this procedure generates the foot inventory of Hayes 1981, stated in (25). However, it is not a mere reformalization, since it has important additional advantages and disadvantages. For arguments on both sides, see HV 19–25; § 6.1.8.5; § 7.1; § 7.4; and § 8.5.4.

In addition to the feet of (25), HV's theory also allows for the equivalent of the moraic trochee. This is done by selecting rhyme segments instead of syllables as the stress-bearing elements (HV 60–63). HV also retains some of the effects of the Obligatory Branching Parameter, in the form of a rule of grid mark projection (HV 12, ex. (12)).

It can be seen that Halle and Vergnaud make assumptions about stress rule typology that are rather different from those adopted here. In order to create the equivalent of the moraic trochee, they assume that the choice of stress-bearing element is language-particular rather than universally being the syllable; for discussion, see § 3.9.1, § 5.6.2. In addition, since their formal mechanisms do not in general distinguish direction, it is reasonable to infer that they assume that the inventory of stress rules in the world's languages is largely symmetrical;[1] this issue is discussed in § 4.2.

Summing up, the theory advocated here attempts to make do with a rather smaller set of primitive metrical units than earlier work. In particular, I propose to dispense with the uneven trochee (25c) and the even iamb (25b). Both these units are posited in the theories of Hayes 1981 and HV. I also propose to do without labeling based on branching (28), the Obligatory Branching Parameter, and the use of any stress-bearing unit other than the syllable.

4.4 FURTHER ARGUMENTS

The fundamental support for the proposal is how it matches up against the typology of bounded stress rules, as discussed in § 4.2. This section presents further arguments. The discussion is in summary, with cross-references to full presentations of individual languages below.

(A) SEGMENTAL PHONOLOGY. As noted in Hayes 1985, 1987; McCarthy and Prince 1986; and Prince 1990, the segmental phonology of a language often appears to be directed toward enforcing the canonical (i.e. maximal) shapes of feet at the surface. The lengthenings and shortenings of syllables carried out by segmental phonology are skewed in the direction of reinforcing the basic durational patterns of feet, those proposed in the inventory of (18). This argument is developed further in § 4.5.3.

(B) FOOT EXTRAMETRICALITY. Foot extrametricality rules are not uncommon. The crucial question here is: What kinds of feet are diagnosed by foot extrametricality? We will see examples here of extrametricality for iambs

1. Under the assumption that all heavy syllables are left-headed, HV's theory generates no mirror image of the moraic trochee. Such feet are generated, however, under the theory of Halle 1990, which adds a further mechanism to the basic HV system.

and for moraic trochees. (The absence of extrametricality for syllabic trochees is unexplained.) Crucially, there are several cases that help decide between the functionally similar moraic trochee and the uneven trochee (27). These parse a final sequence of the form /– ⏑/ (from either direction) as in (29):

(29) a. **Moraic Trochees** (×)
– ⏑ #

b. **Uneven Trochees** (× .)
– ⏑ #

The moraic trochee template must skip over the final light syllable, since /– ⏑/ is excluded as a foot by the two-mora limit.

A crucial difference between (29a) and (29b) is that in (29a) the rightmost foot in the word is not peripheral, since the final stray light syllable intervenes between it and the word boundary. For this reason, the Peripherality Condition (§ 3.11 (47b)) blocks any foot extrametricality rule from applying to the foot in (29a). The non-peripherality of the rightmost foot can be seen more clearly in the "official" (see § 3.5) representation in (30):

(30) ×
(–) ⏑ #

In contrast, foot extrametricality could apply to (29b). This difference has empirical consequences in longer words when the word layer is constructed by End Rule Right: main stress will fall on the penult in (29a), but somewhere to the left of the penult in (29b).

In fact, of the cases discussed here (Palestinian Arabic, § 6.1.1; Egyptian Radio Arabic, § 6.1.2; Maithili, § 6.1.6; Hindi, § 6.1.7; Early Latin, § 6.1.9; Manam, § 6.1.9; and Awadhi, § 6.1.9), a heavy penult always receives the main stress. This observation follows if the uneven trochee is banished from the theory.

(c) Prosodic Morphology. As noted in § 3.8.4, the theory of prosodic morphology posits that rules of non-concatenative morphology may invoke only a restricted set of templates, which may be identified with the categories of prosodic theory. As McCarthy and Prince (1986) argue, the prosodic targets for morphology at the foot level are precisely those posited in the theory here: the syllabic trochee, the moraic trochee, and the iamb. Characteristically, the foot type that is used in a language's prosodic morphological system is the same as that used in its stress system. This correlation appears to be strong, though it is not absolute. For example, in Arabic dialects the morphology of the broken plural is based on the iamb (McCarthy and Prince 1990); but various individual Arabic dialects employ moraic trochees or unbounded feet in their stress systems.

Additional arguments favoring the moraic trochee over the uneven trochee are given in Mester 1992.

4.5 THE RHYTHMIC BASIS OF THE FOOT INVENTORY

So far, the arguments have been entirely empirical: the proposed foot inventory (18) is claimed to match the facts better than alternatives. What I have not addressed yet is the plausibility of (18) on purely theory-internal grounds.

Evaluating theories on their internal merits is often difficult. Most linguists would agree that a theory should predict a wide range of phenomena using a small number of simple principles. In this respect the inventory of (18) may seem defective: since it is completely asymmetrical, its structure must be listed (though see Prince 1990 for a derivation of something like (18) from basic principles, and for discussion Mester 1992). In contrast, the inventory in (25) is easily deducible from two simple parameters (quantity sensitivity and dominance) and is thus fully symmetrical.

In addressing this apparent defect, two points should be considered. First, it is important that enthusiasm for symmetry not cut off empirical inquiry. For example, current theories of syllable structure are focused in large part on accounting for asymmetries within the syllable, both in phonotactic possibilities and in the computation of weight. Were we to limit ourselves to symmetrical theories of the syllable, these interesting lines of inquiry would be cut off. Second, it is possible to seek generality for our theories in more than one way: what may appear as an arbitrary pattern when formally described in isolation may in fact bear interesting connections to other linguistic or even other cognitive domains.

That is the claim to be made here: the foot inventory of (18) is naturally characterizable as the linguistic reflection of a purely rhythmic principle.

4.5.1 The Iambic/Trochaic Law

The principle in question can perhaps best be presented with an example. I invoke a tradition of psychological experiments on rhythmic grouping; see Bolton 1894; Woodrow 1909; and more recent surveys by Woodrow 1951; Fraisse 1974; Allen 1975; and Bell 1977. Bolton and Woodrow's central results have recently been replicated with modern equipment by Rice 1992.

In the crucial experiment, subjects are asked to listen to a sequence of artificially created sounds played in regular rhythm. Alternating sounds are made more prominent, in two different ways: in one set, every other sound is louder (31a), while in another, every other sound is longer (31b).

(31) a. . . . x́ x x́ x x́ x x́ x x́ x x́ x x́ x x́ x x́ x . . .

b. . . . – —— – —— – —— – —— – —— – —— . . .

The judgment that listeners are asked to make is how the sounds are most appropriately grouped in pairs. Naturally, one must control for how the sequences begin; for discussion, see Woodrow 1909, 24–25. Before reading the next paragraph, the reader may wish to compare (31a, b) as a thought experiment.

The usual results are as follows: in the case of intensity contrast, the preferred grouping is with the most prominent element first, as in (32a). In the case of durational contrast, the preferred grouping is with the most prominent element last, as in (32b):

(32) a. . . . [× ×][× ×][× ×][× ×][× ×][× ×][× ×][× ×] . . .

b. . . . [– ——][– ——][– ——][– ——][– ——][– ——] . . .

The preference in the case of (32b) is somewhat stronger.

I assume that this preference does not reflect some mechanical aspect of the perception process, but rather is based on the kinds of groupings people inherently "prefer" to perceive. That is, the perceptual results reflect a law of well-formed rhythmic structure. The statement of this law in (33) paraphrases Bolton 1894, 232:

(33) **Iambic/Trochaic Law**

a. Elements contrasting in intensity naturally form groupings with initial prominence.

b. Elements contrasting in duration naturally form groupings with final prominence.

There is evidence for this law from other rhythmic domains.

Cooper and Meyer (1960) develop a sophisticated account of rhythmic grouping in music, based on the intuitive judgments of musicians. A central postulate of their theory is a version of the Iambic/Trochaic Law: "Durational differences . . . tend to produce end-accented groupings; intensity differentiation tends to produce beginning-accented groupings" (p. 10). They provide numerous examples to support the validity of this law for music.

The rhythmic grouping theory of Lerdahl and Jackendoff 1983 does not posit the Iambic/Trochaic Law as a primitive, but the "preference rules" for grouping that it posits can be applied to (31a, b) to yield the groupings of (32a, b), in conformity with the law.

Rhythmic grouping in verse has been studied by Fant, Kruckenberg, and Nord (1991), who examined the distinction between iambic and trochaic lines in Swedish. They found that the iambic/trochaic distinction had overt correlates both in recitation and in the phonological form of the verse itself. As the Iambic/Trochaic Law predicts, the readers whose renditions were measured produced greater durational contrast in iambic than in trochaic verse, much as in (32). Musicians transcribing verse rhythm in musical notation likewise reflected the pattern of the law in their choice of note values. The same pattern appears to have been respected by the poets themselves, since there is a greater degree of contrast in the phonological length (measured in number of phonemes) of strong and weak syllables in iambic verse than in trochaic verse. Similar work has been carried out for English with parallel results; see in particular Newton 1975, 146–50, and references cited there.

Anecdotally, the iambic/trochaic contrast can be illustrated with the answers that are conventionally given to the question "What does [name of meter] sound like?" For iambic pentameter the answer is something like (34a), while the trochaic tetrameter is usually modeled as in (34b):

(34) a. [də də́:: də də́:: də də́:: də də́:: də də́::]
b. [də́ də də́ də də́ də də́ də]

Before examining the possible linguistic relevance of the Iambic/Trochaic Law, it is worth considering a few other experimental results. First, the rates at which the grouping effects can be observed range from about 0.5 to 5 beats per second. According to Bell 1977, the higher rate encompasses the syllable rate of "ordinary conversational speech." Second, there is a threshold that must be exceeded for durational contrast to have an iambic effect: the long elements must be from 1.5 to 2 times as long as the short ones for iambic rhythmic groupings to be perceived. In fact, Woodrow 1909 found that SMALL differences in duration actually led to trochaic perceptions, with the longer element grouped first. He argues that this effect is psychoacoustic, as longer elements tend to sound louder, all else being equal. The relevance of this duration threshold effect to the realization of stress is made clear below.

4.5.2 The Iambic/Trochaic Law and Foot Inventories

In applying the Iambic/Trochaic Law to stress, my hypothesis is that the set of rhythmic templates for metrical feet is influenced by the law. That is, I posit an EXTRASYSTEMIC motivation, in a law of rhythm, for INTERNAL formal principles of the linguistic system. (For general discussion of dual approaches to explanation in linguistics, see Anderson 1981, 535–36.)

Following the Iambic/Trochaic Law, we would expect the trochaic feet to consist of units roughly equal in duration. There are two ways such feet could be "designed." First, the equal units could be syllables, considered without regard to their length. In this case we motivate the syllabic trochee:

(35) (× .)
σ σ

Second, we can take the equal units to be moras; either in two consecutive light syllables, or within a single heavy syllable. The resulting moraic trochee is shown in (36), with syllable structure included for clarity:

```
(36) (×  .)  or  (×)
      σ  σ        σ
      |  |        |\
      μ  μ        μ μ
```

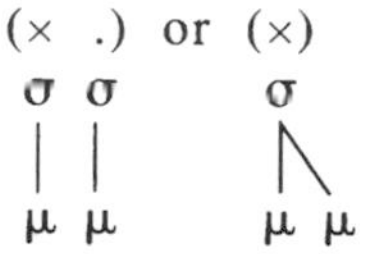

As noted in § 3.9.1, I exclude the possibility of letting a foot contain just one of the two moras of a heavy syllable. For why a single heavy syllable may be considered a trochee, see § 4.1.3.

Logically, we might expect one further structure with equal duration and initial prominence, namely a foot with two heavy syllables:

```
(37) (×     .)
      σ     σ
      |\    |\
      μ μ   μ μ
```

However, it is hard to see how such a foot could play a role in stress systems, since it would leave long strings of light syllables unparsed, often including entire words. In contrast, the feet proposed in the inventory advocated here leave unparsed only single syllables (in syllabic trochee systems) or single light syllables (in quantity-sensitive systems).

A foot with inherent durational contrast can be constructed by concatenating a light syllable with a heavy one. This is the maximal (and, I claim, canonical) form of the iamb.

```
(38) (. ×)
      ⏑ –
```

It is true that the iamb has two smaller versions, consisting of /⏑ ⏑́/ and /–́/. I attribute these to the need to parse most or all of the syllables in the string; the well-formedness of /–́/ as an iambic foot might also be attributed to the inherent stress-attracting properties of heavy syllables. We will see shortly that /⏑ ⏑́/ is unstable, and is often converted to /⏑ –́/.

4.5.3 The Iambic/Trochaic Law and Segmental Rules

It is a recurring theme in this book that metrical structure is not just a means of deriving stress but serves as a general organizing principle for the phonology of a language. This idea has been argued for by Kiparsky 1979; Prince 1980; Hayes 1982a; McCarthy and Prince 1986; and Dresher and Lahiri 1991. The foot structure of a language can govern the prosodic morphology, cause readjustments of stress in response to segmental changes, and also motivate the segmental changes themselves. Here I consider cases of the latter type.

An iamb of the form /⏑ ⏑́/ violates the Iambic/Trochaic Law: it has even duration but final prominence. It can be converted to the canonical /⏑ –́/ by segmental processes that make the stressed syllable heavy, either by lengthening the stressed vowel or by geminating the initial consonant of the following syllable. I refer to such rules as **Iambic Lengthening.**

(39) **Iambic Lengthening**

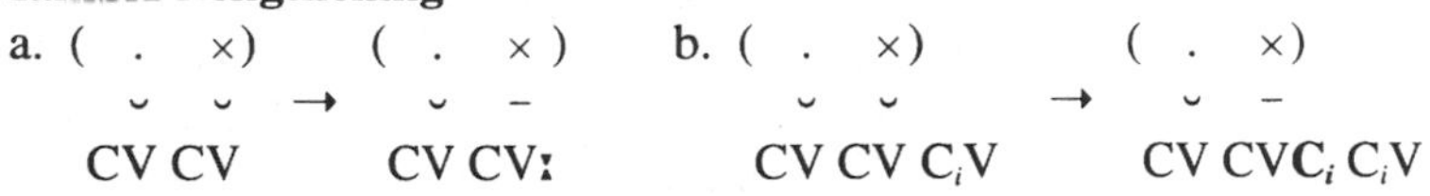

Iambic languages often have such rules, as (40) shows.

(40) a. **Iambic Vowel Lengthening**

Cariban:	Hixkaryana (§ 6.3.1)
	Macushi (§ 6.3.1)
	Surinam Carib (Hoff 1968)
Choctaw/Chickasaw (§ 6.3.2)	
Algonquian:	Menomini (§ 6.3.4)
	Potawatomi (§ 6.3.4)
Lake Iroquoian:	Cayuga (§ 6.3.5)
	Onondaga (§ 6.3.5)
	Seneca (§ 6.3.5)
Eskimo:	St. Lawrence Island Yupik (§ 6.3.8.1)
	Central Alaskan Yupik (§ 6.3.8)
	Pacific Yupik (§ 8.8)
N. California:	Kashaya (§ 6.3.9)
	Maidu (§ 6.3.9)
	Sierra Miwok (§ 6.3.9)
Yidiɲ (§ 6.3.9)	

b. **Iambic Consonant Lengthening**

Algonquian:	Munsee/Unami (§ 6.3.3)
	Menomini (§ 6.3.4)
Eskimo:	Seward Peninsula Inupiaq (Kaplan 1985, 194)
	Central Alaskan Yupik (§ 6.3.8)
	Pacific Yupik (§ 8.8)
Southern Paiute	(§ 6.3.11)

My contention is that the widespread occurrence of lengthening rules in iambic languages is a consequence of their enforcing optimal iambic foot structure.

For this argument to go through, we must also consider lengthening in trochaic languages, which though less common does seem to occur. Examples include Chimalapa Zoque (§ 5.1.9), Icelandic (§ 6.2.2.3), and Mohawk (Michelson 1988), all of which have syllabic trochees and lengthening of stressed vowels. The patterns of length assignment in trochaic languages seem to differ in crucial respects from those of iambic languages, as outlined below.

First, in a number of moraic trochee systems we find actual SHORTENING of stressed vowels. As discussed in § 6.1.5, such shortening is functionally motivated, in that it allows a maximal parse of syllables into perfect moraic trochees, with even duration. Stressed vowel shortening is limited to moraic trochee languages, as one would expect following the Iambic/Trochaic Law: in iambic languages shortening would reduce well-formedness by giving the foot even duration (/⏑ –́/ → /⏑ ⏑́/); and in syllabic trochee languages it would serve no rhythmic function at all, since all syllables are treated as equal.

Second, lengthening in trochaic languages is typically (though not always) phonetic in character, falling short of the duration given to true phonological long vowels. This holds true for languages including Swedish (Bruce 1984), Wargamay (§ 6.1.4), and Tongan (§ 6.1.9). It should be remembered from § 4.5.1 that there is a threshold, around 1.5–2.0, for the duration ratio needed to induce iambic grouping, and that slight durational contrasts in fact enhance trochaic grouping. This suggests that modest degrees of lengthening can safely be used by trochaic languages without disrupting rhythmic structure. The lengthenings measured by Bruce in Swedish generally fall within this range.

Third, trochaic lengthening is often limited to the MAIN-stressed syllable, as in Icelandic or Wargamay. This makes sense if in such cases lengthening is simply a direct manifestation of stress and not an optimization of foot structure.

What we do not find at all, to my knowledge, is trochaic lengthening that both occurs in every foot (as in Swedish) and is fully phonological (as in Mohawk). This contrasts with the iambic cases, where fully phonological lengthening in every foot appears to be common.

Summarizing: the difference in frequency and character between lengthening in iambic languages and lengthening in trochaic languages is a natural consequence of the Iambic/Trochaic Law. In trochaic languages, lengthening is less common, and typically has a lesser status, based on its value as a cue for stress; whereas in iambic languages lengthening is frequent and robust, based on its function in fulfilling a rhythmic target.

A further possible segmental effect of the Iambic/Trochaic Law worth exploring is vowel reduction: by reducing stressless vowels, an iambic language can increase the durational contrast of the foot. Cases of this may be found in the Algonquian languages described in § 6.3.3–4, Choctaw/Chickasaw (§ 6.3.2), Ossetic (§ 6.3.9), Cambodian (§ 6.3.9), Araucanian (§ 6.3.11), and Macushi (§ 6.3.1). A similar process of iambic weakening may be found in Cayuga (§ 6.3.5, § 7.3.1).

There are two difficulties with the hypothesis that vowel reduction represents a response to the Iambic/Trochaic Law. One is that the reduction in duration seen in vowel reduction is not obviously to be represented structurally, though it remains a possibility that the Iambic/Trochaic Law governs phonetic as well as phonological length. The other question is whether vowel reduction might be not specifically iambic, but rather a general trait of languages with distinctive syllable quantity. (Languages without quantity, which typically are stressed with syllabic trochees, quite generally eschew vowel reduction.) My casual impression is that reduction is more common in iambic languages, as the Iambic/Trochaic Law would predict, but it would be difficult to prove this at present.

4.6 SUMMARY

This chapter has focused on the primary empirical area of this book: rhythmic bounded stress systems. I have discussed the role of basic foot inventories in such systems, and have proposed what I take to be the smallest possible inventory, which makes the strongest possible predictions. I outlined the arguments for the inventory, for the most part taking out loans on the data to follow. Finally, I provided external motivation for the inventory by grounding it in the Iambic/Trochaic Law.

The two chapters that follow flesh out the basic proposal. Chapter 5 describes further issues in the theory, making a number of specific proposals with documenting evidence. Chapter 6 is a collection of case studies, ranging from outline sketches to detailed analyses, intended as the main body of evidence in support of the theory.

• 5 •

Further Elements of the Theory

The preceding chapter set out the core of the theory proposed here, namely the basic inventory of bounded metrical feet. In this chapter, I discuss additional principles that will play a role in the analyses that follow.

5.1 DEGENERATE FEET

Do metrical feet have a minimum size? The crucial cases are the so-called **degenerate feet,** defined for the moment as single light syllables in systems that respect syllable weight (iambs and moraic trochees), and single syllables in the quantity-insensitive systems (syllabic trochees). These are the smallest logically possible feet in these systems.

(1) **Degenerate Feet** (preliminary formulation)

a. **Syllabic Trochee**	b. **Moraic Trochee**	c. **Iamb**
(×)	(×)	(×)
σ	˘	˘

If degenerate feet are disallowed, then many words will include unparsable syllables, which will simply be left stray. For example, such syllables will arise for syllabic trochees in words of odd length. If degenerate feet are allowed, such syllables will be footed as in (1).

As Prince 1980; McCarthy and Prince 1986, 1990; and Kager 1989 have argued, there are a number of reasons to favor a ban of some sort on degenerate feet.[1] The difficult issues are: (a) how "degenerate" is to be defined precisely; (b) how general the ban should be. In this section, I present specific proposals concerning these issues, assessing them against cross-linguistic evidence.

Some terminology I will make use of here:

(2)

a. **Degenerate foot**	Logically smallest possible foot
b. **Proper foot**	Any foot which is not degenerate
c. **Canonical foot**	Maximal foot of a given type; i.e. /˘́ ˘/ or /¯́/ for moraic trochees, /˘ ¯́/ for iambs, /σ́ σ/ for syllabic trochees

1. A similar idea has also been proposed within a grid-only framework by Selkirk 1984, 52.

The terms **proper** and **canonical** coincide for moraic and syllabic trochees. I will use "proper" in contexts where it is necessary to emphasize that a foot is not degenerate.

5.1.1 Strong and Weak Prohibitions on Degenerate Feet

My proposal is that languages ban degenerate feet with varying levels of severity: in some languages, the ban is absolute, whereas in others, degenerate feet are allowed if they are metrically strong, that is, dominated by a higher layer grid mark. The latter option could be thought of as follows: the ban on degenerate feet is relaxed if necessary to avoid a violation of the Continuous Column Constraint.

This approach essentially follows that laid out by Kager 1989, though Kager's views about constituency are quite different. My version of the proposal is laid out more precisely in (3):

(3) **Prohibition on Degenerate Feet**
Foot parsing may form degenerate feet under the following conditions:
a. **Strong prohibition** absolutely disallowed
b. **Weak prohibition** allowed only in strong position, i.e. when dominated by another grid mark

We will see in § 5.1.3 that languages may differ almost minimally with respect to this parameter. In particular, Cairene Arabic and Auca form a near minimal pair, in that they both are trochaic and parse feet from left to right, but Cairene invokes the strong and Auca the weak prohibition on degenerate feet.

Various evidence bears on the issue of when in the derivation the weak prohibition applies. Clearly, the removal of weak degenerate feet must be delayed until the construction of the word layer; and I suggest below (§ 5.1.7; § 6.1.6.4; § 7.1.8) that in certain cases it must be delayed until after the application of rules that convert degenerate feet into proper ones. Poser (1989) on Diyari and Halle and Kenstowicz (1991) on enclitic stress in various languages support the view that the prohibition is enforced quite late, at the very end of word stress assignment. In languages with the strong prohibition, degenerate feet are not constructed at all.

I leave it open as to whether the parameter outlined in (3) should include a third value:

(4) **Non-prohibition** Degenerate feet are freely allowed.

The evidence in favor of (4) is surprisingly weak, and will be assessed in more detail below.

5.1.2 Word Minima

I have mentioned already (§ 3.8.5, § 4.1) an important point about degenerate feet made by Prince 1980 and McCarthy and Prince 1986, 1990: a ban on degenerate feet makes predictions about possible word shapes. In particular, assuming that every phonological word must contain at least one foot, and that there are no degenerate feet, then there can be no degenerate-size words.

In the languages examined here, this "minimal word syndrome" is widely attested, as indicated in (5):

(5) a. **Iambs, No /⏑/ Words**

Asheninca (§ 7.1.8)
Cambodian (§ 6.3.9)
Cayuga (§ 6.3.5)
Choctaw/Chickasaw (§ 6.3.2)
Cyrenaican Bedouin Arabic (§ 6.3.7)
Hixkaryana (§ 6.3.1)
Hopi (§ 6.3.9)
Menomini (§ 6.3.4)
Sierra Miwok (§ 6.3.9)
Seminole/Creek (§ 4.1.2)
Southern Paiute (§ 6.3.11)
Winnebago (§ 8.9)
Yidiɲ (§ 6.3.9)
Yupik languages (§ 6.3.8, § 8.8)

b. **Moraic Trochees, No /⏑/ Words**

Awadhi (§ 6.1.9)
Cairene Arabic (§ 4.1.3)
Diegueño (§ 6.1.9)
Fijian (§ 6.1.5)
German (§ 6.1.9)
Hawaiian (§ 6.1.9)
Kawaiisu (§ 6.1.9)
Latin (§ 5.1.4)
Lebanese Arabic (§ 6.1.9)
Maithili (§ 6.1.6)
Mam (§ 7.1.5)
Modern English (§ 6.1.9)
Old English (§ 5.4.4)
Paamese (§ 6.1.9)
Palestinian Arabic (§ 6.1.1)
Sarangani Manobo (§ 6.1.9)
Tongan (§ 6.1.9)
Tümpisa Shoshone (§ 6.1.9)
Wargamay (§ 6.1.4)

c. **Syllabic Trochees, No Monosyllables**

Bidyara/Gungabula (§ 6.2.3)
Cavineña (§ 6.2.3)
Cayuvava (§ 8.2)
Diyari (§ 6.2.3)
Mohawk (Michelson 1988)
Pitta-Pitta (§ 6.2.3)
Wangkumara (§ 6.2.3)

McCarthy and Prince note that the minimal word prediction typically holds only for content words, lexical categories like Noun, Verb, and Adjective. Function words are typically phonologically bound to a neighboring content word and need not be independently footed. Hence function words characteristically may be subminimal, like English *the* /ðə/, *a* /ə/.

Minimal word requirements, though widespread, are hardly universal. Here are examples of languages that allow content words consisting of only a degenerate foot:

(6) a. **Iambs, /⏑/ Words Allowed**

Araucanian (§ 6.3.11)
Dakota (§ 6.3.11)
Munsee/Unami (§ 6.3.3)
Ossetic (§ 6.3.9)

b. **Moraic Trochees, /˘/ Words Allowed**

Romanian (Steriade 1984, p.c.)
Cahuilla (§ 6.1.3)
Inga (§ 6.1.9)
Klamath (§ 7.1.4)
Lenakel (§ 6.1.8)
Sentani (§ 8.7)
Spanish (§ 6.1.9)
Southwest Tanna (§ 6.1.9)
Tol (§ 6.1.9)

c. **Syllabic Trochees, /σ/ Words Allowed**

Auca (§ 6.2.1)
Czech (§ 6.2.3)
Dalabon (§ 6.2.3)
Dehu (§ 6.2.3)
Hungarian (§ 8.6)
Icelandic (§ 6.2.2)
Kela (§ 6.2.3)
Macedonian (§ 6.2.3)
Mae (§ 6.2.3)
Maranungku (§ 6.2.3)
Nengone (§ 6.2.3)
Ono (§ 6.2.3)
Parnkalla (§ 6.2.3)
Piro (§ 6.2.3)
Polish (§ 6.2.3)
Selepet (§ 6.2.3)
Warao (§ 6.2.3)

In the theory assumed here, I propose that such languages invoke only the weak prohibition on degenerate feet. Since the stress of a monosyllable is a main stress, the degenerate foot is dominated by the grid mark of the word layer, and thus is licensed under the weaker ban on degenerate feet (3b):

(7) a. **Moraic Trochees, Iambs**

(×)
(×)
˘

b. **Syllabic Trochees**

(×)
(×)
σ

In the sections that follow, I will consider further consequences of the ban.

5.1.3 Degenerate Feet and Word Layer Labeling

Consider the following scenario: language $\mathscr{L}$ constructs bounded feet of some type from left to right; then a word layer rule amplifies the rightmost foot, making its head the main stress. Alternatively, language $\mathscr{L}'$ constructs feet from right to left, with its word layer amplifying the leftmost foot. In either case, the direction of word layer prominence is opposite to the side from which foot parsing originates.

In systems of this type, we examine words in which the parse of syllables leaves only enough material to form a degenerate foot. The question is whether this material is parsed into a foot or not: if it is, then it will attract the main stress; and if it is not, then there will be no docking site for the grid mark of the word layer (due to the Continuous Column Constraint), and the material will end up stressless. Thus the location of main stress can diagnose whether degenerate feet are tolerated or not.

Cairene Arabic (§ 4.1.3) is a case of this type. The feet are moraic trochees, assigned from left to right. Cairene allows no words consisting of a single light syllable; thus we should expect that the ban on degenerate feet should be absolute. The consequences of this are illustrated by the pair in (8), taken from § 4.1.3.

```
(8) a. ( ×      )      vs.     b. (    ×  )
       ( × .)                     ( ×)(× .)
         ◡ ◡ ◡                      – ◡ ◡
       kátaba                     qattála
```

The sequence /ba/ in (8a) cannot support a foot, and thus rejects main stress. The sequence /tala/ in (8b) can be a proper foot, so /ta/ gets the main stress. Thus the patterning of main stress placement argues that degenerate feet should be disallowed.

Parallel examples may be found in other systems. Seminole/Creek (§ 4.1.2) lacks degenerate words, and demonstrably fails to construct degenerate feet when a single light syllable remains in the iambic parse (see § 4.1.2 (11)). Wargamay (§ 6.1.4) and Paamese (§ 6.1.9) resemble Cairene Arabic in having moraic trochees and in lacking degenerate-size words, but they construct their feet from right to left. Once again, no foot is created when a single syllable is left over in the parse, a point demonstrated in § 6.1.4 (38f, g) and § 6.1.9 (125a, c) respectively.

A useful comparison with the languages just mentioned is Auca (§ 6.2.1). Here, single words CAN consist of a degenerate foot, so we would assume the weak prohibition of (3b), which allows degenerate feet in strong position. Foot construction (at Level 1 of the phonology) involves the creation of syllabic trochees from left to right (9a, b). The weak prohibition on degenerate feet allows such a foot to be created when there is only one syllable left in the parse. This foot survives to the surface, because a later rule of End Rule Right places it in strong position (9c):

```
(9) a.               b.                c.  (    ×)       (= § 6.2.1 (137c))
       (×.)     →       (×.)(×)   →        (×.)(×)
       σσ σ             σσ σ               σσ σ
       mǫi̯ko            mǫi̯ko              mǫ̀i̯kó
```

Cahuilla (§ 6.1.3) is a similar case: words of degenerate length are allowed, and there occur polysyllabic words where the main stress falls on a degenerate foot, as in § 6.1.3 (30a).

The crucial point is that in the cases considered, two phenomena are correlated: the minimal word size, and whether leftover material in a parse may be counted as a foot. Under the theory proposed here, the two phenomena have the same cause, namely whether degenerate feet are strongly prohibited ((3a);

Cairene, Seminole/Creek, Wargamay, Paamese) or weakly prohibited ((3b); Auca, Cahuilla).

5.1.4 Latin and Similar Cases

The well-known stress pattern of Classical Latin was as in (10):

(10) a. Syllable weight: /–/ = CVC, CVV; /⏑/ = CV
 b. Stress the antepenult if the penult is light.
 c. Stress the penult if it is heavy, and in disyllables.
 d. Monosyllables must consist of a heavy syllable, which receives stress.

See § 3.9.2 for examples. The Latin pattern was taken in Hayes 1981 as evidence for the uneven trochee (§ 4.3.1): it can be derived straightforwardly by marking the final syllable extrametrical (§ 3.11) and parsing from right to left with the uneven trochee template (11). Under (12) are the three relevant classes of forms.

```
                                (×  .)                (×)
(11) Uneven Trochee   Form      σ  ⏑     otherwise    σ
```

(12) a. **Heavy Penult** b. **Light Penult, Light Antepenult** c. **Light Penult, Heavy Antepenult**

```
     (×)               ( ×  .)             (×  .)
  ⏑  – 〈–〉            ⏑   ⏑〈–〉        –   –  ⏑〈–〉
  a.míː.kus            sí.mu.laː           do.més.ti.kus
```

Cases (12a) and (12b) also fall out straightforwardly from an analysis in which the foot template is the moraic trochee, since like the uneven trochee the moraic trochee can span the sequences /–/ and /⏑ ⏑/. It is only (12c), with /– ⏑/, that is problematic here.

In this proposal, what induces skipping of the penult in these cases is the need to avoid degenerate feet: the light syllable alone should not be a foot (cf. (13a)), so the parse continues on to the preceding heavy, which forms a proper moraic trochee (13b):

```
(13) a.        (×)              b.     (×)
          –  –  ⏑〈–〉               –  –  ⏑〈–〉
         *do.mes.tí.kus             do.més.ti.kus
```

Latin did not tolerate words of degenerate size; that is, it had no CV content words (Allen 1973, 51). However, see § 5.1.6 for why this is probably not the crucial factor in the /– ⏑/ cases.

The rules for Latin are stated in (14). I leave aside the treatment of secondary stress, since the data are not clearly established.

(14) a. **Syllable Extrametricality** σ → ⟨σ⟩ / ____ $]_{word}$
b. **Foot Construction** i. Form a moraic trochee, going from right to left.
ii. Degenerate feet are banned absolutely.
c. **Word Layer Construction** End Rule Right

Crucially, foot construction is construed as scanning the string leftward until it finds sufficient material for a proper foot. The rules derive (15):

```
(15) a. (    × )        b. ( ×     )        c. (      ×    )
          (× )              ( ×  . )               (×)
       ⌣   – ⟨–⟩             ⌣   ⌣⟨–⟩            –    –   ⌣⟨–⟩
       a.míː.kus            sí.mu.laː          do.més.ti.kus
```

This analysis is in its crucial respects the same as proposed in a quasi-metrical framework by Allen 1973. For more on Latin, see § 5.3 and Mester 1992.

The Latin stress rule occurs in a number of languages, sometimes as the result of inheritance or borrowing. A similar pattern, but without extrametricality, is also attested; see § 6.1.9.

5.1.5 Malayalam and Similar Cases

Mohanan's (1986, 111–15) analysis of stress in Malayalam (Dravidian, South India) also relies on a prohibition against degenerate feet. According to Mohanan, main stress in Malayalam is determined by the rule in (16):

(16) a. Stress the second syllable if the first syllable is light and the second syllable is heavy.
b. Otherwise stress the first syllable.

This pattern cannot be derived by iambs, since a left-edge iamb would assign second-syllable stress to /⌣ ⌣/.

Mohanan's proposal, stated in my terms, is that Malayalam imposes an absolute ban on degenerate feet. As a first approximation, we can suppose that Malayalam employs the moraic trochee, assigning it from the left edge. Applied to the four logical possibilities, this yields the correct result:

```
(17) a.  (× .)   b.    (×)   c.  (×)      d.  (×)
         # ⌣ ⌣       # ⌣ –       # – ⌣        # – –
```

The crucial case is (17b): the sequence /⌣ –/ cannot be parsed as a moraic trochee, nor can /⌣/ form a moraic trochee on its own. Thus the foot parsing algorithm must scan further to the right, making a moraic trochee out of the following /–/. Note that this is much the same phenomenon as seen above for Latin, though the surface consequences for stress placement are quite different.

A full account of Malayalam stress actually must differ somewhat from the above, owing to the patterning of secondary stress. As Mohanan shows, secondary stresses fall on all heavy (= CVV(C)) syllables after the primary. Secondary stresses arguably are metrically genuine, rather than being just an effect of auditory prominence, since they attract pitch accents in the Malayalam intonational system (Mohanan 1986, 113–19).

The analysis I adopt, following Mohanan, invokes the notion of unbounded feet. Unbounded feet were proposed originally by Prince 1976a and are discussed here in § 7.2. In brief, an unbounded foot may include an unlimited number of light syllables, provided that only the head of the foot (leftmost or rightmost, depending on parametric setting) is heavy. For Malayalam, the specific analysis is as in (18):

(18) a. **Foot Construction** i. Form left-strong, unbounded feet.
ii. Degenerate feet are forbidden.
b. **Word Layer Construction** End Rule Left

Parsing representative words (Mohanan 1986, 113) according to these rules, we derive both main and secondary stress correctly:

```
(19) a. (×        )   b. (   ×                        )   c. (×              )
        (×  .  .  )          (×     )(× ) (×  .  .  .  )      (×) (×  .  .  )
         ˘  ˘  ˘            ˘  –   ˘ –   –  ˘  ˘  ˘           –   –  ˘  ˘
        márʲaɳam           angáːrʲasàːt̪mìːkarʲaɳam          páːrʲàːyaɳam
```

In particular, (19b) cannot have initial stress, since the initial light syllable /an/ would form a degenerate foot.

It is not clear what the status of /˘/ content words is in Malayalam; according to Mohanan 1989, underlying CVV and CVCC are allowed, but CV and CVC are forbidden. The resolution of this question turns out not to be crucial, given what is proposed below in § 5.1.6.

The Malayalam main stress pattern has also been reported for Gurkhali (Indo-Aryan, Nepal; Meerendonk 1949), and Yil (Torricelli, Papua New Guinea; Martens and Tuominen 1977).

The Malayalam pattern was used by Hayes 1981 as partial justification for the Obligatory Branching Parameter (§ 4.3.1), which requires that the head of a foot be heavy. The proposal just given is one of the set of reanalyses that permit this parameter to be dispensed with.

5.1.6 Priority in Foot Construction

We have seen so far that a ban on degenerate feet can account for the following:

(a) Restrictions on minimal word length (§ 5.1.2).

(b) A distinction between languages of the Creek/Cairene/Wargamay type and those of the Auca/Cahuilla type (§ 5.1.3): the End Rule has different effects depending on whether degenerate feet are allowed and thus can provide docking sites. This distinction correlates with whether the language in question allows light monosyllables.

(c) The patterning of stress in cases like Latin (§ 5.1.4) and Malayalam (§ 5.1.5): the ban on degenerate feet forces foot construction to skip over light syllables when a heavy syllable is the next one available in the string.

Cases (b) and (c) have a different status. In (b), the parse has reached the end of the string, and we have a choice of WHETHER to create a degenerate foot or not. In (c), the choice is more one of WHERE the foot is to be created: Do we create a degenerate foot at word edge, or do we skip over the light syllable to form a proper foot further on in the string? The choices are illustrated in (20), which for concreteness assumes moraic trochees and left-to-right parsing.

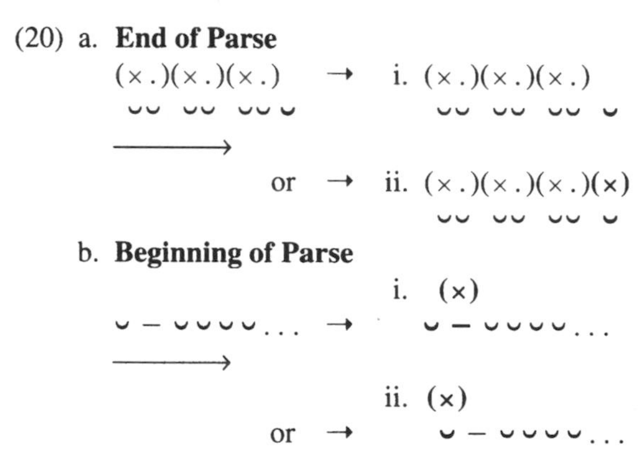

Of these four logical possibilities, what is empirically attested?

For (20a), we have evidence that different languages make different choices: the final foot (will/will not) be constructed when the language in question invokes the (weak/strong) prohibition on degenerate feet (3). This in turn is correlated with whether the language allows words of degenerate size (§ 5.1.3).

For (20b), however, I posit that there are no language-particular choices, and that (20b.i) is the obligatory outcome. This is hardly surprising in the case of languages that ban degenerate feet entirely (i.e. which disallow words of degenerate size). More surprisingly, there are languages that do allow degenerate words but which nonetheless select the option illustrated in (20b.i); that is, they skip over a light syllable to form a canonical foot on the adjacent heavy. For example, in Romanian (Steriade 1984 and p.c.) and Spanish (Harris 1983, 1992), the stress pattern is basically the Latin type, but with degenerate words allowed.[2] The crucial schema is as in (21):

(21)
$$\ldots - \smile \langle\sigma\rangle \# \rightarrow \ldots \overset{(\times)}{-} \smile \langle\sigma\rangle \#, \quad \text{not} \quad {*}\ldots - \overset{(\times)}{\smile} \langle\sigma\rangle \#$$

2. Degenerate words are not particularly common in Spanish, but there is a large class of words with stressed final light syllables. These words, which could be analyzed with a lexically listed final degenerate foot, make essentially the same point.

In fact, it would be surprising if we found languages that went the other way; to do this would go against the general pattern that heavy syllables are stress-attracting.

The upshot of this is that when candidates for degenerate foot status are encountered at the CONCLUSION of the parse, there is language-specific variation in whether a degenerate foot is constructed, correlated with the minimal word requirement (20a). When candidates for degenerate foot status are encountered OTHER THAN at the end of the parse, they are skipped over. Intuitively, this difference arises because the choice of skipping is available only when the parse is not yet finished.

This intuition can be incorporated into the theory as in (22):

(22) **Priority Clause**
If at any stage in foot parsing the portion of the string being scanned would yield a degenerate foot, the parse scans further along the string to construct a proper foot where possible.

This correctly distinguishes the cases of (20a, b): in (20a) the Priority Clause is inapplicable, since there is no more string to scan, and a degenerate foot will be constructed if the language allows degenerate feet at all. This is likewise the case for light monosyllables in languages that allow them. In (20b) (similarly (21)), more of the string is available for parsing, and the Priority Clause enforces skipping of the light syllable. Hence (20b.i) must be the outcome, even if degenerate feet are allowed in other contexts.

Summing up, the basic ingredients of the theory are the prohibition on degenerate feet (3), the link between degenerate feet and minimal word size (§ 5.1.2), and the Priority Clause (22). Taken together, these predict that in surface forms, degenerate feet will only occur in strong position, only at the (right/left) edge of the word where footing is (left-to-right/right-to-left), and only in languages that allow degenerate-size monosyllables. These predictions are largely confirmed in the cases presented in this book.

Other languages whose analyses crucially require the Priority Clause are Malayalam (§ 5.1.5), Hindi (§ 6.1.7), Lenakel (§ 6.1.8), Inga (§ 6.1.9), Tol (§ 6.1.9), Klamath (§ 7.1.4), Turkish (§ 6.3.10), and Sentani (§ 8.7).

5.1.7 Repair of Degenerate Feet

A possibility I include here is that degenerate feet can be created early in the derivation but removed if not **repaired** by other metrical or segmental rules. We will see cases where degenerate feet are repaired (a) by vowel lengthening, as in Chimalapa Zoque (§ 5.1.9) and Badimaya (§ 6.2.3), Hixkaryana (§ 6.3.1), Maidu (§ 6.3.9), Bani-Hassan Arabic (§ 8.10), the cases of § 5.3(a), and the cases of § 5.1.8.1; and (b) by reparsing of foot boundaries. The latter category is illustrated by the cases of § 5.3(b), as well as by the **initial dactyl** effect, to which we now turn.

The **initial dactyl** effect was so named by Prince 1983a, 49. It can be illustrated by the distribution of secondary stresses in Spanish, as described by Harris 1983, 1989 and Roca 1986. Main stress in Spanish is phonemic, though it can be predicted to a fair extent by complex lexical rules, whose character continues to be debated; see Harris 1992 for a clear summary and new proposals. For present purposes, I assume that main stress has already been assigned and consider only secondary stress, which is a simpler, postlexical phenomenon. The domain to be considered is the set of syllables preceding the main stress.

When this domain has an even number of syllables, stress falls in a simple alternating pattern. The examples in (23) are classified by the number of pretonic syllables:

(23)	2:	/σ̀σσ́σ/	*Cònstantíno*	'Constantine' Roca 352
	4:	/σ̀σσ̀σσ́σ/	*cònstantìnopléño*	'Constantinople guy' R 353
	6:	/σ̀σσ̀σσ̀σσ́σ/	*cònstantìnopòlitáno*	'Constantinopolitan' R 353
	8:	/σ̀σσ̀σσ̀σσ̀σσ́σ/	*cònstantìnopòlizàcionísmo*	'Constantinopolizationism' R 353

The odd-syllabled cases prove that the alternation must be computed from right to left. In addition, they show an interesting pattern of systematic variation:

(24)	1:	/σσ́σ/	*constánte*	'constant' R 352
	3:	/σσ̀σσ́σ/	*genèratívo*	'generative'
	or	/σ̀σσσ́σ/	*gèneratívo*	Harris 1983, 85
	5:	/σσ̀σσ̀σσ́/	*natùralìzación*	'naturalization'
	or	/σ̀σσσ̀σσ́/	*nàturalìzación*	H 1989, 1, 5
	7:	/σσ̀σσ̀σσ̀σσ́σ/	*constàntinòpolìtanísmo*	'Constantinopolitanism'
	or	/σ̀σσσ̀σσ̀σσ́σ/	*cònstantinòpolìtanísmo*	H 1989, 1, 5

The variation consists in the possibility of a ternary interval (the "initial dactyl") for odd-syllabled words. The initial dactyl variants are apparently more colloquial, the initial upbeat forms more "rhetorical" in character (Harris 1983, 86).

Hayes 1985 and Harris 1989 suggest an analysis for patterns of this type: the two variants represent alternative outcomes to the resolution of clash. If we allow degenerate feet at an intermediate stage of the derivation, the sort of clash shown in (25) will result. The form given has seven pretonic syllables; similar results would obtain for other odd numbers.

```
(25) (                    ×     )
     ( ×) (×  .)(×  .)(× .)(×   .)
       σ    σ  σ σ  σ σ σ  σ    σ
     constantinopolitanismo
     ←———————————————
```

From this, we may derive the initial dactyl forms by applying a rule of rightward destressing, whose effects are shown in (26a). I also assume that Spanish footing is persistent in the sense of § 5.4.1, so that the destressed syllables are submitted to reparsing (26b).

```
(26)      a. (                    ×     )      b. (                    ×     )
   →         (× )       (×  .)(× .)(×   .) →      (×    .)  (×  .)(× .)(×   .)
             σ     σ  σ σ  σ σ σ  σ    σ           σ    σ  σ σ  σ σ σ  σ    σ
             constantinopolitanismo               cònstantinòpolitanísmo
```

The other option taken is to resolve the clash with leftward destressing:

```
(27) (                    ×     )
          (×   .)(×  .)(× .)(×   .)
       σ    σ  σ σ  σ σ σ  σ    σ
     constàntinòpolìtanísmo
```

If there is just one pretonic syllable, only leftward destressing gets to apply; to destress rightward here would violate the Continuous Column Constraint (§ 3.4.2(d)):

```
(28) (    ×    )        (    ×    )          (     ×   )
     ( × )(×   .)  →         (×  . )          (  ×    .)
       σ   σ   σ         σ   σ   σ               σ   σ   σ
     constante          constante,    not    *constante
```

An observation of Harris 1989 indicates that rightward Destressing is necessary in the phonology of Spanish in any event: *José trabajó* 'José worked' surfaces without secondary stress on /tra/, evidently as a result of destressing triggered by the preceding syllable /sé/.

The crucial point of this analysis is that it relies on a temporary degenerate foot, set up in the middle of the derivation, that either is expanded into a proper foot by destressing and reparsing, or is itself deleted. In neither case does a degenerate foot in weak position survive to the surface; thus the ban on degenerate feet (2) is observed.

The initial dactyl effect is attested in other languages. For instance, it is found in long monomorphemic English words, where secondary stress is not determined by morphology (Hayes 1982b). Initial dactyls are also found in Indonesian (Cohn 1989, 1993), Swedish (Gösta Bruce, p.c.), Polish (Hayes and Puppel 1985), Lenakel (§ 6.1.8.3), and, with some puzzling complications,

Modern Hebrew (Bolozky 1982). The mirror image case, a **final anapest,** is found in Asheninca (§ 7.1.8).[3]

5.1.8 Assessing the Case for Degenerate Feet in Weak Position

To summarize so far: (3) lays out what I take to be the most restrictive theory of the distribution of degenerate feet that is likely to be tenable: on a parametric basis, degenerate feet may be banned entirely or restricted to strong metrical position. A weaker and less interesting version of the theory would allow the additional option under (4), under which degenerate feet are permitted everywhere. Below, I review some of the evidence that would support this option. At present it does not seem fully compelling, though careful study of the relevant languages is needed to resolve the issue.

5.1.8.1 *Degenerate Feet in Clashing Positions*

Consider the case in which trochees (for concreteness, moraic) are assigned from right to left, with a word layer assigned by End Rule Right:

```
(29) (                   ×  )
         (× .) (× .) (× .)
     ˘    ˘ ˘   ˘ ˘   ˘ ˘
```

What is at issue is what happens to leftover single syllables, as in (29).

One attested outcome is just what the theory predicts: no foot is permitted (since it would be both weak and degenerate), and (29) is the surface form. This holds for Fijian (§ 6.1.5), Nengone (§ 6.2.3), Cavineña (§ 6.2.3), Djingili (§ 6.2.3), Inga (§ 6.1.9), and Warao (§ 6.2.3). A different outcome is the initial dactyl effect (§ 5.1.7), which involves a "virtual" degenerate foot that is later repaired.

The outcome that the theory does not predict is the appearance of the degenerate foot in surface form:

```
(30) (                   ×  )
     (×) (× .) (× .) (× .)
     ˘    ˘ ˘   ˘ ˘   ˘ ˘
```

Such configurations are suggested to occur in Maithili (§ 6.1.6), Eastern Ojibwa (§ 6.3.4), Modern Greek (§ 6.2.3), as a marginal variant in Polish (§ 6.2.3), and in Bani-Hassan Arabic (§ 8.10).

A priori, the boldface position in (30) is not a likely location for stress. In particular, it is a light syllable occurring in a clash with the immediately following stress. What is in question here is what might serve phonetically as the cue for stress in this position.

In the case of Bani-Hassan Arabic, the relevant syllable appears to be

3. For an alternative account of the initial dactyl phenomenon see Cohn 1993.

lengthened, which would suggest that the relevant feet are actually / – /, which is non-degenerate. The marginal Polish case also involves lengthening, which I propose to treat as the creation of ad hoc / – / feet in a language that otherwise lacks quantity. Malikouti-Drachmann and Drachmann's description of the analogous cases in Modern Greek (1981, 284, 288) implies the same. For discussion of Maithili and Eastern Ojibwa, see § 6.1.6 and § 6.3.4 respectively.

In my opinion, there are no fully convincing cases where a degenerate foot survives to the surface in clash environments like (30).

Another environment where degenerate feet could in principle arise in clash position is when the main stress is assigned by a non-iterative rule at one edge of the word, with secondary stresses assigned iteratively from the opposite edge. This is shown schematically in (31):

```
(31) (                    ×  )
     (× .) (× .) (× .)  ?  (× .)
      ˘ ˘   ˘ ˘   ˘ ˘   ˘   ˘ ˘
     ————————————→      ←——
```

Here, the outcome in the three cases of which I am aware is to avoid the degenerate foot; see Piro (§ 6.2.3), Garawa (§ 6.2.3), and Lenakel (§ 6.1.8).

5.1.8.2 *Degenerate Feet in Non-clashing Positions*

Consider next the cases where a putative weak degenerate foot is not in a position of clash. This will arise when feet are constructed "tail first," in a direction going away from the main stress. In (32) are the three logically possible cases:

(32)

a. **Iambs, Right to Left End Rule Right**	b. **Moraic Trochees, Left to Right End Rule Left**	c. **Syllabic Trochees, Left to Right End Rule Left**
(×)	(×)	(×)
(×)(. ×)(. ×)	(× .)(× .)(×)	(× .)(× .)(×)
˘̀ ˘ ˘̀ ˘ ˘́	˘́ ˘ ˘̀ ˘ ˘̀	σ́ σ σ̀ σ σ̀

Possible cases of the form (32a) are discussed in § 6.3.10, along with alternative analyses. A case of type (32b) is Cahuilla (§ 6.1.3), and for type (32c), plentiful cases may be found. These cases, where left-to-right alternating stress either is or is not said to leave a secondary stress on odd final syllables, are listed in (33). In cases where morpheme boundaries affect stress, the data listed cover only cases where morpheme boundaries are not relevant.

(33) a. **Final Syllable is Stressed in Odd-syllabled Words**

Czech (§ 6.2.3)	Maranungku (§ 6.2.3)
Hungarian (§ 8.6)	Ono (§ 6.2.3)
Icelandic (§ 6.2.2)	Votic (§ 6.2.3)
Livonian (§ 6.2.3)	

b. **Final Syllable is Sometimes Stressed in Odd-syllabled Words**
Central Norwegian Lappish (§ 6.2.3)
(/σ́ σ σ̀/ or /σ́ σ σ/, /σ́ σ σ̀ σ σ/)
Mansi (§ 6.2.3; sources differ as to whether odd final syllables are stressed)
Selepet (§ 6.2.3; (/σ́ σ σ̀/ or /σ́ σ σ/, no five-syllable forms given)

c. **Final Stress in Odd-syllabled Words Conditioned by Length**
Central Norwegian Lappish = (33b)
Dehu (§ 6.2.3; /σ́ σ σ̀/, but /σ́ σ σ̀ σ σ/)

d. **No Final Stress in Odd-syllabled Words**
Anguthimri (§ 6.2.3)
Badimaya (§ 6.2.3)
Bidyara-Gungabula (§ 6.2.3)
Dalabon (§ 6.2.3)
Diyari (§ 6.2.3)
Karelian (§ 8.6)
Mantjiltjara (§ 8.10)
Mayi (§ 6.2.3)
Pintupi (§ 4.1.1)
Pitta-Pitta (§ 6.2.3)
Walmatjari (Hudson 1978)
Wangkumara (§ 6.2.3)

The number of cases where the alternating pattern has a gap in final position (33d) is fairly striking. But what are we to say about the cases in (33a–c)?

One possibility, obviously, is to give up and allow grammars to elect option (4), which allows degenerate feet to be created in any environment whatever. But it is possible that the relevant descriptions should not be taken at face value.

It is significant that final position is the characteristic environment of word- and phrase-final phonetic lengthening (cf. Klatt 1975; Wightman et al. 1992), which can also make a syllable perceptually prominent. Final lengthening is widespread among the world's languages, and may indeed be a phonetic universal. Supposing it to be the case in the languages of (33a–c), it is easy to see why a linguist might transcribe final stress in odd-syllabled words: the relevant syllable would be subject to final lengthening; and this, together with the natural analytic tendency to extrapolate the binary pattern, could lead to a perception of final stress. (Final lengthening would also affect posttonic final syllables, but given an otherwise alternating pattern, there would be no a priori tendency to hear stress in such cases.)

My own experience in listening to Icelandic utterances is perhaps relevant: the final syllables of odd-length Icelandic polysyllables sound quite ambiguous. On intuitive grounds, it seems difficult to make a firm decision about whether such syllables are stressed or simply phonetically lengthened.

The patterning of certain languages in (33b,c) supports the final-lengthening scenario. In Icelandic, the final stresses disappear in connected speech, where it is known that final lengthening is diminished; plausibly this would also account for the cases of (33b) in which the final stress is said to be optional. In Dehu and Central Norwegian Lappish, the putative final stress appears in three-syllable words but not five-syllable words; since it has been observed that longer words tend to be pronounced at faster tempo (Lindblom and Rapp 1973), this difference could also be attributed to phonetic length.

In making this proposal, I do not intend to impugn anyone's ear. What is at issue is solely the matter of interpretation: whether the perceived prominence should be attributed to a strong position in metrical structure.

Since metrical structure is an abstract phonological entity, what we need in order to test it are cases where the phonological rules of the language refer to stress in final position, with results that are uncontroversially interpretable. Three cases are of interest here. For Icelandic (§ 6.2.2), I argue that the phonological evidence goes against the view that final odd syllables are stressed. Another is Cahuilla (§ 6.1.3), where the facts of vowel allophony and pitch assignment lead to the same conclusion. The pattern in Estonian (§ 8.5.3) involves additional factors (§ 5.1.9), but essentially supports the view adopted here as well.

To sum up, I am proposing that degenerate feet are sharply limited in universal grammar: the languages adopting the strong prohibition on degenerate feet (3) forbid them entirely, while those adopting the weak prohibition allow them only in strong metrical position. Certain cases appear to falsify this, by showing secondary stress that would have to be derived by a degenerate foot. These are concentrated in final position, where a plausible alternative explanation in the form of phonetic final lengthening is also available. The alternative account is supported by cases where phonological diagnostics tell us that the final stresses are actually missing.

5.1.9 Defining "Degenerate" for Syllabic Trochee Systems

The issue of degenerate feet is related to the problem of how heavy syllables are treated in syllabic trochee systems.

Statistically, syllabic trochee languages tend to be languages that have no quantity distinction at all—that is, no vowel length contrast and no phonological rules that distinguish syllable weight. It is plausible to suppose that in such languages, every syllable is monomoraic. As McCarthy and Prince (1986) point out, if it were the case that ALL syllabic trochee languages lacked bimoraic syllables, we could simplify the theory to a considerable extent: we eliminate the syllabic trochee, allowing all its functions to be carried out by the moraic trochee, as in (34).

(34) $\underset{\sigma\ \sigma}{(\times\ .)} = \underset{\mu\ \mu}{(\times\ .)}$, where all σ are of the form $\begin{array}{c}\sigma\\ |\\ \mu\end{array}$

However, the languages in (35) apparently go against the general tendency: they have a vowel length distinction (and in some cases, additional evidence for heavy syllables) but a stress system analyzable with syllabic trochees.

(35) **Syllabic Trochee Systems in Languages with Quantity Oppositions**

Anguthimri (§ 6.2.3)	Finnish (§ 8.6)
Chimalapa Zoque (this section)	Mansi (§ 6.2.3)
Czech (§ 6.2.3)	Pintupi (§ 4.1.1)
Hungarian (§ 8.6)	Piro (§ 6.2.3)
Estonian (§ 8.5)	Votic (§ 6.2.3)

A number of less completely documented cases may be found among the languages listed in § 6.2.3. Spanish (§ 5.1.7) shows quantity-insensitive syllabic trochees for postlexical secondary stress assignment, but the lexical main stress rules diagnose a distinction of quantity (CV vs. CVC).

On the basis of these cases, it appears that we must distinguish moraic and syllabic trochees in the theory. While there may be a tendency for languages having quantity distinctions to "select" quantity-sensitive templates (i.e. iambs and moraic trochees), this is not a requirement.[4] It remains the case that for trochaic languages lacking syllable quantity, the syllabic and moraic trochee foot templates are equivalent. For concreteness, I classify such languages as syllabic trochee languages.

Considering now "quantity-disrespecting" languages, we can ask what constitutes a degenerate foot for purposes of the prohibition on degenerate feet (3). There are three possibilities:

(36) a. Only light syllables are degenerate feet; i.e., $\overset{(\times)}{-}$ is a well-formed syllabic trochee.
b. Any monosyllabic foot is degenerate, even if it is a heavy syllable.
c. The choice between the previous two options is made on a language-particular basis.

The choice adopted here, namely (36a), was first suggested by Prince 1980, for Estonian; see § 8.5. To state this choice explicitly, I adopt (37) as the final definition of degenerate foot:

(37) **Degenerate Foot** $\overset{(\times)}{\smile}$

That is, a degenerate foot is a light-syllable foot. In applying this definition, it should be borne in mind that languages without quantity distinctions are assumed to have only light syllables.

This definition raises the question of how certain sequences are parsed into syllabic trochees. For instance, a left-to-right parse of /– σ/ could in principle create either a monosyllabic or a disyllabic foot. Here I follow the general principle that prosodic structure is created maximally (see Prince 1976a, 1980 for

4. For differing views see Kager 1992a, 1992b; Yip 1992.

metrical structure, J. Ito 1989 for syllable structure). This dictates that the foot be disyllabic, as shown in (38):

```
(38) a. ( ×  . )        not:     b. *( × )
         – σ . . .                    – σ . . .
```

There are two arguments that support (37). First, an adequate account of Estonian stress seems to require it, as Prince showed. Estonian stress basically involves parsing the word from left to right into syllabic trochees (this ignores the complication of optional ternarity, for which see § 8.5). Crucially, if at the end of the parse a heavy syllable is available, it is stressed; whereas a light syllable is not. This is what definition (37) predicts: single heavy syllables are allowed as feet; but they only occur word-finally, since earlier in the word a larger, disyllabic foot is always possible. For reasons discussed in § 8.5, the stresses on odd final heavy syllables in Estonian must be regarded as metrically genuine, and not just the auditory effect of weight.

The other argument for (37) concerns minimal word constraints: syllabic trochee languages characteristically employ a minimal word constraint of the form /–/. To give an example: Pintupi (§ 4.1.1) and Anguthimri (§ 6.2.3) both involve construction of syllabic trochees from left to right. In each, long-voweled syllables are made part of a disyllabic foot where possible; thus the sequence /– σ/ when encountered in the left-to-right parse would be made into a single foot as in (38a). However, in both languages, the minimal content word is either /σ́ σ/ OR /–́/; that is, a single heavy syllable (but not a single light syllable) can form a word. Anguthimri in fact actively imposes this constraint, lengthening the vowels of monosyllabic stems if they have not been amplified by a suffix syllable (Crowley 1981, 154). Similar cases include Badimaya (§ 6.2.3), Djingili (§ 6.2.3), Garawa (§ 6.2.3), Mansi (§ 6.2.3), Votic (§ 6.2.3), Mantjiltjara (§ 8.10), and Icelandic (§ 6.2.2, based on the weight-governed lengthening rule discussed by Kiparsky 1984).

The theory proposed here can account for this minimal word pattern. Where possible, the rules will construct /– σ/, owing to maximality (39a), but when only /–/ is available (39b), it can constitute an acceptable foot under the theory:

```
(39) a. Polysyllables          b. Heavy           c. Light
                                  Monosyllables      Monosyllables
        ( ×  . )( ×  . )( ×  . )   ( × )              *(×)   (excluded
          – σ  σ σ  σ σ . . .       –                  ˘     by (37))
```

This account predicts that final /–/ in an odd-syllabled word will receive stress, as in Estonian. Unfortunately, the prediction is difficult to check: for instance, long vowels in Pintupi occur only in initial syllables, so the crucial cases do not exist; and in many languages the crucial cases may exist but are not listed in the sources.

Chimalapa Zoque (Mixe-Zoque, Mexico; Knudson 1975) provides a different kind of support for the claim that / – / is non-degenerate in syllabic trochee languages. Main stress is penultimate, with an initial secondary stress on all words of three or more syllables. My analysis for this is as in (40):

(40) a. **Foot Construction**
i. Form a syllabic trochee at the right edge of the word.
ii. On the remaining syllables of the word, apply End Rule Left.

b. **Word Layer Construction** End Rule Right

In simple cases, this derives forms like those in (41):

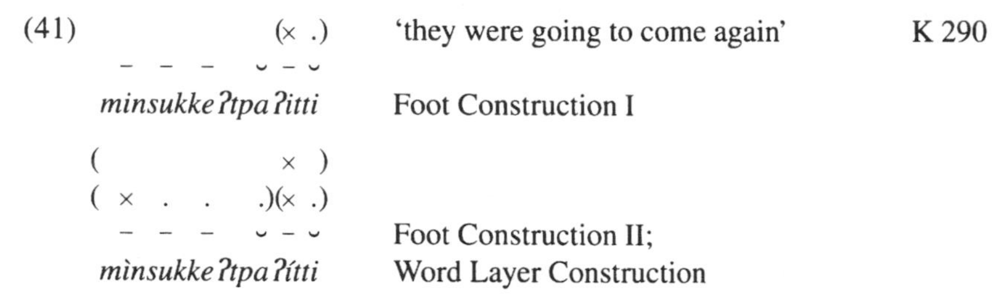

The minimal word in Zoque is not a disyllable but rather a single heavy syllable (Knudson 1975, 292). As before, this follows if Zoque bans degenerate feet in underlying representations and / –́ / is a proper syllabic trochee.

Degenerate feet also arise in Zoque when the domain of the secondary stress rule (40a.ii) is a single light syllable. These are repaired (§ 5.1.7) by a general rule of Zoque that lengthens vowels in stressed open syllables; all long vowels are derived by this rule.

(42)

(×)	(×)	Foot Construction,
(×) (× .)	(×)(× .)	Word Layer Construction
– – ⏑	⏑ ⏑ –	
mìnkéʔtpa	*patanus*	
'he is coming again' K 290	'large cooking banana' K 329	
	(×)	Open Syllable
	(×)(× .)	Lengthening
	– – –	
——	*pàːtáːnus*	

Here again, the notion that lengthening licenses the monosyllabic foot makes sense if we assume that / – / is a proper foot in a quantity-insensitive system.

One further phenomenon remains to be dealt with under the definition of degenerate foot (37): the minimal word constraints of Estonian (§ 8.5) and Nyawaygi (§ 6.1.9). Both languages have heavy syllables but define the minimal word as a disyllable (or for Estonian, its phonological equivalent; see § 8.5.3). In the case of Estonian, we have already seen that, for reasons of secondary stress, /– / must be counted as a proper foot in any event. This suggests that the peculiarity of Estonian and Nyawaygi should be localized in their unusual minimal word constraints: they define the minimal word as a maximal foot (i.e. /σ σ/), rather than the minimal foot that is usually found. A somewhat similar case is Cayuvava (§ 8.2), where the definitions of minimal word and minimal foot must also be separated.

The proposed definition of "degenerate foot" for quantity-insensitive feet suggests that my choice of the descriptive term "syllabic trochee" is not entirely felicitous. More precisely, the constituent is a binary trochee, with freedom of choice as to its terminals (Prince 1980, 528). Possible structures are /σ́ σ/ and /μ́ μ/, where the latter option is empirically distinct when it is instantiated by a heavy syllable. Yet a third possibility, a higher layer trochee whose terminals are feet, also occurs (§ 5.5.1). The general pattern for trochees, then, is that identical elements are grouped together with initial prominence.

5.2 MORE ON EXTRAMETRICALITY

In § 3.11 I laid out the basic principles of extrametricality theory: only constituents can be extrametrical; they may only be extrametrical when located at a specified edge; the unmarked edge is the right one; and extrametricality may not exhaust the domain of the stress rules. This section covers additional matters that arise in the version of extrametricality theory assumed here.

5.2.1 Extrametricality on Multiple Levels

Extremetricality functions on two prosodic levels. At the segmental level, rules of consonant extrametricality describe the commonplace pattern whereby a syllable must contain more segments to count as heavy in final position than nonfinally. At higher levels, extrametricality accounts for cases where a foot is constructed further from word edge than expected, or where the word layer rule selects a nonfinal foot for main stress; these result from syllable or foot extrametricality respectively. A possible intermediate case, mora extrametricality, is plausibly excluded by the ban on foot-splitting (§ 3.9.1).

What happens when there are two kinds of extrametricality at once? In particular, suppose consonant extrametricality has applied, and we wish to determine whether foot extrametricality is applicable:

```
(43)   (×)       (×      .)
        –         ˘      ˘
        σ         σ      σ
      C V C     C V    C V ⟨C⟩
```

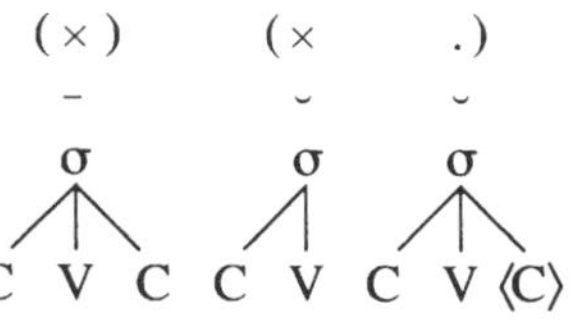

What is at issue is the Peripherality Condition (§ 3.11, (47b)): we want to know whether the rightmost foot in (43) is to be considered peripheral within the word, or whether the extrametrical consonant renders it non-peripheral. If the latter is correct, extrametricality will be blocked, and a word layer rule of the form End Rule Right will derive penultimate stress (44a). If the former is correct, extrametricality will apply, and End Rule Right will derive antepenultimate stress (44b).

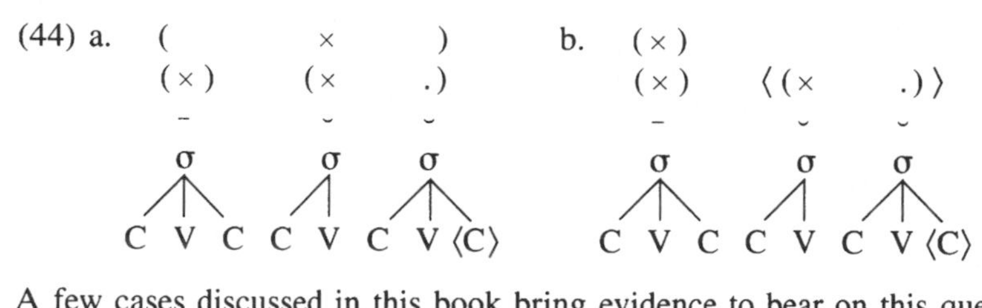

A few cases discussed in this book bring evidence to bear on this question: Palestinian Arabic (§ 6.1.1), Egyptian Radio Arabic (§ 6.1.2), and Bani-Hassan Arabic (§ 8.10). They all come out the same way, namely as in (44b). That is, a foot must be allowed to be extrametrical, even if it includes an extrametrical entity. Thus I propose (45):

(45) Extrametrical higher level constituents may dominate extrametrical lower level constituents.

The case just examined must be carefully distinguished from a separate phenomenon, that of **unsyllabified** consonants. Steriade 1982; Clements and Keyser 1983; Kenstowicz 1986; and others have demonstrated that certain consonants may fall outside the syllable for a large part of the phonological derivation. Typically these are consonants that if syllabified would create syllables that go beyond the central syllable canons of the language. An example would be final /CVCC/ and /CV:C/ syllables in languages where nonfinal syllables must have the structure /CV/, /CVC/, or /CV:/. Extrasyllabic consonants are often resolved by epenthesis or deletion, or they may be incorporated into a neighboring syllable by rules applying late in the derivation.

The crucial point here is that an extrasyllabic consonant is not dominated by a syllable or by a foot. It therefore does not satisfy the description in (45). Suppose, then, that we have a final extrasyllabic consonant in a language with foot extrametricality, as in the schematic form (46):

```
(46)  (×)      (×)
       –        –
       σ        σ
     C V C    C V C C
```

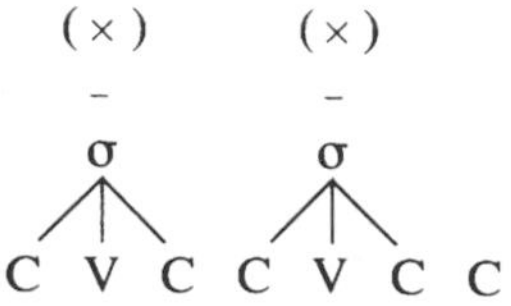

In this case, the rightmost foot is NOT peripheral, since the stray consonant intervenes between it and the right edge of the string. This can be made clearer with the "official" representation of (47):

(47)
```
   ×        ×
   σ        σ
  /|\      /|\
( C V C )( C V C ) C
```

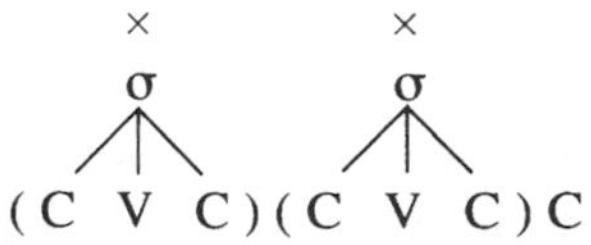

As a result, the rightmost foot cannot be made extrametrical. End Rule Right, applying to such a form, would create final main stress.

One further complication is this: suppose in (47) the final stray consonant is marked as extrametrical. Would the rightmost foot then be peripheral, hence eligible for foot extrametricality? If this is the case, we would have "chained" extrametricality, as shown in (48):

(48)
```
   ×          ×
   σ          σ
  /|\        /|\
( C V C )⟨( C V C )⟩⟨C⟩
```

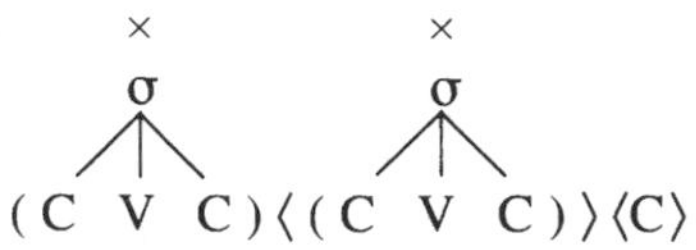

I propose to exclude this configuration, by the constraint in (49):

(49) Extrametricality does not chain; i.e., a constituent followed by an extrametrical constituent is not counted as peripheral.

Following (49), the rightmost foot is regarded as non-peripheral; hence by the Peripherality Condition (§ 3.11), it may not be extrametrical, and End Rule Right would assign it the main stress.

This outcome is widely observed in Arabic dialects, which have final extrasyllabic consonants. Examples presented in this book are Cairene (§ 4.1.3), Palestinian (§ 6.1.1), Negev Bedouin (§ 6.3.6), Cyrenaican Bedouin (§ 6.3.7), and Bani-Hassan (§ 8.10); another possible example is Dakota (§ 6.3.11). The ban on chained extrametricality is also crucial to the prediction made in § 4.4: that in quantity-sensitive languages with End Rule Right, stress may not fall to the left of a heavy penult. If extrametrical feet could occur before extrametrical syllables, this result would be subverted.

At least some kind of ban on chained extrametricality must be assumed in any event if extrametricality theory is to be at all restrictive: if we allow extrametricality to chain freely (essentially as proposed in Lee 1969), then we could write all sorts of exotic, unattested stress rules simply by writing a string of extrametricality rules. For example, the rules in (50) generate preantepenultimate stress in words of four syllables or more, otherwise initial stress.

(50) a. $\sigma \rightarrow \langle\sigma\rangle$ / ____ $]_{word}$
b. $\sigma \rightarrow \langle\sigma\rangle$ / ____ $]_{word}$
c. $\sigma \rightarrow \langle\sigma\rangle$ / ____ $]_{word}$
d. End Rule Right

Summing up, the theory proposed here posits the following principles governing the interaction of extrametricality rules and peripherality: (a) Higher level extrametrical elements may incorporate lower level ones; (b) a stray consonant renders the preceding syllable or foot non-peripheral; (c) extrametricality may not chain.[5]

5.2.2 Extrametricality in Clash

The preceding section was directed toward tightening the theory of extrametricality by fixing the class of possible interactions of extrametricality at different levels. I also propose to loosen the theory a bit, justifying this with the analyses that follow.

I propose that extrametricality rules may refer in limited ways to their left environments (cf. Archangeli 1984). In particular, they may apply when the constituent being marked as extrametrical is **in clash** with a preceding stress. A typical rule of this form would be as in (51):

(51) $\sigma \rightarrow \langle\sigma\rangle \,/\, \times \;\underline{\quad\quad}\;]_{\text{word}}$

The /×/ marks the immediately preceding clashing stress. Cases of extrametricality in clash may be found in Maithili (§ 6.1.6), Awadhi (§ 6.1.9), Manam (§ 6.1.9), Sarangani Manobo (§ 6.1.9, § 6.3.10), Javanese (§ 6.3.10), Malay (§ 6.3.10), and Bani-Hassan Arabic (§ 8.10).

5.2.3 Stray Adjunction?

The convention of **Stray Adjunction** was proposed in Liberman and Prince 1977, 294. In its most general form, it states that once a given layer of metrical structure has been created, any element that is **stray** on that layer is incorporated as a metrically weak member of an adjacent constituent. Thus stray syllables are incorporated into adjacent feet and stray feet into adjacent word layer constituents. Stray Adjunction was particularly important for the theory of

5. As Donca Steriade (p.c.) points out, a potential problem for my view is posed by English words like *gálaxy, álligàtor, éfficacy,* whose stress patterns suggest underlying representations of the form /gælVksy/, /ælVgeytr/, /ɛfıkVsy/ (*SPE* 129–35, 160–1; Liberman and Prince 1977, 295–97). If this analysis is correct, we must assume that the underlying final consonant actually belongs to the preceding syllable, since otherwise it would block syllable extrametricality, leading to **galáxy, *àlligátor, *effícacy.* Steriade points out a similar problem with extrasyllabic consonants and extrametricality in Latin, where words like *múltipleks* would receive final stress if the final /s/ is extrasyllabic.

If we assume extrasyllabicity in such cases, then we need a different approach, namely allowing chained extrametricality in limited circumstances, e.g. when the extrametrical constituents are of different types, as Steriade suggests (cf. also Buckley 1991). To handle Arabic, we would then need to posit that the final stray consonants somehow count as full syllables. Moreover, we would also need a new explanation for why stray syllables block foot extrametricality, as shown in § 4.4(b). I leave this issue open.

Hayes 1981: by adjoining extrametrical units as weak members of the next higher constituent, it guaranteed that extrametrical elements surface as weak.

In the theory proposed here, however, I largely avoid Stray Adjunction, following HV. There are several reasons not to adopt it. (a) The justification for Stray Adjunction under the earlier tree theories just cited does not hold under bracketed grid theory, where no actual structural change is needed to mark a stray element as weak. (b) Stray Adjunction would go against one of the central ideas argued for here: that the "goal" of creating canonical foot structure influences segmental phonology and other areas. A Stray Adjunction convention would create all sorts of non-canonical foot shapes in its output. (c) Stray Adjunction weakens the predictions of the theory, because it expands the size of metrical constituents, and thus increases the set of landing sites available for Move X or the migration of stress under vowel deletion. It also expands the class of feet that are peripheral, thus destroying the argument made in § 4.4(b). (d) Finally, the literature on cyclic stress and the Free Element Condition (§ 5.4.2) crucially depends on there being no Stray Adjunction, as Halle and Kenstowicz 1991, 476, points out.

There is a limited amount of evidence in favor of Stray Adjunction. An empirical argument can be made for Pirahã, discussed in § 7.1.7, where an alternative analysis is also entertained. English phonology has been argued by Kiparsky 1979; Selkirk 1980b; McCarthy 1982; and others to involve segmental rules sensitive to the foot; the relevant feet could be derived in the framework here only by assuming Stray Adjunction, since they are quantity-sensitive and often larger than moraic trochees. But the same phenomena have also been plausibly analyzed with syllable structure rather than foot structure, as in Kahn 1976.

On the theoretical side, Stray Adjunction might be supported by the fact that it enables us to say that prosodic constituents always exhaust the string.[6] The objection to this is that it merely expresses a personal taste unless we make clear what the empirical consequences of exhaustive parsing would be. In syllable theory, we have quite a bit of evidence for exhaustive parsing, since in most languages segments are not allowed to surface unless they are syllabified (Steriade 1982; Harris 1983; Ito 1986), and this requirement often serves to guide much of the phonology. For foot structure, there does seem to be some pressure for all syllables to be incorporated into feet (§ 6.1.5), but this is hardly the pervasive phenomenon that is found in syllabification; moreover, the

6. Hayes 1987 made just this assumption, and therefore treated what I here analyze as stray syllables as the members of **stressless feet.** While this maintains exhaustive parsing, it also provides additional descriptive power: the stressless feet mark when a syllable has been already parsed by the stress rules. Since the need for this power seems doubtful, the present version of the theory jettisons stressless feet in favor of surface stray syllables.

relevant cases could not be derived if Stray Adjunction were an automatic convention.

Comparing further with syllables, note that if Stray Adjunction were to be made part of syllable theory, it would have to be modified so as to respect conditions of syllable well-formedness; otherwise it would overgenerate disastrously. By analogy, Stray Adjunction in metrical theory ought to be limited to cases where adjunction obeys the basic constraints of foot structure. This would greatly limit its scope, since foot construction algorithms seldom place stray syllables adjacent to submaximal feet. Moreover, Bagemihl 1991 and Bates 1992 have argued that at least in some languages, the principle of exhaustive parsing does not hold even for syllables. The upshot seems to be that in our present state of knowledge, it would be aprioristic to adhere firmly to a rigid principle of exhaustive prosodic parsing, especially one that is enforced by an adjunction convention.

5.3 THE UNSTRESSABLE WORD SYNDROME

The preceding sections have discussed the ban on degenerate feet (§ 5.1) and extrametricality rules (§ 3.11, § 5.2). In this section, I discuss a curious phenomenon that results from their conjunction: words of a certain shape become **unstressable.** Consider the following scenario: (a) a given language has quantity-sensitive stress and prohibits degenerate feet (/$\acute{\smile}$/); (b) final syllables are extrametrical; (c) the word being assigned stress has the shape /$\smile$ σ/. Together, these conditions imply a contradiction: no foot can be constructed, since only /$\smile$/ is visible to the stress rules, which cannot make /$\smile$/ into a foot. On the other hand, we have seen (§ 3.1) that languages almost always impose a requirement of "culminativity" on content words; that is, every word must have metrical structure. Thus in an "unstressable" word, four factors—the shape of the input word, foot well-formedness, extrametricality, and culminativity—are in conflict.

It is precisely in such words that we find a remarkable range of outcomes across (and even within) languages. It appears that languages having the above set of conflicting properties solve the contradiction in essentially ad hoc fashion. The wide range of outcomes is unappealing, since we would like to be able to propose a single convention that predicts the correct resolution for all unstressable words. But the theory proposed here does have the merit of singling out this particular case as one where peculiar things happen.

Here is a survey of the strategies that languages use to deal with unstressable words.

(A) LENGTHENING. Hixkaryana (§ 6.3.1), with iambs and final syllable extrametricality, lengthens the vowel of an initial light syllable in disyllabic words. This renders it capable of bearing a foot, so the word can be stressed:

(52) (×) = § 6.3.1 (171a)

˘ ⟨˘⟩ → – ⟨˘⟩ → – ⟨˘⟩

kwaya *kwa:ya* *kwá:ya*

Dutch, in the account of van der Hulst and van Lit 1988, is a similar case. Sporadically, lengthening in unstressable words also occurs in English: in dialectal *police* [pó:lì:s], *Detroit* [dí:trɔ̀it], *cement* [sí:mɛ̀nt] (Flexner 1987), and *Arab* [é:ræ̀b] (Kenyon and Knott 1944), the initial vowel is lengthened to enable it to bear a foot (final syllables are extrametrical in English nouns; Hayes 1982b).

The lengthening strategy can sometimes be seen in systems without extrametricality. In Fijian (§ 6.1.5), Anguthimri (§ 5.1.9), and Badimaya (§ 6.2.3), words of the underlying form /˘/ are lengthened by rule to / – /, a proper moraic trochee. For some English speakers, the citation forms of *the* and *a* are [ðʌ:], [ʌ:]. The long [ʌ:] vowels are segmentally aberrant, but fulfill a more demanding prosodic requirement for isolation forms (compare the similar facts of Seminole/Creek; § 4.1.2). In Mohawk (Michelson 1988), words of the underlying form /σ/ are lengthened by epenthesis to form /σ σ/, a proper syllabic trochee.

(B) INCORPORATION OF EXTRAMETRICAL MATERIAL. Latin (§ 5.1.4) assigns stress to a moraic trochee at the right edge of a word, with final syllables extrametrical. Since Latin forbids degenerate /˘/ feet entirely, words of the shape /˘ σ/ are unstressable. I propose that they receive surface stress by a process of **incorporation,** that is, a degenerate foot is constructed, but immediately repaired by adjoining the extrametrical syllable to it. This creates a non-canonical /˘́ – / foot. By violating canonical foot form, this process preserves culminativity and the ban on degenerate feet: /˘́ – /, though non-canonical, is not degenerate.

Incorporation can also arise when stress is assigned from left to right. Hopi (§ 6.3.9) forms an iamb at the left edge of a word. Final syllables are extrametrical, and degenerate words are forbidden. The resulting "unstressable" words are treated by forming a degenerate foot and incorporating the final extrametrical syllable:

(53) (×) → (× .) 'wood' Jeanne 1982, 254

˘ ⟨˘⟩ ˘ ˘

koho *kóho*

Similar patterns are found in Southern Paiute (§ 6.3.11) and Asheninca (§ 7.1.8).

The incorporation process can be characterized explicitly as follows: it is triggered by the presence of an illegal degenerate foot, and consists of the addition to that foot of a weak member consisting of a neighboring stray syllable.

Since degenerate feet typically do not arise next to stray syllables (cf. the Priority Clause (22)), incorporation is a relatively rare phenomenon; and the constraints on foot form posited here are not often violated by it. See § 5.4.4 for incorporation in another context.

(C) SHORTENING. The non-canonical /◡́ –/ feet created by incorporation in Latin display an interesting property: the segmental rules of the language often repair them. In particular, the so-called **Iambic Shortening** rule (Allen 1973, 179–85) converts /◡́ –/ words to the canonical /◡́ ◡/ by shortening the stressless vowel: *égo:* 'I' becomes *égo.* For discussion of Iambic Shortening see § 5.6.1, which also describes the similar rule of English.

(D) REVOCATION OF EXTRAMETRICALITY. In English, the final syllables of nouns are normally extrametrical (Hayes 1982b). However, as Liberman and Prince (1977) point out (citing unpublished work by R. Oehrle), disyllables of the "unstressable" form /◡ –/ often show final stress; cf. words like *políce, canóe, ballóon, manúre, ravíne* (see also Ross 1972). I suggest that this reflects a strategy for making such words stressable: by avoiding syllable extrametricality, these words provide enough room for a proper moraic trochee:

(54) (×)
(×)
◡ –
pV̆li:s

It can be seen that English invokes multiple solutions to the unstressable word problem. This reflects the lexicalized character of the English stress system (Hayes 1982b, 236–37): since English stress is lexically listed, the stress rules serve essentially as redundancy rules, describing the set of possible (as well as unmarked) patterns. A word has a well-formed stress contour if it is stressable by some derivation provided by the rules.

Allen 1973, 186–89 suggests that in certain positions, /◡ –/ words in Latin bore final stress. The account would be the same as above.

There are many cases in which revocation of extrametricality cannot be distinguished from incorporation. For example, either strategy derives initial stress on /◡ ◡/ nouns in English, such as *mánna, cíty,* or *grótto.*

(E) VIOLATION OF CULMINATIVITY. One further way to deal with unstressable words is not to stress them. This appears to be the case in Central Sierra Miwok (§ 6.3.9). In this language, the stress rule assigns iambs, degenerate words are forbidden, and final syllables are extrametrical. According to Freeland 1951, 8, when /◡ –/ words "are spoken alone there is a secondary stress on the final syllable. . . . In connected speech, however, either syllable may be stressed, or the word may be entirely without stress."

My interpretation of this is that unstressable words in Sierra Miwok indeed receive no metrical structure within the word phonology. Phrasally, they are incorporated into the metrical structure of surrounding words, with optional Beat Addition (see § 9.2.4) giving them weak prominence contours. For the isolation forms with "secondary stress," I conjecture that extrametricality is revoked (see (d) above), and that a normal iamb is formed on them postlexically. This allows culminativity to be respected at the phrasal level. In this sense, violations of culminativity in Sierra Miwok are only temporary: all phrases have culminative metrical structure. Sierra Miwok differs from other languages only in that some content words receive no metrical structure in their word level derivations.

Freeland's claim that /◡ – / words bear only secondary stress in isolation is most likely to be explainable with reference to the intonational system. Broadbent 1964, 17–18, discussing intonation in closely related Southern Sierra Miwok, gives a plausible rationale for why monopod words could sound like they bear only secondary stress.

A case somewhat similar to Sierra Miwok is Turkish, in the analysis of Barker 1989. Here, the unstressable words are simply the monosyllables; a rule of syllable extrametricality renders them unstressable in Level I of the morphology, thus sending them on to later levels where extrametricality is not in effect.

Summing up, the unstressable word syndrome gives rise to one of four basic strategies: (a) lengthening and shortening, (b) incorporation, (c) revocation of extrametricality, or (d) blockage of foot construction. These reflect which one of the four competing factors in an unstressable word is forced to yield: option (a) sacrifices the input word shape, (b) sacrifices foot form, (c) sacrifices extrametricality, and (d) sacrifices culminativity.

5.4 MODES OF PARSING

Given a particular foot template, there are various possible algorithms by which it can be used to parse a word. I assume here two basic parsing parameters: (left to right/right to left) and (iterative/noniterative). The latter is the subject of a controversy (cf. HV; Halle 1990; Hammond 1990b; and Blevins 1990), to which I have nothing to contribute; while the analyses below occasionally make use of non-iterative footing, they could be re-expressed using the more complex mechanisms invoked by HV in order to dispense with the iterativity parameter.

Below, I consider further parsing parameters and other issues in foot parsing.

5.4.1 Persistent versus Non-persistent Footing

Surface well-formedness conditions in phonology can be thought of in two ways. Under one approach (e.g. that of *SPE*), they are regarded essentially as theorems: the well-formedness conditions observed on the surface are the result of underlying well-formedness conditions plus phonological rule application. An opposing view regards surface well-formedness conditions as in some sense more fundamental than the rules themselves: the rules can be thought of as relatively ad hoc ways of insuring that the representations obey the well-formedness conditions.

In general, I am sympathetic to the second view. For example, the account below of segmental rules that create canonical foot forms (§ 4.5.3) and the discussion of the "unstressable word syndrome" in § 5.3 both presuppose it. A large body of work (e.g. Kisseberth 1970; McCarthy and Prince 1986; Paradis 1988; and references cited below) has demonstrated the centrality of surface well-formedness conditions to phonology.

One way to enforce surface well-formedness conditions is to suppose that the rules achieving well-formedness apply **persistently,** rather than just once in the derivation. This is the view, for example, of McCarthy 1979b and Ito 1986 on syllabification: the outputs of phonological rules are persistently adjusted to conform to a language's syllable canons. Stress rules are assumed to be generally persistent in HV (p. 135), and Myers 1991 has argued that many phonological processes are also persistent in this sense.

I conjecture that stress rules are variably persistent, on a language-specific basis. Below, I discuss languages in which foot well-formedness conditions continue to be enforced even after the application of phonological rules that follow basic footing. On the other hand, there appear to be languages in which metrical structure is assigned just once and can be grossly deformed after it is assigned. I will say that languages of the former category have **persistent** stressing and the latter **non-persistent.** Examples of languages with persistent stress include Spanish (§ 5.1.7), Maithili (§ 6.1.6), Fijian (§ 6.1.5), Lenakel (§ 6.1.8.3), Icelandic (§ 6.2.2), the Yupik dialects described in § 6.3.8, Asheninca (§ 7.1.8), Estonian (§ 8.5.3), Finnish (§ 8.6), Karelian (§ 8.6), Hungarian (§ 8.6), Winnebago (§ 8.9.5), Auca (§ 8.10), Bani-Hassan Arabic (§ 8.10), and Latin (Mester 1992). Languages with non-persistent stress include Cahuilla (§ 6.1.3 (30b, c)), Wargamay (§ 6.1.4 (40)), Negev Bedouin Arabic (§ 6.3.6), Cyrenaican Bedouin Arabic (§ 6.3.7), Cayuvava (§ 8.2), Sentani (§ 8.7), Mantjiltjara (§ 8.10), and English.[7] A problematic case of "semi-persistent" stress occurs in Pacific Yupik (§ 8.8).

7. The relevant data for English are words in which rules of destressing create syllable sequences that could in principle be reparsed, but are not: e.g. *Òkefenókee* /–̀ ◡ ◡ –́ ◡/ (Hayes 1982b, 257), or *règularizátion* /◡̀ ◡ ◡ ◡ –́ ◡/ (Liberman and Prince 1977, 285–86). Also included would be sequences created by late rules of glide vocalization: e.g. *présidency* /◡́ ◡ ◡ ◡/ (*SPE*, 160–61).

In my account, persistent stressing consists of two things. First, when segmental processes produce a metrical structure that is ill formed (either degenerate or else larger than the foot template allows), that structure is deleted, creating stray syllables. Second, stray syllables (both from foot loss and those left stray earlier in the derivation) are resubmitted to the foot construction algorithm. Persistent footing is assumed to obey the Free Element Condition, stated in (55):

(55) **Free Element Condition** (Prince 1985)
Rules of primary metrical analysis apply only to free elements.

Because of the Free Element Condition, persistent footing (construed here as primary metrical analysis) may apply only to stray syllables. More precisely, I define persistent footing as in (56):

(56) **Persistent Footing**
a. Single stray syllables are adjoined to existing feet if the result is well formed.
b. Otherwise, sequences of stray syllables may be converted into feet.

Provision (56a) is taken from Steriade 1988b; it is ranked over (56b) when both are applicable. For discussion and examples, see § 5.1.7, § 6.1.6.4, § 6.1.8.3, § 6.2.2.4, § 6.3.8.3, and § 7.1.8.4.

5.4.2 More on the Free Element Condition

The Free Element Condition makes possible a simple account of exceptionally stressed syllables: I assume that they come with lexically listed feet. Due to the Free Element Condition, the ordinary stress rules cannot override the lexically listed structure. For discussion, see Kiparsky 1979, 1982a.

The Free Element Condition (55) also forms the basis of much insightful work on cyclic stress, notably Steriade 1988b; Poser 1989; Halle 1990; and Halle and Kenstowicz 1991. For a number of languages, these authors argue that when stress is assigned cyclically, the only syllables left over from an old cycle that may be footed on a new cycle are those that the previous cycle leaves stray, usually because of extrametricality. This contrasts with the view expressed by Kiparsky 1982a on the basis of English evidence, whereby foot construction may rebracket syllables that have been footed on a previous cycle, when (and only when) this is needed to foot a syllable added on the new cycle, as in *solid* [(sáləd)] ~ *solidify* [sə(**lídə**)fay]. The analyses in this book for the most part do not bear on this issue, but I tentatively assume that cyclic stress systems may vary parametrically along this dimension. For further discussion, see § 6.2.1.[8]

8. Kenstowicz 1991 observes that the Free Element Condition approach to cyclic stress bears on the choice of a foot inventory. In Latin, various cyclically derived forms appear to support the existence of uneven trochees (§ 4.3.1). In (i.a), with uneven trochees, we get the correct stress on

5.4.3 Top-Down versus Bottom-Up Stressing

In all the analyses given so far, bracketed grids have been constructed from the bottom up. For example, Pintupi (§ 4.1.1), constructs syllabic trochees from left to right, then forms a word layer by End Rule Left above the feet. This yields main stress on the initial syllable, with secondary stress on subsequent odd-numbered nonfinal syllables.

```
(57)                                                        (×               )
                              (×  .)(×   .)(×  .)           (×  .)(×   .)(×  .)
     σ σσ  σ σ σ      →       σ σσ  σ σ σ         →         σ σσ  σ σ σ
  tʲámulìmpatʲùŋku         tʲámulìmpatʲùŋku              tʲámulìmpatʲùŋku
```

Van der Hulst 1984, 178–82, suggested a less obvious procedure: assign the main stress first. In the formalism adopted here, this would involve applying End Rule Left, as in (58b), then "tucking under" a foot layer, as in (58c):

```
(58) a.                              b. ( ×  . .   .  .   .)
       σ σσ  σ σ σ      →              σ σσ  σ σ σ      →
     tʲamulimpatʲuŋku                tʲámulimpatʲuŋku

     c. (×               )
        (×  .)(×   .)(×  .)
         σ σσ  σ σ σ
       tʲámulìmpatʲùŋku
```

---

the second cycle, since only the (originally extrametrical) syllable /na/ is stray. In (i.b), with moraic trochees, the whole sequence /na + kʷe/ is available for footing, and we would wrongly derive antepenultimate stress.

```
(i) a.                         (      ×)
            (×  .)             ( ×  .)(×)                 Kenstowicz 1991, ex. (4)
            –  ᴗ(ᴗ)      →      –  ᴗ ᴗ  (ᴗ)
          liːmina+kʷe         liːminákʷe                  'and thresholds'

   b. not:                     (   ×  )
            (×)                ( ×)(× .)
            –  ᴗ(ᴗ)      →      –  ᴗ ᴗ  (ᴗ)
          liːmina+kʷe         *liːmínakʷe
```

Mester 1992 replies to this point, arguing first that the analysis Kenstowicz assumes faces its own empirical problems (i.e., it derives the wrong output for disyllabic clitics attached to monosyllabic stems); and second that the basic data on Latin enclitic accentuation are quite difficult to establish firmly in any event. Mester's own analysis of Latin clitic analysis is quite different, based on a rule of clitic extrametricality, plus End Rule Right. Mester's approach is made especially plausible by the substantial amount of evidence he presents from other areas that Latin prosody is based on the moraic trochee, not the uneven trochee.

In this case, the foot layer is constrained to contain a mark underneath the grid mark for the main stress, in order to satisfy the Continuous Column Constraint. For Pintupi, the foot construction rule does this trivially. In § 9.6.1, I propose other mechanisms for satisfying the Continuous Column Constraint in top-down creation of metrical structure.

Which of these two options (bottom up vs. top down) is correct? The answer appears to be that it depends on the language in question. For example, in Swedish (Bruce 1984), it is clear that primary stress must be assigned before secondary stress, since primary stress assignment is lexical, secondary postlexical. Churchyard 1990, 1992 argues for top-down construction in Tiberian Hebrew on similar grounds. Here I argue for top-down stressing in Cahuilla (§ 6.1.3), Tümpisa Shoshone (§ 6.1.9), Czech (§ 6.2.3), Mayi (§ 6.2.3), Old English (§ 5.4.4), Cayuvava imperatives (§ 8.2), Estonian (§ 8.5.3), in English phrasal stress (§ 9.6.1), and in a number of examples given in § 6.3.10. On the other hand, I discussed several cases in § 3.4.2 (e.g. Cairene Arabic or Seminole/Creek) in which applying main stress first would require an extremely complex main stress rule; for these languages the bottom-up analysis is far more straightforward and permits a far more constrained general theory.

For the majority of alternating stress languages, it is apparently impossible to determine the answer, because the main stress usually falls on the point of origin of the alternating count (Hammond 1984b). In such cases, it is usually straightforward to express the stress rules either way. For concreteness, I express stress rules in bottom-up fashion here, except where the facts require otherwise.

5.4.4 Unstressable Words from Top-Down Stressing

Suppose a metrical system involves the rules in (59):

(59) a. **Word Layer Construction** End Rule Left
(ordered first; i.e. top down)

b. **Foot Construction**
i. Form moraic trochees from left to right.
ii. Degenerate feet are banned entirely.

Such a system runs into an interesting problem when it encounters words beginning with /˘ – /:

(60)

a.	b.	c.
*(×)	*(×)	(×)
(×)	(×)	(× .)
˘ – . . .	˘ – . . .	˘ – . . .

Outcome (60a) includes a degenerate foot, violating (59b.ii); outcome (60b) violates the Continuous Column Constraint. Given the rules of (59), words beginning with /˘ – / are thus "unstressable," in the sense of § 5.3.

My suggestion is that the contradiction may be resolved much as it is in

Latin (§ 5.3(b)): the heavy syllable is incorporated into a degenerate foot, yielding (60c). This scenario plausibly constitutes the foot construction system for Old English, as discussed by Tanaka 1990 and Lahiri and van der Hulst 1988; the use of top-down construction for Old English originates with the latter. A ban on degenerate feet in Old English can be justified by the absence of light-syllable content words (Russom 1987, 12).

The central evidence for the feet derived in this way is that they govern a rule of High Vowel Deletion. Keyser and O'Neil (1985) formulate this rule as follows: delete high vowels in open syllables that directly follow a foot. I adopt this formulation, though the feet proposed here are left-strong, and thus unlike Keyser and O'Neil's can account for the stem-initial main stress as well.

Dresher and Lahiri (1991) propose a rather different foot structure for Old English, consisting of the feet noted above augmented by a following weak light syllable. While the analysis is carefully argued, it involves a substantial unacknowledged cost: to account for High Vowel Deletion, it must posit vowels that are stressless but not metrically weak. These occur sometimes as the weak half of a "disyllabic strong branch" (Dresher and Lahiri ex. (9)), sometimes as the result of destressing in final position (exx. (12c) and (16)). Dresher and Lahiri note that their proposal can also provide accounts of metrical resolution and secondary stress. With the feet proposed here, resolution can be described by allowing a strong foot to occupy a single metrical position. Scholars differ on the secondary stress facts (cf. Dresher and Lahiri vs. Tanaka); in my analysis either pattern can be derived, the difference hinging on whether Old English employs strong versus weak local parsing (chapter 8).

It is interesting to compare Old English with a metrically similar language, namely Cahuilla (§ 6.1.3). Unlike Old English, Cahuilla allows light monosyllables; I interpret this as a choice of the weak prohibition on degenerate feet in Cahuilla versus the strong prohibition in Old English. Since Cahuilla allows degenerate feet in strong position, it can stress initial /◡ –/ sequences without using incorporation:

```
(61) (×          )
     (×)( × )
      ◡   –   . . .
```

Another interesting comparison is with Malayalam (§ 5.1.5). Here, the metrical structure is created bottom up rather than top down. This allows the initial light syllable of /◡ –/ words to be skipped, so that such words receive second-syllable stress:

```
(62)    ( × )
      ◡   –   . . .
```

Tümpisa Shoshone (§ 6.1.9) represents an intermediate case, with free variation between the Cahuilla and Malayalam outcomes. I interpret this as free selection of top-down versus bottom-up stressing.

5.5 THE NUMBER OF GRID LAYERS

5.5.1 Cola

There appears to be nothing sacred about having just two layers in the metrical representation, although two layers typically suffice for word stress. Phrasal stress, discussed in chapter 9, characteristically involves multiple layers, and some word stress systems involve three. For these systems, Halle and Clements 1983 suggest the term **colon** to designate a constituent of the intermediate layer. Cola seems to come in two varieties: unbounded (created by the End Rule), and binary with initial prominence (apparently an analogue of the syllabic trochee). In this book, cola are invoked for Maithili (§ 6.1.6), Garawa (§ 6.2.3), Malecite-Passamaquoddy (§ 6.3.4), Eastern Ojibwa (§ 6.3.4), Tiberian Hebrew (§ 6.3.10 and Dresher 1980), Asheninca (§ 7.1.8), and Hungarian (§ 8.6).

5.5.2 Conflation

In a number of languages, foot construction appears to serve primarily as a computational device: it is crucial to an account of main stress placement, but the feet are not (as far as we know) reflected in secondary stress or any other phonetic correlates. To account for this, we can make one of two assumptions.

First, we might suppose that the phonetic and phonological rules of the language just happen not to provide any means of manifesting foot structure. This solution is viable, given what we have seen (chapter 2) concerning the language-specific phonetic realization of stress.

The other possibility is that the feet are actually removed by phonological rule, so that in principle they cannot have phonetic consequences. The latter solution is advocated by HV (50–55), who provide a mechanism called **conflation** to remove the unnecessary structure. The following version of conflation is proposed by Halle and Kenstowicz 1991, 462:

(63) **Conflation** Remove the lowest line of the grid.

The requirement that only the lowest line be removed prevents conflation from violating the principle of culminativity (§ 3.1).

It is difficult to prove the existence of Conflation, given the subtlety of the data and the alternative of accidental phonetic non-interpretation; for extended discussion, see Blevins 1990. A relevant example is Seminole/Creek (§ 4.1.2), treated by HV 59–60 with conflation: Martin 1992 suggests that metrically weak feet in Creek do have subtle phonetic correlates and therefore must persist into surface representations. Other cases in this book where the possibility of conflation arises include Cairene Arabic (§ 4.1.3), Lenakel (§ 6.1.8), Paamese (§ 6.1.9), Negev Bedouin Arabic (§ 6.3.6), Cyrenaican Bedouin Arabic (§ 6.3.7), Mam (§ 7.1.5), and many of the examples in § 7.2.

5.6 FURTHER ISSUES IN SYLLABLE THEORY

The basic elements of the theory of syllable structure adopted here, that is, moraic theory, were laid out in § 3.9.2. This section adds further discussion relevant to the case studies in the next chapter.

5.6.1 The Quantity of CVC Syllables

Universally, CV counts as a light syllable. It is probable that long-voweled syllables universally count as heavy; see § 7.4. But CVC syllables vary: in some languages they are heavy, in others light. As noted in § 3.9.2, these patterns are predicted by the moraic theory of syllable structure, in which CVː must be represented as bimoraic, CV as monomoraic, but CVC has both monomoraic and bimoraic representations:

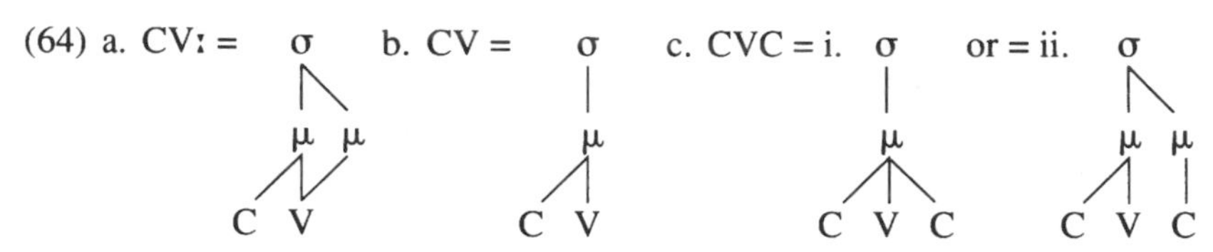

In most previous literature, it is assumed that the choice between (64c.i) and (64c.ii) is made by individual languages. However, as Kager 1989 points out, it is logically possible that the choice of (64c.i) versus (64c.ii) might vary within a single language, determined by the phonological context.

The Iambic Shortening rule of Latin, discussed in § 5.3, supports this view. I noted that in "unstressable" /⏑ –/ words, extrametricality plus incorporation create an aberrant /⏑́ –/ foot, which is often repaired by shortening the final vowel: *égoː* → *égo* 'I', *máleː* → *mále* 'bad'. The same phenomenon can be found with secondary stress, where the aberrant foot may be created to avoid clash: *àmiːkítiam* → *àmikítiam* (cf. *amíːkus* 'friend, kind'; examples from Allen 1973, 179, 181). We can express Iambic Shortening formally as in (65):

(65) **Iambic Shortening**

```
(×   .)        (×   .)
 σ   σ     →    σ   σ
 |   |\         |   |
 μ   μ  μ       μ   μ
```

An interesting aspect of Iambic Shortening is that it also applied to CVC syllables. This had no effect on segment count, but its effects can be observed in Latin quantitative verse, which depended on the light/heavy distinction. The following are typical cases, taken from Allen 1973, 182–83:

(66)	a.	b.	c.	
	⏑ ⏑ – – –	⏑ ⏑ – –	⏑ ⏑ – –	metrical scansion
	***úter** vostróːrum*	***dédit** dóːnoː*	*gù**berná**ːbunt*	verse text

Assuming that Latin allowed light CVC syllables, such scansions are nonproblematic, as they fall out from the same rule of Iambic Shortening (65) that applies to long vowels:

(67)
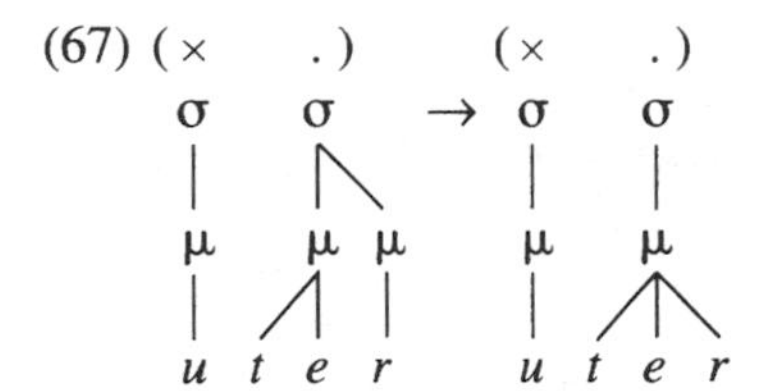

We need only assume that the consonant stranded by mora deletion is reaffiliated within its own syllable.

Iambic Shortening is not limited to Latin; as Allen 1973, 191–99 points out, the so-called "Arab" Rule (Ross 1972) of present-day English has essentially the same function: CVC is almost always stressless (and, I would claim, monomoraic) after stressed light syllables, and CVː is very rare in the same context. An example is the word *presentation,* whose second syllable must be stressless if the first is light: [prɛ̀zəntéyšən], [prìːzɛ̀ntéyšən], *[prɛ̀zɛ̀ntéyšən].

Other cases of variable quantity for CVC are presented elsewhere in this book: Cahuilla (§ 6.1.3), Palestinian Arabic (§ 6.1.1), Eastern Ojibwa (§ 6.3.4), and the Yupik languages (§ 6.3.8, § 8.8).

5.6.2 Assessing the Evidence for Syllable-splitting Foot Construction

I claimed in § 3.9.1, following earlier work, that the syllable is universally the stress-bearing unit. Here I consider possible counterexamples to this claim, in which stress is said to be assigned to moras, sometimes dividing a heavy syllable into two feet.

One possible case of this sort is Southern Paiute (§ 6.3.11). Here, iambs are formed from left to right across what superficially appear to be moras; the word layer is formed by End Rule Left. In (68), /./ marks syllable division, long vowels are represented as VV for clarity, and the underlying representation follows Harms 1966.

```
(68) (        ×                                   )
     (.       ×)(.    ×)  (.     ×)  (.     ×)(.    × )
     μ        μ μ     μ    μ     μ    μ     μ μ     μ     ⟨μ⟩
     t ɨ . xʷ i i . n a . t ɨ . β i . č u . xʷ a i . ʔ i . ŋʷ a   →
```

[tɨχʷíːnːàtːɨβʷìčuχʷàiʔìŋʍḁ] 'go and ask him to tell a story'
Sapir 1930, 81

If the second syllable /xʷii/ were treated as a single unit, we would derive the wrong result:

```
(69)  (      ×                    )
      ( .    ×) ( . ×) ( .  ×) (×) (×)
        ˘    –    ˘  ˘   ˘  ˘   –   ˘   ⟨˘⟩
      tɨ.xʷiː.na.tɨ.βi.ču.xʷai.ʔi.ŋʷa   →   *[tɨχʷíːnːḁtɨ̀βʷɪčùχʷàiʔìŋʷḁ]
```

In light of this, I consider here the strength of the Southern Paiute evidence for mora stressing. First, it is not a surface phenomenon, since the sequences /V́V/ versus /VV́/ derived by the stress rule in different contexts are not distinct from one another phonetically.[9] More important, there is evidence that surface CVV syllables in Southern Paiute are represented at an earlier stage of the phonology as disyllabic CV+V, as in (70):

```
(70)  (          ×                               )
      ( .        ×)( .    ×) ( .    ×) ( .       ×)( .   ×)
        ˘        ˘  ˘     ˘    ˘    ˘    ˘       ˘  ˘    ˘     ⟨˘⟩
      t ɨ . xʷ i . i . n a . t ɨ . β i . č u . xʷ a . i . ʔ i . ŋʷ a
```

In particular, certain allophonic rules of Southern Paiute apply to one "half" of a VV sequence. For example, a rule of word-final vowel devoicing applies to single final vowels and to half of a final long vowel; other examples are cited in Hayes 1981, 53–54. This pattern goes against the characteristic "inalterability" of long segments (Schein and Steriade 1986; Hayes 1986a) if we regard long vowels as unitary, but follows straightforwardly if the long vowels are two-vowel sequences (hence disyllabic) at the stage where the rules apply.

The proposal, then, is that both stress assignment and vowel allophony in Southern Paiute take place at a stage of the derivation in which surface long vowels are disyllabic sequences, and only later are these converted to single syllables. Precedent for the latter process may be found in Kalkatungu (Blake 1969, 1979a), Pitta-Pitta (Blake 1979b), Wangkumara (McDonald and Wurm 1979), and Diyari (Austin 1981), where it is reported that the merger of adjacent identical vowels into a single long vowel, with inheritance of stress, is an optional phonological process. A similar case, with apparent obligatory syllable merger, may be found in Cayuga (Foster 1982; § 6.3.5). Under this account, there is no need to add the possibility of stress-bearing moras to metrical theory.

A second case in which it is claimed that stress is assigned to moras in Winnebago, as analyzed by Hale and White Eagle 1980; HV; and Halle 1990.

9. For some rare surface contrasts of CVː vs. CVV́, where the latter is derived from a different source, see Harms 1966, 232.

An alternative account of Winnebago under which only syllables bear stress is presented in § 8.9.

Other than a lack of compelling cases, there is a more direct argument against the view that metrical structure can split syllables. Were this possible, we would logically expect to find it in languages where (unlike in Southern Paiute) CVC counts as heavy. Such a pattern could be clearly diagnosed as follows:

(71) a. (× .)(× .)(× .)
μ μ μμ μ μ
CV́.CV.**CV́C**.CV́.CV . . .

b. (× .)(× .)(× .)(× .)
μ μ μ μμ μ μ μ
CV́.CV.CV́.**CV́C**.CV.CV́.CV . . .

CVC would be stressed in either of these abstract forms (assuming the syllable inherits the stress of either of its parts), but the alternating count would be preserved after CVC.

Such cases, which appear to be unattested, are excluded by the theory. The reason is that the consonant mora of CVC cannot be a syllable on its own, so there can be no early stage of the derivation at which CVC is disyllabic.

Another possible source of counterexamples to the view that footing does not split syllables may be found in the theory of Prosodic Morphology, as developed by McCarthy and Prince 1986, 1990, and others. Here, we find that the parsing out of bases for nonconcatenative morphological processes freely splits up syllables. For example, McCarthy and Prince's (1990) derivation of the Arabic broken plural of /ħaluːb/ 'milch-camel' takes the following form: (a) circumscription of a moraic-trochee base: /**ħalu**/ + /ub/; (b) remapping of the base onto an iambic template: /**ħaluː**/ + /ub/; (c) vowel overwriting: /**ħalaː**/ + /**i**b/; (d) glottal epenthesis in onset position: [ħalaːʔib]. The first stage of the derivation appears to construct a moraic trochee over the first syllable plus the first mora of the second syllable of /ħaluːb/.

To regard such cases as counterexamples to the syllable-splitting prohibition would be to misconstrue the nature of prosodic morphology. As Marantz 1982 originally showed, rules of nonconcatenative morphology characteristically parse out ENOUGH MATERIAL TO FORM a syllable, foot, or minimal word, rather than parsing out a syllable, foot, or minimal word PER SE. This is seen, for instance, in heavy syllable reduplication, where the syllabification of the base is characteristically ignored; cf. Ilokano *pus.pú.sa* 'cats' (Hayes and Abad 1989, 357), where reduplication gets enough material to form a heavy CVC syllable by ignoring the syllabification of the base. The Arabic example works the same way: the prosody of the base in /ħaluːb/ is ignored when the broken plural rule parses out enough material to form a moraic trochee. Since the rule of moraic trochee formation is not actually assigning a foot to the base, but is only extracting enough material from the base to form a foot, it cannot be said to be splitting syllables in the relevant sense.

5.7 SUMMARY

This chapter has outlined the following topics: (a) prohibitions on degenerate feet; (b) the theory of extrametricality; (c) the "unstressable word syndrome" as it results from (a) and (b); (d) modes of foot parsing; (e) variable syllable weight; and (f) the integrity of syllables with respect to metrical structure. With this apparatus and that of the preceding chapters, we are now ready to consider the exemplifying analyses that follow.

•6•

Case Studies

6.1 MORAIC TROCHEES

Taking into account the discussion of degenerate feet from the preceding chapter, the moraic trochee foot template can be stated as in (1):

(1) Form $\overset{(\times\ .)}{\smile\smile}$ or $\overset{(\times)}{-}$. Where degenerate feet are allowed, form $\overset{(\times)}{\smile}$.

Below, I present instances of the moraic trochee.

6.1.1 Palestinian Arabic

The stress pattern of Palestinian Arabic has been described and analyzed by Brame 1973, 1974; Kenstowicz and Abdul-Karim 1980; Kenstowicz 1981, 1983; and others. The facts treated here are taken from the latter three references, as well as from Abu-Salim 1980.

The relevant syllable types in Palestinian are as in other dialects of Arabic: light (/CV/), heavy (/CVC, CV:/), and superheavy (/CVCC, CV:C/, notated /=/). In final position, we find an aberrant pattern: superheavy syllables attract stress, /CV/ and /CVC/ syllables do not attract stress, and /CV:/ syllables do not occur (at the level where the stress rules apply; see below).

In analyzing these facts, I make the following assumptions. First, final consonants are extrametrical, at the level of syllabification. This means that in final /CVC/, the final consonant will not receive a mora, so /CVC/ is structurally a light syllable (§ 5.2.1). Second, I follow Kenstowicz 1986 in assuming that the final consonant of a superheavy syllable is not syllabified at the initial stage. Thus we have the structures in (2):

(2) a. Final /CV/: Light b. Final /CVC/: Light

c. Final /CVCC, CV:C/: Heavy + Stray C

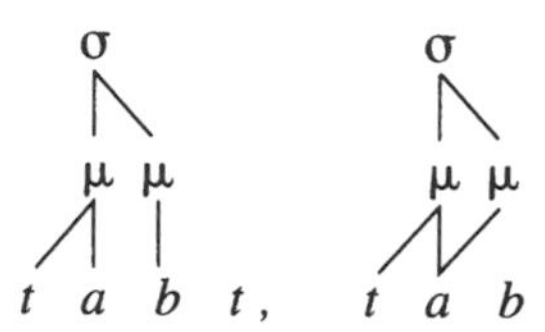

Note that in (2b), the final consonant is made part of its syllable, so that the sole effect of consonant extrametricality is to exempt this consonant from the assignment of prosodic weight. In the data that follow, I designate final /CVC/ as light /⌣/, assuming these syllabic representations.

The stress pattern of Palestinian is illustrated in (3)–(5).

(3) **Final Superheavy Syllables are Stressed**

/⌣ =́/	a. *darást*	'I studied'	K 1983, 212
/– =́/	b. *ba:bé:n* (→ *babé:n*)	'two doors'	K 1983, 208
	c. *dukká:n*	'shop'	K 1983, 212
/– ⌣ =́/	d. *mo:ladé:n*	'two feasts'	K 1983, 208

(4) **Heavy Penults are Stressed**

/–́ ⌣/	a. *bá:rak*	'he blessed'	K+AK, 73
	b. *máktab*	'office'	K+AK, 73
	c. *yírši*	'bribe (3 m.sg. imperf.)'	K 1981, 2
/⌣ –́ ⌣/	d. *katábna*	'we wrote'	AS 1980, 1
/– –́ ⌣/	e. *maktábna*	'our office'	K+AK, 73
	f. *mo:ládna* (→ *moládna*)	'our feast'	K 1983, 208
/⌣ ⌣ –́ ⌣/	g. *bakarítna*	'our cow'	K+AK, 73
/⌣ – –́ ⌣/	h. *maka:tíbna* (→ *makatíbna*)	'our offices'	K 1983, 208
/– ⌣ –́ ⌣/	i. *ba:rakátna*	'she blessed us'	K+AK, 73
	j. *maktabítna*	'our library'	K+KA, 73

(5) **Words Ending in /⌣ ⌣/**

i. **Heavy Antepenults are Stressed**

/–́ ⌣ ⌣/	a. *bá:rako*	'he blessed him'	K+AK, 73
	b. *ʕállamat*	'she taught'	K+AK, 73
	c. *ʔídfaʕu*	'pay (pl. imper.)'	K 1981, 2
/⌣ –́ ⌣ ⌣/	d. *maká:tibi* (→ *maká:tbi*)	'my offices'	K 1983, 208

ii. **Disyllables and Trisyllables: Initial Stress**

/⌣́ ⌣/	e. *ʔána*	'I'	K 1981, 16
	f. *kátab*	'he wrote'	K+AK, 55
/⌣́ ⌣ ⌣/	g. *kátabu*	'they wrote'	K 1983, 210

(plus (5i. a–c))

iii. **Longer Words Ending in /⌣ ⌣ ⌣/**

A. **Four Moras Total: Preantepenultimate Stress**

/◡́ ◡ ◡ ◡/ h. *bákarito* (→ *bákarto*)	'his cow'	K+AK, 73
i. *ḍárabato* (→ *ḍárbato*)	'she hit him'	K+AK, 73
j. *šáǰaratun* (Classical)	'a tree'	K 1981, 15

B. **Five Moras Total: Antepenultimate Stress**

/– ◡́ ◡ ◡/ k. *ʕallámato*	'she taught him'	K 1981, 13
l. *maktábito* (→ *maktábto*)	'his library'	K+AK, 73
/◡ ◡ ◡́ ◡ ◡/ m. *šaǰarátuhu* (Classical)	'his tree'	K 1981, 15

Some comments on the data are in order. The basic stress patterns are sometimes obscured by the operation of later phonological rules. These include the deletion of stressless high vowels in open syllables (5d, h, l); deletion of /a/ in open syllables in certain environments (5i); and shortening of long vowels in immediately pretonic position (3b), (4f, h). The forms marked "Classical" are renderings by Kenstowicz's consultant of Classical Arabic forms that escape the colloquial syncope rules.

In formulating an analysis, I take the last group of data as crucial. In particular, one should ask: (a) What can give rise to a pattern in which stress always falls on one of the last FOUR syllables (cf. (5iii.A))? This seems particularly strange from the viewpoint of the theory assumed here, in which metrical rules in general cannot count higher than two. (b) Why is the four-syllable limit achieved only in (5iii.A) and not the forms of (5iii.B), which also end in three light syllables?

My proposal for how the theory can impose a four-syllable limit is: count to two twice. In particular, we can set up binary trochaic feet, and posit that preantepenultimate stress reflects the presence of two binary feet in final position, with the second extrametrical. For (5h) this yields (6):

```
(6)  (×       )
     (×    .)⟨( ×    .)⟩
      ◡    ◡    ◡    ◡
     b á k a r i t o
```

This tacit binary pattern can also account for why preantepenultimate stress occurs in four-mora words but not five-mora words ((5iii.A) vs. (5iii.B)). To capture the distinction, I assume that the feet are moraic trochees and that parsing goes from left to right:

(7) **Five-Mora Words Ending in /◡ ◡ ◡/**

```
a. (    ×    )              b. (  ×      )
   (× .)(× .)                  (×)(×   .)
   ◡  ◡  ◡  ◡  ◡               –   ◡   ◡ ◡
   šaǰarátuhu = (5m)           ʕallámato = (5k)
   ———————→                    ———————→
```

In (7), the rightmost foot is not extrametrical, because it is not peripheral (§ 3.11), a stray syllable intervenes between it and the right word boundary. I assume that the rightmost syllables are stray, rather than degenerate feet, because degenerate-size words are absent in Palestinian.

The rules are stated together in (8):

(8) a. **Consonant Extrametricality** C → ⟨C⟩ / ___ $]_{word}$
b. **Foot Construction** Form moraic trochees from left to right. Degenerate feet are forbidden absolutely.
c. **Foot Extrametricality** Foot → ⟨Foot⟩ / ___ $]_{word}$
d. **Word Layer Construction** End Rule Right

I will now show that the analysis correctly generates the remaining cases in the data.

Case (4), heavy penults: a heavy penult will invariably form a moraic trochee on its own. This foot is protected from foot extrametricality by the following stray syllable; hence it receives stress by End Rule Right:

```
(9) (     ×  )
    (×  .)(×)
    ᴗ  ᴗ  –  ᴗ
    bakarítna = (4g)
```

Case (5.i), /. . . – ᴗ ᴗ/: the heavy antepenult forms a moraic trochee, as does the following /ᴗ ᴗ/ sequence. Foot extrametricality results in stress being awarded to the antepenult by End Rule Right.

```
(10)  (    ×)
        ( ×) ⟨( ×   .)⟩
      ᴗ   –    ᴗ   ᴗ
      m a k á : t i b i      = (5d)
```

The analysis proposed here for Palestinian is almost exactly the same as the analysis for Cairene Arabic (§ 4.1.3). The crucial difference is that Palestinian has a rule of foot extrametricality, which results in stress appearing one foot to the left in words that end in /– ᴗ ᴗ/. The same holds for cases like (6), with /ᴗ ᴗ ᴗ ᴗ/. In all other cases, foot extrametricality is blocked, and the two languages have the same stress pattern.

Disyllables and trisyllables composed of all lights (5ii) receive just one foot, which takes the stress by default. In disyllables, the single foot escapes Foot Extrametricality, since extrametricality rules cannot exhaust the stress domain (§ 3.11):

```
(11) a. (×      )           b. (×    )
        (× .)                  (× .)
        ˘  ˘  ˘                ˘  ˘
        kátabu = (5g)          ʔána = (5e)
```

The behavior of words ending in a superheavy syllable (3) is accounted for under the assumption that their final extrametrical consonants are unsyllabified. These stray consonants prevent the rightmost foot from being peripheral. The rightmost foot therefore escapes Foot Extrametricality by the Peripherality Condition and receives main stress. To illustrate this, (12) gives both schematic and "official" representations (§ 3.5) for (3a):

```
(12) (   × )                  ×
     (× )       =             ×
     ˘  –               σ     σ
     daras⟨t⟩           |     |\
                        μ     μ μ
                       /|    /| |
                    (d a  (r a s)) ⟨t⟩
```

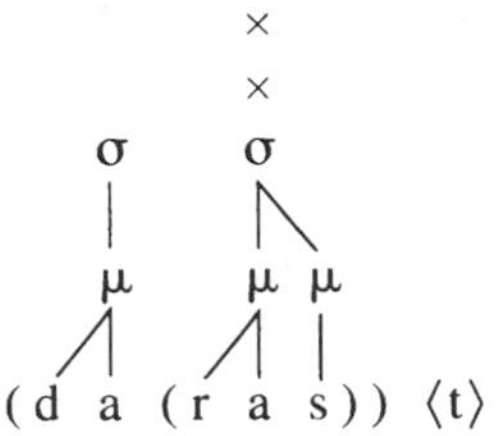

This should be contrasted with final CVC syllables: following § 5.2.1, such consonants are extrametrical but syllabified, and hence belong to the rightmost foot. The rightmost foot is therefore peripheral, and extrametricality may apply:

```
(13) (×)                         ×
     (×)⟨(×  .)⟩     =           ×        ×
     –   ˘   ˘                   σ        σ    σ
     ʕallama⟨t⟩                  |\       |    |
                                 μ  μ     μ    μ
                                /|   \   /|   /|\
                             ((ʕ a))⟨((l a m a⟨t⟩))⟩ = (5b)
```

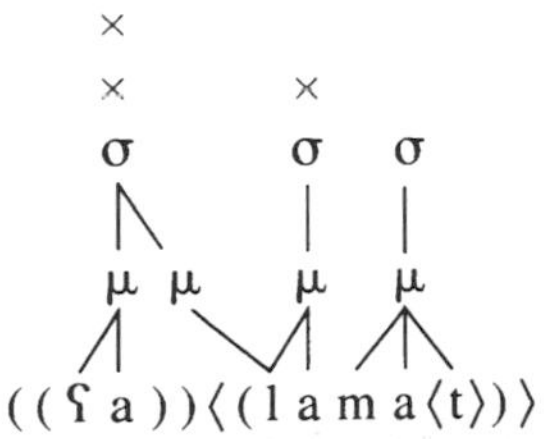

A final item to note is the existence of word-final long vowels in surface forms such as *dawáː* 'his medicine' (Kenstowicz and Abdul-Karim 1980, 74). Kenstowicz and Abdul-Karim derive these cases as follows. The underlying form is /dawa+u/ (cf. *dáwa* 'medicine', *bírak-u* 'his pools' K+AK 58), and a general rule of pre-suffix lengthening gives *dawaːu* (syllabic pattern /˘ – ˘/), which is stressed penultimately by the regular rules (cf. (4d), *katábna*). The suffix *-u* is then deleted postvocalically.

This completes the basic analysis. It should be further noted that Palestinian Arabic stress is assigned cyclically (Brame 1974; Kenstowicz and Abdul-Karim 1980), with the cyclic domains determined by loosely bound affixes such as the object suffixes. Cyclic stressing may provide alternative ways of stressing some of the examples above; however, the analysis proposed here appears to be the only one that generalizes to Classical forms pronounced in Palestinian style.

An important point is that this analysis would be impossible to replicate using a theory that uses uneven (/σ́ ⏑/) trochees (§ 4.3.1) instead of moraic trochees. In particular, an uneven trochee analysis cannot distinguish /⏑ ⏑ ⏑ ⏑/ from /– ⏑ ⏑ ⏑/:

```
(14) a.  (×      )                  b.  (×        )
         (×    .)⟨( ×    .)⟩            (×      .)⟨( ×    .)⟩
          ⏑    ⏑   ⏑    ⏑                –      ⏑    ⏑    ⏑
       b á k a r i t o   = (5h)      *ʕ á l l a m a t o   = (5k)
```

The latter form is in fact stressed *ʕallámato*. This follows from the moraic trochee analysis (7b), in which the alternating count begins immediately after the heavy syllable.

My analysis of Palestinian is inspired by the early metrical account of Kenstowicz (1981), who was the first to notice the alternating, left-to-right nature of Palestinian stress. For Kenstowicz, the binary count was carried out by a language-particular rule of foot splitting, which applied only in five-mora words. My analysis derives comparable effects with elements posited to be part of the general theory, namely moraic trochees and extrametricality. Subsequent analyses that incorporated Kenstowicz's insight were Hayes 1987 (essentially the same as this one) and Halle and Kenstowicz 1989.

6.1.2 Egyptian Radio Arabic

I have analyzed Cairene and Palestinian Arabic as almost identical systems, differing only in a rule of final foot extrametricality. The pattern of Egyptian Radio Arabic, discussed by Harrell 1960, essentially combines the possibilities of Cairene and Palestinian in free variation. Harrell's data came from extensive listening to the shortwave transmissions of the Egyptian Broadcasting Service. The language is the modern-day Classical language (Modern Standard Arabic), as spoken by Egyptian news broadcasters. It appears that the same pattern may characterize the Classical speech of Mitchell's (1960) consultant "D," from Qena in Upper Egypt.

Harrell found the stress patterns of (15) in his data (pp. 11–13):

(15) a. Final superheavy syllables are stressed:

salá:m	'peace'
dimášq	'Damascus'

b. Otherwise, stress falls on a heavy penult:

qàddámna	'we presented'
dàwlí:ya	'international'

c. Penultimate stress in /⏑ ⏑/: (N.B. final /CVC/ counted as light)

málik	'king'
húna	'here'

d. Antepenultimate stress in /˘ ˘ ˘/:
ʔábadan ‘never’
šárika ‘company’

e. For /˘ ˘ ˘ ˘/, penultimate or preantepenultimate stress in free variation:
kàtabáhu, kátabàhu ‘he wrote it’

f. For /– ˘ ˘/, antepenultimate or penultimate stress in free variation:
háːðihi, hàːðíhi ‘this’
múškila, mùškíla ‘problem’

g. For /– ˘ ˘ ˘/, antepenultimate stress:
mùxtálifa ‘different’

The secondary stresses noted in (15e) are from Harrell. He also notes that in general, heavy syllables receive secondary stress, and this is transcribed above. Unlike Mitchell 1960, Harrell does not explore the range of words with longer strings of light syllables.

The simplest hypothesis for what is going on here would be that the radio announcers vary freely between applying their own colloquial stress rule (§ 4.1.3) and applying the ancient stress rule of Classical Arabic, which is: stress the rightmost nonfinal heavy syllable, otherwise the initial (McCarthy 1979a; see § 7.2 for analysis of unbounded systems of this type). That this is not the case can be seen from two facts. Forms like *muxtalifa* (15g) do not show free variation, but may only be stressed on the antepenult (the Classical stressing would be *múxtalifa*). In addition, the Classical stress rule would have no means of deriving the posttonic secondary stress in (15e), *kátabàhu.*

The rules proposed here are much closer to those of the Cairene colloquial. They assume, as before, that final consonants are extrametrical.

(16) a. **Foot Construction** Form moraic trochees from left to right.
b. **Foot Extrametricality** Foot → ⟨Foot⟩ / ___$]_{word}$ (optional)
c. **Word Layer Construction** End Rule Right

This is almost exactly the same as the analysis of Palestinian Arabic, except that foot extrametricality is optional. Here are representative output forms. In the following examples, foot extrametricality is inapplicable, either because the foot is not peripheral (17a, b, d, e) or because extrametricality would exhaust the word ((17c); see § 3.11 (47d)):

(17)	a.	b.	c.	d.	e.
	(×)	(×)	(×)	(×)	(×)
	(×)	(×) (×)	(× .)	(× .)	(×)(× .)
	˘ –	– – ˘	˘ ˘	˘ ˘ ˘	– ˘ ˘ ˘
	dimáš.q	*dàwlíːya*	*húna*	*šárika*	*mùxtálifa*

Cases of free variation are the result of making foot extrametricality optional:

```
(18) a. (     ×  )    (×  )            b. (  ×  )   (×)
        (× .)(× .)    (× .)⟨(× .)⟩        (×) (× .)  (×)⟨(× .)⟩
        ᴗ  ᴗ  ᴗ  ᴗ    ᴗ  ᴗ  ᴗ  ᴗ          –   ᴗ  ᴗ    –  ᴗ  ᴗ
```

kàtabáhu, kátabàhu *hà:ðíhi, há:ðihi*

Notice that the pattern of secondary stress is also largely accounted for by this analysis. The only exception is the second form of (18b), where a rule of destressing in clash must be posited to eliminate the posttonic secondary stress. The rule may take the simple form in (19), since all possible cases of overgeneration (i.e. eliminating a main stress) are blocked by the Continuous Column Constraint (see § 3.4.2).

(19) **Destressing in Clash** $\times \rightarrow \emptyset / \times$ ____

Concerning the origin of this pattern, my conjecture is as follows: the announcers do not fully shift to authentic Classical Arabic stress because this pattern, being based on unbounded rather than binary feet, is rhythmically alien to their native speech. Instead, the announcers "classicize" their speech with a less dramatic change that remains more faithful to the basic rhythmic structure of Cairene: they simply allow the final foot of the word to be optionally extrametrical. This shifts the stress pattern in the general direction of Classical Arabic, while retaining the basic Cairene foot structure. This creates a stress rule that happens to allow as an option the characteristic stressings of Palestinian Arabic.

Summing up, the Egyptian Radio Arabic stress pattern provides support for the basic foot structures (i.e. moraic trochees) I have posited for Cairene and Palestinian Arabic. As before, an analysis using uneven trochees would be unworkable. The foot structure is manifested quite clearly in Egyptian Radio Arabic for two reasons: in certain words, more than one foot is eligible to serve as the docking site for main stress; and most feet that do not receive main stress are manifested on the surface by secondary stress.

6.1.3 Cahuilla

Cahuilla is a Uto-Aztecan language of Southern California, investigated in detail by Hansjakob Seiler. The data and crucial generalizations below are from Seiler 1957, 1965, 1967, 1977, and p.c.; and Seiler and Hioki 1979; a metrical analysis for Cahuilla has previously been proposed by Levin 1988b.

6.1.3.1 *Data*

Stress in Cahuilla is based on a distinction between heavy and light syllables. Heavy syllables are those with long vowels or diphthongs, as well as syllables closed by glottal stop. I assume here that Weight by Position (§ 3.9.2) applies only to /ʔ/. A process of expressive gemination described below also creates heavy syllables.

In most Cahuilla words, main stress falls on the initial syllable of thc root, although a few forms have irregular root-internal stress that must be lexically listed. To a first approximation, secondary stress follows a binary alternating count of moras. Thus if all syllables of the root and the following suffix string are light, stress falls on every odd-numbered syllable:

(20) a. /⏑́ ⏑ ⏑̀ ⏑/ *tákalìčem* 'one-eyed ones' S 1977, 27
b. /⏑́ ⏑ ⏑̀/ *táxmuʔàt* 'song' S 1965, 57

When the initial syllable of the root is heavy, the immediately following syllable is also stressed, and the alternating count continues:

(21) a. /–́/ *mú:t* 'owl' S 1977, 32
b. /–́ ⏑̀/ *páʔlì* 'the water (objective case)' S 1977, 28
c. /–́ ⏑̀ ⏑/ *qá:nkìčem* 'palo verde (pl.)' S 1977, 27
háʔtìsqal 'he is sneezing' S 1965, 52

An alternating count also occurs in prefixes. In (22), the boundary between the prefix string and the root is marked with /#/:

(22) a. /⏑ # –́/ *ne#yú:l* 'my younger brother' S 1977, 33
b. /⏑̀ ⏑ # ⏑́ ⏑ ⏑̀ ⏑ ⏑̀/ *pàpen#túleqàlevèh* 'where I was grinding it' S 1965, 52

6.1.3.2 *Analysis*

Consider now how these patterns might be derived in a metrical analysis. To begin, I follow Levin's (1988b) account of bidirectionality. Levin proposes a derivation of the type shown in (23):

(23) a. **Morphology** Attach all suffixes to the root.
b. **Stress** Construct feet from left to right.
c. **Morphology** Attach all prefixes to the root.
d. **Stress** Construct feet from right to left over unfooted syllables.

As for the feet, I propose that they are moraic trochees; this accounts for the appearance of stress immediately after either a heavy syllable or a /⏑́ ⏑/ string. I assume the weak version of the prohibition on degenerate feet (§ 5.1.1): degenerate feet are allowed, but are deleted at the end of the word phonology if they do not surface in strong position. Finally, I assume that Cahuilla is a "top-down" system (§ 5.4.3), with the word layer constructed first. Summing up, the analysis is as in (24):

(24) a. On the cycle consisting of root + suffixes:

i.	**Word Layer Construction**	End Rule Left
ii.	**Foot Construction**	Form moraic trochees from left to right, invoking the weak prohibition on degenerate feet.

b. Add prefixes and foot them as in (a.ii), only right to left.

Here are some representative derivations. The first set of cases is forms with no prefixes, where only the first application of footing (24a.ii) is relevant:

(25)	a.	b.	
	(×)	(×)	Word Layer Construction
	˘ ˘ ˘ ˘	– ˘ ˘	
	takaličem = (20a)	*qa:nkičem* = (21c)	
	(×)	(×)	Foot Construction
	(× .)(× .)	(×) (× .)	
	˘ ˘ ˘ ˘	– ˘ ˘	
	tákalìčem	*qá:nkìčem*	

The following derivations for (22a) and (22b) are more complex, involving separate cycles for suffixation and prefixation, and (in (26b)) loss of a weak degenerate foot.

(26)	a.	b.		
	yu:l	*tuleqaleveh*	**Morphology**	suffixation
	(×)	(×)	**Phonology**	Word Layer Construction
	–	˘ ˘ ˘ ˘ ˘		
	yu:l	*tuleqaleveh*		
	(×)	(×)		Foot Construction
	(×)	(× .)(× .)(×)		
	–	˘ ˘ ˘ ˘ ˘		
	yu:l	*tuleqaleveh*		
	(×)	(×)	**Morphology**	prefixation
	(×)	(× .)(× .)(×)		
	˘ –	˘ ˘ ˘ ˘ ˘ ˘ ˘		
	ne-yu:l	*papen-tuleqaleveh*		
	(×)	(×)	**Phonology**	Foot Construction
	(×)(×)	(× .)(× .)(× .)(×)		
	˘ –	˘ ˘ ˘ ˘ ˘ ˘ ˘		
	neyu:l	*papentuleqaleveh*		
	(×)	(×)	**Word Level**	loss of weak degenerate feet
	(×)	(× .)(× .)(× .)		
	˘ –	˘ ˘ ˘ ˘ ˘ ˘ ˘		
	neyú:l	*pàpentúleqàleveh*		

Note that this analysis does not permit feet on weak light syllables in final position, despite Seiler's transcription of stress there. This aspect of the analysis is defended in § 6.1.3.3.

Consider now the following issue: When a heavy syllable is not root-initial, how is it affected by the alternating count? This can be divided into two cases. First, there are examples in which the first mora of the heavy syllable is "in phase," that is, in even position with respect to the alternating count. These are stressed as we might expect:

(27) ⏑ ⏑ – ⏑
táxmuʔàʔtì 'the song (objective case)' S 1977, 33

(×) Word Layer Construction
(× .)(×)(×) Foot Construction
⏑ ⏑ – ⏑
táxmuʔàʔtì

(×) loss of degenerate feet in weak position
(× .)(×)
⏑ ⏑ – ⏑
táxmuʔàʔtì (for final stress, see § 6.1.3.3)

Next, consider cases where a heavy syllable is "out of phase," with its initial mora in even position. All examples I have located involve a heavy syllable that immediately follows an initial light.[1]

(28) a.	/⏑ – ⏑/	*súkàʔtì*	'the deer (objective case)'	S 1977, 28
b.	/⏑ – ⏑ ⏑/	*púkawtèmih*	'gopher snakes (obj. pl.)'	S 1965, 53
		nèsun káviːčì-wen	'I was surprised'	S+H, 73
	or	*nèsun kávìːčì-wen*		
		kíhmay-ʎù-qal	'wonder why'	S+H, 78
		pálaw-wènet	'that which is beautiful, pretty'	S+H 140
c.	/⏑ – – ⏑/	*héʔi kákawlàː-qà*	'his legs are bow-shaped'	S+H, 74
d.	/⏑ # ⏑ – ⏑ ⏑/	*tax#kíʎiw-kàtem*	'companions'	S+H, 79
		pen#péníːčì-ni-qà	'translate'	S+H, 148
	or	*pen#péniːčì-ni-qà*		

1. The stressings for the first two forms in (28) are from the sources as indicated. For the remaining examples, the patterns indicated were provided by Hansjakob Seiler (p.c.).

These pattern as follows. The heavy syllables vary in terms of whether they receive stress; Seiler notes (p.c.) that stress in this position was difficult to transcribe consistently, and the variation in the data is likely to be fairly random. Below, I attribute this variation to an optional destressing rule. More crucially, Seiler observes that the alternating count for stress is always RESTARTED immediately after a heavy syllable. This follows from the proposal: since /˘ –/ cannot form a moraic trochee, it is parsed as /˘/ (a degenerate foot) plus the possible moraic trochee /–/. The count then continues with a foot formed on the following syllables. The degenerate foot is allowed under the theory since it bears main stress; compare (24a.ii).

To account for the variation in whether the heavy syllable of /˘́ –/ bears stress, I posit the optional destressing rule in (29):

(29) **Destressing in Clash**

× → ∅ / × ____ (applies optionally, from left to right)
˘ –

Here are derivations for representative forms of (28):

(30)	a.	b.	c.	
	(×) ˘ – ˘ *suka*ʔ*ti*	(×) ˘ – ˘ ˘ *kavi:čiwen*	(×) ˘ – – ˘ *kakawla:qa*	Word Layer Construction
	(×) (×)(×) (×) ˘ – ˘ *suka*ʔ*ti*	(×) (×)(×) (× .) ˘ – ˘ ˘ *kavi:čiwen*	(×) (×)(×) (×) (×) ˘ – – ˘ *kakawla:qa*	Foot Construction
	(×) (×)(×) (×) ˘ – ˘ *suka*ʔ*ti*	(×) (×) (× .) ˘ – ˘ ˘ *kavi:čiwen*	(×) (×) (×) (×) ˘ – – ˘ *kakawla:qa*	Destressing in Clash
	(×) (×)(×) ˘ – ˘ *suka*ʔ*ti*	—	(×) (×) (×) ˘ – – ˘ *kákawlà:qa*	Loss of degenerate feet in weak position
	*súkà*ʔ*tì*	*kávì:čìwen*, *kávi:čìwen*	*kákawlà:qa*	output

These derivations show why I assume that Cahuilla is a "top-down" stress system: the initial main stress forces a degenerate foot to be constructed on the initial syllable. This overrides the Priority Clause (§ 5.1.6), which would require that the initial syllable be skipped to create a proper foot on the second syllable.

Summing up, I claim that the moraic trochee appropriately formalizes Seiler's view of Cahuilla stress (1977, 29) as "a mixed system with alternation based partly on the mora, partly on the syllable." The foot template counts moras, but the system is syllable-based in that syllables can never be divided between feet (§ 3.9.1, § 5.6.2).

6.1.3.3 *Degenerate Feet in Cahuilla*

To complete the analysis, I must defend the discrepancy between the stresses Seiler transcribes on final light syllables and the absence of a foot on them, which is required by the ban on weak degenerate feet (§ 5.1.1). My suggestion is that while such syllables do sound prominent (and thus could reasonably be assigned an accent mark), this can be attributed to phonetic final lengthening, rather than to a metrical foot in this position.

In support of this we may note the following: (a) The rise in pitch that ordinarily accompanies secondary stress does not occur on final syllables (Seiler 1965, 52). (b) In words with sequences containing /a/, this phoneme appears as back [ɑ] in syllables with primary or secondary stress, and as central [a] in stressless syllables, as in /aʔamnaʔwet/ = [ɑ́ʔamnɑ̀ʔwet] 'big (personal pl.)' S 1957, 214. The form [tɑ́lalqal] = /talalqal/ 'to snore', with final [a], suggests that final light syllables are not stressed. (c) Similar conclusions follow from Seiler's discussion (1957, 213–14) of the allophones of /e/.

Other evidence also bears on the status of degenerate feet in Cahuilla. Monomoraic words are fairly common, for example *nét* 'ceremonial chief' S 1977, 32, *pál* 'water' S 1965, 53, *máx* 'to give' S 1967, 142. These are allowed, since the degenerate foot is in strong position. When a monomoraic noun stem is preceded by a monosyllabic prefix, main stress falls on the prefix, as shown in (31) (examples from Seiler 1977, 39):

(31)	a. *-na*	'father'	compare:	b. *-yu:l*	'younger brother'
	čém-na	'our father'		*ne-yú:l*	'my younger brother'

This suggests that where a degenerate foot would result on the root syllable, it is "repaired" where possible by incorporating the prefix into the initial stress domain. This again suggests a tendency to avoid degenerate feet.

6.1.3.4 *Theoretical Consequences*

(A) MORAIC TROCHEES VERSUS UNEVEN TROCHEES. A significant aspect of the Cahuilla pattern is the appearance of secondary stress immediately after a heavy syllable. This can be patently seen in (21c), and less directly (due to destressing) in the forms of (28). The appearance of stress on post-heavy light syllables is the automatic consequence of an analysis using moraic trochees, but could not be derived using uneven trochees (§ 4.3.1). The uneven trochee foot template restarts the binary count TWO light syllables after a heavy, and thus derives the wrong output in numerous forms, such as (32):

(32) (×   )   = (21c)
     (×   .)(×)
     –   ⏑ ⏑

qa:nkičem → **qá:nkičèm* if degenerate feet allowed, otherwise **qá:nkičem*

The correct output is *qá:nkìčem,* derived in (25b).

(B) SPLITTING SYLLABLES INTO FEET. Consider the Cahuilla pattern from the viewpoint of a theory that lacks moraic trochees. In such a theory, the most plausible approach would be to suppose that the terminals of metrical structure are not syllables, but moras. An account along these lines is presented by Levin 1988b. To a rough approximation, we can say that stress falls on the root-initial mora and on alternating moras before and after. This works with all lights (20), and where heavy syllables occur "in phase" ((21), (27)). In forms like (28a) *súkàʔtì,* an alternating mora rule would stress the mora containing /ʔ/, and the phonetic appearance of stress on the preceding vowel plausibly reflects a shift from a segment that cannot bear phonetic stress to one that can; the final stress could be rationalized as in § 6.1.3.3 above.

Where this approach fails is in the forms of (28), in which syllables with long vowels and diphthongs appear out of phase. If Cahuilla feet could actually group moras together freely, without regard to syllable structure, then we would expect the patterns in (33) for such forms:

(33) *ka vi i či wen* — string of moras — = (28b)

(× .)(× .) (×) — stress on alternating moras
ka vi i či wen

**káviìčiwèn* — (correct outcome is *kávì:čiwen* or *kávi:čìwen*)

The output in (33) would involve a long vowel with a rising stress contour: [iì]. Given the phonetics of stress in Cahuilla (Seiler 1957), it is easy to reconstruct what such a rising-stress vowel would sound like in terms of pitch and vowel quality. But according to Seiler (p.c.), rising-stress long vowels never occur. The analysis could be partially repaired by positing a rule merging /i + í/ to [í:]. But this would still place secondary stress in the wrong position AFTER the long vowel, as (33) shows.

The point is that the two moras of a long vowel are unable to sustain the alternating count, which always starts over after a long vowel. I take this to be evidence for the claim made in § 3.9.1: syllables cannot be split into separate feet.

6.1.3.5 *Multiple Weights for CVC Syllables*

I suggested in § 5.6.1 that in some languages CVC syllables can vary in their weight, depending on context. Cahuilla seems to be a plausible case of this sort. While in general CVC is light in Cahuilla, a morphological rule of "intensification" (Seiler 1977, 58) creates heavy CVC. The rule applies to stems beginning with CVCV or CVCCV. In either case, the second consonant receives greater length.

(34) a. *čéxiwèn* 'it is clear' b. *wélnet* 'mean one'
čéxxìwen 'it is very clear' *wéllnèt* 'very mean one'

Seiler describes the change as follows: "As an invariant we find that the initial syllable receives an extra mora, thus a total value of two morae. As a consequence, the second vowel of the sequence will bear secondary stress."

I formalize this account as follows: intensification applies before stress assignment and adds a mora to the initial syllable. Where the syllable is open, the mora is linked to the initial consonant of the next syllable, as in (35a); where it is closed, the mora lodges on the syllable-final consonant, as in (35b):[2]

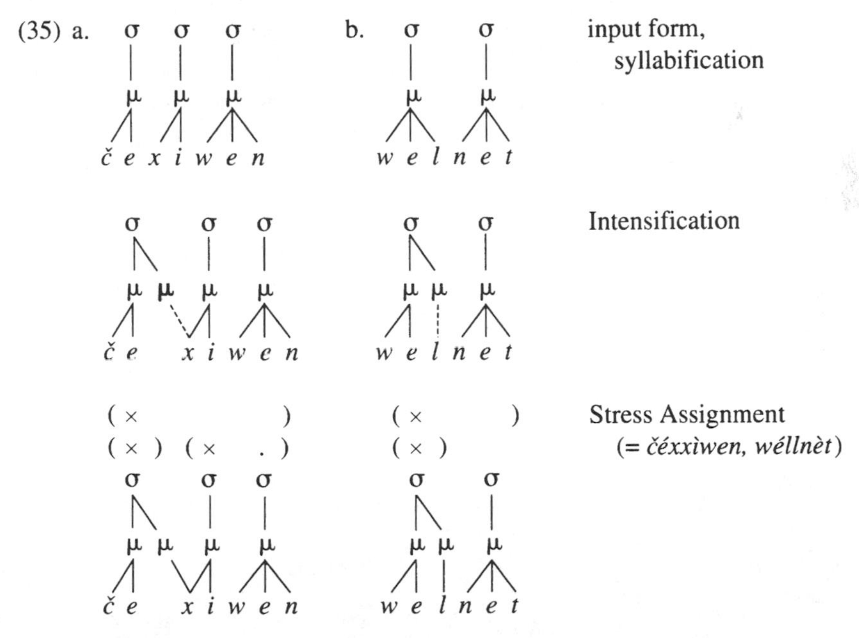

Under this account, what Seiler transcribes as syllable-final [ll] is moraic /l/, whereas his single [l] is /l/ without its own mora; compare (35b). These are essentially what was proposed in Hayes 1989b as UNDERLYING representations for geminate versus non-geminate consonants; Cahuilla is unusual in showing the distinction in surface forms.

2. In the Mountain Cahuilla dialect, the inserted mora links to the vowel: hence the intensive forms corresponding to (34) are *čeːxiwe* and *weːlnet* (Pamela Munro, p.c.).

Intensified versus non-intensified *welnet* form a minimal pair for heavy versus light CVC. The crucial evidence for the claim that intensification creates heavy syllables, rather than just segmental length, is that secondary stress falls immediately after the lengthened syllable: *čéxxìwen* is stressed like *qá:nkìčem* (21c), showing that its quantitative pattern is /– ⌣ ⌣/.

6.1.3.6 *Conclusion*

Cahuilla stress provides support for three theoretical proposals adopted here: (a) metrical theory should include the moraic trochee and not the uneven trochee; (b) the stress-bearing unit is the syllable; and (c) CVC may occur as light or heavy, as determined by context. I have also suggested an interpretation of the facts under which Cahuilla does not violate the proposed universal ban on degenerate feet in weak position.

6.1.4 Wargamay

Wargamay is a Pama-Nyungan language of Australia (North Queensland), described and analyzed by Dixon 1981. In Wargamay, long-voweled syllables are heavy, and all others (including short-voweled closed syllables) are light. In simplex words, which form the domain of the stress rule, long vowels may only occur in the initial syllable. The description in (36) of the stress pattern follows Dixon 1981, 20, closely:

(36) a. If the initial syllable is heavy, it receives the main stress:

/–́ ⌣/	*mú:ba*	'stone fish'
/–́ ⌣ ⌣/	*gí:baɽa*	'fig tree'

b. Otherwise, stress is on the second syllable in words with three or five syllables:

/⌣ ⌣́ ⌣/	*gagára*	'dilly bag'
/⌣ ⌣́ ⌣ ⌣̀ ⌣/	*ɟuɽágay-mìri*	'Niagara-Vale-from'

c. . . . and on the initial syllable in words with two or four syllables:

/⌣́ ⌣/	*báda*	'dog'
/⌣́ ⌣ ⌣̀ ⌣/	*gíɟawùlu*	'freshwater jewfish'

d. Secondary stress falls on alternating syllables after the primary stress, except that it may never fall in final position (*ɟuɽágay-mìri*, *gíɟawùlu*).

Monosyllabic content words are permitted; in the absence of any indication to the contrary, I assume they are stressed: *má:l* 'man', *yá:* 'top of a tree' D 18. All monosyllables have a long vowel; there are no */(C)V(C)/ words.

In the theory proposed here, the analysis of Wargamay stress would proceed as follows. First, the distinction between (36b) and (36c) shows that stress is computed from right to left; only by counting in this way can the odd–even distinction result in a difference between initial and second-syllable stress.

Since the final syllable is always stressless, the feet should be trochaic. The special behavior of long vowels indicates that the feet are quantity-sensitive, hence moraic trochees.[3] Finally, since the leftmost stress of a two-stress word is the strongest, the word layer must be constructed by End Rule Left. The absence of words consisting of a single light syllable shows that the ban on degenerate feet (§ 5.1.1) is absolute in Wargamay.

This analysis is summarized and illustrated in (37) and (38).

(37) a. **Syllable Weight** — Long voweled syllable = /–/, others /◡/
b. **Foot Construction** — Form moraic trochees from right to left. Degenerate feet are banned absolutely.
c. **Word Layer Construction** — End Rule Left

(38)

a.
(×)
(×)
–
má:l

b.
(×)
(×)
– ◡
mú:ba

c.
(×)
(×) (× .)
– ◡ ◡
gí:baɽa

d.
(×)
(× .)
◡ ◡
báda

e.
(×)
(× .)(× .)
◡ ◡ ◡ ◡
gíɟawùlu

f.
(×)
(× .)
◡ ◡ ◡
gagára

g.
(×)
(× .) (× .)
◡ ◡ ◡ ◡ ◡
ɟuɽágaymìri

All forms are derived correctly except (38c). Since this is the only form with a stress clash, the destressing rule in (39) can plausibly be posited:

(39) **Destressing** × → Ø in clash

The left-to-right direction of Destressing is not written into the rule, since it is the consequence of the Continuous Column Constraint (§ 3.4.2). In the only relevant forms it is the weak stress that deletes:

(40)
(×) → (×)
(×) (× .) → (×)
– ◡ ◡ → – ◡ ◡
gí:baɽa → *gí:baɽa*

The crucial part of the analysis for present purposes is the behavior of odd-syllabled words: under the assumption that degenerate feet are forbidden, a light initial syllable cannot be footed. Thus by the Continuous Column Constraint it cannot serve as a docking site for the × assigned in the word layer.

3. See, however, § 5.1.9, which if correct permits an analysis with syllabic trochees as well; I include Wargamay among the moraic trochee languages, since such an analysis works independently of the assumptions made in § 5.1.9.

The ban on degenerate feet is independently supported by the absence of words consisting of a single light syllable.

An analysis in which degenerate feet were freely allowed would face difficulties in assigning the position of the main stress. If the initial syllable of an odd-syllabled word is footed, then End Rule Left would assign it main stress, as in (41):

```
(41) a.  (×      )       b.  (×              )
         (×)(× .)            (×)(× .) (× .)
          ˘  ˘  ˘             ˘  ˘  ˘   ˘  ˘
         *gágara             *ɟúɽagaymìri
```

The alternatives would be either (a) to allow main stress rules to refer to branching, an option which appears to be dispensable (§ 3.11, § 3.12); or (b) to posit initial extrametricality in clash, which would be very unusual typologically (§ 3.11, § 5.2.2). Given that the prohibition on degenerate feet is independently motivated by the minimal word constraint, the analysis here seems preferable.

6.1.5 Trochaic Shortening in Fijian and Other Languages

Fijian (Austronesian, Fiji) has a fairly straightforward moraic trochee stress system. The interest of this language here lies in a segmental rule of Trochaic Shortening, which bears on a number of general issues in metrical stress theory. The data and generalizations that follow are taken from the Standard Fijian grammar of Schütz 1985, as well as Dixon's (1988) description of the Boumaa dialect.

6.1.5.1 *Fijian Stress*

Fijian syllables take the form (C)V or (C)VV. C can be a prenasalized consonant: [m͡b], [n͡d], [ŋ͡g], or [n͡r], where the last is a prenasalized flap or trill, and VV may be a long vowel or a diphthong. CVV syllables count as heavy, CV as light. The stress pattern of Fijian is as stated in (42):

(42) a. If the final syllable is light, main stress falls on the penult.
b. If the final syllable is heavy, main stress falls on the final syllable.
c. Secondary stress falls on remaining heavy syllables, and on every other light syllable before another stress, counting from right to left.

In (43) are some native Fijian words illustrating the distribution of main stress.

(43)	/˘́ ˘/	*láko*	'go'	S 489
		tálo	'pour'	S 489
	/˘ ˘́ ˘/	*βináka*	'good'	S 489
	/˘ –́/	*seŋái*	'no'	S 489
		kilá:	'know'	S 490
	/–̀ –́/	*n͡rè:n͡ré:*	'difficult'	S 497

Since long native stems are uncommon, and since morphology affects stress (by determining the formation of phonological words; see Dixon 1988, chap. 3), it is easiest to check secondary stress in borrowings. The examples below are from the list of loans in Schütz 1978:

(44)			
a.	/⏑ ⏑́ ⏑/	*atómi*	'atom'
	/⏑̀ ⏑ ⏑́ ⏑/	*n͡dìkonési*	'deaconess'
	/⏑ ⏑̀ ⏑ ⏑́ ⏑/	*perèsitén͡di*	'president'
	/⏑̀ ⏑ ⏑̀ ⏑ ⏑́ ⏑/	*m͡bàsikètepólo*	'basketball'
	/–̀ ⏑́ ⏑/	*m͡bè:léti*	'belt'
	/⏑ –̀ ⏑́ ⏑/	*taràusése*	'trousers'
	/⏑̀ ⏑ –̀ ⏑́ ⏑/	*m͡bèlem͡bò:tómu*	'bellbottoms'
	/⏑ –̀ ⏑ ⏑́ ⏑/	*parò:karámu*	'program'
	/–̀ ⏑̀ ⏑ ⏑́ ⏑/	*mì:sìniŋ͡gáni*	'machine-gun'
b.	/⏑ –́/	*m͡basá:*	'bazaar'
	/⏑̀ ⏑ –́/	*n͡dòketá:*	'doctor
	/⏑ ⏑̀ ⏑ –́/	*palàsitá:*	'plaster'
	/⏑̀ ⏑ ⏑̀ ⏑ –́/	*mìnisìtirí:*	'ministry'
	/⏑ ⏑̀ ⏑ ⏑̀ ⏑ –́/	*terènisìsitá:*	'transistor'
	/⏑ –̀ ⏑ –́/	*paràimarí:*	'primary'
	/–̀ ⏑̀ ⏑ –́/	*n͡dàirèkitá:*	'director'

The analysis of such a system is straightforward:

(45) a. **Foot Construction** Form moraic trochees right to left. Degenerate feet are prohibited.

b. **Word Layer Construction** End Rule Right

```
(46) a. (          × )       b. (      ×  )      c. (          ×)
        (× .)(× .)(× .)            (×) (× .)         (  ×)(× .)(×)
         ⏑ ⏑ ⏑ ⏑ ⏑ ⏑               ⏑ –  ⏑ ⏑             –  ⏑ ⏑ –
        m͡bàsikètepólo             taràusése          n͡dàirèkitá:

     d. ( ×  )      e. (          ×)      f. (×  )
        (× .)          (× )(× .)(×)          (× .)
         ⏑ ⏑ ⏑          ⏑ ⏑ ⏑ ⏑ ⏑ –            ⏑ ⏑
        βináka         terènisìsitá:         láko
```

This is to a large extent a translation of the analyses in Scott 1948 and Schütz 1985. These scholars parse the word into "groups" (Scott) or "measures" (Schütz), which are the same as moraic trochees, except that they may optionally incorporate a stray light syllable at the beginning of the metrical unit: thus *taràu-, βináka,* and *palàsi-* in the examples above would be counted as trimoraic units. There appear to be no data within Fijian that could decide between these two views of metrical constituency, though only moraic trochees would generalize to other languages.

The strict prohibition on degenerate feet in (45a) is reflected in the minimal word constraint for Fijian: all independent words (in fact, all roots) have at least two moras. Schütz 1985, 490, and Dixon 1988, 25, give evidence that the bimoraic minimum is ironclad: bound monomoraic forms are lengthened when pronounced alone as citation forms. For example, when the monomoraic ablative marker /-i/ is pronounced in a context like "Did you say ____?" it is pronounced [íː] to conform to the word minimum.

While rule (45) states that foot parsing proceeds from right to left (following Schütz and Dixon), this is actually a difficult issue. The rightmost, main stressed foot is indeed always in final position. But other loanwords indicate that secondary stress can sometimes be assigned from left to right:

(47)	/⏑̀ ⏑ ⏑ –́/	*mìnisitáː*	'minister'	*kònitaráki*	'contract'
		òtakarísi	'watercress'	*pàrakaráβu*	'paragraph'
	/–̀ ⏑̀ ⏑ ⏑ –́/	*kèːmìsitiríː*	'chemistry'		

The motivation for secondary stress placement in these forms should be clear: it falls either on a syllable that bears main stress in the source (e.g. ***mì**nisitáː* 'mínister') or on a vowel that corresponds to a vowel in the source, as opposed to being inserted to create lawful syllables: *kèː**mì**sitiríː* 'chemistry'. The forms in (44), with right-to-left assignment, also fit into these categories. Thus it appears necessary (Schütz 1985, 534) to posit phonemic secondary stress in Fijian, treated with lexically listed foot structure. With Icelandic (§ 6.2.2), Fijian shares the distinction of being a language with predictable primary stress but phonemic secondary stress. The phonotactic laws that hold absolutely are (a) there must be a complete parse into moraic trochees (i.e., sequences of stray light syllables are disallowed); (b) the main-stressed foot must be word-final.

To determine whether there is in any sense a default pattern of right-to-left foot assignment for secondary stress, one must check non-borrowed forms in which 3, 5, 7, . . . light syllables precede the main stress. Dixon 1988, 24, 30, notes cases like *liŋà-mun͡ráu* 'arm-2 DUAL POSSESSOR', suggesting that right-to-left assignment is the default at least in the dialect he describes. Scott 1948, 745, however, gives an entirely parallel form with left-to-right stressing: *ùlu-mun͡ráu* 'head-2 DUAL POSSESSOR'. The issue thus remains open.

Suffixes form part of the domain of stress assignment, so we find alternations like those in (48):

(48)					
a.	*máta*	'eye'	*matá-ŋ͡gu*	'my eye'	S 478
b.	*m͡butáko*	'steal'	*m͡bùtakó-ða*	'steal it'	S 482
c.	*n͡dalíŋa*	'ear'	*n͡dàliŋá-na*	'her ear'	S 478
d.	*kám͡ba*	'climb'	*kam͡bá-ta*	'climb it'	S 483
			kàm͡ba-tá-ka	'climb with it'	S 496

Many prefixes and preceding grammatical words also cohere phonologically with a stem and thus participate in the alternating count of secondary stresses:

(49) *mè+kilá:* 'that he might know it' S 477
ì-leβúka 'to Levuka' S 477
kì-na-βále 'to the house' S 480

6.1.5.2 *Trochaic Shortening*

The theoretically most interesting part of the system is a phonological rule that shortens long vowels in penultimate syllables when the final vowel is short; informally, V: → V / ___ CV#. Because of this rule, there are no stems in Fijian that end on the surface in CV:CV. Moreover, borrowings are brought into conformity with the rule; thus English *pepper* [pέpr̩] and *paper* [pé:pr̩] are both borrowed into Fijian as *pépa* (Schütz 1978). Suffixation can bring underlying long vowels into penultimate position, where they regularly are shortened:

(50) *m͡bú:* 'grandmother' *m͡bú-ŋ͡gu* 'my grandmother' S 528
tá: 'chop' *tá-y-a* 'chop-TRANS.-3 SG. OBJ.' S 528
n͡ré: 'pull' *n͡ré-ta* 'pull-TRANS.' D 26

Dixon 1988, 26–27, shows that we cannot take a form like /m͡bu/ to be the underlying form and derive [m͡bu:] by lengthening, because polysyllabic forms like *ðaðá:* 'lots of bad things' also undergo shortening before a monomoraic suffix. Dixon also notes that alternations like *síβi* 'exceed' ~ *sì:βí-ta* 'exceed-TRANS.' can only be derived by shortening, from underlying /si:βi/.

In the standard dialect described by Schütz, the shortening rule applies even to diphthongs, as in *rai* 'see' ~ *răĭ-ða* 'see it'. A diphthong like [ăĭ] is phonetically shorter than [ai], and its /a/ component undergoes a greater degree of phonetic assimilation to the following segment (Schütz 1985, 545).

The structural change of shortening can be treated as loss of a mora. Where the input is a long vowel, this shortens it and nothing further happens. Where the input is a diphthong, the stranded vowel segment adjoins to the surviving mora, forming a short diphthong:

(51)
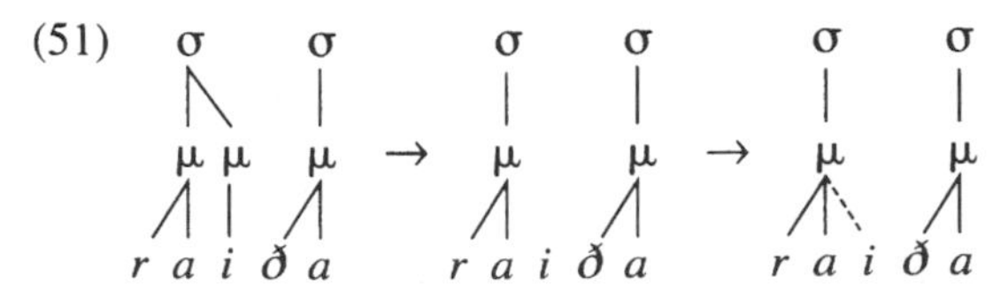

Geraghty 1983, 69, 169–70, notes interesting variations on this theme: some Fijian dialects fully delete the stranded vowel; others monophthongize shortened /ai, au/ to [e, o]; still others avoid the problem entirely by breaking the diphthong into a disyllabic sequence: *ra.í.ða.*

Shortening does not take place when a long penult precedes a heavy syllable: *n͡rè:n͡ré:* 'difficult' (43); nor does it take place before a sequence of two light syllables: *m͡bà:léti* 'belt' (44a).

The central question is why long vowels and diphthongs should be shortened in the / ____ CV# context. This is hardly to be expected a priori; the shortened vowel bears main stress, and the languages of the world generally prefer to LENGTHEN main-stressed vowels rather than shorten them.

Following a strategy similar to that of Prince 1990, I propose that the motivation of Fijian stressed vowel shortening is that it permits a more complete parse of the word into metrical feet. This assumes that there is a general pressure for syllables to be parsed into feet, analogous to the (much stronger) pressure for segments to be parsed into syllables (§ 5.2.3). Trochaic Shortening permits the sequence of (52a) (foot plus stray) to be converted into (52b), a canonical moraic trochee.

(52) a. (×)   b. (× .)
   – ᴗ   ᴗ ᴗ

Based on this account, I call the rule **Trochaic Shortening,** formulating it as in (53):

(53) **Trochaic Shortening**

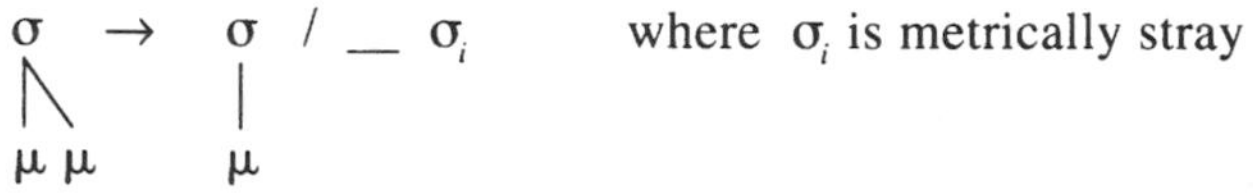

```
σ   →   σ  / __ σᵢ        where σᵢ is metrically stray
|\      |
μ μ     μ
```

The rule itself just alters quantity, and the actual adjustment of foot structure shown in (52) is attributed to persistent footing.

A sample derivation appears in (54). The basic rules of foot construction yield (54a); note that the ban on degenerate feet (§ 5.1.1), as well as the Priority Clause (§ 5.1.6), have resulted in the stress rule skipping over the final light syllable. The resulting representation is altered to (54b) by Trochaic Shortening, and the surface form is then created by persistent footing and the linking of stray /i/ (54c).

```
(54) a. (×       )   b. (×       )   c. (×       )
        (×)             (×)             (×     .)
         σ     σ         σ     σ         σ     σ
         |\    |         |     |         |     |
         μ μ   μ         μ     μ         μ     μ
        /| |  /|        /|    /|        /|\   /|
        r a i ð a       r a i ð a       r a i ð a
```

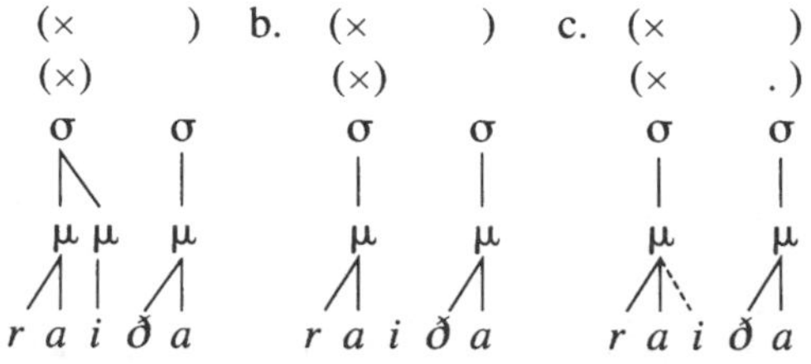

When a long vowel is followed by another long, or by two shorts, then the condition on Trochaic Shortening that σ_i be stray is not met, and Shortening does not apply:

(55) a. (×) b. (×)
(×) (×) (×)(× .)
– – – ⏑ ⏑
n͡rè:n͡ré: 'difficult' *m͡bè:léti* 'belt'

The argument is slightly obscured by the fact that all pretonic long vowels in Fijian may shorten optionally (Schütz 1985, 527); the crucial point is that shortening is an option in (55) but a rigid requirement in (50) and similar forms.

Long vowels may appear before metrically stray syllables in positions other than penultimate. According to Schütz, Trochaic Shortening may apply to them as well, though in such cases it is not obligatory:

(56) a. (×) (×)
(×) (× .) (× .)(× .)
– ⏑ ⏑ ⏑ ⏑ ⏑ ⏑ ⏑
ká:+m͡baláβu → *kàm͡baláβu* 'thing + long' = 'long thing' S 480

b. (×) (×)
(×) (×) (× .)(×)
– ⏑ – ⏑ ⏑ –
sa:+kila: → *sàkilá:* 'she knows it' S 525

A caveat here is that we must distinguish Trochaic Shortening from the rule just alluded to that shortens any pretonic long vowel. It is plausible to infer from Schütz's description (1985, 527–29) that a pretonic long vowel is more likely to shorten if it also meets the requirements of Trochaic Shortening; but the text does not make this fully clear.

6.1.5.3 *Discussion*

The analysis of Trochaic Shortening raises a number of theoretical issues.

(A) FORMALIZING THE RULE. The formulation in (53) does not directly incorporate the rule's "goal" of creating an exhaustive parse of moraic trochees. Instead, the rule simply adjusts quantity, and I attribute to persistent footing the improvements in syllable parsing that result. An alternative approach would be to attempt a rule that simultaneously shortens the penult and incorporates the final syllable. This would have the advantage of expressing the teleology of the rule in the formalism, but would have the disadvantage of allowing phonological rules to have multiple simultaneous actions. The rule as given reflects the view that phonological rules may take on relatively ad hoc forms but are valued within the grammar according to whether they create well-formed structures.

(B) THE GENERALITY OF TROCHAIC SHORTENING. Since Trochaic Shortening is attributed here to the general Iambic/Trochaic Law, we would expect it to occur elsewhere among trochaic languages. Some examples are given below.

Trochaic Shortening is found in Hawaiian (Elbert and Pukui 1979, 14) and in Tongan (§ 6.1.9), both of which belong together with Fijian in the Central Pacific subgroup of Austronesian. The Hawaiian rule is like that of Fijian, except that it applies only when the long vowel is followed by /ʔ/. Tongan Trochaic Shortening also resembles Fijian, except that it applies only sporadically, with /V́ːCV/ sequences are usually repaired by breaking the long vowel into two syllables: [**V.V́.**CV].

Prince 1990 points out that the well-known Trisyllabic Shortening process of English (productive in Middle English; currently a lexical rule) can also be regarded as creating canonical moraic trochees. The difference between Fijian and English is that English has final extrametricality; thus the sequence that is made to conform to the trochaic template consists of the antepenult and the penult, rather than the penult and the final. This is illustrated in the following derivation of *sanity* [sǽnɪti] from /séːn+ɪti/ (the issue of vowel quality shift is ignored here):

```
(57) a. (×   )      Trochaic      b. (×   )
        (×)         Shortening       (× .)
        –  ⌣⟨⌣⟩        →               ⌣  ⌣⟨⌣⟩
        seːnɪti                      sænɪti
```

Various dialects of Italian also exhibit Trisyllabic Shortening: Rohlfs 1949, 65, reconstructs it at an early stage for most southern dialects on the basis of the distinct qualitative evolution of long and short vowels. It is also observed as a synchronic phenomenon in certain Abruzzese dialects by Fong 1979, 285. Since the relevant stress pattern here is essentially the same as in the English case, we can again attribute the rule to the goal of achieving an exhaustive parse of moraic trochees.

Prince 1990 points out an interesting counterpart to Trisyllabic Shortening: lengthening of stressed vowels in penultimate, but not antepenultimate position. This is found in Chamorro (Chung 1983) and Italian. Prince suggests that the mechanism in such cases is a late application of extrametricality, which forces lengthening to preserve a well-formed foot (cf. § 5.1.7):

```
(58)     (× .)   →     (×)       →     (×)
       . . . ⌣ ⌣      . . . ⌣ ⟨⌣⟩      . . . – ⟨⌣⟩
```

This mechanism cannot derive lengthening in an antepenultimate syllable, since to make the rightmost two syllables extrametrical would require illegal chained extrametricality (§ 5.2.1).

The interesting aspect of Prince's treatment is that under it, the lengthening of stressed vowels in trochaic languages does not counterexemplify the Iambic/

Trochaic Law. The lack of lengthening in antepenults indicates that the lengthening of the penult is not due to pressure to create an uneven trochee (i.e. /◡́ ◡/ → /–́ ◡/), but rather to the maintenance of a well-formed foot under extrametricality.

(C) TROCHAIC SHORTENING AND FOOT INVENTORIES. The existence of trochaic shortening rules is relevant to the choice of a basic foot inventory. Under the "uneven trochee" (§ 4.3.1), there is no reason to expect shortening in a /–́ ◡/ sequence, since the input to the rule already forms a canonical uneven trochee:

(59) (×   )
     (× .)
      – ◡
     raiða

Such a theory is therefore at a loss to explain why rules of trochaic shortening should exist at all.

(D) UNEVEN TROCHEES AS SUBOPTIMAL FEET. The account of trochaic shortening in Prince 1990 adopts a different strategy from that employed here: first, suboptimal /–́ ◡/ trochees are created, then they are brought into conformity with the Iambic/Trochaic Law by shortening the heavy syllable. My analysis relies instead on a hypothesis that (at least in some languages) metrical parses are preferred that include as many syllables as possible. Additional support for this hypothesis may be found in Yidiɲ (§ 6.3.9; Hayes 1982a; Kirchner 1992) and Maniwaki Ojibwa (§ 6.3.4), which delete unfooted vowels; and in Latin, whose phonology operates to avoid unfooted syllables in a number of ways (Mester 1992). The hypothesis of maximizing the incorporation of syllables into metrical structure is analogous to maximal syllabification of segments, which is very widely observed (J. Ito 1989). For discussion of the relative merits of this approach with respect to that of Prince 1990, see Mester 1992.

6.1.6 Maithili

Maithili is an Indo-Aryan language spoken in India and Nepal. The variety of Maithili discussed here is described in the work of Subhadra Jha: a phonetic description (1940–44) and a grammar and historical phonology (1958). Jha describes a rather archaic form of speech. Other dialects are treated in Davis 1984 and Yadav 1984; these have lost phonemic vowel length, apparently with drastic effects on the stress system. All discussion here should be taken as referring only to Jha's dialect.

6.1.6.1 *Data*

Jha marks primary stress on a subset of the data. In the forms cited below, stress locations determined by applying Jha's rules are cited with the location in parentheses; i.e. "(G)" for forms from the grammar (Jha 1958) and "(P)" for forms from the phonetic description (Jha 1940–44). Jha marks secondary stress in just a few forms, but he asserts that secondary stress falls on every initial syllable that does not bear primary stress. I insert this secondary stress in the transcriptions below, following Jha's rule.

It can be argued that Maithili also has tertiary stress. This is not transcribed by Jha, but can be inferred by a phonological diagnostic, namely its effect in blocking a rule of Vowel Reduction. In certain positions fully stressless /a, ɒ, i, u/ are realized as extremely short, and apparently may even be deleted. When pronounced, they largely retain their basic vowel quality, except that /a/ and /ɒ/ merge as [ə]. Since the position of stress within the word varies, the vowel reduction rule creates alternations, such as those in (60). Reduced vowels are transcribed here with /˘/; reduced /i, u/ are /ĭ, ŭ/ and reduced /a, ɒ/ are /ə̆/. Above each syllable is marked the level of stress assumed in the analysis here.

(60)

	1 3 0		1 0	
a.	*gáːč^h**a**-kə̆*	'tree-GEN.'	*gáːč^{h}ə̆*	'tree-(dir. base)' (G 277)
	1 3 0		1 0	
b.	*máːṭ**i**-kə̆*	'earth-GEN.'	*máːṭĭ*	'earth (dir. base)' (G 277)
	1 3 0		2 0 1	
c.	*déːk^h**ɒ**b-ə̆*	'seeing-DIR. BASE'	*dèk^{h}ə̆b-áː*	'seeing-OBL. BASE' (G 279)
	2 1 0		2 0 1	
d.	*bìs**ɒ́**r-ĭ*	'forget-CONJ.	*bìsə̆r-úː*	'forget-IMP. 2 HON.' (G 453)

Vowel Reduction in Maithili can be argued to be stress-sensitive. In particular, vowels bearing primary or secondary stress never undergo it. Moreover, in long words reduced vowels are distributed rhythmically: *bùɽ^{h}ə̆nagə̆ríː* 'name of a village' (G 60), much as in English and other languages with alternating stress. I therefore assume that unreduced short vowels, such as the first and third vowels of the form just given, bear a weak stress, which protects them from reduction: *bùɽ^{h}ə̆nagə̆ríː* has the stress contour 20301.

Vowel Reduction usually does not apply to vowels that are adjacent to another vowel. For example, the second vowel of the stem /guru/ 'teacher' reduces in the isolation form *gúrŭ* (G 205), but not in the derived form *guru-áːi* 'teachership' (G 205), despite the fact that (on analogy with forms without hiatus) its stress pattern must be *gùru-áːi*. Similarly, the stressless final /i/ of *gùru-áːi* fails to reduce because it is preceded by /aː/. The failure of vowel reduction in hiatus may reflect an additional rule that merges adjacent vowels into a single syllable prior to reduction: transcribed *gùruáːi,* for instance, should perhaps be treated phonologically as *gùrwáːy.* Alternatively, we could simply be dealing with an idiosyncrasy of the Vowel Reduction rule. It is hard to tell from the data, and resolving this question will not be crucial.

Syllable weight in Maithili is not straightforward; for the moment I will assume that only long-voweled syllables are heavy, but I adjust this assumption in § 6.1.6.3.

The placement of main stress in Maithili can be summarized as follows. If a word has a long vowel in the penult, it will receive penultimate stress:

(61) a.	/–́ ⏑/	*máːʈĭ*	'earth'	P 437
b.	/⏑ –́ ⏑/	*kìšáːnə̆*	'a cultivator'	G 61
c.	/⏑ ⏑ –́ ⏑/	*àdʰə̆láːhə̆*	'bad'	G 60, P 440
d.	/⏑ ⏑ ⏑ –́ ⏑/	*dàhinə̆báːrĭ*	'the right one'	(G 258)
e.	/–́ –/	*sáːɽiː*	'woman's garment or cloth'	P 437
f.	/⏑ –́ –/	*tèlíːbaː*	'belonging to an oilman'	(G 275)
g.	/⏑ ⏑ –́ –/	*àgə̆ráːhiː*	'setting fire to a thing'	G 60

Words ending in /⏑ –/ bear final stress:

(62) a.	/⏑ –́/	*sàkʰáː*	'issue'	G 60
		bʰàʈʈíː	'a big oven'	G 60
b.	/⏑ ⏑ –́/	*pàʈə̆híː*	'thin'	G 60
c.	/⏑ ⏑ ⏑ –́/	*kùʈilə̆táː*	(no gloss)	(G 61)
d.	/⏑ ⏑ ⏑ ⏑ –́/	*bùɽʰə̆nagə̆ríː*	'name of a village'	G 60
		pʰùlə̆kumə̆ríː	(proper name)	G 62
		hàrĭbolə̆báː	(no gloss)	(G 83)

The stress assigned to words ending in /⏑ ⏑/ depends on what precedes this sequence. If there are only two syllables, we get penultimate stress, as in (63a). If a final /⏑ ⏑/ is preceded by a short-voweled syllable we also get penultimate stress, as in (63b). But if the final /⏑ ⏑/ sequence is preceded by a long vowel, we get antepenultimate stress (63c):

(63) a. **Disyllables**

/⏑́ ⏑/	*pámʰə̆*	'little whiskers'	P 449

b. **Final / . . . ⏑ ⏑ ⏑/**

/⏑ ⏑́ ⏑/	*sùnnŏ́rə̆*	(no gloss)	G 60
	bìndúlə̆	'a fabulous horse'	P 456
	pàtítə̆	(no gloss)	G 60
/⏑ ⏑ ⏑́ ⏑/	*dʰànə̆hŏ́rə̆*	(no gloss)	(G 191)
	bʰìnə̆sŏ́rə̆	(no gloss)	(G 220)
	čʰùčʰunnŏ́rĭ	(no gloss)	G 60

c. **Final /– ⌣ ⌣/**

/–́ ⌣ ⌣/	*gá:hʰinə̆*	'pregnant'	P 447
	dí:ɒṭĭ	(no gloss)	G 60
/⌣ ⌣ –́ ⌣ ⌣/	*kàkə̆ɻó:hɒrĭ*	'residence of a crab'	G 60
	mànə̆mó:hɒnə̆	(proper name)	G 60
/⌣ ⌣ ⌣ –́ ⌣ ⌣/	*ǰìmutə̆bá:hɒnə̆*	(proper name)	G 62

Lastly, monosyllables are stressed; according to Jha (P 457), only long voweled monosyllables occur, e.g. *kí:* (no gloss) G 59, *sǽ̃:* 'husband' P 440, *ǰʰá:* 'proper name'.

As noted earlier, all initial syllables not bearing primary stress are claimed by Jha to bear secondary stress.

6.1.6.2 *Analysis*

It is useful to focus on the two contrasts in (64):

(64) a. **Penultimate Stress** vs. **Final Stress**
. . . –́ – # . . . ⌣ –́ #

b. **Antepenultimate Stress** vs. **Penultimate Stress**
. . . –́ ⌣ ⌣ # . . . ⌣ ⌣́ ⌣ # and # ⌣́ ⌣ #

The point is that a final heavy syllable, or a final /⌣ ⌣/ sequence, can attract stress just in case it is not preceded by a heavy syllable. Suppose we organize the syllables from right to left into moraic trochees. Then the relevant examples will be footed as in (65):

(65) a. **Penultimate Stress** vs. **Final Stress**

```
    (×)(×)              .)(×)
. . .  –  –  #        . . .⌣  –  #
```

b. **Antepenultimate Stress** vs. **Penultimate Stress**

```
    (×)(× .)            .)(×  .)            (×  .)
. . .  –  ⌣⌣ #        . . . ⌣  ⌣  ⌣  #  and  # ⌣  ⌣ #
```

To pick out the right /×/ for the main stress, it is sufficient to mark the final foot as extrametrical when it is preceded by a heavy syllable. This syllable will necessarily be stressed, so this is essentially a case of extrametricality in clash (§ 5.2.2). If we then create a word layer with End Rule Right, the correct stressings will result:

(66) a. **Penultimate Stress** vs. **Final Stress**

```
     ×)                   ×)
    (×)⟨(×)⟩            .)(×)
. . .  –́  –  #        . . .⌣ –́ #
```

```
b. Antepenultimate Stress     vs.     Penultimate Stress
      ×)                                   × )           (× )
   (×)⟨(× .)⟩                             .)(× .)        (× .)
. . .  –́   ᴗ ᴗ  #                     . . .ᴗ  ᴗ́ ᴗ  #  and  #  ᴗ́ ᴗ  #
```

For reasons to be made clear below, Extrametricality in Clash must be restricted to cases where the clashing stress is constituted by a long-voweled syllable. With this addition, the rules for Maithili can be stated as in (67):

(67) a. **Foot Construction** — Form moraic trochees from right to left. Degenerate feet are allowed in strong position.
b. **Extrametricality in Clash** — Foot → ⟨Foot⟩ / $\overset{\times}{\text{V}}$: ____ $]_{word}$
c. **Word Layer Construction** — End Rule Right
d. **Beat Addition** — Form a new layer below the word layer, and apply End Rule Left.

In this analysis, rule (67a) sets up the metrical feet, and rules (67b) and (67c) select a foot for main stress. Rule (67d), Beat Addition, promotes the foot on the initial syllable. This accounts for the distinction between secondary stress (noted by Jha) and tertiary stress (diagnosed by the absence of vowel reduction). The Beat Addition rule must be assumed to add a grid mark to retain the main stress in its original position, as shown in stage (d) of the representative derivation in (68):

```
(68) a. (×)(× .)(×) (×  .)       Foot Construction
         ᴗ  ᴗ ᴗ  –   ᴗ ᴗ
        ǰimutaba:hɒna            = (63c)

     b. (×)(× .)(×)⟨(×  .)⟩      Extrametricality in Clash
         ᴗ  ᴗ ᴗ  –   ᴗ ᴗ
        ǰimutaba:hɒna

     c. (        ×)              End Rule Right
        (×)(× .)(×)⟨(×  .)⟩
         ᴗ  ᴗ ᴗ  –   ᴗ ᴗ
        ǰimutaba:hɒna

     d. (        ×)              Beat Addition
        (×      )(×)
        (×)(× .)(×)⟨(×  .)⟩
         ᴗ  ᴗ ᴗ  –   ᴗ ᴗ
        ǰimutaba:hɒna

     e. [ǰìmutə̆bá:hɒnə̆]           Vowel Reduction
```

For an explicit proposal concerning how Beat Addition works, see § 9.6.1, where the process is discussed as part of the theory of phrasal stress.

In (69)–(73) are further metrical structures generated by the rules of (67).

(69) **Final Heavy, No Extrametricality: Final Stress**

```
a. (×)     b. (    ×)     c. (     ×)     d. (       ×)     e. (              ×)
   (×)        (×) (×)        (×    )(×)      (×     )(×)      (×            )(×)
   (×)        (×) (×)        (× .  )(×)      (×)(× .)(×)      (×   .)(×   .)(×)
    –          ⌣   –          ⌣  ⌣  –         ⌣  ⌣  ⌣ –        ⌣   ⌣   ⌣   ⌣  –
```

a. *kíː* b. *sàk^{h}áː* c. *pàʈə̆híː* d. *kùʈilə̆táː* e. *bùɽ^{h}ə̆nagə̆ríː*

(70) **Words Ending in / – ⌣/: Penultimate Stress**

```
a. (×  )     b. ( ×   )     c. (      ×   )     d. (        ×   )
   (×  )        (×)(×)         (×     )(×)         (×       )(×)
   (×)          (×)(×)         (×    .)(×)         (×)(×  .)(×)
    –  ⌣         ⌣  –  ⌣        ⌣   ⌣  –  ⌣          ⌣  ⌣  ⌣  –  ⌣
```

a. *máːʈĭ* b. *kìsáːnə̆* c. *àd^{h}ə̆láːhə̆* d. *dàhinə̆báːrĭ*

(71) **Words Ending in / – – /: Extrametricality, Penultimate Stress**

```
a. (×)           b. (   ×)              c. (          ×)
   (×)              (×) (×)                (×     ) (×)
   (×) ⟨(×)⟩        (×) (×)  ⟨(×)⟩         (×    .) (×)  ⟨(×)⟩
    –    –           ⌣   –     –            ⌣    ⌣    –      –
```

a. *sáːɽiː* b. *tèlíːbaː* c. *àgə̆ráːhiː*

(72) **Words Ending in / – ⌣ ⌣/: Extrametricality, Antepenultimate Stress**

```
a. (×)              b. (      ×)
   (×)                 (×  .)(×)
   (×) ⟨(×  .)⟩        (×  .)(×)⟨(×  .)⟩
    –    ⌣  ⌣           ⌣   ⌣  –   ⌣  ⌣
```

a. *gáːb^{h}inə̆* b. *kàkə̆ɽóːhɒrĭ*

(73) **Other Words Ending in /⌣ ⌣/: Penultimate Stress**

```
a. (×     )     b. ( ×    )     c. (       ×  )
   (×     )        (×)(×   )       ( ×  )(×   )
   (×    .)        (×)(×  .)       ( ×  .)(× .)
    ⌣    ⌣          ⌣   ⌣  ⌣          ⌣   ⌣  ⌣  ⌣
```

a. *pám^{h}ə̆* b. *pàtítə̆* c. *d^{h}ànə̆hɒ́rə̆*

As can be seen, these structures also account for the pattern of vowel reduction: vowels reduce when their syllables are not dominated by /×/; that is, when they are completely stressless. This is of course the expected environment for reduction rules. In many Maithili words, the basic trochaic quantity-sensitive alternating structure is readily apparent on the phonetic surface, owing to reduction.

6.1.6.3 *The Quantity of CVC Syllables*

The rules so far have assumed that only CVV syllables are heavy. In fact, there is evidence that CVC should be counted as heavy as well. The relevant facts involve vowel reduction. As Jha notes (P 457), vowels are never reduced in closed syllables:

(74)	*č^{h}ùč^{h}**unn**ɒrĭ*	(no gloss)	G 60
	*čà**ma**tká:rə̆*	'cunning'	(G 635)
	*bèb**as**t^{h}á:*	(no gloss)	(G 125)
	*pà**ri**čč^{h}á:*	'shaking of the whole body by a goat before sacrifice'	(G 636)

Assuming that CVC is heavy, this is just what the analysis would predict. The closed syllable must form a foot on its own, which blocks reduction:

```
(75) (      ×   )
     (×   )(×)
     (×)(×)(×)
      ˘  –  –  ˘
     čàmatká:rə̆
```

Moreover, when there are three syllables before the main stress, the pattern of vowel reduction again depends on the distribution of open and closed syllables. If the syllable before the main stress is open, it undergoes reduction, as shown in (76a). But if this syllable is closed, and the preceding syllable is open, that syllable undergoes reduction, as in (76b):

(76)	a.	*àdinə̆tá:*	'misfortune'	(G 633)
		dàhinə̆bá:rĭ	'the right one'	(G 258)
		ǰìmutə̆bá:hɒnə̆	(proper name)	G 62
	b.	*gòrə̆minṭí:*	'conspiracy'	(G 638)
		èkə̆hatt^{h}á:	'a man with one hand'	(G 62)
		bèṭĭbeččá:	'seller of a daughter'	(G 82)
		màsalə̆gaṭṭí:	'a kind of sweetmeat'	P 455

This is again a direct consequence of the analysis, if CVC is counted as heavy:

```
(77) a. (           ×)        b. (             ×)
        (×         )(×)          (×          ) (×)
        (×) (×   . )(×)          (×    . ) (×)  (×)
         ˘    ˘   ˘   –            ˘    ˘    –    –
         a d i n ə̆ t á :          g o r ə̆ m i n ṭ í :
```

In (77b), the heavy syllable /min/ forms a foot on its own, placing the syllable /ra/ in metrically weak instead of strong position.

CVC syllables are rare in positions other than immediately before the main stress (none occurs finally). But such examples as there are resist vowel reduction, as the analysis would predict:

```
(78)  (        ×)      'magistrateship'                    (G 224)
      (×      )(×)
      (×)(×)   (×)
       ⏑  –    ⏑ ⏑
      majǐšṭə̆ri:
```

The blockage of vowel reduction in closed syllables is completely exceptionless. Thus it seems safe to conclude that, insofar as vowel reduction is a diagnostic for metrical structure, CVC counts as heavy in Maithili.

However, there is one respect in which CVC syllables do not count as heavy: they do not trigger Extrametricality in Clash (67b). This is shown by the examples like those in (79), from (62a) and (63b):

```
(79) a.  (   ×)        b.  (   ×  )
         (×)(×)            (×) (×  )
         (×)(×)            (×) (× .)
          –  –              –   ⏑ ⏑
         bʰàṭṭí:           sùnnɓrə̆
```

If CVC could trigger Extrametricality in Clash, the rightmost foot in these examples would be extrametrical, and stress would fall one syllable further to the left. The condition that only CV: triggers Extrametricality in Clash also prevents penultimate stress in /⏑ –/ forms and antepenultimate stress in /⏑ ⏑ ⏑/, as can be seen in (69b) and (73b) above.

There is one class of forms, namely those ending in /. . . CV:CVCCV/, where in principle the weight of CVC could be diagnosed by main stress placement. In such cases, the foot dominating the CVC syllable would not be word-final and thus would be prevented by the Peripherality Condition from being marked as extrametrical in clash:

```
(80)  (    ×     )
      (×) (×     )
      (×) (×)
       –   –   ⏑
      CV:CVCCV
```

This may be compared with /– ⏑ ⏑/ (e.g. (72a)), where the final two syllables form a single foot, which becomes extrametrical.

Unfortunately, this prediction concerning /CV:CVCCV/ forms is difficult to test. In particular, CVCCV is rare in word-final position. It appears that such sequences may occur underlyingly, but they are normally eliminated by a process of degemination and compensatory lengthening. This rule is stated informally in (81), and is illustrated by the alternations of (82).

(81) **Compensatory Lengthening** VCCV# → Vː CV#

(82) a.	/ɖ^{h}aʈʈ^{h}-a/	→	*ɖ^{h}áːʈ^{h}ə̆*	'bamboo fence'	G 63
	/ɖ^{h}aʈʈ^{h}-aː/	→	*ɖ^{h}àʈʈ^{h}áː*	'bamboo fence (long form)	G 63
b.	/hatt^{h}-a/	→	*háːt^{h}ə̆*	'hand'	(G 353)
	/eːk-a-hatt^{h}-aː/	→	*èkə̆hatt^{h}áː*	'man with one hand'	(G 62)
c.	/nakk-a/	→	*náːkə̆*	'nose'	(G 227)
	/nakk-uː/	→	*nàkkúː*	'long-nosed'	(G 227)
	/nakk-ula/	→	*nàkkúlə̆*	'crooked-nosed'	(G 229)

It would be feasible to order stress assignment after compensatory lengthening, so most of the relevant cases simply cannot test the hypothesis. Only words that exceptionally fail to undergo Compensatory Lengthening are candidates for testing the predictions of the analysis concerning /CVːCVCCV/.

In addition to this problem, there is an additional difficulty: if (as the analysis predicts) /CVːCVCCV/ is assigned penultimate stress, the initial long vowel would normally then undergo a rule of Pretonic Shortening (P 457), which shortens all long vowels to the left of the main stress.

(83) **Pretonic Shortening**

```
                 ×
                 ×
                 ×
Vː → V / ___ . . . V
```

(84) a.	***báːǰ-alə̆***			'speak-PAST PART.'	(G 453)
	báːǰ-uː			'speak-2 HON.-IMP.'	
	/baːǰ-itɒt^{h}i/	→	***bàǰĭtɒ́t^{h}ĭ***	'speak-3 FUT.'	
b.	***kúːɽ-aː***			'a big jar'	(G 201)
	/kuːɽ-aniː/	→	***kùɽə̆níː***	'a small jar'	(G 201)
c.	***ǰamíːnə̆***			'earth'	(G 255)
	/ǰamiːna-daːra/	→	***ǰàminə̆dáːrə̆***	'landlord'	(G 255)
d.	***déːk^{h}ɒb-ə̆***			'seeing-DIR. BASE'	(G 279)
	/deːk^{h}-aːra/	→	***dèk^{h}áːrə̆***	'seen'	(G 214)

Pretonic Shortening is not surface-true; it has exceptions in borrowed words, and it fails to apply in loose-knit compounds and in words formed by "stress-neutral" affixation processes. But the rule does hold for the bulk of Maithili morphology, and a large number of morphemes containing underlying long vowels alternate according to the rule.

Consider now the issue of determining whether CVC is heavy. As noted, we would like to see if in a word ending in /CVːCVCCV/, the (claimed to be heavy) penult can take main stress in perference to the antepenult. The problem is that if it does, the surface form will end in [CVCV́CCV], owing to Pretonic Shortening. This means that such forms are not easily spotted in Jha's grammar; one needs an alternating form to identify the underlying long vowel.

The following cases bear on this issue. From underlying /galaː/ 'neck' is derived the form *gàlə̆bɒ́nnə̆* 'muffler' (G 257), which bears penultimate stress as the analysis would predict. The derivation is as in (85):

```
(85) (       ×      )    stress rules
     (×  ) (×)
     (×)(×) (×)
      ˘  –   –    ˘
     g a l a ː b ɒ n n a

     (       ×      )    Pretonic Shortening
     (×  ) (×       )
     (×)(×) (×)
      ˘  ˘   –    ˘
     g a l a b ɒ n n a

     (       ×      )    Loss of degenerate feet in weak position
     (×  ) (×       )    persistent footing (see below)
     (×  .) (×)
      ˘  ˘   –    ˘
     g a l a b ɒ n n a

     [gàlə̆bɒ́nnə̆]          Vowel Reduction
```

The form *àːdɒ́ttĭ* (no gloss) G 63 (Jha's main stress mark), a borrowing from Persian, also shows stress on the penult in a /VːCVCCV/ form. It is a lexical exception to both Compensatory Lengthening and Pretonic Shortening. However, the form *čáːrillo* 'fourth' G 261 (his stress mark), derived from *čáːrĭ* 'four' G 262, has antepenultimate stress. The data are thus contradictory.

I conjecture that the stress on *čáːrillo* is aberrant. This word is part of a sequence of ordinal numbers ending in the suffix *-llo* (i.e. *čáːrillo, pã́ːčillo* . . . 'fourth, fifth, . . .'). A parallel situation in English induces aberrant stress; we say *thírteen, fóurteen, fífteen* . . . , despite the isolated pronunciations *thirtéen* etc. The repeated suffix is deaccented in favor of the changing (and semantically focused) stem. It is possible that the same phenomenon is involved in Maithili.

In any event, the data do not seem strong enough to decide the issue either way. The evidence of vowel reduction constitutes stronger grounds for believing that CVC counts as heavy in Maithili.

6.1.6.4 *Persistent Footing*

In derivation for *gàl*ə̆*bɓ́nn*ə̆ under (85), the foot created over the (originally heavy) second syllable is lost in weak position, following the general principles proposed above (§ 5.1.1) governing weak feet. Similar derivations occur in which a weak degenerate foot is created immediately to the left of a stray syllable, as shown in (86):

(86) /ǰami:na-da:ra/ = (84c)

```
( × )          Creation of metrical structure
(× )(× )       (/na/, /ra/ cannot be footed, by the Priority
(×)(×) (×)     Clause, § 5.1.6)
 ˘ – ˘ – ˘
ǰami:nada:ra

( × )          Pretonic Shortening (83)
(× )(× )
(×)(×) (×)
 ˘ ˘ ˘ – ˘
ǰami:nada:ra
```

What emerges on the surface is *ǰàminădá:ră,* with reduction of the third, but not the second, syllable. This would result naturally from the assumption that footing in Maithili is persistent, in the sense defined in § 5.4.1. Following the conventions laid out there, the stray syllable /na/ would be reparsed into the preceding foot, forming the correct surface metrical representation:

```
(87) ( × )          persistent footing
     (× )(× )
     (×)(× .)(×)
      ˘ ˘ ˘ – ˘
     ǰàminə̆da:rə̆
```

In effect, the adjoined syllable "saves" the foot over /na/ from the ban on weak degenerate feet.

Similar pairs from which the same point could be made are listed in (88). In each, the boldface vowel does not reduce and therefore must have been placed in strong position by persistent footing:

(88)

a.	*tàmá:s-a:*	'amusement'	(G 249)
	*tàm**a**s-*ə̆*gí:r*ə̆	'those who come to witness a performance'	(G 249)
b.	*sìpá:hi:*	'peon'	(G 249)
	*sìp**a**hĭ-gí:ri:*	'profession of a peon'	(G 249)
c.	*gàrí:b*ə̆	'poor'	(G 253)
	*gàr**i**b*ə̆*-tá:i*	'poverty'	(G 253)

6.1.6.5 *Discussion*

The analysis bears on a number of theoretical issues.

(A) EXTRAMETRICALITY IN CLASH. The analysis provides support for this rule type, proposed in § 5.2.2. A final foot of the form CV: attracts stress only if it does not clash with a preceding stress on another CV: syllable. The same holds for a final foot of the form CVCV. Maithili has numerous words having the prosodic shape CV:CVCV̆. These show clearly that the clash results in extrametricality rather than destressing: the penultimate syllable is subordinated in stress to the preceding heavy syllable, but it retains a weak stress and thus escapes reduction.

What is odd about the analysis is that only CV: syllables induce this clash-governed extrametricality. One would like to relate this to the typological observation that CV: syllables somehow count as "heavier" than CVC; that is, there are many rules that count CV: as heavy and CVC as light, but no rules that go the other way (chap. 7). I have no concrete suggestions for doing this.

(B) MORAIC VERSUS UNEVEN TROCHEES. The Maithili facts also support the moraic trochee over the uneven trochee (§ 4.4(b)). It is useful to compare the three cases in (89), which are parsed according to the analysis above (forms omit the middle layer of the grid, derived later by Beat Addition):

(89) a.
                (×)   = (61e)
(×)⟨(×)⟩ → (×)⟨(×)⟩
sa:ɻi: → *sa:ɻi:*

b.
                (×)   = (63c)
(×) ⟨(× .)⟩ → (×) ⟨(× .)⟩
ga:b^h*ina* → *ga:b*^{h}*inə̆*

c.
                (×)   = (84d)
(×) (×) → (×) (×)
de:k^h*a:ra* → *de:k*^{h}*a:rə̆*   (→ *dek*^{h}*á:rə̆* by Pretonic Shortening)

In (89a) and (89b), the rightmost foot in the word follows a CV: syllable. This foot becomes extrametrical in clash, and we get stress on the penultimate foot. But in (89c), the rightmost foot is not peripheral, being followed by a stray syllable. Extrametricality is blocked by the Peripherality Condition, and main stress is penultimate.

Now consider the same examples footed with uneven trochees. Examples (89a, b) will look the same, but (89c) will be as in (90):

(90)
                (×)
(×)⟨(× .)⟩ → (×)⟨(× .)⟩
de:k^h*a:ra* → **dé:k*^{h}*a:rə̆*

Under this template, final /CV:CV/ ends up as a single foot, rather than a two-foot sequence. If we mark this foot as extrametrical, then we incorrectly derive stress on the antepenult, as shown. Only moraic trochees permit a straightforward account of the system.

(C) VOWEL REDUCTION. A further question to be considered is why a rule of vowel reduction should occur in a trochaic language, given the prediction of the Iambic/Trochaic Law that segmental phonology should preserve the even timing of the foot (§ 4.5.3). One possibility to consider is that what Jha transcribes as reduction is in fact vowel DELETION, creating syllable-final consonants. Here, the "vowel" would be regarded as merely a consonant release. Where the lost vowel is /i/ or /u/, the preceding consonant is palatalized or labialized respectively. Under this interpretation, the process of vowel deletion might be expressed formally in (91) (a schematic form):

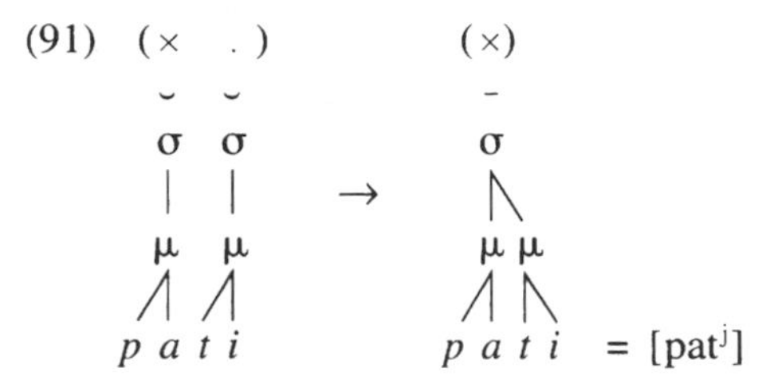

In this case, there is no violation of the Iambic/Trochaic Law, since the output is a well-formed moraic trochee, just like the input.

A number of passages in Jha's grammar support this interpretation. For example, he implies (G 65) that reduced vowels do not form syllables for purposes of verse composition. Further, in colloquial speech, metatheses of the form VC*i*#, VC*u*# → V*i*C#, V*u*C# are observed (G 120; Davis 1984, xiv–xv). These are plausibly either a way of transcribing the palatalized and labialized consonants, or a subsequent historical development from them. Finally, the form *pàɳɖitá:in𝑎̆* 'wife of a *paɳɖita*' (G 205) is interesting in this respect. Superficially it appears to be an exception to Vowel Reduction (note the boldface vowel), but if we construe Vowel Reduction as deletion, it falls into place: the triple cluster [ɳɖʲt] that would result from deletion is more complex than anything actually tolerated in the language. (Such forms are rare, since CVC syllables seldom occur early enough in the word to precede a medial reduction site.)

(D) DEGENERATE FEET. Maithili has a potential bearing on the question of whether degenerate feet can ever surface in weak position (§ 5.1): there are many possible examples of this, for instance (70b) *kìsá:n𝑎̆*. This issue can be addressed on various fronts.

First, it appears that monomoraic words do not occur in Maithili. By itself, this fact would suggest that degenerate feet should be banned entirely. This makes it all the more puzzling that degenerate feet should occur in weak position within the word.

Second, it is cross-linguistically very unusual for secondary stresses to occur on a light syllable, directly clashing with the main stress (e.g. as in *kìsáːnə̆*). Jha does not give phonetic correlates for secondary stress, so at present most of the evidence available is intuitive. What is particularly remarkable is the claimed audibility of secondary stress when even primary stress is relatively weak: "Though stress exists in Maithili, it is very weak and causes no semantic difference. It is not so prominent as in English or Bengali" (G 63).

The most clearly documented evidence of a more objective character is that Vowel Reduction is blocked in initial syllables. One is led to wonder whether there is some other reason for this. We might suppose, for example, that reduction is blocked here because it would ordinarily create an unsyllabifiable consonant cluster: **ksáːna*. We have just seen evidence that the avoidance of complex clusters is a factor in Maithili phonology.

The analysis proposed conservatively assumes that Jha's views on secondary stress are correct. In this analysis, the degenerate feet of Maithili are not strictly speaking in weak position, since they are dominated by a higher grid mark. For example, the structure proposed for (61f) *tèlíːbaː* is as in (92):

```
(92) (  ×      )
     (×)(×)
     (×)(×)  (×)
      ˘  –    –
     t è l í ː b a ː
```

The second-layer structure is needed independently to account for the distinction claimed by Jha between secondary and tertiary stress. However, it seems here that the second-layer structure is itself degenerate and in weak position. Thus although the formal principle has not been violated, one has a sense that a loophole has been taken advantage of. Clearly, further research to resolve this issue would be desirable.

6.1.7 Hindi

The stress pattern of Hindi has been described in a fair number of articles in the literature; see M. Ohala 1977 for a review. As Ohala makes clear, this topic has its empirically dismaying aspects: the published descriptions almost all disagree with one another, and seldom mention the disagreement.

The most interesting analysis in the literature, Pandey 1989, illustrates the difficulties. Pandey's data consist of his own speech plus elicitations from thirty other Eastern Hindi speakers. He counted a stressing as "regular" if seventy percent of the consultants provided it, and any stressings provided by fewer than forty percent of the consultants were thrown out (p. 39). There must be doubt about whether under such conditions any really firm conclusions about Hindi stress can be maintained. The very complex stress pattern Pandey describes exceeds the capacity of the theory proposed here, though a rough resemblance to the pattern assumed below can be seen.

There is good reason to expect that there would be variation in Hindi stress. Stress is a non-phonemic feature of words in Hindi, and the language is spoken by a very large population, many of whom are native speakers of some other Indo-Aryan language. Some careful dialectology might help to clarify the currently murky picture.

Fairbanks 1981, 1987a, 1987b, has brought relatively objective evidence to bear on the issue by examining the patterning of stress in Hindi verse, both modern and from the sixteenth century. The diagnostic Fairbanks used was to examine which syllable of a word poets most frequently place in the metrically strong positions of the line. To a rough approximation, the line structure can be depicted as in (93):

(93) ◡́ ◡ ◡ ◡   ◡́ ◡ ◡ ◡   ◡́ ◡ ◡ ◡   ◡́ ◡ ◡ ◡

That is, the line contains four feet of four moras each. Any two consecutive /◡/ positions may be realized as such, or by a single / – /. The strong positions, marked /◡́/, are normally filled by a stressed light syllable or by either mora of a stressed heavy, as in (94):

(94) ◡́ ◡◡ ◡ ◡́◡ ◡ ◡ ◡́ ◡◡ ◡ ◡́ ◡◡ ◡   meter (F 1981,184)

béːgi áːnu ǰala páːya pakʰáːruː   line

With such metrical data, Fairbanks motivates a stress rule ((96) below) proposed originally by Grierson 1895; this rule also agrees with the intuitions of two Hindi-speaking linguists Fairbanks consulted. For some syllable patterns, the predictions of the rule are unanimously corroborated by the metrical data. For others, the data suggest additional, less common stress patterns. I postpone discussion of these cases, treating for the moment only the normal outcome.

The Grierson/Fairbanks stress rule refers to syllable weight (as do other rules proposed for Hindi). Hindi syllables can be light (CV), heavy (CVC, CVː), or superheavy (CVːC, CVCC, and anything longer). The superheavy syllables derive historically from / – ◡/ sequences, and pattern in the same way synchronically. To treat this fact, I make two assumptions. First, such syllables are trimoraic (for a defense of trimoraic syllables, see Hayes 1989b, 291–97). Second, I assume with Pandey 1989 that trimoraic syllables are assigned a metrical form parallel to that of / – ◡/ sequences. In particular, the third mora is syllabified as a kind of degenerate syllable (95a), up to a point late in the derivation when it is adjoined to the preceding heavy, with inheritance of stress (95c):

(95) 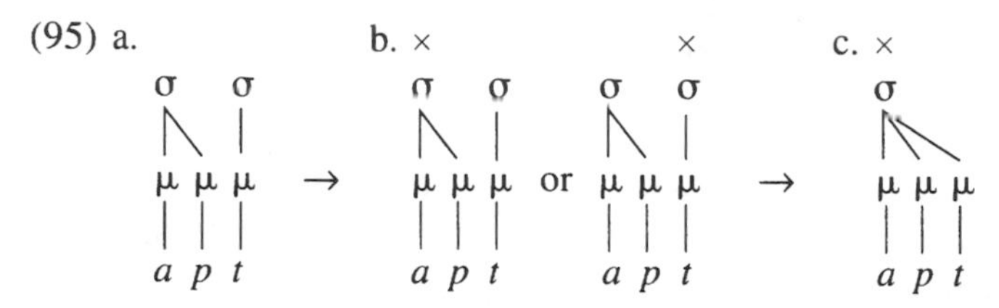

This scheme enables the syllabification algorithm to recapitulate history, without positing an actual phantom vowel. With this proviso, we may treat the superheavy syllables of Hindi in the same way as the rest of the data. Some precedents provide justification for this approach: Estonian (§ 8.5) treats trimoraic syllables in a similar way, and cases where syllables merge with inheritance of stress are cited in § 3.8.1 and § 5.6.2.[4]

With this background, the Grierson/Fairbanks stress rule may be stated as in (96):

(96) a. Stress the initial syllable of a disyllabic word:

/◡́ ◡/	*bál(a)*	'force'	G 139
/◡́ –/	*kála:*	'art'	F 1987a, 5
/–́ –/	*čú:ṛa:*	'bangle'	G 144

b. Weight-based rules:

i. Stress a heavy penult:

/◡ –́ –/	*asú:*$\check{j}^h$*a:*	'invisible'	G 139

ii. Otherwise, stress a heavy antepenult:

/–́ ◡ ◡/	*bánd*^{h}*an(a)*	'binding'	G 139

iii. Otherwise (i.e. words ending in /◡ ◡ σ/), stress

1. the preantepenult if the final is light:

/◡́ ◡ ◡ ◡/	*ánumati*	'approval'	F 1987a, 4

2. the antepenult if the final is heavy; or in trisyllables:

/◡ ◡́ ◡ –/	*titáliya:*	'butterfly (long form)'	G 140
/◡́ ◡ ◡/	*áditi*	(proper name)	F 1987a, 4
/◡́ ◡ –/	*ámita:*	(proper name)	F 1987a, 5

This complex pattern has two striking aspects: the limitation of stress to one of the last four syllables in a word (cf. § 6.1.1, Palestinian Arabic) and the manner in which the choice of preantepenultimate versus antepenultimate stress (96b.iii) depends on the weight of the FINAL syllable. Both oddities fall into place in much the same way as in Palestinian if we suppose that there is a tacit foot layer, with main stress often assigned to a nonfinal binary foot. The main difference with Palestinian is that Hindi foot construction goes from right to left.

4. My account of superheavy syllables in Estonian is rather different; conceivably it is applicable to Hindi as well, but this is hard to determine in the absence of sufficient data.

A specific analysis is given in (97). Under (98)–(101), I give representations derived by the rules for representative schematic forms.

(97) a. **Foot Construction** Form moraic trochees from right to left. Degenerate feet are allowed in strong position.

b. **Foot Extrametricality** Foot → ⟨Foot⟩ / ___ $]_{word}$

c. **Word Layer Construction** End Rule Right

(98) **Two Syllables**

```
a. (      )      b. (×)           c. (×)
   (×   .)          (×)⟨(×)⟩         (×)  ⟨(×)⟩
    ˘́   ˘             ˘́   –            –́     –
   b á l a          k á l a ː        č ú ː ɽ a ː
```

(99) **Heavy Penults** (100) **Heavy Antepenults**

```
(99)                    (100)
 (  ×)                   (×)
   (×) ⟨(×)⟩             (×)  ⟨(×  .)⟩
 ˘  –́    –                –́     ˘  ˘
 asúːǰʰaː                 bándʰana
```

(101) **Remaining Cases**

```
a. (×      )           b. ( ×  )            c. (×)            d. (×     )
   (×   . ) ⟨(×  .)⟩      (×)(×  .)⟨(×)⟩       (×)⟨(×  .)⟩        (×   .)⟨(×)⟩
    ˘́   ˘    ˘   ˘         ˘   ˘́   ˘   –         ˘́   ˘   ˘          ˘́    ˘    –
   á n u m a t i          t i t á l i y a ː     á d i t i          á m i t a ː
```

Some notes on this analysis: (a) The preantepenultimate maximum occurs in cases with two disyllabic feet, of which the second is extrametrical (101a). (b) The dependency of preantepenultimate (101a) versus antepenultimate (101b) stress on the weight of the final syllable is determined by the right-to-left creation of moraic trochees. (c) Degenerate feet are necessary to derive initial stress in (98b) /˘́ –/ and (101c) /˘́ ˘ ˘/. Such feet are licensed by their strong metrical position. (d) The stress pattern supports the Priority Clause proposed in § 5.1.6: degenerate feet may be created only when the parsing algorithm cannot create a proper foot by extending the parse. Were this not the case, we would derive penultimate stress in words ending in /– ˘ –/, since the penult would attract a degenerate foot. (e) Forms ending in /– – ˘/ receive penultimate stress, since the rightmost foot (i.e. the heavy penult) is nonperipheral and thus cannot be marked as extrametrical.

The analysis just given incorporates an insight of Fairbanks (1987a), who describes the pattern as involving (in the central core of cases) stress on the fourth mora from the end of the word. Where the analysis might be said to improve on Fairbanks's account is as follows: (a) Stress on words ending in /–́ ⌣/ is on the third mora from the end. This follows automatically from the metrical analysis, but requires a special stipulation in Fairbanks's account. (b) The rules involved in the metrical analysis do not actually count to four, but rather count to two twice, once in constructing bimoraic feet and once in locating the penultimate foot. Thus they are compatible with the general goal of imposing locality constraints (here, counting limitations) on phonological rules.

As Fairbanks notes, there are exceptions in the data from poetry on which the analysis is based, namely, cases where the poet places a word so that some syllable other than its main-stressed syllable (as the rules would predict) occurs in strong metrical position. These exceptions turn out to be of interest in their own right. The data given in (102) show the basic distribution of regular cases versus exceptions, based on my own counts of the two hundred lines scanned in Fairbanks 1981. The number below each syllable represents the number of times it appeared in metrically strong position in the Fairbanks sample. Above each syllable sequence is placed the foot structure assigned to it by the rules above.

(102) a. **Forms to which Foot Extrametricality is Inapplicable**

```
(× .)          (×)
⌣ ⌣      . . .  – ⌣
71 0           153 0
```

b. **Forms to which Foot Extrametricality is Applicable**

```
(×) ⟨(×   .)⟩    (× .)⟨(× .)⟩    (×) ⟨(×)⟩    (×) ⟨(×)⟩
 ⌣    ⌣   ⌣      ⌣ ⌣   ⌣ ⌣        –    –       ⌣    –
55   26   0      120    30       128   11      33   11

(× .) ⟨(×)⟩   (×)   ⟨(×)⟩     (× .)⟨(×)⟩        (×) ⟨(×   .)⟩
⌣ ⌣    –      –   ⌣   –     ⌣  ⌣ ⌣   –     . . .  –    ⌣   ⌣
230   14      6   0   0     0   90   0            51   27   0
```

The data may be summed up as follows. As already established, the first choice for the metrical placement of a word is determined by its main stress. There is only one second choice available, namely, the syllable that would have borne main stress if foot extrametricality had not applied. A reasonable surmise is that in the dialect of Hindi spoken by the poets, foot extrametricality was optional but preferred. This yields the correct pattern of frequencies in the data of (102b), and moreover correctly predicts the absence of variation in the forms of (102a), where foot extrametricality is inapplicable. More generally, this account correctly predicts that syllables that are not foot heads can never occur in strong metrical position.

The analysis given above also can serve as the basis of an account of cross-linguistic and cross-dialectal variation. For Hindi, Fairbanks notes that several published descriptions of Hindi stress assign the main stress to the "second choice" syllables marked in (102b). The various authors are almost unanimous, however, in assigning penultimate stress to the word shapes of (102a). Going beyond Hindi to related languages, it can be noted that the Maithili system (§ 6.1.6) is of very much the same character: there is the same foot pattern, but with the extrametricality rule rendered categorical and limited to a particular clash environment. Similar stress systems also occur in Awadhi (§ 6.1.9) and Bhojpuri (§ 7.1.2).

A theoretical result of this analysis is further support for the moraic trochee. Were we to employ the uneven trochee (§ 4.3.1) instead, then final /– ⌣/ would constitute a foot. It thus would be marked extrametrical, and we would wrongly derive antepenultimate or preantepenultimate stress in words ending in /– ⌣/.

6.1.8 Lenakel

Lenakel is an Austronesian language spoken on the island of Tanna in Vanuatu. Its intricate phonology is described and analyzed in Lynch 1974, 1977, 1978. The Lenakel stress system has been analyzed metrically by Hammond 1986, 1990b; HV; and Blevins 1990. Hammond has made an interesting argument on the basis of Lenakel data for a revised Obligatory Branching Parameter (§ 4.3.1); one of my interests here is in motivating a plausible alternative account, since I have proposed to dispense with this parameter.

6.1.8.1 *Data*

Stress in Lenakel is assigned fairly late in the phonological derivation, following rules of Glide Formation and Epenthesis. The transcriptions below reflect the application of these rules, but are otherwise phonemic.

Syllable quantity in Lenakel is based solely on vowel length, so CVː(C) is heavy, while CV and CVC are light. In this scheme, glides count as consonants; that is, CVG syllables are light. Long vowels are fairly rare, and are almost entirely limited to stem- and suffix-final syllables. Their surface phonetic status is unclear. Lynch notes that the vowels he phonemicizes as long resist laxing rules that apply to short vowels, and unlike short vowels, long vowels can participate in stress clashes: *nɨkìːnílar* 'their (pl.) hearts' L 1974, 198. Both factors strongly suggest underlying vowel length, though Lynch notes (1974, 200; 1978, 14) that phonetically there is no clear length difference. One possibility is that long vowels, which are always stressed, are phonetically indistinguishable from short vowels that have been phonetically lengthened in stressed position.

Words containing only light syllables show the pattern given in (103). Main stress is normally penultimate. There are a few cases of antepenultimate stress, analyzed by Blevins 1990 with syllable extrametricality; and words with final stress are discussed below. Secondary stress respects a distinction between nouns and verbs/adjectives. In nouns, it falls every other syllable to the left of the main stress (going from right to left):

(103)	a.	/◡́/	*nám*	'fish'	L 1974, 182
	b.	/◡́ ◡/	*nápuk*	'song'	L 1978, 19
	c.	/◡ ◡́ ◡/	*tɨkómkom*	'branches'	L 1978, 13
	d.	/◡̀ ◡ ◡́ ◡/	*nèluyáŋyaŋ*	'twig'	L 1978, 19
	e.	/◡ ◡̀ ◡ ◡́ ◡/	*tupʷàlukáluk*	'lungs'	L 1978, 19
	f.	/◡̀ ◡ ◡̀ ◡ ◡́ ◡/	*lètupʷàlukáluk*	'in the lungs'	L 1974, 183

In adjectives and verbs, secondary stress falls on the initial syllable and every following syllable going from left to right, except that secondary stress may never occur on a light syllable immediately before the main stress. This avoidance of clash can be seen in (104b, d, f).

(104)	a.	/◡́ ◡/	*éheŋ* 'to blow the nose'	L 1978, 16
	b.	/◡ ◡́ ◡/	*rɨmáwŋɨn* 'he ate'	L 1978, 19
	c.	/◡̀ ◡ ◡́ ◡/	*rɨ̀molkéykey* 'he liked it'	L 1978, 19
	d.	/◡̀ ◡ ◡ ◡́ ◡/	*nɨ̀marolkéykey* 'you (pl.) liked it'	L 1978, 19
	e.	/◡̀ ◡ ◡̀ ◡ ◡́ ◡/	*nɨ̀mamàrolkéykey* 'you (pl.) were liking it'	L 1978, 19
	f.	/◡̀ ◡ ◡̀ ◡ ◡ ◡́ ◡/	*tɨ̀nakàmarolkéykey* 'you (pl.) will be liking it'	L 1978, 19
	g.	/◡̀ ◡ ◡̀ ◡ ◡̀ ◡ ◡́ ◡/	*nàtyakàmetwàtamnɨ́mʷan* 'why am I about to be shaking?'	L 1974, 184

Comparing (103) with (104), we see that because of the difference in direction of assignment, nouns are stressed differently from adjectives and verbs when there are five (or, in principle, seven) syllables.

6.1.8.2 *Analysis*

The pattern just described can be derived by the rules in (105):

(105) a. **Foot Construction**
   i. Form a moraic trochee at the right edge of the word.
   ii. Verbs and adjectives: form moraic trochees from left to right.
   iii. Nouns: form moraic trochees from right to left.
   Degenerate feet are permitted in strong position.

b. **Word Layer Construction** End Rule Right

Degenerate feet must be permitted in strong position, because Lenakel freely tolerates light-syllable words, such as *nú* 'yam', *rhó* 'he hit it' L 1978, 10.

In (106) are schematic derivations for long nouns and verbs of even and odd length:

(106) a. **Nouns**

```
            (× .)              (× .)       Foot Construction:
 ˘   ˘ ˘ ˘ ˘       ˘ ˘   ˘ ˘ ˘ ˘         right edge
 tupʷalukaluk      letupʷalukaluk

 (×)  (× .)(× .)   (× .) (× .)(× .)       Foot Construction
 ˘   ˘ ˘ ˘ ˘       ˘ ˘   ˘ ˘ ˘ ˘         right to left
 tupʷalukaluk      letupʷalukaluk

 (        ×  )     (         ×  )        Word Layer Construction
      (× .)(× .)   (× .) (× .)(× .)       Loss of degenerate feet
 ˘   ˘ ˘ ˘ ˘       ˘ ˘   ˘ ˘ ˘ ˘         in weak position
 tupʷàlukáluk      lètupʷàlukáluk
```

b. **Verbs**

```
                (×   .)                         (×   .)   Foot Construction:
 ˘ ˘ ˘ ˘ ˘ ˘ ˘       ˘ ˘ ˘ ˘ ˘˘ ˘ ˘                  right edge
 tɨnakamarolkeykey   natyakametwatamnɨmʷan

 (× .)(× .)(×)(× .)  (× .)(× .) (× .) (× .)          Foot Construction:
 ˘ ˘ ˘ ˘ ˘ ˘ ˘       ˘ ˘ ˘ ˘ ˘˘ ˘ ˘                  left to right
 tɨnakamarolkeykey   natyakametwatamnɨmʷan

 (            ×  )   (                    ×  )        Word Layer
 (× .)(× .)  (× .)   (× .)(× .) (× .) (× .)             Construction,
 ˘ ˘ ˘ ˘ ˘ ˘ ˘       ˘ ˘ ˘ ˘ ˘˘ ˘ ˘                  Loss of degenerate
 tɨ̀nakàmarolkéykey   nàtyakàmetwàtamnɨ́mʷan               feet in weak
                                                        position
```

These derivations are approximate, in that below I suggest a destressing rule that (outside of its crucial function) has the vacuous effect of removing the degenerate feet in these forms, before the general convention eliminating such feet could affect them.

The trochees of Lenakel must be moraic, not syllabic. This can be seen from (107), in which an underlying long vowel interrupts the right-to-left count:

```
(107) (          ×  )
      (×   .) (×)(× .)
      ˘   ˘   –  ˘ ˘
      nɨ̀mʷansìːnílar        'their (pl.) buttocks'        L 1974, 198
```

In (108), the Priority Clause (§ 5.1.6) ensures that the final syllable is skipped over rather than attracting a degenerate foot:

(108) (   )
  (×)
  ˘ – ˘
  nɨkí:tar 'our (incl.pl.) hearts' L 1974, 197

It is worth noting that although Lenakel has three foot construction rules, all three form moraic trochees. That is, as HV suggest, foot type seems to be a property of the metrical layer rather than of individual rules.

6.1.8.3 *Finally Stressed Words and Their Junctural Properties*

The central perplexity of Lenakel stress concerns the fairly small set of words that are stressed finally. Of these, a large number are verbs ending in the transitive suffix *-in,* for example *ɨŋn-ín* 'be afraid of it', *rhìnat-ín* 'he knew it' L 1974, 190.

An adequate analysis of such forms must take into account the JUNCTURAL properties of *-in.* As Lynch argues, *-in* does not fully cohere phonologically with the preceding stem. In particular, a rule of Glide Formation that normally applies between morphemes is blocked across the juncture preceding *-in: aru-in* 'to throw', not **arwin* (Lynch 1974, 188). However, certain other rules that are blocked across word boundary do apply before *-in.* These include a low-level palatalization rule ($t \rightarrow t^j$ / ___ *i*) and a rule of medial stop voicing, cf. /rhinat-in/ → [r̥ìnadʲín]. Because of the behavior of these rules, we cannot treat *-in* as separate word.

In Lynch's view, *-in* is a "postclitic": it is separated from the preceding stem by a juncture that is weaker than word boundary but stronger than morpheme boundary. In his *SPE*-style analysis, Lynch posits that a /=/ boundary, intermediate in strength between # and +, separates *-in* from the preceding stem. The = boundary plays a dual role: it accounts for the junctural effects, and (given Lynch's formulation of the stress rules) it induces final stress.

The analysis I propose basically follows Lynch in this respect. However, I translate his account of juncture into a more recent framework. In the theory of the Prosodic Hierarchy (Selkirk 1980a; Nespor and Vogel 1986), words and phrases may be parsed into prosodic constituents that form domains of rule application. This theory has been extended to word-internal contexts by Booij and Rubach 1984; Cohn 1989; and Inkelas 1989. Adapting these proposals to Lenakel, I assume two prosodic layers within the word, which following Inkelas I call α and β. Thus the representation in (109) indicates that the word consists of one loosely cohesive domain β, divided into two tightly cohesive domains α.

(109) $[[\text{aru}]_\alpha[\text{in}]_\alpha]_\beta$

The morpheme *-in* is constrained to begin its own α-domain. Rules such as Glide Formation are confined to α-domains; other rules such as $t \rightarrow t^j / \_\_ i$ apply within β-domains. The representation in (109) translates Lynch's *#aru=in#*.

Forms with *-in* are not the only words that contain more than one α-domain, though they are the most common. Other examples may be found in compounds of a relatively frozen character:

(110)

$[[\text{nɨm}^{w}\text{a}]_{\alpha}[\text{mɨ́v}]_{\alpha}]_{\beta}$[5]	'cover-*mɨv* (kind of heliconia)' = '*mɨv* leaf'	L 1977, 79
$[[\text{nɨ̀m}^{w}\text{a}]_{\alpha}[\text{nár}]_{\alpha}]_{\beta}$	'cover-thing' = 'leaf used as container' (*nɨm^w a* is a bound morpheme.)	L 1977, 79; L 1974, 223
$[[\text{nɨmop}]_{\alpha}[\text{tɨ́n}]_{\alpha}]_{\beta}$	'(bound form)-land' = 'earth'	L 1977, 78
$[[\text{nɨp}^{w}\text{aŋnhaŋ-ɨ}]_{\alpha}[\text{tɨ́n}]_{\alpha}]_{\beta}$	'nose-land' = 'riverbank'	L 1977, 81
$[[\text{nɨviŋ-ɨ}]_{\alpha}[\text{nár}]_{\alpha}]_{\beta}$	'outside, cover-thing' = 'cup'	L 1977, 83
$[[\text{tɨn}]_{\alpha}[\text{múrh}]_{\alpha}]_{\beta}$	'land-(bound form)' = 'neighboring island'	L 1977, 105
$[[\text{twi}]_{\alpha}[\text{tɨ́n}]_{\alpha}]_{\beta}$	'top-land' = 'the world'	L 1977, 106
$[[\text{nelɨm}]_{\alpha}[\text{núw}]_{\alpha}]_{\beta}$	'arm-yam' = 'long yam'	L 1977, 71

Further, in Lynch's (1974) analysis, various reduplicated forms that bear final stress (not all do) are treated as pseudo-compounds, even in cases where the reduplicated sequence does not occur as an isolation form (Lynch 1974, 216):

(111)

$[[\text{m}^{w}\text{àtik}]_{\alpha}[\text{tík}]_{\alpha}]_{\beta}$	'Mauitikitiki'	(**tik*)
$[[\text{àriŋ}]_{\alpha}[\text{rɨ́ŋ}]_{\alpha}]_{\beta}$	'to wash'	(**rɨŋ*)
$[[\text{àkɨl}]_{\alpha}[\text{kɨ́l}]_{\alpha}]_{\beta}$	'to be thin'	(*kɨl* occurs, but means 'flying fox')

Consider now the accentual behavior of words containing a separate final monosyllabic α-domain. (Data are available only for *-in* forms, but Lynch's rules imply that the other cases would behave the same way.) These words exhibit final stress. Further, the forms in this class that have six or eight syllables show an additional peculiarity in their secondary stress pattern, which here will be called **aberrant antepenultimate stress.** This aberrant stress can be seen in the six- and eight-syllable forms of (112e, g) (all forms in (112) from Lynch 1974, 190). Since these are verb forms, we expect alternating secondary stress from left to right. But the rightmost secondary stress shows up in antepenultimate position, rather than the expected four from the end.

5. Secondary stresses are not marked in Lynch 1977.

(112) a. /◡ ◡́/ *ɨŋnín* 'be afraid of it'
b. /◡̀ ◡ ◡́/ *rhìnatín* 'he knew it'
c. /◡̀ ◡ ◡ ◡́/ *rɨ̀mhinatín* 'he knew it'
d. /◡̀ ◡ ◡̀ ◡ ◡́/ *kɨ̀marhìnatín* 'they (pl.) knew it'
e. /◡̀ ◡ ◡ ◡̀ ◡ ◡́/ *kɨ̀mamarhìnatín* 'they (pl.) were aware of it'
f. /◡̀ ◡ ◡̀ ◡ ◡̀ ◡ ◡́/ *tɨ̀repàwkɨrànɨm*^{w}*ín* 'he'll subsequently drown it'
g. /◡̀ ◡ ◡̀ ◡ ◡ ◡̀ ◡ ◡́/ *nàtyepàyukɨrànɨm*^{w}*ín* 'we (excl. pl.) will be ready to drown it'

The situation is actually even more puzzling than this: given the behavior of eight- and six-syllable forms, we would also expect aberrant antepenultimate stress in four-syllable forms (i.e. /◡ ◡̀ ◡ ◡́/), but do not get it: the actual outcome is /◡̀ ◡ ◡ ◡́/, as in (112c).

An analysis of these facts can be developed from an idea advanced tentatively by Lynch 1974, 224: the material preceding *-in* is stressed almost exactly like a word in isolation, abstracting away from the primary–secondary distinction. To integrate this idea into the analysis, and relate it to the junctural facts already described, I make two basic assumptions: (a) the stress rules apply within α-domains; (b) words consisting of two α-domains have an additional metrical layer constructed within β-domains. This constituent is labeled by End Rule Right.

Under this account, aberrant antepenultimate stress is simply the normal PENULTIMATE stress, as assigned to the nonfinal α-domain. The derivation in (113) illustrates this:

(113)

```
(               ×  )    (×)       Stressing of α-domains
(×  .)(×  .)(×)(×  .)   (×)
 ◡   ◡   ◡   ◡  ◡  ◡     ◡
[[natyepayukɨranɨm^w]α[in]α]β

(                        ×)       End Rule Right (β-domain)
(               ×  )    (×)
(×  .)(×  .)(×)(×  .)   (×)
 ◡   ◡   ◡   ◡  ◡  ◡     ◡
[[natyepayukɨranɨm^w]α[in]α]β

(                        ×)       Loss of degenerate feet in weak
(               ×  )    (×)       position
(×  .)(×  .)   (×  .)   (×)
 ◡   ◡   ◡   ◡  ◡  ◡     ◡
[[natyepayukɨranɨm^w]α[in]α]β
```

To complete the analysis, the lack of a distinction between ternary and secondary stress on the surface must be explained. Here, we can appeal to a rule of Conflation (§ 5.5.2), eliminating the second row of the grid.

(114)
```
(                         ×)      Conflation
(×  .)(× .)  (× . )      (×)
˘   ˘ ˘ ˘ ˘  ˘  ˘         ˘
```
[[nàtyepàyukɨrànɨm^{w}]$_{\alpha}$[ín]$_{\alpha}$]$_{\beta}$

Alternatively, we might suppose that the structural distinction between secondary and tertiary stress is not phonetically realized.

What remains now is to account for the ABSENCE of aberrant antepenultimate stress in quadrisyllabic forms, noted above. This can be done with a rule of Move X (§ 3.4.2), shifting secondary stress leftward. The language-specific conditions on this rule are (a) it moves stress only onto initial syllables; (b) it moves stress only the distance of one syllable (cf. a similar distance restriction in Tiberian Hebrew; McCarthy 1979b). These conditions limit the rule to quadrisyllabic forms:

(115) a. **Move X** Move a second-layer grid mark leftward.
Distance limit: one syllable
Target must be initial.

b.
```
(            ×)          (            ×)
(    ×   )  (×)          (×        )  (×)
(×)  (× .)  (×)    →     (×)  (× .)   (×)
 ˘    ˘ ˘    ˘            ˘    ˘ ˘     ˘
```
[[rɨmhinat]$_{\alpha}$[in]$_{\beta}$ → [[rɨmhinat]$_{\alpha}$[in]$_{\beta}$

Following Move X, stresses are removed from light syllables in clashing positions by (116). Assuming that persistent footing then applies (§ 5.4.1), we derive the correct output form:

(116) a. **Destressing in Clash** × → Ø / ___ in clash; default direction is
˘ right to left

b.
```
(            ×)      Destressing
(×       )  (×)
(×)         (×)
 ˘    ˘ ˘    ˘
```
[[rɨmhinat]$_{\alpha}$[in]$_{\alpha}$]$_{\beta}$

```
(            ×)      Persistent footing
(×       )  (×)
(×   .)     (×)
 ˘    ˘ ˘    ˘
```
[[rɨmhinat]$_{\alpha}$[in]$_{\alpha}$]$_{\beta}$

```
(            ×)      Conflation
(×   .)     (×)
 ˘    ˘ ˘    ˘
```
[[rɨ̀mhinat]$_{\alpha}$[ín]$_{\alpha}$]$_{\beta}$

As with other destressing rules (§ 3.4.2), the structural description is simplified by relying on the Continuous Column Constraint to rule out destressing of stronger stresses by weaker ones. Thus in (116b), Destressing must apply left to right, since right-to-left destressing would violate the Continuous Column Constraint. The default right-to-left direction is imposed in order to account for the clashing forms in (106), which under a Destressing analysis must lose their degenerate feet by the Destressing rule, rather than by the general convention of § 5.1.1.

My use of destressing to account for the quadrisyllabic forms basically follows Hammond 1986, 215. The proposed analysis differs from Hammond's in decomposing the destressing process into two rules; this enables the analysis to express the rules as local operations, rather than having a single destressing rule that refers to the entire four-syllable string.

To sum up so far: I have suggested that the puzzling aberrant antepenultimate stresses in Lenakel do not reflect some special rule of antepenultimate stress; rather, they result from the fact that the words in question consist of two phonological domains, which are stressed separately. The analysis is independently motivated by the junctural properties of the relevant words, as diagnosed by segmental phonology. Following Hammond, I have adopted a destressing analysis to explain the absence of the phenomenon in four-syllable words.

6.1.8.4 *Long Vowels*

The remaining words in Lenakel that receive final stress are those with long vowels in their final syllables. These attract final stress, as one might expect. Oddly, this is not straightforwardly attributed to their length, because just like pseudo-compound structures, forms ending in long vowels exhibit the aberrant antepenultimate stress syndrome. The data in (117) are from Lynch 1974, 196:

(117) a.	/⏑ –́/	*rɨmá:mh*	'he saw it'
b.	/⏑̀ ⏑ –́/	*rɨ̀mamá:mh*	'he was looking at it'
c.	/⏑̀ ⏑ ⏑ –́/	*tɨ̀nakamwá:mh*	'you two will see it'
d.	/⏑̀ ⏑ ⏑̀ ⏑ –́/	*tɨ̀nakàmwapké:n*	'you two will be jealous'
e.	/⏑̀ ⏑ ⏑ ⏑̀ ⏑ –́/	*nàtyaka**mày**pɨké:n*	'we (excl. pl.) are going to be jealous'
f.	/⏑̀ ⏑ ⏑̀ ⏑ ⏑̀ ⏑ –́/	*tyèsamàrolkèykeyá:n*	'we (excl. pl.) won't like it'
g.	/⏑̀ ⏑ ⏑̀ ⏑ ⏑ ⏑̀ ⏑ –́/	*nàtyesàmarol**kèy**keyá:n*	'we (excl. pl.) aren't about to be liking it'

The pattern here is exactly the same as with *-in* (112).

My suggestion here follows an idea proposed for Estonian by Prince 1980; namely, that the aberrantly stressed words are divided into more than one phonological domain. That is, final long-voweled syllables are split off as in (118):

(118) *natyakamaypɨkeːn* → $[[\text{natyakamaypɨ}]_\alpha[\text{keːn}]_\alpha]_\beta$

Assuming such a structure, the rules already given will derive the correct results.

Such pseudo-morphological structures arguably are present in other languages. For example, there is no morphological support for regarding such English words as *gobbledy-gook, helter-skelter,* or *bibbity-bobbity-boo* as consisting of more than one phonological word, but native intuition seems to support such a view, as does the phonology: these words have stress patterns that resemble those of compounds or phrases (Hayes 1982b, 264), and some contain tense short [i] (a vowel normally confined to word-final position) just before the "boundary" spelled with a hyphen. For Estonian, Prince 1980, following Hint 1973, suggests that the borrowed words that bear non-initial stress and/or overlong quantity are integrated into the native vocabulary by treating them as pseudo-compuonds, on analogy with native forms. Under this analysis, the normal initial stress rule and quantitative patterns of Estonian are greatly regularized. A similar analysis has also been proposed for Icelandic (§ 6.2.2.6).

The Lenakel situation is analogous to the Estonian and Icelandic cases: given the presence of many finally stressed words that have compound structure, it is plausible that the remaining cases would similarly be assigned compound status. In essence, this is a more formal statement of a traditional "analogical" account: words with long vowels in final syllables are assigned secondary stress by analogy with forms containing an internal juncture.

The plausibility of this account is increased by two parallels between the clitic *-in* and final long vowels. First, long-voweled suffixes pattern like *-in* in that they fail to condition Glide Formation: note the prevocalic high vowels in *rɨ̀s**ɨni**-áːn* 'he didn't say', *nàr**u**-áːn* 'the making of grass skirts' (Lynch 1974, 162). This follows if the long vowels are parsed with the same structure given to *-in:* $[[\text{rɨsɨni}]_\alpha[\text{aːn}]_\alpha]_\beta$, $[[\text{naru}]_\alpha[\text{aːn}]_\alpha]_\beta$.

Second, Lynch formulates two rules of /a/ Raising: one raises /a/ to [e] before *-in,* the other raises /a/ to [e] before the long-voweled suffixes. The examples in (119) illustrate this.

(119) a. $[[\text{r-am-aklha}]_\alpha[\text{in}]_\alpha]_\beta$ → $[[\text{ràmaklh}\mathbf{e}]_\alpha[\text{ín}]_\alpha]_\beta$ 'he is stealing it' Lynch 1974, 168

b. $[[\text{n-aklha}]_\alpha[\text{aːn}]_\alpha]_\beta$ → $[[\text{nàklh}\mathbf{e}]_\alpha[\text{áːn}]_\alpha]_\beta$ 'theft' Lynch 1974, 161

Although more data would be needed to check the claim thoroughly, it would appear that these rules can be collapsed: that is, /a/ is raised to [e] in the environment/ $\_\_\,]_\alpha[_\alpha V$; that is, prevocalically across an α-juncture. Again, the facts suggest that final syllables containing long vowels are treated as separate domains.

Finally, it should be noted that there are words in Lenakel with underlying long vowels in consecutive syllables. These obey an alternating pattern whereby only odd-numbered long vowels survive to influence the stress contour. This is illustrated in (120) (Lynch 1974, 194–95):

(120) a. /r-ɨm-ety**aː**w/ → *rɨ̀mety**áː**w* 'he arrived'
b. /r-ɨs-ety**aː**w-**aː**n/ → *rɨsety**aː**wan* → *rɨ̀sety**áː**wan* 'he didn't arrive'
c. /r-ɨs-ety**aː**w-hy**aː**i̯-**aː**n/ → *rɨsety**aː**why**a**i̯-**a**ːn* → *rɨ̀sety**à**ːwhy**a**i̯**á**ːn*
'he didn't arrive in the north'

I follow here the analysis proposed by Lynch 1974, 194–95: long vowels that follow a long vowel in the preceding syllable are shortened. This rule can be applied either right to left, as Lynch suggests, or alternatively it can be cyclic, since all long vowels appear to be stem- or suffix-final.

6.1.8.5 *Earlier Analyses*

I take the central problem of Lenakel accentology to be that of accounting for aberrant antepenultimate stress. To conclude, I consider earlier approaches to this problem.[6]

Lynch's (1974) analysis of Lenakel stress, expressed in *SPE* formalism, is notably simple and concise, given the complexity of the data. Lynch's approach to aberrant antepenultimate stress is to posit a special Antepenultimate Stress rule, which applies to words bearing final stress. My efforts have been directed toward eliminating this special rule, based on taking seriously Lynch's suggestion that the material preceding a final stressed syllable itself resembles a normally accented word. The advantage of this approach is that it relates the antepenultimate stress pattern to the aberrant JUNCTURAL properties of the words involved, rather than just describing it on an ad hoc basis.

Hammond (1986) has addressed the aberrant antepenultimate stress puzzle within a metrical framework.[7] His solution is to expand the universal foot inventory to allow "Revised Obligatory Branching" feet. For Lenakel these take the form of (121):

(121) Form $\overset{(\times\ .)}{-\ \sigma}$ or $\overset{(\times)}{-}$

Applied from left to right, these feet account for the vowel shortening paradigm shown in (120). Since the surface difference between long and short vowels is

6. HV also analyze Lenakel but do not address the antepenultimate stress facts.

7. See also Hammond 1990b for a somewhat different version of the same analysis, based on different general theoretical assumptions; the remarks here would mostly apply to both analyses.

subtle, stress-based and length-based accounts give essentially equivalent results here.

Hammond's other rules are roughly along the lines of (105) above. Under his analysis, aberrant antepenultimate stresses are assigned by the equivalent of (105a.i)—that is, the rule that in simple words assigns the MAIN stress. The reason is that where there is a final long vowel, an Obligatory Branching foot has already been formed on the final syllable. This is illustrated by the derivation (122) for (117e):

(122) a. (×) Foot Construction: Obligatory Branching Feet
˘ ˘ ˘ ˘ ˘ –
natyakamaypɨke:n

b. (× .)(×) Foot Construction: single foot, right to left
˘ ˘ ˘ ˘ ˘ –
natyakamaypɨke:n

c. (× .)(×)(× .)(×) Foot Construction: iterative, left to right
˘ ˘ ˘ ˘ ˘ –
natyakamaypɨke:n

With destressing, this gives the correct surface form *nàtyaka**màyp**ɨké:n*. Note that the non-iterative rule applying at stage (122b) applies to the RIGHTMOST UNFOOTED SYLLABLE SEQUENCE; hence in short-voweled words that lack Obligatory Branching feet, this rule also serves to assign the penultimate main stress.

Hammond gives a similar account for the suffix *-in:* it is lexically footed, so words in *-in* are derived much as in the last two steps of (122). To account for the absence of aberrant antepenultimate stress in quadrisyllables, he posits a destressing rule, analogous in its effects to (115)–(116).

At this point I will summarize what might be taken as the main defects of both metrical proposals.

Hammond's account fails to link the accentual behavior of *-in* to its aberrant junctural behavior; that is, because *-in* is assumed to be lexically footed, there is no reason to expect that this suffix should also exhibit aberrant junctural effects as well, as outlined above. In the proposal made here, the stress pattern and the junctural effects have exactly the same origin.

Another defect of Hammond's proposal is that it makes it an accident that lexical footing should in almost all cases be limited to monosyllabic morphemes, such as the postclitic *-in* or the second members of compounds, as in (110). The analysis proposed here need not assume such a coincidence, since it is based on juncture rather than lexical feet. Cross-linguistically, clitic and compound structures are precisely where strong junctures are ordinarily found.

The last objection is entirely theoretical in character: Hammond's account requires the introduction of Revised Obligatory Branching into the theory. I believe this is not needed for other languages, and elsewhere provide or cite alternative analyses for Hammond's other cases; see § 7.2 for Khalkha Mongolian, § 6.3.10 for Turkish, and § 7.1.4 for Klamath. Moreover, if feet of the form (121) are admitted into the theory, the grounding of foot theory in the Iambic/Trochaic Law will no longer go through; there is no reason why the law would favor (121) as a foot template.

It is worth noting further that a slight modification of Hammond's analysis brings it much closer to the confines of the general theory proposed here. If we introduce Lynch's vowel shortening rule into Hammond's account, then the Obligatory Branching parameter can be simplified to (123), a rule type first proposed in Prince 1983a:

(123) Form a moraic trochee over a heavy syllable.

Hammond's analysis can then be retained, otherwise unchanged. The change in the general theory allows Obligatory Branching but in the much less ambitious form "stress heavy syllables." Such a move would extend the theory by adding a new foot construction algorithm, but it does not expand the actual set of metrical structures that can be created. Note that if this move is correct, it provides support for HV's view (cf. § 4.3.2) that the basic mechanism by which stress rules refer to quantity is "stress a heavy syllable."

The proposal made here goes out on a limb in its liberal assignment of juncture: that is, in positing a rule that places $]_\alpha[_\alpha$ junctures before every final long-voweled syllable. The main justification made for this is that final long vowels are treated by analogy with the cases where the juncture is morphologically motivated. But this is clearly taking out a loan on our meager understanding of the phenomenon of morphologically unmotivated juncture.

6.1.9 Other Moraic Trochee Systems

(A) PAAMESE (Austronesian, Vanuatu; Crowley 1982) normally has antepenultimate stress but with preantepenultimate stress in exceptional forms. It can be treated with an analysis similar to that proposed above for Hindi (§ 6.1.7):

(124) a. **Syllable Extrametricality** $\sigma \rightarrow \langle\sigma\rangle$ / ___ $]_{word}$
b. **Foot Construction** Form moraic trochees from right to left. Degenerate feet are forbidden absolutely.
c. **Foot Extrametricality** $F \rightarrow \langle F\rangle$ / ___ $]_{word}$
d. **Word Layer Construction** End Rule Right

In regular words (e.g. (125a)), the extrametricality of the final syllable means that the rightmost foot is not peripheral, hence Foot Extrametricality cannot apply, and we get antepenultimate stress. But in forms that are lexical exceptions to Syllable Extrametricality (e.g. (125b)), the rightmost foot will be peripheral, so we get preantepenultimate stress.

(125)	a.	b.	c.	d.
	(×)	(×)	(×)	(×)
	(× .)	(× .)⟨(× .)⟩	(× .)	(× .)⟨(× .)⟩
	ᴗ ᴗ ᴗ⟨ᴗ⟩	ᴗ ᴗ ᴗ ᴗ	ᴗ ᴗ ᴗ	ᴗ ᴗ ᴗ ᴗ
	visókono	*ná-tahosi*	*tahósi*	*mátu-vaa*
	'morning'	'I am good'	'he is good'	'we (paucal excl.) went'
	C 29	C 29	C 29	C 29

Where there are not enough syllables to form two feet, exceptional stems like /tahosi/ 'good' display penultimate stress (125c). This is as expected, since the single syllable /ta/ could form only a degenerate foot, and these are banned absolutely in Paamese. The ban is supported by the absence of (C)V content words; see Crowley 1982, 24, 36, 51, for discussion.

Some indication that this approach is correct can be seen in the behavior of disyllabic stems, as in (125d): when preceded by two syllables, these regularly induce preantepenultimate stress. This can be accounted for if we make the plausible assumption that stems must contain at least one foot; cf. similar requirements in St. Lawrence Island Yupik (§ 6.3.8.1), Winnebago (§ 8.9.5), and English (Liberman and Prince 1977, 306).

Paamese is treated by Goldsmith 1990 as an exception to the Peripherality Condition (§ 3.11): in his account, antepenultimate vowels, and only such vowels, may be exceptionally extrametrical. The above analysis would permit the original Peripherality Condition to be retained.

The treatment of Paamese as involving moraic rather than syllabic trochees is tenative, and hinges on the analysis (discussed by Crowley and by Goldsmith) of surface glides, long vowels, and diphthongs in this language.

(B) AWADHI (Indo-Aryan, India and Nepal; Saksena 1971) and SARANGANI MANOBO (Austronesian, Philippines; variety described by DuBois 1976). Here the pattern is as follows: words ending in the sequence /. . . ᴗ –/ bear final stress; otherwise, stress falls on the penult. Heavy syllables are CVː and CVC in Awadhi, full-voweled syllables in Sarangani Manobo. In both languages the minimal word is / – /. This can be analyzed as in (126):

(126) a. **Foot Construction** — Form moraic trochees from right to left. Degenerate feet are forbidden.

b. **Syllable Extrametricality** $\sigma \rightarrow \langle\sigma\rangle \ / \ \times \ \_\_\_\_]_{word}$
(i.e., syllables are extrametrical in clash)

c. **Word Layer Construction** End Rule Right

These rules derive the basic pattern as in (127):

(127)

```
a.     × )        b.       ×)       c.     ×  )       d.     ×)
      (× .)              .)(×)            (×)                (×)(×)
   . . .  ˘ ˘ #       . . . ˘  – #      . . .  –  ˘ #      . . .  – ⟨–⟩ #
```

(C) TÜMPISA SHOSHONE (Uto-Aztecan, Eastern California; Dayley 1989a, 1989b) has a stress pattern quite similar to that of Cahuilla (§ 6.1.3), to which it is distantly related. A difference is that words beginning /˘ – . . ./ normally receive second syllable stress, though they can also be stressed initially, as in Cahuilla. It appears that Tümpisa Shoshone normally constructs metrical structure in bottom-up fashion, as in Malayalam (§ 5.1.5), with the top-down pattern of Cahuilla as an additional option. Degenerate words: no /˘/ content words occur, but /˘/ as a function word may be stressed.

(D) EARLY LATIN. A stress pattern much the same as that discussed in § 6.1.2 for Egyptian Radio Arabic (i.e. moraic trochees from left to right, with optional foot extrametricality) is posited for early stages of Latin by Allen 1973, 188–90. Later, Latin shifted to right-to-left stress, as discussed in § 5.1.4.

The theory here offers an answer to a puzzle raised by McCarthy (1979a), who asks why the shift in direction should have been simultaneous with a change in the foot template (in my terms, from moraic to uneven trochees). In the account given here, the template was always the moraic trochee, and only the direction of assignment shifted. A similar directional shift seems to have taken place for Lebanese Arabic ((128e) below), assuming it arose from a Palestinian-like pattern.

(E) NYAWAYGI (Pama-Nyungan, Australia; Dixon 1983) has the same stress pattern as neighboring Wargamay (§ 6.1.4). For the minimal word in Nyawaygi, see § 5.1.9.

(F) SOUTHWEST TANNA (Austronesian, Vanuatu; Lynch 1982) has the same basic stress pattern as neighboring Lenakel (§ 6.1.8).

(G) LATIN-LIKE STRESS. A number of languages, listed in (128), have essentially the stress pattern of Classical Latin, discussed in § 5.1.4. While in the first four cases the resemblance is due to inheritance or borrowing, this is not so for the last two. Many of these languages have additional complexities not treated here; in particular, stress often falls to the right of where it would in Latin. Usually, the location of stress in Latin establishes the leftmost limit of lawful stress placement. For Klamath, the deviation from Latin is quite different; see § 7.1.4. All these languages use a Latin-like quantity distinction, with both CVV and CVC heavy.

(128) **Languages with Latin-like Stress**

a. **Spanish** (Harris 1983, 1992). No minimal word constraint.
b. **Romanian** (Steriade (1984). No minimal word constraint.
c. **English** (Chomsky and Halle 1968). Minimal word = / – /.
d. **German** (Giegerich 1985; Hayes 1986c). Minimal word = / – /.
e. **Lebanese Arabic** (Kenstowicz and Abdul-Karim 1980; Kenstowicz 1981). Minimal word = / – /.
f. **Klamath** (§ 7.1.4). No minimal word constraint.

To the list might be added Bedouin Hijazi Arabic, as described by Al-Mozainy, Bley-Vroman, and McCarthy 1985. However, the Bedouin Hijazi pattern is also analyzable using the left-to-right footing of Palestinian (§ 6.1.1), at least for the available data.

(G) LATIN WITHOUT SYLLABLE EXTRAMETRICALITY. When the foot construction rules of Latin are employed without syllable extrametricality, final stress is derived when the final syllable is heavy, otherwise penultimate stress. We have seen this in Fijian (§ 6.1.5); other instances include Mam (§ 7.1.5) and those listed in (129):

(129) **Languages with Fijian-like Stress**

a. **Ancient Greek**, in the approach taken in Sauzet 1989 and Golston 1989. Their derivations include a considerable tonal component as well, as Ancient Greek was a pitch accent language. For an alternative, purely metrical approach, see Steriade 1988b.
b. **Bergüner-Romansh** (Romance; Kamprath 1987). / – / = CVː, CVC
c. **Diegueño** (Yuman, California and Baja California; Langdon 1970; Couro and Hutcheson 1973). / – / = CVː, CVC. Cases of CV́CəC appear to involve epenthetic [ə]. Minimal word = / – /.
d. **Hawaiian** (Austronesian; Elbert and Pukui 1979). Essentially the same as Fijian (§ 6.1.5), including secondary stress. Minimal word = / – /.
e. **Inga** (Quechuan, Columbia; Levinsohn 1976). / – / = CVC_i, where C_i is [+sonorant] (long vowels do not occur). Secondary stress on alternating syllables before the main stress (i.e. syllabic trochees from right to left). No minimal word constraint.
f. **Kawaiisu** (Uto-Aztecan, Central California; Zigmond, Booth, and Munro 1990). / – / = CVV. Minimal word = / – /, though most words are at least disyllabic.

g. **Manam** (Austronesian; Manam Island, New Guinea; Lichtenberk 1983; Chaski 1985; L. Ito 1989; Halle and Kenstowicz 1991). / – / = CVC, possibly also CVV. / – ⏑ ⏑/ words receive antepenultimate stress, requiring a rule of extrametricality in clash or equivalent. No minimal word constraint.
h. **Tol** (Hokan, Honduras; Fleming and Dennis 1977). / – / = CVC (long vowels do not occur). No minimal word constraint.
i. **Tongan** (Austronesian, Tonga; Churchward 1953; Feldman 1978). / – / = CVV (CVC does not occur). Minimal word = / – /.

Other moraic trochee systems appear elsewhere in this book and in the literature: Tiberian Hebrew (§ 6.3.10), Dutch (§ 7.4), Sentani (§ 8.7), Bani-Hassan Arabic (§ 8.10), and Japanese (Poser 1990 and references cited there).

6.2 SYLLABIC TROCHEES

The syllabic trochee foot template is as in (130):

(130) (× .)
σ σ

Under the hypothesis about degenerate feet made in § 5.1.9, / –́ / also qualifies as a proper syllabic trochee for those syllabic trochee languages that have heavy syllables.

Stress systems based on the syllable trochee tend to be fairly simple. This section presents two of the more interesting and well documented systems, then a cross-linguistic survey.

6.2.1 Auca

The stress pattern of Auca (unclassified, Ecuador) is analyzed in Pike 1964; further information on Auca phonology can be found in Saint and Pike 1962. The Auca pattern is of interest for its bidirectionality, its interaction with morphology, and its treatment of degenerate feet.

In describing the Auca facts it is useful to treat morphologically complex words as consisting of a **stem** plus a suffix sequence. The stem may itself be complex, consisting of one or more roots. The phonological justification for this is that a different alternating stress pattern is found in each domain. Within the stem, there is stress on odd-numbered syllables, counting from left to right; whereas in the suffix sequence there is stress on the penult and (with a complication to be discussed later) every other syllable before it going from right to left. These two "stress trains" collide in the middle of a word, with various adjustments taking place.

To illustrate this, (131) lists all the available cases of stem and suffix trains, organized by what kind of collision, if any, takes place in the middle of a word.

In the examples, /]/ is used to mark the right edge of the stem, nasality is transcribed with / ̨ /, and stop consonants are written phonemically as in Saint and Pike 1962.

(131) a. **No Suffixes**

σ́]	*bǫ́*	'cotton bird'	SP 6
	mǫ́	'leaf'	SP 6
σ́ σ]	*bádą*	'mother'	P 426
σ́ σ σ́]	*mǫ́įkó*	'blanket'	SP 24
σ́ σ σ́ σ]	*bódæpóka*	'anthill'	P 426

b. **Trains are in Phase**

i. **Both Trains Even in Length**

σ́ σ] σ́ σ	*kǽga] ką́ba*	'his tooth hurts'	P 427
σ́ σ σ́ σ] σ́ σ	*pǽdæpǫ́nǫ] ɲą́ba*	'he handed it over'	P 426

ii. **Both Trains Odd in Length**

σ́] σ	*gó] bo*	'I go'	P 425
σ́] σ σ́ σ	*gó] tabópa*	'I went (declar.)'	P 426
σ́ σ σ́] σ	*kíwęɲǫ́] ŋa*	'where he lives'	P 426
σ́ σ σ́] σ σ́ σ	*ápæ̨né] kądápa*	'he speaks'	P 428

c. **Trains Produce a Clash**

i. **Odd Stem Train + Even Suffix Train**

σ́] σ́ σ	*gó] bópa*	'I go (declar.)'	P 425
σ́ σ σ́] σ́ σ	*yíwæ̨mǫ́] ŋą́ba*	'he carves; he writes'	P 427
σ́] σ́ σ σ́ σ	*gó] támǫnápa*	'we two went'	P 426
σ́ σ σ́ σ σ́] σ́ σ	*tíkawódǫnó] ką́ba*	'he lights'	P 427

ii. **Even Stem Train + Odd Suffix Train**

σ́ σ́] σ	*wódǫ́] ŋą*	'she hangs up'	P 427
σ́ σ́] σ σ́ σ	*ę́ɲá] kądápa*	'he was born'	P 427
σ́ σ σ́ σ́] σ	*gą́næǽ̨mǽ̨] ŋą*	'he raised up his arms'	P 427

The alternating pattern is stress-first from left to right, stressless-first from right to left, suggesting syllabic trochees. When both stem and suffix trains are odd, a foot must span the boundary between the two (131b.ii). Where stem and suffix train are odd and even respectively, then the output is clashing stresses straddling the boundary (131c.i); and where stem and suffix trains are even and odd respectively, clashing stresses appear before the boundary (131c.ii).

Pike does not designate any particular stress as the strongest, but an argument can be made from his description of intonation (p. 430): "The most frequent [intonational pattern] has high pitch on stressed syllables, with stressed syllables of the suffix train cascading a bit lower in a contour 'fade'." Cross-linguistically, one of the strongest diagnostics for main stress is that it serves as the principal turning point for intonation contours (§ 2.3.1). This suggests that we analyze the contour with an initial H boundary tone, a H* pitch accent, and a final L boundary tone (see Bruce 1977; Pierrehumbert 1980; and chap. 2). The H* pitch accent must be associated to the rightmost stress assigned within the stem domain, as in (132):

```
(132) [  [σ̀ σ σ̀ σ σ́]stem σ̀ σ ]
      |           |        |
      H           H*       L
```

In this respect it closely resembles a contour of English, where the position of the main stress is uncontroversial:

(133)
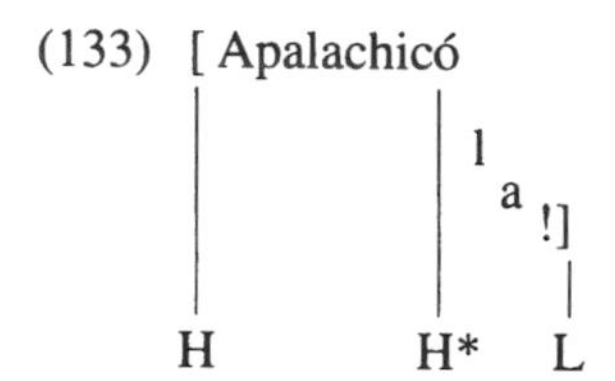

In this account, the pitch peaks that occur on secondary stresses would be regarded as phonetic perturbations due to stress; that is, the tonal specification describes only the main outlines of the contour. The crucial point is that the rightmost stem stress serves as the docking site for H*, which suggests that it is the metrically strongest of the word.[8]

How should the two stress trains be described in a formal analysis? First, there is good evidence that the stem stress train must be assigned before the suffix train. Pike notes (p. 430) that vowel sequences are treated differently in the two. Within the stem, each vowel counts as a separate syllable, even where two identical vowels occur in a row: *wòǫǫ́] ŋą̨dàpa* 'he blew his blow gun' (p. 430). But in the suffix train, identical vowels are collapsed together into a single vowel (134a), and adjacent distinct vowels fuse into single-syllable nuclei (134b):

(134)	a. /æ̨] mi – i/ → *ǽ̨] mi* (not **ǽ̨] mìi*)	'take] 2 SG.-SUBJUNCTIVE'	P 430
	b. *wǽ̨] kį̨mòį̨ba* (not **wǽ̨] kį̨moį̨ba*)	'I will cry'	P 430

8. Against this goes the discussion by Saint and Pike (1962, 20) of stressed vowels in hiatus, where they judge a suffix-chain stress as "slightly more intense" than a stem stress. Clearly, more information would be needed to settle the question of Auca main stress in a satisfactory way.

The ligature in (134b) follows Pike's notation; he does not specify whether the diphthong created by fusion is long or short.

Since the underlying form of a short vowel is irrecoverable after collapsing has taken place, the rules must be ordered as follows: (a) stress assignment within the stem; (b) collapsing of vowel sequences in the suffix train; (c) stress assignment within the suffix train. A suitable formalization of this procedure is available within a version of Lexical Phonology (e.g. Kiparsky 1982a; Mohanan 1982), if the morphological levels and rules are arranged as in (135):

(135) a.	*Level I:*	stem formation (includes compounding) stress assignment (left to right)
b.	*Level II:*	suffixation collapsing of adjacent vowels into single syllables stress assignment (right to left)

In this arrangement, Level II stress assignment must take place after all the suffixes are added, rather than occurring each time a new suffix is attached. In the proposal of Halle and Mohanan 1985, Level II of Auca would be a noncyclic lexical level. Other instances of the same phenomenon have been documented by Brame 1974 for Palestinian Arabic; Pulleyblank 1986a for Kikuyu; Jackson 1987 for Seminole/Creek; Odden 1987 for Kimatuumbi; Levin 1988b and § 6.1.3 for Cahuilla; and § 6.3.8.7, § 8.8.5 for Yupik.

Within Level I, stress assignment may be formulated as in (136):

(136) a.	**Foot Construction**	Form syllabic trochees from left to right. Degenerate feet are permitted in strong position.
b.	**Word Layer Construction**	End Rule Right

Since the weak prohibition on degenerate feet is invoked, monosyllabic words should be possible; this is confirmed by (131a).

The effects of these rules on all forms can be illustrated with unsuffixed words, which are derived entirely at Level I:

(137)	a. (×)	b. (×)	c. (×)	d. (×)
	(×)	(× .)	(× .)(×)	(× .)(× .)
	σ	σ σ	σσ σ	σ σ σ σ
	bǫ́	*bádą*	*mǫ̀įkó*	*bòdæpóka*

The crucial aspect here is that only the weak ban on degenerate feet is imposed. Since final degenerate feet in odd-syllabled stems are placed in strong position by word layer construction, they may persist to the surface.

When a word receives suffixal material at Level II, any adjacent vowels in the suffixes are assigned to a single syllable, as noted above. Parsing into syllabic trochees again takes place, this time from right to left. It is not clear what the status of Word Layer Construction (136b) at Level II should be; here I assume for concreteness that this rule is restricted to Level I; hence suffixes do not induce a shift in the main stress. Alternatively, we could construct a higher grid layer at Level II, vacuously amplifying the existing main stress.

The crucial aspect of the analysis at this stage is what determines the assignment of extra stem-final stresses seen on the surface in the forms of (131c.ii), such as the boldface syllable of *gą́næę̨́***mǽ̨**] *ŋą*. Plausibly, these stresses represent a continuation of the alternating count of the suffix train, continued for the distance of one syllable into the stem. This idea can be implemented formally by following the view of Kiparsky 1982a that at least in some languages, cyclic stress rules are permitted to analyze material belonging to previous cycles, provided this is necessary in order to assign metrical structure to material added on the current cycle; for discussion see § 5.4.2. Under the assumption that cyclic stress in Auca is of the restructuring type, the crucial derivations will appear as in (138):

```
(138) a. (×   )        (×     )          (     ×     )      Output of
         (×  .)        (×   . )          (×  .)(×   . )     Level 1
          σ σ ] σ       σ σ ] σ σ σ       σ σσ  σ ] σ
         wodǫ ] ŋą     ęɲa ] kądapa      gąnææ̨mæ̨ ] ŋą

      b. (×)          .(×)               (     ×)           Level II
         (×)(×    .)  (×)(×    .)(× .)   (×  .)(×)(×    .)  footing
          σ σ ]  σ     σ σ ] σ σ σ        σ σσ  σ ]  σ
         wódǫ̀ ] ŋą    ę́ɲà ] kądàpa      gą̀næǽ̨mæ̨̀ ] ŋą
```

In the forms of (131b.ii), parsing at Level II will vacuously restress the final syllable of the stem:

```
(139) (      )   (     ×     )   (    )         (     ×      )
      (×    .)   (×  .)(×   .)   (×   .)(× .)   (×  .)(×   .)(× .)
       σ ] σ      σ σ σ ] σ       σ ] σ σ σ      σ σ σ ] σ σ σ
      gó ] bo    kìwęɲǫ́ ] ŋa     gó ] tabòpa    àpæ̨né ] kądàpa
```

For concreteness, I assume that the word layer is adjusted to incorporate the newly constructed foot, though evidence on this question is lacking.

In all remaining forms, the suffix train has an even number of syllables; thus by the principles outlined above, Level II footing cannot alter the content of the stem train:

(140) a. **Both Trains Even in Length (131b.i)**

```
(×  )              (       ×  )
(× .) (× .)        (×  .)(× .) (× .)
 σ σ ] σ σ          σ σ σ σ ] σ σ
kǽga ] kąba        pæ̀dæpǫ́nǫ ] ɲą̀ba
```

b. **Odd Stem Train + Even Suffix Train (131c.i)**

(×)　　　　　(×)
(×)　(×　.)　　(×)　(×　.)(×　.)
σ]　σ　σ　　σ]　σ　σ　σ
gó] bòpa　　*gó] tàmǫnàpa*

(　　×)　　　　(　　　×)
(×　.)(×)　(×　.)　　(×　.)(×　.)(×)　(×　.)
σ　σ　σ]　σ　σ　　σ σ　σ　σ　σ]　σ　σ
yìwæ̨mǫ́] ŋą̀ba　　*tìkawòdǫnó] ką̀ba*

This is the core of the analysis; two details follow.

According to the analysis, the reparsing of stem material with an odd-syllabled suffix train should not affect main stress, since word layer construction is confined to Level I. That this is correct is suggested by Pike's observation (p. 428) that forms of the type *1 + 2* (no reparsing of the stem syllable; cf. (141a)) are prosodically the same as forms of the type *2 + 1* (final stem syllable prosodically reparsed at Level II; cf. (141b)). The analysis assigns the same structure to both:

(141) a. **1 + 2**　(×)
　　　　(×)　(×　.)
　　　　σ]　σ　σ
　　　　gó] bòpa

b. **2 + 1**　(×)
　　　　(×)(×　.)
　　　　σ　σ]　σ
　　　　wódǫ̀] ŋą

This conclusion is important, since it makes it possible to maintain the view that degenerate feet are allowed on the surface only in strong position.

A second detail concerns the stress pattern of five-syllable suffix trains; remarkably, these deviate from alternating stress, having stresses on their first and penultimate syllables: *gó] kæ̨̀dǫmǫnà͡i̧ba* 'we two would have gone' (p. 426). This can be accounted for under the theory of ternary alternation developed in chapter 8; for discussion, see § 8.10.

Auca has been analyzed earlier in a metrical framework by Halle and Kenstowicz 1991. These authors assume a theory in which cyclic stress may never restructure existing feet. To permit stem-final syllables to be stressed, they mark the stem-final syllable as extrametrical, so that it will be free to be footed at the non-cyclic level where suffixes are stressed. In addition, to cover cases where the latter rule generates excess stresses, a destressing rule is posited. Their analysis does not address unsuffixed odd-syllabled forms like (131a) ***mǫ́i̧kó:*** these receive final stress, despite the putative extrametricality of the final syllable; Kenstowicz (p.c.) notes that one might stress this syllable later in the derivation, if extrametricality is not counted at the non-cyclic level.

The Halle/Kenstowicz analysis makes differing predictions from the present one in cases where a polysyllabic stem is followed by four suffix syllables: the proposal made here predicts e.g. /σ́ σ] σ́ σ σ́ σ/; whereas for Halle and Kenstowicz the prediction would be /σ́ σ́] σ σ σ́ σ/, since for them the suffix train has five syllables. Pike's verbal description implies that /σ́ σ] σ́ σ σ́ σ/ is correct, but unfortunately the sources contain no actual examples.

The decision as to whether cyclic stress may restructure existing feet relates to other formal questions. The theory assumed by Halle and Kenstowicz (see Halle and Vergnaud 1987a; HV) can dispense with the parameter that permits restructuring under cyclic stress, but at the cost of adding new apparatus to the theory: multiple stress planes and rules that copy grid marks between planes. These devices are not needed in a theory that allows restructuring, but a restructuring theory may also need to allow rules of metrical structure removal ("deforestation") for some cyclic stress systems, to the extent that such systems cannot be treated as tonal. For discussion of these issues, see Steriade 1988b; Poser 1989; Kager 1989, 150–53; Dresher 1989; and Kiparsky 1993.

In summary, the Auca facts and analysis support the following theoretical notions.

First, Auca constitutes an argument in favor of constituency in stress assignment: only by expressing stress assignment as foot parsing can the common basis of left-to-right stressing and right-to-left stressing be captured (§ 3.8.3).

Second, Auca supports a particular view of the weak prohibition on degenerate foot construction: the deletion of degenerate feet cannot take place until word layer construction is complete, since in forms with an odd syllable count in the stem train, the final degenerate foot must be present to serve as the docking site for the word layer grid mark. In this respect, Auca forms a minimal contrast to languages like Wargamay, Seminole/Creek, or Cairene Arabic (see § 5.1.3): in these, the ban on degenerate feet is absolute, and syllables left unfooted because of the ban cannot serve as docking sites for main stress.

6.2.2 Icelandic

Stress in Icelandic is analyzed in detail by Árnason 1985, drawing on earlier work, especially Gussmann 1985. Árnason's analysis is based on the metrical framework of Giegerich 1983, 1985; here, I propose an analysis that captures Árnason's generalizations and insights using the framework proposed in this book. The central theoretical issue at stake is the status of weak degenerate feet on word-final syllables, created by alternating stress rules. This problem was discussed in § 5.1.8.2, where it was proposed that such word-final syllables sound prominent not because they are footed (this is forbidden by the theory), but because of phonetic final lengthening. Icelandic stress brings evidence to bear on this issue.

6.2.2.1 *Data*

There is a certain amount of disagreement among linguists concerning Icelandic stress. The disagreement usually does not concern how a particular word is to be categorized, but rather how we should label the various distinguishable patterns in terms of stress levels. My own judgments given below are based on listening to the speech of Sigriður Sigurjonsdottir,[9] a native speaker, as well as to tapes of spoken Icelandic from the UCLA Phonetics Laboratory archives. With a single exception discussed below, these judgments agree with Árnason's.

Icelandic stress is based on both morphological and rhythmic principles. The simplest pattern is attested in "simplex" words, that is, non-compound words with only "cohering" affixes (Árnason 1985, 105–11). In the latter, according to Árnason, primary stress falls on the initial syllable and secondary stresses fall on alternating syllables thereafter, including the final syllable in odd-syllabled words. In the data below, examples are given in Icelandic orthography. Since the orthography uses accent marks to distinguish different vowel qualities, IPA notation ([ˈ, ˌ]) is used to mark primary and secondary stress. All page references are to Árnason 1985.

(142)				
a.	/σ́/:	*ˈJón*	'John'	A 97
b.	/σ́ σ/:	*ˈtaska*	'briefcase'	A 99
c.	/σ́ σ σ̀/:	*ˈhöfðingˌja*	'chieftain (gen. pl.)'	A 94
d.	/σ́ σ σ̀ σ/:	*ˈakvaˌrella*	'aquarelle'	A 98
e.	/σ́ σ σ̀ σ σ̀/:	*ˈbíóˌgrafíˌa*	'biography'	A 98

The crucial stresses for our purposes are the final secondaries in (142c, e): either we attribute them to phonetic final lengthening, or else they would have to be represented with degenerate feet.

What do such syllables sound like? If they are stressed, the stress is not strong; in fact, Gussmann 1985 believes that such syllables are completely stressless. To my ears, these syllables do sound slightly prominent, particularly when phrase-final; the percept is ambiguous between secondary stress and final phonetic lengthening. In connected speech, these syllables can sound completely stressless when another word follows in the phrase; in fact they can sometimes be very short and have a schwa-like vowel.

The sound of final odd syllables is compatible with either theoretical view. We could say that the prominence is due to phrase-final lengthening, in which case the lack of prominence in phrase-medial position is automatically accounted for. Alternatively, if one adopts the other interpretation of the facts, one could extend the rule of Destressing ((152) below) to apply at the phrasal level.

9. To whom many thanks for her patient and thoughtful help, and especially for her own linguist's judgments of Icelandic stress patterns. Some areas where we have disagreed on the judgments are reported below.

6.2.2.2 *Analyses*

The pattern of (142) clearly involves syllabic trochees, created from left to right, with the word layer created by End Rule Left. Under the assumption that degenerate feet may be freely created, the structures in (143) are derived:

(143) a. (×) b. (×) c. (×)
(×) (× .) (× .) (×)
ˈJon *ˈtaska* *ˈhöfðingˌja*

d. (×) e. (×)
(× .)(× .) (×.) (× .)(×)
ˈakvaˌrella *ˈbíóˌgrafíˌa*

Under the theory proposed here, in which degenerate feet may not occur in weak position, the analysis would be the same, except that the ban on degenerate feet is invoked. (The choice of the weak vs. the strong ban would depend on whether the lengthening rule discussed below in § 6.2.2.3(b) is considered to "rescue" degenerate feet or not.) The analysis attributes the prominence of final syllables to phonetic final lengthening rather than to metrical structure. Examples (143c) and (143e) would differ as shown in (144), while the other examples would be the same.

(144) c. (×) e. (×)
(× .) (× .)(× .)
ˈhöfðingja *ˈbíóˌgrafía*

6.2.2.3 *Stress in Icelandic Compounds*

To understand whether the final syllables are footed or not, we first need to explore the more complex aspects of Icelandic stress assignment. Árnason notes that the binary alternating pattern seen in (142) can be disturbed in compounds. To a first approximation, compound stress works as in many other languages: the components of the compound are stressed separately, then the second member is metrically subordinated to the first. In the transcriptions below, the simplex forms that make up a compound are separated by /#/.

(145) a. *ˈforustu#ˌsauður* 'leadership (gen.) # sheep' = 'leading sheep' A 95
b. *ˈFramsóknar#ˌflokkur* 'progress (gen.) # party' = 'Progressive Party' A 95
c. *ˈhöfðingja#ˌvald* 'chieftain (gen. pl.) # power' = 'aristocracy' A 94

Were these words to be stressed as single units, we would incorrectly derive **ˈforusˌtu#sau(ˌ)ður*, **ˈFramsókˌnar#flok(ˌ)kur*, and **ˈhöfðingˌja#vald.*

The same stress pattern may be found in words containing a "noncohering" suffix: these suffixes act as if they were compound members, disrupting the alternating count:

(146)	*ˈsnúninga#ˌsamur*	'bothersome'	A 107
	ˈaumingja#ˌlegur	'miserable looking'	A 107
	ˈmiskunnar#ˌfullur	'merciful'	A 107

This reflects the historical sources of these suffixes, which at an earlier stage of the language were separate words. The suffixes may be treated as members of pseudo-compounds, forming a separate phonological domain from the stem to which they attach. Below, the category "compounds" is intended to include suffixed forms of this type.

For the prosodic structure of compounds, there are two basic possibilities: either all the feet from both words are joined together under a single word layer (147a), or else each simplex word has its own word layer, with an additional higher compound layer (147b).

(147) a. **Single Word Layer** b. **Extra Compound Layer**

```
a.                                  b.
                                    (×                          )
(×                          )       (×           )  (×          )
(×  .)(×  .)    (×  .)(×  .)        (×  .)(×  .)    (×  .)(×  .)
[ σ  σ  σ  σ ] [ σ  σ  σ  σ ]       [ σ  σ  σ  σ ] [ σ  σ  σ  σ ]
```

There are three arguments that favor the compound layer hypothesis.

(a) Orešnik 1971, 55, notes that the complex compound *skóla#bóka#safn* 'school book collection' is stressed differently depending on whether it means 'collection of school books' ([[school book] collection]) or 'school library' ([school [book collection]]). Assuming (as in Sigurjonsdottir's speech) that the difference is as in (148), we can derive it only if compound structure is interpreted metrically with layers in the grid:

```
(148) a. ( ×              )   'collection of    b. ( ×              )   'school
         ( ×         )         school books'       ( ×  )(×         )    library'
         (  × )(×   )(× )                          ( ×  )(×   )(× )
         (  × .)(× .)(× )                          ( × .)(×  .)(× )
         [[skólabóka]safn]                         [skóla[bókasafn]]
```

(For the theory of phrasal stress rules assumed here, see chap. 9.) Since extra compound layers are needed for complex cases like (148), it is reasonable to suppose that they are also constructed in simple two-word compounds.

(b) Icelandic has a rule that lengthens vowels in stressed open syllables, with final consonants counted as extrametrical (Kiparsky 1984). This rule applies to vowels that bear the strongest stress of a word within a compound, but not to vowels receiving mere alternating stress, insofar as this can be determined from Árnason's (1980, 43–52) description. To be able to express this distinction, we need a separate word layer for each simplex word.

(c) The intuitions of various linguists also support the view that each simplex word in a compound has its own word layer. For example, Jóhannsson 1924, cited in Árnason 1985, 94, stresses the compound formed from *'forða* 'supplies (gen.)' + *búr* 'pantry' as *'forða#'bur* 'storage room', but simplex *hamingja* 'happiness' as *'hamin,gja.* These are also Sigurjonsdottir's judgments.[10] Árnason hears both as ['σ σ ,σ]. Gussmann (1985) hears a final secondary stress in forms like *'forða#,bur* but none at all in simplex forms like *'hamingja.* My own opinion is that compounds like *forða#bur* have a clear secondary stress on their final syllable, even when they are not phrase-final (ruling out the account of prominence based on final lengthening), whereas *'hamin,gja* has only the ambiguous final prominence that might be attributed to final lengthening. Summing up, everyone but Árnason hears forms like *forða#bur* and *hamingja* as different, with a greater degree of stress on the final syllable of the /σ σ # σ/ compound than on the final of the simplex form. This suggests that *forða#bur* and *hamingja* should have distinct metrical representations.

For these reasons, I assume the compound layer hypothesis, as in the representation in (149) for *'forða#,bur:*

```
(149) ( ×          )     compound layer
      ( ×     ) (× )     word layer
      ( ×    .) (× )     foot layer
       forða#bur
```

Under this analysis, the deviation from binary alternation in examples like (145) and (146) is straightforwardly accounted for, since each half of the compound is footed separately. Under the two general hypotheses we are considering, the representations would be as in (150):

(150) a. **Degenerate Feet Freely Allowed**

```
(×                  )  compound layer
(×         ) (×     )  word layer
(× .)(×)     (×    .)  foot layer
forustu#sauður
```

b. **Degenerate Feet Only in Strong Position**

```
(×                  )  compound layer
(×         ) (×     )  word layer
(× .)        (×    .)  foot layer
forustu#sauður
```

10. It is possible to argue against the view that there is level stress in Icelandic compounds (which sound much like English compounds): if they had genuinely level stress, the distinction of (148a) vs. (148b) would not be audible.

At first blush, it looks as though the facts already favor the analysis in which degenerate feet are restricted, since almost everybody agrees that there is no secondary stress on the third syllable of *ˈforustu#ˌsauður.* But we will see shortly that there is reason in any event to posit a rule of Destressing, which would convert (150a) to the correct result. Thus we do not yet have a true test of the two hypotheses.[11]

6.2.2.4 *Destressing*

An important surface generalization in Icelandic word stress is that except in deliberate speech, there are never two adjacent stresses within a word. When compounding or attachment of non-cohering affixes would create adjacent stresses, one of the two is deleted in fluent speech by a rule of Destressing (Gussmann 1985, 84). Árnason 1985, 94, provides clear phonetic justification for Destressing: destressed syllables can undergo reduction or even deletion.

Destressing sometimes eliminates the stress on the right (151a, b), sometimes the stress on the left (151c):

```
(151) a. (×       )        (×       )
         (× ) (× )         (×       )
         (× ) (× )   →     (×     . )
         fisk#fars         ˈfisk#fars        'fish # paté'                A 102

      b. ( ×          )      ( ×           )
         ( × )( ×     )      ( ×       )
         ( × )( ×  .) →      ( ×     . )     'strain # briefcase'        A 94
         stress#taska        ˈstress#tas(ˌ)ka  = 'businessman's briefcase'

      c. (×                  )       (×                  )
         (×             )            (×             )
         (×     )( × )(×  )          (×     )      (×  )
         (×   .)( × )(× .)   →       (×   .)       (× .)     'dog # shit # like'
         hunda#skíts#legur           ˈhunda#skítsˌ#legur     = 'like dog shit'
                                                                          A 108
```

The stress on the left wins if it is the main stress of the compound (151a, b), otherwise the stress on the right wins (152c). Formally, Destressing may be stated as in (152):

(152) **Destressing** $\times \rightarrow \emptyset$ in clash (all layers, applies in fluent speech)

11. Sigurjonsdottir hears this form as *ˈforusˌtu#ˌsauður,* and similarly for other forms. This is not unreasonable, since her judgments are based on quite slow and careful speech (i.e. suitable for demonstrating pronunciation to a foreigner). In such a speaking style, there could be phonetic final lengthening (alternatively, blockage of Destressing) at the boundary between the two compound members.

I assume that metrical structure creation in Icelandic is persistent (§ 5.4.1), so that constituency is reassigned to the sequence affected by destressing. Where the clashing stresses include the initial main stress, the main stress must "win," since otherwise there would be a violation of the Continuous Column Constraint (§ 3.4.2). For cases like (151c), it may be necessary to add a default direction to the structural description of the rule. Alternatively (and speculatively), we might rely on general principles of rhythmic well-formedness (e.g. Hayes 1984) to guarantee the correct output: in *ˈhunda#skíts#ˌlegur* the two remaining stresses are separated further apart than in hypothetical *ˈ*hunda#ˌskíts#legur*. Resolving this issue will not be crucial in what follows.

6.2.2.5 *The Argument Against Degenerate Feet*

The rules set up so far are listed in (153):

(153)			
	a.	**Foot Construction**	Within simplex words, form syllabic trochees from left to right.
	b.	**Word Layer Construction**	End Rule Left, within simplex words
	c.	**Compound Layer Construction**	End Rule Left, within compounds
	d.	**Destressing**	= (152)

Consider now an ordering argument: (154) and similar examples show that Foot Construction must occur before Destressing.

```
(154) (×              )       (×                 )
      (×) ( ×         )  →    (×        )
      (×) ( ×    .)(× .)      (×    .  ) (× . )
       ó#krambúleruð          'ó#krambúˌleruð       'un#damaged (f. sg.)'
                                                                  A 120
```

That is, the secondary stress on the penult reflects an alternating count starting from *kram*. It must therefore be assigned before *kram* is destressed.

A priori, we might have expected a different system, based on top-down application of the rules (§ 5.4.3): the End Rules would assign the major stresses, Destressing would apply, and then Foot Construction would create the alternating secondary stresses. But this alternative is excluded empirically, since it would wrongly derive *ˈ*ó-kramˌbule(ˌ)ruð*.

The derivation in (154) establishes that Foot Construction must be allowed to apply before Destressing. The analysis thus places the rules in this order. However, since Foot Construction is assumed to be persistent (see preceding section), it may also apply after Destressing. Further evidence for this assumption is presented shortly.

We can now consider the forms that bear on the issue of whether final degenerate feet are allowed in Icelandic. Consider the data in (155):

(155)	*ˈstress#töˌskuna*	'businessman's briefcase (acc. fem. sg.)'	A 116
	ˈó#þoˌlandi	'unbearable'	A 116
	ˈlífs#fögˌnuður	'joy of life (nom. m. sg.)'	A 116

We will examine derivations for these forms under both possible assumptions about degenerate feet made above. First, assuming that Icelandic allows degenerate feet only in strong position, the derivation for *stress#töskuna* would be as in (156):

```
(156)  ( ×              )     Foot Construction,
       ( ×   )(×        )     Word Layer Construction
       ( ×   )(×    )         Compound Layer Construction
       stress#töskuna

       ( ×              )     Destressing
       ( ×         )
       ( ×       . )
       stress#töskuna

       ( ×              )     Foot Construction (persistent)
       ( ×              )
       ( ×       . )(× .)
       ˈstress#tösˌkuna
```

Here, the assumption that Foot Construction is persistent makes it possible to derive the correct output.

Next consider how the derivation will work if footing is allowed to create a degenerate foot (marked in boldface), over the final syllable:

```
(157) a. ( ×              )     Foot Construction,
         ( × ) (×         )     Word Layer Construction,
         ( × ) (×     )(×)      Compound Layer Construction
         stress#töskuna

      b. ( ×              )     Destressing
         ( ×        )
         ( ×      . ) (×)
         stress#töskuna         = *ˈstress#töskuˌna
```

This output is incorrect. Moreover, it appears that additional rule application would not save the derivation: one might assign additional metrical structure to promote the stress on the syllable *-ku-* (i.e. *ˈstress#tösˌkuˌna*), but this would still not eliminate the unwanted stress on *na:* recall from (151c) that destressing in this context proceeds from right to left. Thus the only workable analysis is the one in which weak degenerate feet are forbidden in the first place.

To return to the main point, what we were trying to determine was whether the prominence heard on final odd syllables of simplex words in Icelandic is to be attributed to phonological representation, that is, to a degenerate metrical foot, or to the phonetic effects of final lengthening. The way to test this is to see whether the putative foot has effects deeper in the phonology. For the binary secondary stress feet that are agreed upon, such effects are apparent in derivations like (154), where due to destressing they produce stress patterns that contrast minimally with the binary norm. On parallel grounds we would expect the putative degenerate foot in representations like (157b) also to make its presence felt phonologically. It does not; instead, the derivation comes out right only if we assume that there is never any foot present in final position. This follows straightforwardly from the theory proposed here, which forbids such feet on general grounds.

6.2.2.6 *Remaining Forms*

To complete the proposed account, below are derivations for all the possibilities given by Árnason for the various combinations of syllable count and division of the full word into simplex words. (Simplex words of one to five syllables were already shown in (143)–(144)).

(158) **Complex Words, Two Segments**

a. **1 + 1** (151a) b. **1 + 2** (151b) c. **1 + 3** (157) d. **1 + 4** (154)

```
e. 1 + 5 ( ×                 ) Foot Construction,            'undamaged'
         ( ×) ( ×            ) Word Layer Construction        A 120
         ( ×) ( ×   .)(× .)    Compound Layer Construction
           ó#krambúleraður

         ( ×                 ) Destressing
         ( ×                 )
         ( ×    .)   ( × .)
          'ó#krambú,leraður

f. 2 + 1   g. 2 + 2  ( ×            )     h. 2 + 3  (×              )
   (149)             ( ×    ) ( ×   )               (×    ) (×      )
                     ( ×   .) ( ×  .)               (×   .) (×   .)
                     'barna#,skapur                 'undir#,leikari
                     'childish#ness' A 101          'accompanist' A 107

i. 3 + 1   (×              )          j. 3 + 2  (×              )
           (×          )(× )                    (×        )(×    )
           (×  . )     (× )                     (× .)     (×   .)
           'höfðingja#,vald = (145c)            'forustu#,sauður = (145a)
```

(159) **Complex Words, Three Segments**

a. **1 + 1 + 1** b. **1 + 1 + 2** c. **2 + 1 + 2** (151c)

```
(×              )     (×            )     Foot Construction,
(×         )          (×)(×         )     Word Layer Construction,
(× ) (×   )(× )       (×)(×) (×     )     Compound Layer
(× ) (×   )(× )       (×)(×) (× .   )     Construction
þing#hüss#hurd        ó#boð#legur
```

þing#húss#hurd 'parliament # house # door' A 94

ó#boð#legur 'unpresentable' A 114

```
(×              )     (×            )     Destressing
(×         )          (×       )
(×         ) ( × )    (×       )(×  )
(×         ) ( × )    (×       )(× .)
'þing#húss#ˌhurd      ó#boð#ˌlegur[12]
```

One other fact about Icelandic stress is relevant: in borrowings, there can arise secondary stresses that idiosyncratically fall on even syllables, as in *'alternaˌtor* 'alternator', *'Rakmaniˌnoff* 'Rachmaninov' A 112. Other borrowed forms have invariant stress on odd syllables: *'almaˌnak* 'calendar' A 98. That this is a true instance of final secondary stress can be seen from the compound *'plat#almaˌnak* 'make-believe almanac' A 121, which forms a minimal pair for stress with (156) *'stress#tösˌkuna*. These borrowed words may be treated as pseudo-compounds, similar to (158i) and (149), so that their final monosyllabic feet will be in strong metrical position, just like the feet of monosyllables. This analysis has some phonological support, in that the borrowed forms are treated like compounds by the rule of /u/ Umlaut (Árnason 1985, 121; see Anderson 1972 for the relevant phonology):

(160) a. **Borrowed Forms** /alma#nak + um/ → *alma#nökum* 'calendar (dat. pl.)' A 121

b. **True Compound** /matar#gat + um/ → *matar#götum* 'food hole' = 'big eater' A 121

c. **Simplex Word** /fatnað + um/ → *fötnuðum* 'suit of clothes (dat. pl.)' Anderson 1972, 16

6.2.2.7 *Conclusion*

The crucial point of the discussion above is that what Árnason transcribes as final stress in odd-syllabled simplex forms is arguably not represented in the phonology, being instead the effect of phonetic final lengthening. I take this as

12. This derivation presupposes left-to-right Destressing, as predicted by Hammond's (1984a) principle of Trigger Prominence; the results with right-to-left destressing followed by persistent footing would be essentially the same.

evidence that, as argued in § 5.1.8.2, putative cases of weak degenerate feet in final position should not necessarily be taken at face value.

Discussing the Icelandic stress rules in general terms, Árnason (121–24) notes an interesting "conspiracy": irrespective of the morphological structure of the input, stressed syllables are never followed by more than two stressless syllables. In the analysis proposed here, this surface pattern is the result of three basic postulates: (a) the limitation of feet to binary size; (b) the persistence of foot construction, which forces the outputs of Destressing to undergo further footing; and (c) the ban on degenerate feet, which prevents single stray syllables from being footed. Together, these three ingredients (only (b) particular to Icelandic) suffice to enforce the conspiracy. In general, we expect to find "conspiratorial" behavior whenever structure-creating rules apply persistently.

6.2.3 Other Syllabic Trochee Systems

The following lists reflect a somewhat conservative procedure for determining whether a language has syllabic trochees. In particular, a number of languages (e.g. Modern Hebrew (Bolozky 1982) and Spanish (§ 5.1.6, § 5.1.7)) have secondary stresses on alternating syllables before a main stress that is determined by lexical or other principles. These are fully compatible with a syllable trochee analysis, but could conceivably be reanalyzed on other terms, for instance with top-down even iambs (§ 4.3.1):

```
(161) (                              ×     )
      (. ×)(. ×)(. ×)(. ×)(. ×)(. ×)
       σ σ̀  σ σ̀  σ σ̀  σ σ̀  σ σ̀  σ σ́ σ . . .
```

The lists below exclude such cases.

(162) **Left-to-Right Systems**

a. **Anguthimri** (Crowley 1981) Paman, Cape York Peninsula, Australia
Stress on odd nonfinal syllables, e.g. 1, 10, 100, 1010
No distinctions marked between primary and secondary stress.
Vowel length is phonemic; minimal word is disyllabic or /CVː/.
Analysis: Syllabic trochees from left to right; degenerate feet forbidden
No word layer
See § 5.1.9 for nondegenerate status of /CVː/.

b. **Badimaya** (Dunn 1988) Pama-Nyungan, Western Australia
Main stress is initial.
Secondary stress on odd nonfinal syllables; e.g. 100, 1020, 10200
No vowel length distinction, but all vowels in monosyllables are lengthened
Analysis: Syllabic trochees from left to right
Word Layer: End Rule Left
Degenerate feet created, then "repaired" in strong position by vowel lengthening (§ 5.1.9)

c. **Bidyara/Gungabula** (Breen 1973) Pama-Nyungan, South Queensland, Australia
Main stress is initial.
Secondary stress on odd nonfinal syllables, e.g. 10, 100, 1020, 10200, 102020
Morphology influences stress.
Minimal word is disyllabic; no phonemic vowel length.
Analysis: Syllabic trochees from left to right; degenerate feet forbidden
Word Layer: End Rule Left
After stress assigned, certain vowel sequences may coalesce to a single syllable, forming long vowels.

d. **Central Norwegian Lappish** (Itkonen 1955) Finno-Ugric
Main stress is initial.
Secondary stress on odd nonfinal syllables, except that in trisyllables a weak secondary stress may appear (see § 5.1.8.2).

e. **Dalabon** (Capell 1962a) Guwinyguan, Arnhem Land, Australia
Main stress is initial.
Secondary stress on odd nonfinal syllables, e.g. 10, 100, 1020
Vowel length is phonemic; no minimal word constraint.
Analysis: Syllabic trochees from left to right
Word Layer: End Rule Left
Capell mentions attraction of stress by final CVʔ syllables, for which see chap. 7.

f. **Dehu** (Tryon 1967b) Austronesian, Loyalty Islands
1, 10, 102, 1020, 10200
No minimal word constraint; vowel length is phonemic.
Analysis: Syllabic trochees from left to right
Degenerate feet allowed only in strong position
Word Layer: End Rule Left
See § 5.1.8.2 for final secondary stress.

g. **Diyari** (Austin 1981; Poser 1989; Hammond 1990b) Pama-Nyungan, South Australia
Main stress is initial.
Secondary stress on (a) the third syllable of quadrisyllabic morphemes
(b) the first syllable of disyllabic and longer suffixes
Vowel length is not phonemic.
Only degenerate word is *ya* ‘and’ (Poser, p. 144).
Analysis (adapted from Poser):
Cyclically form syllabic trochees from left to right.
Word Layer Construction (postcyclic): End Rule Left
Degenerate feet in weak position lost following Word Layer Construction; this replaces the language-specific rule posited by Poser, p. 122.

h. **German** (Giegerich 1985; Hayes 1986c)
Main stress calculated from right edge
Secondary stress: 01 . . . , 201 . . . , 2001 . . . , 20201 . . .
Vowel length is phonemic; but in syllables lacking main stress, a tenseness contrast is substituted for the length contrast.
Minimal word is a heavy syllable.
Analysis: Syllabic trochees from left to right, degenerate feet forbidden

i. **Hungarian** (§ 8.6)

j. **Livonian** (Sjögren 1861; Kettunen 1938) Balto-Finnic, Latvia
Main stress is initial, secondary on third syllable, tertiary on fifth.
Vowel length is phonemic; very few CV̌ words.
Analysis: Syllabic trochees from left to right
Word Layer: End Rule Left
See § 5.1.8.2 for final secondary stress.
Secondary–tertiary distinction possibly to be attributed to intonational downdrift (cf. Hayes 1981, 120–23)

k. **Mansi** (Lakó 1957; Kálmán 1965; Rombandeeva 1973) Uralic, Western Siberia
Main stress is initial; secondary stress on subsequent odd-numbered syllables. Sources differ on whether final syllables are stressed.
Vowel length contrast is present, mostly in initial syllables.
Minimal word is a heavy syllable.
Analysis: Syllabic trochees from left to right
Degenerate feet prohibited (see § 5.1.8.2 for final secondary stress)
Word Layer: End Rule Left

l. **Maranungku** (Tryon 1970) Daly, Australia
Main stress is initial, secondary stress on odd non-initial syllables (examples only given for trisyllables).
No minimal word constraint, no phonemic vowel length
Analysis: Syllabic trochees from left to right
Word Layer: End Rule Left
See § 5.1.8.2 for final secondary stress.

m. **Mayi** (Breen 1981) Queensland, Australia
10, 100 with occasional 120, 1020, 10200; morphology influences stress.
Analysis: Syllabic trochees from left to right
Degenerate feet allowed only in strong position
Word Layer: End Rule Left
120 may diagnose "top-down" stressing (§ 5.4.3), with occasional right-to left foot construction.

n. **Ono** (Phinnemore 1985) Western Huon, New Guinea
Main stress is initial; secondary on odd numbered syllables.
1, 10, 102, 1020, 102020

No minimal word constraint, no phonemic vowel length
Analysis: Syllabic trochees from left to right
Word Layer: End Rule Left
See § 5.1.8.2 for final secondary stress.
Comment: According to Phinnemore, where there are two secondaries, one may be omitted. If this is to be treated in the theory proposed here, we must assume that this reflects variation in the relative prominence of the secondaries (via Beat Addition, chap. 9), with only the stronger of the two transcribed.

o. **Piro** (Matteson 1965; Blevins 1990) Arawakan, Eastern Peru
Main stress is penultimate; secondary stress is initial.
Tertiary stress on 3, 5, 7, 9, 11 in sufficiently long words, but not on the syllable preceding the main stress
Vowel length is present; no minimal word constraint.
Analysis: Form a syllabic trochee at the right edge of the word.
Form syllabic trochees from left to right.
Degenerate feet allowed only in strong position, hence no stress on syllables immediately preceding the main stress
Word Layer: End Rule Right
Secondary vs. tertiary distinction: Beat Addition (§ 9.6.1)

p. **Pitta-Pitta** (Blake 1979b) Pama-Nyungan, Southwest Queensland, Australia
Main stress is initial; secondary stress on third syllable of quadrisyllables; morphology influences stress.
Minimal word is disyllabic; vowel length is phonemic but limited to two morphemes.
Analysis: Syllabic trochees from left to right
Degenerate feet forbidden
Word Layer: End Rule Left

q. **Selepet** (McElhanon 1970) Western Huon, New Guinea
1, 10, 102 ~ 100, 1020
No minimal word constraint, no phonemic vowel length
Analysis: Syllabic trochees from left to right
Degenerate feet allowed only in strong position
Word Layer: End Rule Left
See § 5.1.8.2 concerning final secondary stress.

r. **Votic** (Ariste 1968) Balto-Finnic, Northwest Russia
Main stress is initial; secondary stress on subsequent odd numbered syllables, but not on case suffixes.
Vowel length is phonemic; minimal word is $CVVC_0$.

Analysis: Syllabic trochees from left to right
Word Layer: End Rule Left
See § 5.1.8.2 concerning final secondary stress.
Case suffixes attached on a later morphological level

s. **Wangkumara** (McDonald and Wurm 1979) Southwest Queensland, Australia
Main stress normally on first syllable
Secondary stress normally on 3rd and 5th syllables
Final syllables are stressless in connected speech (but see MW 5 for post-lexically derived final stressing).
No phonemic vowel length
Monosyllabic words absent. Identical vowels can merge to form a long vowel.
Analysis: Syllabic trochees from left to right
Degenerate feet forbidden
Word Layer: End Rule Left

(163) **Right-to-Left Systems**

a. **Cavineña** (Key 1968) Tacanan, Bolivia
Main stress is penultimate.
Secondary stress on alternating syllables before the penult, e.g. 02010
Key's glossary lists no monosyllabic forms except bound roots.
Vowel length is not phonemic; adjacent vowels may coalesce into single nuclei in fast speech.
Analysis: Syllabic trochees from right to left
Degenerate feet forbidden
Word Layer: End Rule Right

b. **Djingili** (Chadwick 1975) West Barkly, Northern Territory, Australia
Main stress is penultimate (a few words: antepenultimate, final)
Secondary stress two syllables before the penult, e.g. 10, 010, 2010, 002010
Vowel length is phonemic; minimal word is CVV.
Analysis: Form up to two syllabic trochees from right to left.
Degenerate feet prohibited (see § 5.1.8).
Some words have syllable extrametricality or lexical final feet.
Word Layer: End Rule Right
Comment: 002010 is very unusual; one wonders whether it is actually 202010, with some phonetic effect diminishing the first secondary stress.

c. **Garawa** (Furby 1974) Karawic, Northern Territory and Queensland, Australia
10, 100, 1020, 10020, 103020, 1003020, 10303020, 100303020, 1030303020

No phonemic vowel length
Minimal word is CVC.
Analysis: Form an initial syllabic trochee.
Syllabic trochees from right to left
Degenerate feet forbidden in weak position
Word Layer: End Rule Left
Beat Addition (§ 9.6) applies within the word.

d. **Malakmalak** (Birk 1976) Daly, Northern Territory, Australia
1, 10, 010 or 102, 1020, 01020, 102020, 0102020
No minimal word constraint, no phonemic vowel length
Analysis is problematic from viewpoint of constraints on degenerate feet; see Goldsmith 1990, 173–77, for a possible account.

e. **Nengone** (Tryon 1967a) Austronesian, Loyalty Islands
1, 10, 010, 2010, 02010
Vowel length is phonemic; no minimal word constraint.
Analysis: Syllabic trochees from right to left
Word Layer: End Rule Right

f. **Warao** (Osborn 1966) Paezan, Venezuela
Main stress is penultimate; exceptions with antepenultimate and final stress.
Secondary stress every other syllable before penult; e.g. 10, 010, 02010, 20202010, 020202010
No minimal word constraint, no phonemic vowel length
Certain vowel sequences apparently treated as diphthongs; not clear from data if these are stress-attracting
Analysis: For exceptional words: final syllable extrametricality (antepenultimate)
listed final degenerate feet (final)
Syllabic trochees from right to left
Degenerate feet allowed only in strong position
Word Layer: End Rule Right

(164) **Other Syllabic Trochee Systems**

a. **Czech** (Jakobson 1962)
10, 102, 1020, 10202 or 10020
According to Chlumský 1928, presence of secondary stress is optional.
Vowel length is phonemic, no minimal word constraint.
Analysis: Word Layer: End Rule Left (top down construction)
Optionally parse the word into feet, obeying the Continuous Column Constraint and avoiding degenerate feet.
For final secondary stress, see § 5.1.8.2.

b. **Gugu-Yalanji** (Oates and Oates 1964) Pama-Nyungan, Queensland, Australia
10, 100, 1020, 10020, 100020
Phonemic status of vowel length is unclear.
Very few words of degenerate (CV) length
Analysis: Form an initial syllabic trochee.
Form a final syllabic trochee.
Degenerate feet are forbidden.
Word Layer: End Rule Left

c. **Modern Greek** (Malikouti-Drachmann and Drachmann 1981)
Main stress is lexically determined; limited to one of last three syllables.
Secondary stress (/2:/ = lengthened syllable bearing secondary stress):
01 . . . , 2: 1 . . .
001 . . . , 201 . . .
0001 . . . , 2001 . . . , 0201 . . . , 2:201 . . .
202001 . . . , 020201 . . . (analogous cases implied)
No minimal word constraint, no phonemic vowel length
Analysis: Main Stress: syllabic trochee at right edge of word; some words have final syllable extrametricality or lexically listed final degenerate feet.
Secondary Stress: Form syllabic trochees in either direction on the layer containing the main stress foot. Weak degenerate feet may be licensed in initial position by lengthening (§ 5.1.7, § 5.1.8.1).

d. **Polish** (Comrie 1976; Hayes and Puppel 1985; Hayes 1985; Franks 1985, 1991; Booij and Rubach 1985; HV; Hammond 1989; Inkelas 1989)
Main stress is penultimate, antepenultimate in exceptional words
Secondary stress: as in Modern Greek (164c), but with left-to-right parsing preferred (or obligatory for some speakers)
No minimal word constraint, no phonemic vowel length
Analysis: see refs. for various analyses of main stress;
(164c) for secondary stress

A very common pattern not yet mentioned is simple penultimate stress, which plausibly results from the construction of a single syllabic trochee at the right edge of a word. Alternatively, if one invokes final syllable extrametricality, such cases could be treated as final stress, as HV (18–19) do for Warao. Some typological support for the syllabic trochee analysis can be found in the propensity of penultimate-stress languages to tolerate exceptional words (e.g. borrowings) with antepenultimate stress, as in Polish (164d), Lenakel (§ 6.1.8), Djingili (163b), Warao (163f), Yawelmani (Archangeli 1984), Chamorro (Chung 1983), and Swahili (Wald 1987). This pre-empts extrametricality for purposes of accounting for the basic stress pattern. For a listing of penultimate stress languages, see Hyman 1977b, 62–63.

Less common is the pattern of regular antepenultimate stress, which is analyzed here with syllable extrametricality and a syllable trochee (§ 3.11). The languages in (165) show antepenultimate stress, with initial stress on shorter words.

(165) a. **Macedonian** (Lunt 1952; Comrie 1976; Franks 1987, 1989) Slavic
No phonemic vowel length; no minimal word requirement
b. **Parnkalla** (Schürmann 1844) Pama-Nyungan, South Australia
nouns only
No phonemic vowel length; no minimal word requirement
c. **Kela** (Apoze dialect, Collier and Collier 1975) Austronesian, Papua New Guinea
No phonemic vowel length; no minimal word requirement
d. **Mae** (Capell 1962b) Austronesian, Vanuatu
Vowel length is phonemic; no minimal word requirement.
e. **Cayuvava** (§ 8.2)

For discussion of the mechanisms allowing such languages to assign initial stress to shorter words, see § 5.3.

6.3. IAMBS

Incorporating the discussion of degenerate feet from chapter 5, I restate the iambic foot template in (166):

(166) Form $\overset{(.\ \times)}{\smile\ \sigma}$ or $\overset{(\times)}{-}$. Where degenerate feet are allowed, form $\overset{(\times)}{\smile}$.

The canonical version of the iamb is /⏑ –́/, following the Iambic/Trochaic Law (§ 4.5.2).

6.3.1 Hixkaryana and Other Cariban Languages

Hixkaryana is a Cariban language of Northern Brazil, described by Derbyshire 1985, from which all data and generalizations have been taken. An earlier metrical analysis of Hixkaryana was proposed by Blevins 1990.

Hixkaryana has no length contrast in its vowel inventory, which is /e, æ, o, u, ɯ/. For legibility, I use Derbyshire's transcription of these vowels, which is /e, a, o, u, ɨ/. There is a weight distinction, between heavy /CVC/ and light /CV/. This is unusual, in that the great majority of languages with a weight distinction also have phonemic vowel length. /CVC/ syllables do not occur finally, except as the output of an Apocope rule (Derbyshire 1985, 179) whose interaction with stress is not described.

To a first approximation, stress in Hixkaryana falls on heavy syllables and on the nonfinal even members of a string of light syllables. Stressed short vowels in nonfinal open syllables undergo a rule of lengthening. In (167) are some examples (Derbyshire 1985, 181) chosen to show the same morpheme (boldface) in different contexts, with different vowels receiving stress and/or lengthening. Stress marks should be interpreted in light of the discussion below.

(167)

a. /⏑ ⏑́ –́ ⏑/   /kɨ-**hananɨhɨ**-no/ → *khananɨhno* → [k**haná:nɨ́h**no]
1SUBJ.,2OBJ.-teach-IMMED. PAST   'I taught you'

b. /⏑ ⏑́ ⏑ –́ ⏑/   /mɨ-**hananɨhɨ**-no/ → *mɨhananɨhno* → [mɨ**há:nanɨ́h**no]
2SUBJ.,3OBJ.-teach-IMMED. PAST   'you taught him'

c. /–́ ⏑ ⏑́ ⏑/   /owto-**hona**/ → [ówto**hó:na**]   'to the village'
village-to

d. /–́ ⏑ ⏑́ ⏑ ⏑/   /tohkur^{j}e-**hona**/ → [tóhkur^{j}é:**hona**]   'to Tohkurye'
Tohkurye-to

e. /–́ ⏑ ⏑́ ⏑ ⏑́ ⏑ ⏑́ ⏑/   /tohkur^{j}e-**hona**-hašaka/ → [tóhkur^{j}é:**honá:**hašá:ka]
Tohkurye-to-finally   'finally to Tohkurye'

This stress pattern can be derived by left-to-right parsing into iambs, following the rules given in (168):

(168) a. **Syllable Weight**   /–/ = CVC, /⏑/ = CV

b. **Syllable Extrametricality** $\sigma \rightarrow \langle\sigma\rangle$ / ______]$_{\text{phonological word}}$

c. **Foot Construction**   Form iambs from left to right; degenerate feet are prohibited absolutely.

The extrametricality rule (168b) prevents final syllables from receiving metrical prominence, hence length. The assumption in (168c) that degenerate feet are prohibited accords with the apparent absence of monosyllabic content words in the language (a /CVC/ word would fulfill the minimum, but no word may end underlyingly in a consonant in any event).

The vowel lengthening rule may be stated as in (169):

(169) **Iambic Lengthening**

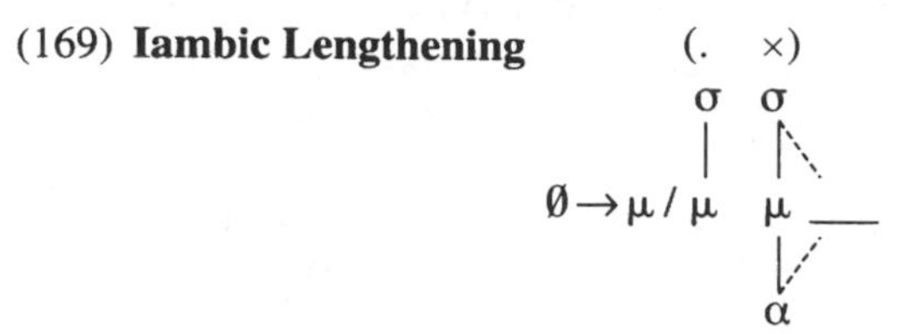

Iambic Lengthening is so named, and is written in moraic notation, to emphasize its property of creating canonical iambic feet; i.e. /μ μ́μ/ = /⏑ –́/. The derivations in (170) illustrate the rules.

(170) a. *kɨ-hananɨhɨ-no* b. *mɨ-hananɨhɨ-no*

khananɨhno	*mɨhananɨhno*	segmental rules
(. ×)(×) ⏑ ⏑ – ⟨⏑⟩ *khananɨhno*	(. ×)(. ×) ⏑ ⏑ ⏑ – ⟨⏑⟩ *mɨhananɨhno*	Syllable Extrametricality, Foot Construction
(. ×) (×) ⏑ – – ⟨⏑⟩ *khan**a**ːnɨhno*	(. ×) (. ×) ⏑ – ⏑ – ⟨⏑⟩ *mɨh**a**ːnanɨhno*	Iambic Lengthening
c. (×) (. ×) – ⏑ ⏑ ⟨⏑⟩ *owtohona*	d. (×) (. ×) – ⏑ ⏑ ⏑ ⟨⏑⟩ *tohkur^j^ehona*	Syllable Extrametricality Foot Construction
(×) (. ×) – ⏑ – ⟨⏑⟩ *owtoh**o**ːna*	(×) (. ×) – ⏑ – ⏑ ⟨⏑⟩ *tohkur^j^**e**ːhona*	Iambic Lengthening
e. (×) (. ×)(. ×)(. ×) – ⏑ ⏑ ⏑ ⏑ ⏑ ⏑ ⟨⏑⟩ *tohkur^j^ehonahašaka*		Syllable Extrametricality Foot Construction
(×) (. ×) (. ×) (. ×) – ⏑ – ⏑ – ⏑ – ⟨⏑⟩ *tohkur^j^**e**ːhon**a**ːhaš**a**ːka*		Iambic Lengthening

Note that because Hixkaryana completely disallows degenerate feet, words may end in two unfooted syllables (e.g. (170d)): the final is extrametrical, and the penult has insufficient bulk on its own to form a foot.

The most puzzling aspect of the Hixkaryana data is the position of the main stress. This appears to be utterly independent of metrical structure, and depends instead on the intonation pattern. Derbyshire lists five intonation contours (p. 182); in four of them, main stress falls on the final syllable, and in the remaining one, it falls on the penult. To a rough approximation, Derbyshire records main stress where a particular intonation pattern places high pitch, though in declarative intonation he places main stress on the falling-pitch final syllable.

Dependence of stress on intonation (rather than vice versa) is unusual, and little theoretical work has been done in this area. If Hixkaryana truly displays this phenomenon, an appropriate account would have to specify how intonationally determined stress overrides the inherent metrical structure of words.

The crucial point is that main stress seems to have nothing to do with the metrical structure of individual words in Hixkaryana. For this reason, I will retain the analysis above, which treats the word-internal metrical system without regard to the intonationally governed main stress.

Two facts support this decision. First, final syllables do not undergo Iambic Lengthening, a fact which follows if (a) they are not metrically strong at the word level, prior to intonation assignment; and (b) Iambic Lengthening is a word-level rule. Second, there is an important deviation in the lengthening pattern, not yet mentioned, that also supports the analysis. Derbyshire notes (p. 181) that words of the underlying shape /CVCV/ (i.e. /⏑ ⏑/), which would be expected to have no lengthening at all, in fact undergo lengthening of their INITIAL vowels:

(171)	a. /kwaya/	→	[kwaːya]	'red and green macaw'	D 181
	b. /kana/	→	[kaːna]	'fish'	D 177
	c. /tuna/	→	[tuːna]	'water'	D 177

This lengthening is a puzzle, given the overall iambic pattern of the language. Note that /CVCCV/ words do not lengthen: *arko* 'take it' D 177.

A solution is possible under the assumption that Hixkaryana displays the "unstressable word syndrome" (§ 5.3): if final syllables are extrametrical, and degenerate feet are forbidden, then words of the form /⏑ σ/ need special help to obtain metrical structure and thus respect culminativity. The lengthening of the initial syllable of /CVCV/ serves just this function:

(172)	a. **Without Lengthening**	?? ⏑ ⏑ CV ⟨CV⟩	b. **With Lengthening**	(×) – ⏑ CVː⟨CV⟩

If this account is correct, we have additional evidence for the view that main stress assignment is intonational in character and independent of the word-internal metrical system. Were we to incorporate the main stress system at the word level, we would have to allow for final stress at this level, and there would be no explanation for initial lengthening.

Clearly, the interpretation of main stress in Hixkaryana is controversial, and would benefit from phonetic studies of stress and intonation in this language. For our purposes, however, the crucial facts are the alternating pattern of prominence and lengthening within words. These form a canonical example of iambic stress, and of segmental rules that reinforce the inherent durational contrast of iambic rhythm.

Two languages related to Hixkaryana within the Cariban family have quite similar iambic patterns: Carib of Surinam (Hoff 1968) and Macushi (Hawkins 1950; Abbott 1991).

6.3.2 Rhythmic Lengthening in Choctaw and Chickasaw

Choctaw and Chickasaw are closely related languages that form the Western branch of the Muskogean family. Data and analysis have been drawn here from Nicklas 1972, 1975; Munro and Ulrich 1984; and Ulrich 1986. Earlier metrical accounts of these languages include the latter two references, as well as Lombardi and McCarthy 1991.

I have already discussed another pair of Muskogean languages, namely Creek and Seminole (§ 4.1.2). Choctaw and Chickasaw share with Creek and Seminole the following characteristics: they are pitch accent languages, with certain syllables in the word marked for H tone. All four languages can be argued to have metrical systems in which iambs are constructed left to right, with heavy syllables defined as /CVː/ or /CVC/. Remarkably, the effects of metrical structure are completely different in the two groups. In Creek/Seminole, the feet apparently have no segmental phonetic correlates, and serve only to locate H tones. In Choctaw and Chickasaw, the feet appear to have ONLY segmental phonetic correlates, with the H tones assigned by quite different principles (Ulrich 1986, 67–72, 220–27).

According to Munro and Ulrich 1984, 193–94, the heads of feet in Choctaw and Chickasaw are "perceptually prominent, and could be characterized as stressed." Vowels in the weak syllables of a foot are slightly centralized, and in certain segmental contexts they are deleted. Nonfinal short vowels in metrically strong open syllables undergo a rule called Rhythmic Lengthening, which converts feet of the form /⏑ ⏑́/ to the canonical /⏑ –́ / shape. This is essentially the same rule of Iambic Lengthening found in other iambic languages (§ 4.5.3); the name of the rule follows Muskogeanist usage.

The rules are stated explicitly in (173):

(173) a. **Syllable Weight** /–/ = CVː, CVC /⏑/ = CV

b. **Foot Construction** Form iambs from left to right; degenerate feet are forbidden.

c. **Rhythmic Lengthening**

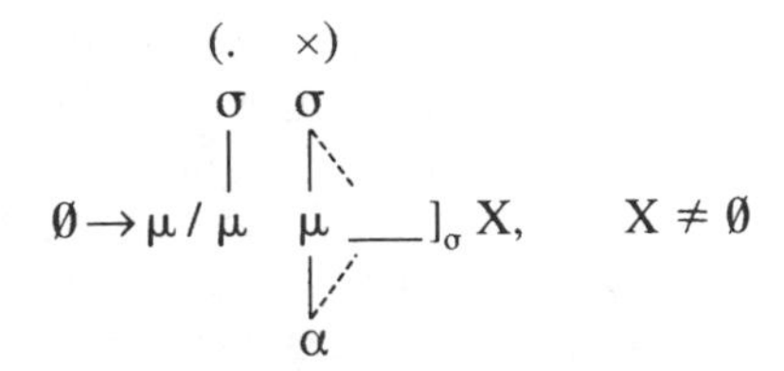

For (173c), Lombardi and McCarthy 1991 propose to subsume the condition X ≠ Ø under a general and persistent prohibition against final [Vː]. However, Chickasaw has acquired new phonemic final long vowels from historically earlier /VhV/ (P. Munro, p.c.). I include a blocking condition in the Rhythmic Lengthening rule to cover the Chickasaw case.

In (174) are examples (from Choctaw dialects) of the alternations that Rhythmic Lengthening creates. References labeled N are to Nicklas 1975.

(174)

a. /litiha-tok/ 'it was dirty' MU 192	b. /sa-litiha-tok/ 'I was dirty' MU 192	c. /okča-li-li-h/ 'I woke him up' U 54	
(. ×) (. ×)	(. ×)(. ×)(×)	(×) (. ×)(×)	Foot
˘ ˘ ˘ –	˘ ˘ ˘ ˘ –	– ˘ ˘ –	Construction
litihatok	*salitihatok*	*okčalilih*	
(. ×) (. ×)	(. ×) (. ×) (×)	(×) (. ×) (×)	Rhythmic
˘ – ˘ –	˘ – ˘ – –	– ˘ – –	Lengthening
*lit**i:**hatok*	*sal**i:**tih**a:**tok*	*okčal**i:**lih*	
d. /pisa-li/ 'see-I (subj.)' N 242	e. /či-pisa-li/ 'you (obj.)-see-I (subj.)' N 242	f. /či-habina-či-li/ 'you (obj.)-receive a present-CAUS.-I (subj.)' N 242	
(. ×)	(. ×)(. ×)	(. ×)(. ×)(. ×)	Foot
˘ ˘ ˘	˘ ˘ ˘ ˘	˘ ˘ ˘ ˘ ˘ ˘	Construction
pisali	*čipisali*	*čihabinačili*	
(. ×)	(. ×) (. ×)	(. ×) (. ×) (. ×)	Rhythmic
˘ – ˘	˘ – ˘ ˘	˘ – ˘ – ˘ ˘	Lengthening
*pis**a:**li*	*čip**i:**sali*	*čih**a:**bin**a:**čili*	

Some further notes on this system:

(a) The metrical rules of Choctaw and Chickasaw interact in an intricate way with morphological structure: they apply cyclically, based on a set of morphological levels. For details, see Munro and Ulrich 1984, Ulrich 1986.

(b) Although the above representations assume that final degenerate feet are forbidden, the facts do not permit clear conclusions in this area. Pamela Munro (p.c.) observes that the vowels of final syllables are considerably less tense than those of rhythmically lengthened syllables, but slightly more tense than the weak vowels of disyllabic feet. This fact could be attributed either to a final degenerate foot in cases like (174d), or to the effects of phonetic final lengthening (§ 5.1.8.2). As noted above, the selection of which syllable is most prominent in the word as a whole is determined by the pitch accent system, which is entirely independent of metrical structure.

Degenerate feet arguably are forbidden, since there are no content words in Choctaw/Chickasaw consisting of a single light syllable. This constraint holds of nouns at the underlying level (Nicklas 1972, 22), and is enforced for verbs by a rule of Prothesis (Nicklas 1975, 240).

(c) According to Munro and Ulrich 1984, 195, Rhythmic Lengthening is non-neutralizing; that is, the long vowels it creates are phonetically not as long as underlying long vowels. Munro and Ulrich show that Rhythmic Lengthening applies in the "deep" phonology, which makes it unlikely that it adds mere phonetic, non-structural length.[13] It is conceivable that Choctaw and Chickasaw work like St. Lawrence Island Yupik (§ 6.3.8.1): in stressed open syllables, the underlying contrast of one versus two moras is realized on the surface as two versus three.

6.3.3 Unami and Munsee Delaware

Unami and Munsee are mutually unintelligible "dialects" of Delaware, an Eastern Algonquian language. Their highly complex phonology is described and analyzed in Goddard 1979 (hereafter G), 1982. I begin with a cautionary note: Goddard observes that many of the alternations conditioned by stress in these dialects are being leveled out, and that the synchronic productivity of the rules he posits for descriptive purposes is questionable. The rules are more likely to be valid for an earlier stage of the language.

Goddard sets up as the basic vowel inventory of Munsee and Unami the following: long /iː, eː, oː, aː/ and short /ə, a/. In addition, there are limited occurrences of phonemic short /i, e, o/. Syllable weight distinguishes heavy CVː, CVC from light CV.

The metrical structure of Munsee and Unami is reflected in what Goddard calls "weak" and "strong" vowels: "A vowel in an odd-numbered syllable in a string of (one or more) open, short-voweled syllables is WEAK; other vowels are STRONG" (G xi). Main stress operates off this distinction: "the last non-ultimate V^{ST} [= strong vowel] is stressed . . . or a final-syllable V^{ST} if it is thc only one in the word" (G 21). Since the strong vowels that lack main stress resist reduction processes, it is plausible to suppose that they bear secondary stress; cf. Goddard 1982, 19.

In the theory proposed here this pattern can be analyzed as in (175):

(175) a. **Foot Construction** — Forms iambs from left to right. Degenerate feet are allowed in strong position.

b. **Foot Extrametricality** — Foot → ⟨Foot⟩ / ___ $]_{word}$

c. **Word Layer Construction** — End Rule Right

The left-to-right iambs account for the alternating strong and weak vowels noted by Goddard, in that they assign some degree of stress to all heavy syllables and to the even syllables of a light-syllable string. The foot extrametricality rule reflects Goddard's observation that only "nonultimate" strong

13. As a reviewer notes, this is important, in that it indicates that the influence of the Iambic/Trochaic Law can extend fully into the phonology and is not just a phonetic effect.

syllables bear main stress, unless the word is so short that there is only one strong syllable; in the latter case, Foot Extrametricality is blocked by the constraint that it must not exhaust the stress domain (§ 3.11). Examples (176)–(177) illustrate the analysis.

(176) **Unami**

a. '(vegetable) gum'
G xi
(×)
(. ×)
˘ –
pəkəw
[pkó]

b. 'he is weak'
G xiv
(×)
(. ×)⟨(×)⟩
˘ ˘ –
šawəsəw
[šawə́so]

c. 'I am weak'
G xiv
(×)
(. ×)⟨(. ×)⟩
˘ ˘ ˘ –
nəšawəsi:
[nšáwsi]

d. 'he is red'
G xx
(×)
(×)⟨(. ×)⟩
– ˘ –
maxkəsəw
[máxkso]

e. 'when he found me'
G xiii
(×)
(×) (. ×)⟨(. ×)⟩
– ˘ – ˘ –
e:nta-maxkawi:t
[entamáxkai:t]

f. 'the little ones starved to death'
G xiii
(×)
(×) (. ×) (×)⟨(. ×)⟩
– ˘ – – ˘ –
ša:wala:mwi:təwak
[ša:ɔla:mwí:ttowak]

(177) **Munsee** (Goddard 1982)

a. 'I am well' p. 35
(×)
(. ×)(. ×)⟨(. ×)⟩
˘ ˘ ˘ ˘ ˘ –
nəwəlamaləsi:
[no:lamálsi]

b. 'he is well' p. 35
(×)
(. ×) (. ×)⟨(×)⟩
˘ ˘ ˘ ˘ –
wəlamaləsəw
[wə̆lamalə́səw]

c. '(in) heaven' p. 30
(×)
(. ×)(×) ⟨(. ×)⟩
˘ ˘ – ˘ ˘
awasáhkame:w

I assume that Munsee and Unami tolerate degenerate feet, at least in strong position, since segmental rules often create monosyllabic words of degenerate size, such as (176a). Other than in monosyllables, the Munsee/Unami evidence concerning the status of degenerate feet is difficult to interpret. Final "leftover" syllables undergo reduction and deletion processes, but not the same processes as weak syllables within the word. Goddard notes (G xi) that "the alternations [due to deletion] have been leveled out in many morphemes; hence some short vowels in open syllables must be marked as basically STRONG." In addition, apparent weak degenerate feet are often created by segmental rules applying after stress. Whether such cases require us to posit degenerate feet in weak position is hard to determine, since the synchronic productivity of the system is in question.

Both Unami and Munsee have a fair number of segmental rules that are based on metrical structure. Among the more interesting is a rule which has the effect of increasing the durational contrast of the iambic foot: in Unami, a voiceless consonant other than /h/ is geminated after a strong vowel (G xii, 22). In Munsee, the rule applies to all consonants, but only after short vowels (G 27; gemination not reflected in Goddard's transcriptions). In (178) is an Unami example of gemination:

(178) (×) (×) 'I follow a trail' G xiii

(. ×)⟨(. ×)⟩ → (. ×) ⟨(. ×)⟩

⏑ ⏑ ⏑ – → ⏑ – ⏑ –

nəmətəme: → *nəmətteme:* → [nəmə́ttəme]

compare /mətəme:w/ → [mətə́me:(w)] 'he follows a trail' G xiii

Where the target of this rule is a light syllable, it has the effect of creating a canonical /⏑ –́ / iamb, following the Iambic/Trochaic Law.

Another iambic effect on segmental phonology is a decrease in the duration of the weak syllable of a foot, by vowel reduction. In various contexts the reduction in fact is extended to full deletion. Deletion of course actually destroys canonical feet; this might be viewed as the anti-canonical result of taking the reduction process too far.

One final point of interest is an apparent shift of stress under vowel deletion. Goddard (G xvi) states the following rule for Unami: "A short vowel is syncopated before a cluster with first member *x, s,* or *š,* except after a weak-vowel syllable." In (179) and (180) are some examples, with their metrical structures:

(179) **Weak-Vowel Syllable Precedes: No Deletion**

a. (×)

(. ×) ⟨(. ×)⟩

⏑ – ⏑ –

/kət**ə**xkwəsəw/ → [kt**ó**xkwso] 'he crawled out' G xvi

```
b.  (         ×)
    (.  ×) (.  ×)⟨(×)⟩
     ⌣  –   ⌣ –  –
```

/nə-p**ax**kəši:kan/ → [mp**ax**kší:kkan] 'my knife' G xxi

(180) **No Preceding Weak Vowel: Deletion**

```
a.  (    ×  )
    (.  ×)(×  ) ⟨(. ×)⟩
     ⌣  ⌣ –      ⌣ –
```

/nəkət**ə**xkwəsi:/ → [nkə́txkwsi] 'I crawled out' G xvi

```
b.  (       ×)
    (×) (.  ×)⟨(×)⟩
     –   ⌣  –  –
```

/p**ax**kəši:kan/ → [kší:kkan] 'knife' G xxi

The first point to observe is that the vowels of the underlying syllables /təx/ and /pax/ are undeletable where deletion would destroy a canonical /⌣ –́ / iambic foot, as in (179). Also of interest is the surface position of main stress in (180a), which falls on what was originally the antepenultimate foot, violating the ordinary pattern of the language. What is odd about this is that we cannot derive it by ordering deletion before stress assignment, since deletion is itself dependent on metrical structure.

To account for this, I posit that deletion does indeed take place after stress assignment, with the fate of the "stranded" stress determined by metrical structure, as laid out in § 3.8.1. In (181) is a representation of the output of vowel deletion, applied to the forms of (180):

```
(181) a. (          ×     )                b. (              ×)
         (.   ×) (×      )⟨(.    ×)⟩           (×     ) (.    ×) ⟨(×)⟩
          σ    σ    σ        σ    σ             σ        σ    σ     σ
          |    |    |\       |    |\            |\       |    |\    |\
          μ    μ    μ μ      μ    μμ            μ μ      μ    μμ    μ μ
          nə   kə   t    x   kwə  si:           p   x    kə   ši:   ka n
```

The newly vowelless syllables are ill-formed, and assuming that syllabification is persistent (see e.g. Ito 1986), their prosodic structure must delete:

```
(182) a. (          ×     )                b. (              ×)
         (.   ×) (×      )⟨(.    ×)⟩           (×     ) (.    ×) ⟨(×)⟩
          σ    σ             σ    σ                      σ    σ     σ
          |    |             |    |\                     |    |\    |\
          μ    μ             μ    μμ                     μ    μμ    μ μ
          nə   kə   t   x    kwə  si:           p   x    kə   ši:   ka n
```

The grid marks serving as heads of the maimed feet could in principle migrate to another docking site, but by the Faithfulness Condition (§ 3.7) they cannot move outside their own constituents. I assume, therefore, that grid marks and foot structure must also be deleted:[14]

```
(183) a. (          ×    )               b.        (   ×)
         ( .   ×)        ⟨⟨( .   ×)⟩⟩              ( .   ×) ⟨⟨(×)⟩⟩
          σ   σ             σ   σ                   σ   σ    σ
          |   |             |   |\                  |   |\   |\
          μ   μ             μ   μμ                  μ   μμ   μ μ
         nə  kə t  x      kwə  si ː           p  x kə  ši ː ka  n  → [kšíːkkan]
```

For (183b), the derivation is completed by syncope of schwa (G xiv) and simplification of the resulting /pxkš/ cluster to [kš] (G xxi). The case of (183a) is more interesting: here the grid mark of the word layer has been rendered floating by the loss of its supporting foot. The geometry of the metrical representation is such that it can dock leftward, respecting the Faithfulness Condition. This derives the correct output:

```
(184) (     ×     )
      ( .   ×)     ⟨⟨( .   ×)⟩⟩
       σ   σ           σ   σ
       |   |           |   |\
       μ   μ           μ   μμ
      nə  kə tx  kwə   si ː  (→ [nkə́txkwəsi])
```

If this account, or something along these lines, is correct, Unami can be added to the list of cases where stress migrates under deletion of stressed vowels. It differs interestingly from other cases in that the actual foot of the deleted vowel disappears entirely; migration shifts only the main stress.

Summing up, Munsee and Unami appear to be highly characteristic iambic languages, with segmental phonologies that are "tilted" toward creating and preserving canonical iambic structure.

6.3.4 Further Algonquian Languages: Malecite-Passamaquoddy, Eastern Ojibwa, Menomini, Potawatomi

Iambic stress patterns can be observed in other Algonquian languages besides Munsee/Unami.

(a) **Malecite** and **Passamaquoddy** are two names for essentially the same language, spoken by different groups in Maine and New Brunswick. Like Munsee and Unami, Malecite-Passamaquoddy is Eastern Algonquian. Its stress pattern is analyzed in Stowell 1979, citing data and a metrical analysis from Philip LeSourd. Brief sketches of the phonology appear in Teeter 1971 and Teeter and LeSourd 1983.

14. A useful comparison is Cyrenaican Bedouin Arabic (§ 6.3.7), which at the corresponding stage of its derivations intervenes with vowel epenthesis. This provides a docking site for the stranded grid mark and saves the foot.

Malecite-Passamaquoddy stress is basically the same as in Munsee/Unami, but with two differences. First, CVC syllables count as light, not heavy. Second, there is an alternating pattern of prominence among stressed syllables, going from right to left before the main stress. Following Stowell, I attribute this to an intermediate layer of cola (§ 5.5.1), which are binary and left-strong. In (185) is a sample representation:

```
(185) Word layer:   (           ×      )    'those (animate) who  S 57, 59
      Colon layer:  (×       )(  ×     )     must have been chosen'
      Foot layer:   (×)(.  ×)(.  ×)(.  ×)
                     –  ˘  –  ˘  ˘  ˘  –              2  0 3   1 0 3
                    me:kənu:təsəpəni:k      →   [me:kənu:tsəpəni:k]
```

As in Munsee/Unami, there are rules deleting weak vowels in certain segmental environments, such as the fourth vowel of (185).

Malecite-Passamaquoddy distinguishes three surface degrees of vowel length. Underlyingly, the short vowel /ə/ is opposed to /i: e: a: o:/, which are somewhat longer phonetically (Stowell 1979, 57). Rules presented by Teeter and LeSourd 1983 derive yet longer variants of the underlying long vowels. I conjecture that these fully long vowels are trimoraic, derived by adding a mora to underlying bimoraic /i: e: a: o:/.

(b) **Eastern Ojibwa** is a Central Algonquian language, whose dialects go by various names. Data and analysis for its prosody are found in Bloomfield 1956, Kaye 1973, Piggott 1980, and Piggott 1983; see the last and HV for earlier metrical accounts.

Eastern Ojibwa feet are iambs, assigned from left to right. CVC is counted as a light syllable; compare *mindídò* 'he is big', *niškádizìw* 'he is angry' (Piggott 1983, 14, 18), where initial CVC is skipped over by the stress rule. Severe reduction processes apply to the weak syllables of feet, following the general pattern of increasing durational contrast in an iambic system.

Rhodes (1985) contends that historically, reduction has been restructured as deletion: in his data, ultrashort vowels in weak position are no different in their quality or distribution from the excrescent vowels inserted into underlying consonant clusters. Thus Rhodes favors an analysis with full deletion, followed by later epenthesis. Rhodes also notes that the alternations created by deletion are starting to be leveled out, and he questions the synchronic productivity of the iambic system.

Accounts of main stress vary. Bloomfield 1956 does not describe any differences in prominence among multiple stresses. In the Odawa dialect (Kaye 1973; Piggott 1980), main stress falls on the head of the antepenultimate foot, otherwise the initial foot in shorter words. In the theory proposed here, this can be derived on the model of antepenultimate-syllable stress (§ 3.11): I posit foot extrametricality, plus a rule that creates a colon layer, constructing on it a trochee whose terminals are feet, as in Malecite-Passamaquoddy. For concreteness I add a word layer, though no evidence bears on whether this is necessary. This analysis is essentially that suggested by Piggott 1983, 47–50. In (186) are two sample representations:

```
(186) a. Word layer:   (  ×            )          'we (excl.) sit'  P 1980, 69
         Colon layer:  (  ×            )
         Foot layer:   (. ×) (.  ×)⟨(.  ×)⟩
                        ˘  ˘   ˘  ˘   ˘  ˘
                       ninamadabimi    →   [nnámdàbmì]

      b. Word layer:   (      ×          )        'a store'        P 1980, 84
         Colon layer:        (×      )
         Foot layer:   (. ×) (×) (. ×)
                        ˘  –   –  ˘ ˘ ˘
                       oda:we:wigamigw   →   [dà:wé:gàmìk]
```

In (186b), extrametricality is blocked by the Peripherality Condition, so stress resides on the penultimate foot. I propose to attribute the final secondary stress heard by Piggott to phonetic final lengthening, as discussed in § 5.1.8.2.

A small set of words having the shape CVNCV(:), where N is a nasal consonant, bear initial stress (Kaye 1973, 45; Piggott 1980, 84–85). To account for this, I adopt (translated into moraic theory) Piggott's proposal (1983, 42): a special rule of Weight by Position assigns a mora to the nasal, making the initial syllable heavy (see § 5.6.1), and the normal stress rules are applied to the result.

The issue of degenerate feet in Eastern Ojibwa is a difficult one. I assume tentatively that the language forbids them entirely. The deeper phonology supports this assumption. For example, the shortest underlying content word types are CVCV and CV:, both proper iambs (Piggott 1980, 307). Moreover, a rule of final short vowel deletion (Bloomfield 1956, 16; Piggott 1980, 306–7) is blocked in input forms of the shape CVCV—precisely when it would create a degenerate foot.

Where the degenerate foot prohibition faces more difficulty is in the phenomena associated with vowel reduction/syncope. Here, apparent degenerate feet abound on the surface. For example, [mwi] 'he sleeps' (Piggott 1980, 122) apparently violates the minimal word constraint; it is derived from /mawi-w/ by syncope and loss of the final glide. The final two feet of (186a) are likewise rendered degenerate in the same way.

I see two possibilities for coping with such examples. One is to suppose that syncope remains a rule of phonetic implementation, in the sense of Pierrehumbert 1980. Such a rule would assign metrically weak vowels very short durations, or even "hide" them under overlapping consonant articulations; but it would not alter the original metrical structure. If syncope turns out on the other hand to be categorical rather than phonetic, there is another possibility: to assume with Rhodes 1985 that the rule has led to a massive restructuring of the system, and that the iambic pattern is now of only historical interest.

The fact that syncope does not apply to final light syllables does not seem to be compelling evidence that such syllables form degenerate feet. In various Arabic dialects (see § 6.1.1; § 6.3.6; § 6.3.7; and references cited there), syncope is likewise blocked in final syllables, yet the final syllable is uncontroversially stressless. It is plausible to suppose that phonetic final lengthening can prevent reduction/syncope from gaining a hold in this location.

In the Maniwaki dialect of Ojibwa (Piggott 1978, 173–74), the evidence against final degenerate feet is clear, since certain suffixes (e.g. *-wag,* 3 pl.) undergo a kind of attrition whenever they form a final stray syllable:

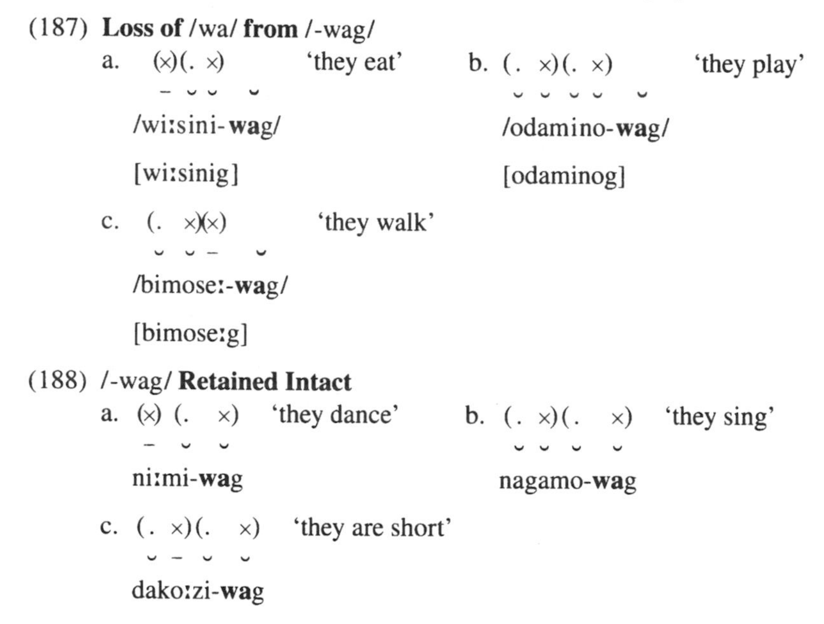

The pattern here would hardly make sense if a degenerate foot were to be constructed over the final light syllables of these forms.

(c) **Menomini** is (like Eastern Ojibwa) a Central Algonquian language, described by Bloomfield 1939, 1962 (hereafter B), 1975; and Hockett 1981. A metrical analysis of Menomini was first proposed by Pesetsky 1979.

Menomini assigns stress by forming iambs from left to right, with CVV(C) counted as heavy and CV(C) as light. The minimal content word is a nondegenerate iamb, that is, CVV(C) or CVCV(C), and this constraint is actively enforced by a phonological rule: underlying short-voweled monosyllables undergo vowel lengthening. Menomini increases the durational contrast of the iamb by reducing vowels in the weak syllable of a foot, but not to the same degree as the other Algonquian languages just discussed.

There are two rules in Menomini that lengthen the head of a disyllabic iambic foot, following the expected /◡ ◡́/ → /◡ –́/ pattern. When a word begins with two light syllables, the vowel of the second syllable is lengthened. This can be seen, for example, in different inflected forms of the same verb:

(189) a. (. ×)(×) → (. ×) (×) 'he is fed' B 57

˘ ˘ – → ˘ – –

*ahs**a**ma:w* → *ahs**a:**ma:w*

b. (. ×) (. ×) → (. ×) (. ×) 'I feed him' B 57

˘ ˘ ˘ – → ˘ – ˘ –

*net**a**hsama:w* → *net**a:**hsama:w*

Bloomfield also notes (B 17) that "between short vowels, when the preceding one is even-numbered in a sequence of short vowels (and accordingly has secondary stress), an *n* is often lengthened." His example is (190):

(190) (. ×) (. ×) → (. ×) (. ×) 'I don't know'

˘ ˘ ˘ ˘ ˘ → ˘ – ˘ – ˘

*kan nekɛhkena**n**an* → *kan nekɛ:hkena**nn**an*

For discussion of the claimed heavy CVC quantity created here, see § 7.3.

The phonology of Menomini includes another rule that refers to metrical structure, but in a quite unusual way. Whereas initial feet show ordinary iambic lengthening, non-initial feet display the following pattern: when occurring as the head of a non-initial disyllabic foot, a vowel must be short in an open syllable and long in a closed syllable (B 91, with corrections by Goddard, Hockett, and Teeter 1972). In the determination of whether a syllable is closed, final consonants must be counted as extrametrical, so that final CVC is a position of shortening, not lengthening.

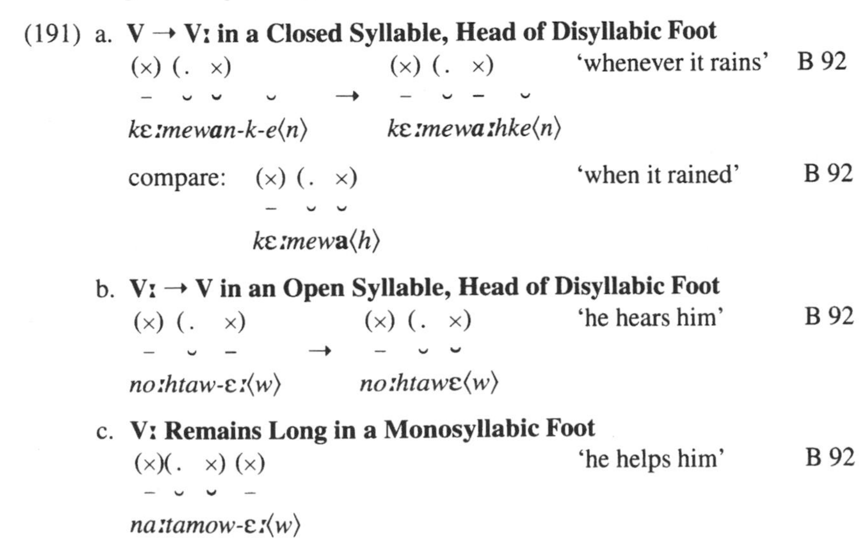

The lengthening in (191a) is expected, since it takes /⏑ ⏑́/ to /⏑ –́ /. But the shortening in (191b) is anti-iambic, taking /⏑ –́ / to /⏑ ⏑́/, and thus constitutes a problem for the theory. Note first that this rule is odd in more than one respect: the natural pattern, found in a great number of languages, is for vowels to shorten in closed syllables and to lengthen in open ones, not the other way around. Thus the Menomini facts are a problem for theories of syllable markedness as well as for the theory of foot structure proposed here.

The rule governing length in non-initial feet seems to be a "crazy rule," in the sense of Bach and Harms 1972. I believe the best account of such rules often is to reconstruct their diachronic origins, explaining them away as the synchronically unnatural result of a sequence of natural changes. Here is a possible scenario of this sort.

Stage 1: Menomini lengthens short vowels in the heads of disyllabic feet; that is, at this stage the rule for initial feet, illustrated in (189), applies across the board. As a result of this lengthening, the contrast between long and short vowels is neutralized in strong syllables.

Stage 2: Long vowels in non-initial feet are somewhat reduced in their phonetic duration. Possibly this was due to assignment of main stress to the initial foot, though this is purely conjectural.

Stage 3, restructuring: A new generation acquires the language and must assign the phonetically shortened long vowels of non-initial feet to a phonemic category, either short or long. Given that there is no contrast of length in this position, and that the vowels are phonetically intermediate in value, which phonemic category is selected?

It is known that across languages, vowels characteristically are phonetically shorter in closed syllables (Maddieson 1985). Assuming that this pattern is encoded as part of universal grammar, children constructing a phonology plausibly would adopt the following tacit reasoning: a semi-long vowel in an open syllable sounds like a phonemic short vowel made somewhat longer under stress; whereas a semi-long vowel in a closed syllable sounds like a phonemic long vowel, slightly shortened in the closed-syllable environment. Thus the semi-long vowels induce a phonemic restructuring of the phonetic facts. Note that long vowels in non-branching feet would be unlikely to be restructured, even if phonetically shortened: the language learner can tell that they are long since they attract stress, as an underlying short vowel would not attract stress in the same position.

Stage 4: The restructured phonemic categories are grammaticalized; that is, the morphophonemic length rule derives long vowels in closed syllables and short vowels in open ones.

This is my conjecture of how Menomini could have acquired its "crazy" length rule. Crucially, the shortening part of the rule was not in itself an added innovation, but the result of restructuring. The only actual foot-based sound change was a normal iambic lengthening.

To sum up, the rule for non-initial feet in Menomini is an apparent counterexample to my claim that quantitative phonology characteristically creates canonical /⏑ –́/ feet in iambic languages. But like many puzzling phenomena, the rule is a counterexample to more than one principle; it also violates the characteristic pattern of lengthening in open syllables and shortening in closed. The conjecture of historical restructuring accounts for both anomalies.

The pattern of main stress in Menomini (B 19–21) is an odd one: for example, words without long vowels apparently have no primary stress. Given this, I conjecture that Bloomfield's "main stress" may actually be a form of pitch accent, but I will not attempt to analyze it here.

(d) **Potawatomi** (Hockett 1948), like Eastern Ojibwa and Menomini, is a Central Algonquian language. Hockett's analysis of Potawatomi morphophonemics clearly diagnoses a left-to-right iambic system, with a syllable weight criterion of CV(C) versus CVː(C) and an underlying vowel system we may interpret as /iː, eː, oː, aː, ə, o/. (This is not Hockett's interpretation; he uses no vowel length distinction, and uses abstract morphophonemes for the deletable vowels /ə, o/.) Loss of weak vowels is obligatory, except in singing. Undeleted /o/ (in strong position) is phonetically identical to underlying /oː/, suggesting that it undergoes iambic lengthening. Information on how the iambic system is related to surface stress, if at all, is not presented.

To conclude, I will sum up what emerges from the study of the six iambic Algonquian languages examined here.

First, segmental rules in all the cases considered have the properties characteristic of iambic rhythm: they either lengthen the strong syllables of a foot, shorten the weak syllables, or both.

Second, it is striking that two of the Algonquian languages assign to their FEET prominence patterns that quantity-insensitive languages often assign to SYLLABLES: Eastern Ojibwa has antepenultimate-foot stress, and Malecite-Passamaquoddy has alternating stress on even feet from the end. The reason may be this: both languages have severely eroded the weak syllables of iambic feet; that is, they have taken the natural iambic process of weak syllable reduction so far that by now, most feet are monosyllabic. Rather than give up alternating rhythm entirely, these languages have created new rhythmic alternation on a higher grid layer. Menomini, with less phonetic erosion of weak syllables, has not adopted a syllable-like rhythm at higher layers.

The common pattern of left-to-right iambs in both Central and Eastern Algonquian languages leads one to wonder whether the pattern was a property of Pronto-Algonquian. Cowan 1982, 43, states that "stress is not reconstructed for Proto-Algonquian," but Hockett 1942, 538, posits an iambic pattern for "at least those dialects of [Proto-Central Algonquian] which underlay Potawatomi, southern Ojibwa, and Mohican." Goddard (p.c.) is cautious on constructing iambic stress for Proto-Algonquian, noting among other things the difference in the treatment of CVC syllables between Munsee/Unami (CVC = /–/) and the other languages (CVC = /⏑/). In addition, a fair number of Algonquian languages do not show iambic patterns. Thus one cannot rule out the possibility that iambic stress is not a Proto-Algonquian inheritance, but has arisen separately in more than one branch of Algonquian.

6.3.5 Cayuga and Other Lake Iroquoian Languages

Accent in the Lake Iroquoian languages is documented and analyzed in Chafe 1977; Foster 1982; Michelson 1983, 1988; and references cited there. The westernmost three of these languages, Cayuga, Onondaga, and Seneca, are noteworthy for having evolved iambic accentual systems out of an original penultimate stress pattern. I concentrate here on Cayuga.

Early Cayuga inherited from Proto–Lake Iroquoian a rule that stressed the penultimate syllable. If the penult was open (according to the rules of syllabification that prevailed at the time), its vowel was lengthened. I will call the synchronic reflex of this rule **Tonic Lengthening I.**

(192) **Tonic Lengthening I**
V → Vː / ____σ $]_{word}$ (with "early" syllabification)
Condition: V ≠ /a/.

The curious condition in (192) has a historical explanation (Foster 1982, 66): the rule originally avoided lengthening instances of /a/ that were derived by a rule of epenthesis; later, this restriction was generalized to all /a/.

Subsequently, the stress pattern of Cayuga changed dramatically, becoming left-to-right iambic. For the modern system, I posit the following analysis, based on ideas of Chafe 1977; Michelson 1983; Prince 1983a; and Benger 1984. These rules must be ordered after Tonic Lengthening I, since lengthened vowels attract stress.

(193) a. **Foot Construction** Form iambs from left to right (/ – / = CVː(C)). Degenerate feet are prohibited.

b. **Foot Extrametricality** Foot → ⟨Foot⟩ / ____$]_{word}$

c. **Word Layer Construction** End Rule Right

According to Foster 1982, 61, the iambic metrical pattern is realized directly as stress. In the examples in (194), the long penultimate vowels are derived by Tonic Lengthening I. Vowel nasality is transcribed with /˛/.

(194)

a. **Odd Heavy Penult**	b. **Even Heavy Penult**	c. **Odd Light Penult**	d. **Even Light Penult**
(×)	(×)	(×)	(×)
(. ×)(×)	(. ×)(. ×)	(. ×)(. ×)⟨(. ×)⟩	(. ×)
˘ ˘ – ˘	˘ ˘ ˘ – ˘	˘ ˘ ˘ ˘ ˘ ˘	˘ ˘ ˘
hę̨nato:was	*ę̨hę̨nató:wat*	*tewakatáwę̨nyeʔ*	*kanéstaʔ*
→ [hę̨na:tó:was]			
'they are hunting' F 63	'they will hunt' F 61	'I'm moving about' F 63	'board' F 62

Example (194a) shows that main stress can fall on an odd-numbered syllable only when the penult is heavy. Otherwise, stress falls either on an even penult (194d) or on an even antepenult (194c). Foot Extrametricality applies only in (194c); elsewhere, it is blocked because the rightmost foot is non-peripheral.

Most of the long vowels that attract stress in Cayuga are derived by Tonic Lengthening I. However, Foster 1982, 62, 66, points out some penultimate long vowels that synchronically are unpredictable and must be regarded as phonemic. These long vowels attract stress just like derived long vowels.

The ban on degenerate feet is based on the absence of degenerate-size content words, inferred from the glossary in Mithun and Henry 1982.

Several rules in Cayuga amplify the durational contrast of an iambic foot, following the Iambic/Trochaic Law (§ 4.5.3). A rule of Laryngeal Metathesis, discussed in Foster 1982, 68–71, and § 7.3.1, shortens syllables in the weak position of a foot; and there are two rules that lengthen the strong position. To make sense of the first of these, one must know that the syllabification principles of modern Cayuga are different from those of the ancestor language, allowing for more cases of /V.CCV/ and hence more open syllables; for the two systems of syllable division, see Foster 1982, 60, 62. A penultimate syllable assigned main stress by the rules of (193) lengthens its vowel when it is open by the modern criterion, even if it was not lengthened earlier:

(195) a. **Tonic Lengthening II**

$$V \rightarrow V\!: / \underset{}{\overset{\times}{\_\_\_\_}}\ \sigma\ ]_{\text{word}}$$ (with "modern" syllabification)

b.
(×)
(. ×)(. ×)(. ×)
˘ ˘ ˘ ˘ ˘ ˘ ˘

ę̨kà ta tǫ̨ kwʔe tǫ̨́ nyę̨ʔ → [ę̨kàtatǫ̨kwʔetǫ̨́:nyę̨ʔ]

'I will make some people for myself' F 61

In other words, a penultimate vowel in Cayuga has two chances to get lengthened: early on, if its syllable is open by the ancestral criterion; and later, if its syllable receives main stress by the alternating count and is open by the more liberal modern criterion. In addition, while Tonic Lengthening I does not apply to /a/, Tonic Lengthening II does; compare *hotíyaneʔ 'chiefs'* F 64 (odd /a/) with *hoyá:neʔ* 'chief' F 64 (even /a/).

Puzzlingly, Tonic Lengthening II usually does not apply in antepenultimate syllables; compare (194c), as well as Foster 1982, 65, for isolated cases of lengthening in this position. Such cases are problematic for the theory of trisyllabic shortening discussed in § 6.1.5.2; for an alternative account, see Kager (1993).

The other iambic lengthening rule of Cayuga (Foster 1982, 63) applies when a /⏑ ⏑́/ foot ending in an open syllable immediately precedes the main stress, as in (194a). Here, the /⏑ ⏑́/ foot /henà/ is lengthened to /henà:/. Prince 1983a attributes to this rule the function of alleviating a clash between adjacent stressed syllables.

Some additional puzzles arise within the metrical phonology of Cayuga.

(a) Cayuga has changed its syllabic and metrical structures over time, leading to problems in the formulation of its synchronic grammar. For instance, the two iambic lengthening rules are based on different syllable divisions from the historically earlier rule of Tonic Lengthening I. To the extent that the syllabification algorithm must be altered in the course of the phonological derivation, we cannot make use of invariant, persistent syllabification to give "coherence" to the phonology, in the sense of Dresher and Lahiri 1991. The alternative is to analyze Tonic Lengthening I with a purely linear structural description, with loss of generality. Likewise, the earlier shift of Cayuga from syllabic trochees (which governed Tonic Lengthening I) to iambs raises the same issue: whether Tonic Lengthening I should presently be governed by a syllabic trochee, or instead should be assigned a relatively ad hoc, purely segmental structural description. If a syllabic trochee is invoked, it must be abstract in character; that is, it must be overridden by the later assignment of iambs, so that some penults (194c) can surface unstressed.

(b) Following the introduction of iambic stress, Cayuga acquired new long vowels and diphthongs from original /VCV/ sequences. The accentual behavior of these newer VV is hard to establish. VV separated by an earlier /r/ count as disyllabic for purposes of stress, as (196) shows.

```
(196)  (              ×)               'he planted the word again'
       (.  ×)(.       ×)⟨(.    ×)⟩                 Michelson 1988, 103
        ⏑   ⏑   ⏑      ⏑   ⏑    ⏑

       s a h a r i h w a y ę t h o ʔ  →  [sahaihwáyętho ʔ]
                                    (not *[sahaihwayę́:thoʔ])
```

In many cases the underlying consonant is synchronically recoverable. For example, the /r/ in (196) is part of the underlying stem /-rihw-/ 'word, matter', which shows up overtly in other forms, such as *takatatríhǫnyęʔs* 'read it for me' (Mithun and Henry 1982, 561).

There may also be other VV sequences that behave accentually in the opposite fashion, as single heavy syllables. One argument in favor of this view is that it would help to explain why the lengthening rules are blocked in hiatus (Benger 1984, 45–49). A basic CVV syllable, being already bimoraic, would resist further lengthening:

```
(197) (            × )        'the food/meal will settle'        B 64
      (.  ×)(.     × )
       ˘   ˘  ˘     –  ˘
      haʔtękakhwaętaʔ         (not *[haʔtękakhwaː.ę́ːtaʔ])
```

(c) According to Chafe 1977, 176, Cayuga is a counterexample to the principle of culminativity (§ 3.1): most words that (in my terms) consist of just a single foot "have no accent at all," as in *kanyoːʔ* 'wild animal'. To account for this, I adopt the approach of Prince (1983a), who hypothesizes that Lake Iroquoian accent is partly metrical and partly tonal. In Cayuga, I posit that the H tone that docks to the strongest syllable of a word is normally prohibited from attaching H to a final foot. Chafe's accentless words are thus analyzed as bearing all low tone. The use of tone to account for Lake Iroquoian accent is supported by the phonetic correlates of accent in these languages: see Prince 1983a, 84–85), as well as Chafe 1977, 169, who notes that accented syllables have invariant high pitch, a typical characteristic of pitch accent (§ 3.9.1).

Below, I briefly summarize iambic patterns elsewhere in Lake Iroquoian.

(a) **Seneca**'s system is fairly close to Cayuga's, with rules corresponding to Tonic Lengthening I and II, plus some interesting complications. Prince 1983a argues that Seneca accent results from the interaction of metrical structure and tone: high tone docks first onto the rightmost nonfinal closed syllable, then may flop leftward onto an adjacent metrically strong syllable. In words lacking nonfinal closed syllables, H tone cannot dock at all; these words are described by Chafe 1977, 180, as accentless. The presence of iambs can nonetheless be demonstrated, because the iambically based rule of Tonic Lengthening II applies, as in the low-toned word *hata**ː**khe**ʔ*** 'he's running' (Chafe 1977, 179). Other accounts of Seneca (Stowell 1979; Levin 1985a; HV) have attempted to cover the data using only metrical structure. As Prince points out, such analyses fail because they cannot account for Tonic Lengthening II.

Late historical developments have considerably obscured the Seneca system. As in Cayuga, loss of intervocalic consonants has produced long vowels and diphthongs that pattern as disyllabic, documented in Michelson 1989. Further, loss of syllable-final /h/ with compensatory lengthening has created long vowels that behave as if they were short for accentual purposes. However, as Michelson 1984 and Levin 1985a show, these long vowels do not force us to assume that Seneca foot construction is quantity-insensitive; using a form of "even iamb" in the sense of § 4.3.1. The reason is that for purposes of the H tone docking rule mentioned above, their syllables pattern like CVC—that is, as light. Since tone docking must follow foot construction, such syllables count as light at the stage where feet are created. In contrast, heavy penults, with long vowels derived deeper in the phonology, attract stress, showing that the iambs are truly sensitive to quantity.

(b) **Onondaga** retains the penultimate stress pattern of Proto–Lake Iroquoian to a greater degree, but nonetheless also has iambic characteristics: vowels in even open pretonic syllables are lengthened, much as in Cayuga, and in long words the second vowel in a word lengthens before certain consonant clusters (Chafe 1977, 173–75; Michelson 1988, 94–98).

6.3.6 Negev Bedouin Arabic

The Negev Bedouin dialect of Arabic is described by Blanc 1970; the analysis here borrows from Kenstowicz 1981, 1983, whose rules are stated in a different metrical framework. Stress in Negev Bedouin Arabic is fairly regular, though Blanc notes a fair number of lexical exceptions. His data include a wide variety of syllable patterns. I give only examples where Blanc actually includes a stress mark; however, his verbal description of stress for the unmarked examples (p. 4) is consistent with what is given below.

(198) a. Stress a final superheavy (CVCC, CV:C) syllable:

/⏑ ⏑ ⏑ =́/:	*gahawatí:(h)*	'my coffee'	B 142

b. Otherwise, stress a heavy (CVC, CV:) penult:

/–́ –/:	*šárgiy*	'eastern'	B 122
/⏑ –́ ⏑/:	*ɣanámna*	'our sheep'	B 120
/⏑ ⏑ –́ –/:	*taħatá:niy*	'lower'	B 126
/– – –́ –/:	*baššibríyyih*	(no gloss)	B 140

c. Otherwise, stress a heavy antepenult:

/–́ ⏑ –/:	*áttifag*	'to agree'	B 117
/– –́ ⏑ –/:	*astáfhamah*	'he queried him'	B 132

d. In disyllables beginning with /⏑/, stress the final syllable:

/⏑ ⏑́/:	*biná*	'he built'	B 124
/⏑ –́/:	*ǰimál*	'camel'	B 121

e. Otherwise, stress the penult or the antepenult, whichever is separated by an odd number of light syllables from the nearest preceding heavy syllable, or in the absence of such a syllable, from the beginning of the word:

i. Penult	/⏑ ⏑́ ⏑/:	*aʔáma*	'blind'	B 124
	/⏑ ⏑́ –/:	*gaháwah*	'coffee'	B 126
	/– ⏑ ⏑́ –/:	*ankitálaw*	'they were killed'	B 121
ii. Antepenult	/⏑ ⏑́ ⏑ –/:	*zalámatak*	'your man'	B 121

Except for deviations involving syllable structure, this is exactly the same stress pattern as in Munsee/Unami (§ 6.3.3) and Cayuga (§ 6.3.5). I propose the rules in (199), giving sample representations in (200).

(199)

a.	**Syllable Weight**	CVC, CVː = /–/, CV = /˘/. Final C is unsyllabified in CVCC, CVːC.
b.	**Foot Construction**	Form iambs from left to right: degenerate feet are permitted in strong position.
c.	**Foot Extrametricality**	Foot → ⟨Foot⟩ / ____ $]_{word}$
d.	**Word Layer Construction**	End Rule Right

(200)

```
a. (     ×)             b. ( ×)     c. ( ×)     d. (  ×)
   (×)(×)⟨(. ×)⟩           (. ×)       (. ×)       (. ×)⟨(×)⟩
    –  –   ˘  –             ˘  ˘        ˘  –        ˘  ˘   –
```

a. *astáfhamah* b. *biná* c. *ǰimál* d. *gaháwah*

```
f. (        ×)          g. ( ×)                h. (  ×    )
   (×)  (. ×)⟨(×)⟩         (. ×)⟨(. ×)⟩             (×)
    –    ˘  ˘   –           ˘  ˘   ˘  –           ˘  –    ˘
```

f. *ankitálaw* g. *zalámatak* h. *ɣanámna*

As expected (§ 3.11), Foot Extrametricality is blocked when the rightmost foot is non-peripheral (200h) or is the only foot in the word (200b, c).

For words ending in a superheavy syllable, rule (199a) posits that the final consonant is unsyllabified until after the stress rules apply, just as in other Arabic dialects (§ 5.2.1). The unsyllabified consonant prevents the rightmost foot from being peripheral, so extrametricality is blocked and the rightmost foot receives the main stress:

(201)

```
(         ×)
(. ×)(. ×)
 ˘  ˘  ˘ –
```

gahawatiː⟨h⟩

As Blanc mentions no secondary stress, I assume that metrical feet in weak position either are removed by Conflation (§ 5.5.2) or simply receive no phonetic interpretation.

In general, Negev Bedouin Arabic is a less spectacular example of iambic stress than its Amerindian counterparts. The reason is that as with other modern Arabic colloquials, long strings of light syllables do not occur. Thus the "odd number" mentioned in (198e) is in fact always just one; the description is worded to emphasize the parallel.

Kenstowicz 1981, 1983 notes that the Negev dialect also shows accentual parallels with neighboring Arabic dialects: in terms of the present theory, it shares its direction of foot construction, its rule of foot extrametricality, and its End Rule with Palestinian and Egyptian Radio Arabic (§ 6.1.1, § 6.1.2). I conjecture that Negev once had the stress pattern still found in the latter languages, and underwent a shift from moraic trochees to iambs (perhaps by Move X, § 3.4.2), preserving the remaining parameter values.

Negev Bedouin Arabic seems to impose a weaker prohibition on degenerate feet than other Arabic dialects (McCarthy and Prince 1990). There are a number of content words that must count as monomoraic if (as is usual in Arabic) final consonants are extrametrical: *mliy* 'full' B 123, *miy* 'water' B 123, *gum* 'get up' B 137, *šil* 'pick up' B 137.[15] Moreover, application of a rule of epenthesis following stress assignment creates degenerate feet in strong position, as in ***addáluw*** 'the pail', from /al-dálw/ (cf. *dálw-ih* 'his pail' B 120–21). Degenerate feet also appear to be needed to account for exceptional stressing, as in *wáladah* 'his boy', in free variation with regular *waládah* B 121; or *ħamṛá* 'red (f. sg.)' B 115.

Blanc notes (p. 121) that words having the shape /– ⏑ ⏑ –/, as in *ankitálaw* above, have systematically variable stress: although *ankitálaw* is the most common stressing, it is also possible to say *ánkitalaw,* and similarly for parallel forms. I postpone analysis of this variation until § 7.1.6, where an account is presented based on the theory of syllable prominence.

6.3.7 Syncope and Stress in Cyrenaican Bedouin Arabic

The Cyrenaican Bedouin dialect of Arabic, spoken in Eastern Libya, is notable for its segmental rules, which can drastically alter the shape of a morpheme in different contexts. Here I focus on the (basically iambic) stress pattern and the way in which it is affected by syncope.

In certain Cyrenaican derivations, ALL the vowels of a foot may be deleted, resulting in a "floating" stress. A rule of Epenthesis provides a new syllable for the consonants stranded by syncope, and the floating stress docks onto this syllable. As I will show, the migration of floating stress is governed by the Faithfulness Condition, as laid out in § 3.8.1. This general type of analysis was first proposed for the Arabic dialect of Tripoli, Lebanon, by Kenstowicz and Abdul-Karim 1980, using an early tree-based metrical framework (§ 3.5). The bracketed grid account given below for Cyrenaican would apply straightforwardly to the Tripoli data as well.

My analysis of Cyrenaican is primarily based on the data and insights of Mitchell 1960. Additional material on a related dialect, with a different analysis, may be found in Owens 1980, 1984. Langendoen 1968 provides a useful discussion of Mitchell's work within the framework of early generative phonology.

15. Also *jṛa* 'puppies' B 124, which may be a typographical error.

6.3.7.1 *Segmental Phonology*

The dialect described by Mitchell apparently has the following vowel phonemes: /i, a, ɑ, ɨ, iː, aː, ɑː, uː/. /ɨ/ is realized as [u] in final position, and in certain contexts /aː/ is diphthongized to [ɪɛ]. Vowel backness is affected by a complex set of harmony rules (Owens 1984, 36–43; Mitchell 1975b). It appears that the backness of epenthetic vowels can be predicted, by assuming a kind of floating backness of the type proposed by Clements 1976 and applied to Arabic by Kenstowicz 1979.

A short vowel is maintained in its underlying form in two contexts: word-finally, and in closed syllables. In nonfinal open syllables, short vowels undergo raising and deletion processes, which I will now describe. My rules roughly follow the statements in Langendoen 1968. Stress marks are included in the transcriptions, but discussion of how stress is assigned is delayed to a later section.

A short high vowel in a nonfinal open syllable is deleted.

(202) **High Vowel Syncope** $\begin{bmatrix} \text{V} \\ \text{+high} \end{bmatrix} \rightarrow \emptyset \,/\, \underline{\quad\quad} \text{CV}$

The effects of this rule can be seen in the alternations in (203); page numbers refer to Mitchell 1975a (hereafter M).

(203) a.	*gássim*		'divide up!'	M 85
b.	*gassím-ha*		'divide it (f.) up!'	M 85
c.	/gáss**i**m-ih/ →	[gássmih]	'divide it (m.) up!'	M 85

There is also a rule of Raising, which applies to short low vowels in the same context, as illustrated in the examples of (205).

(204) **Raising** V → [+high] / ____ CV

(205) a.	*gássam*		'he divided up'	M 85
b.	*gassám-ha*		'he divided it (f.) up'	M 85
c.	/gass**a**m-ih/ →	[góss**i**mih]	'he divided it (m.) up'	M 85

Raising is blocked by neighboring guttural consonants: ***ħ**aší:š* 'grass' M84.

Since high vowels derived by Raising are not deleted, Raising must be ordered after High Vowel Syncope. Moreover, since non-alternating surface forms like *kitáb* 'he wrote' M 84 and *gitál* 'he killed' M 85 do not undergo Syncope, they must be phonemicized as /katab/, /gatal/, with underlying low vowels. This is supported by the absence of actual surface forms like **katáb*.

High Vowel Syncope often creates long consonant clusters, which are resolved by a rule of Epenthesis. I state Epenthesis below in a syllabic format, assuming a canonical syllable pattern of the form CV(V)(C) (/ay, aw/ are treated as VV, M 87). Any consonant that cannot be incorporated into this pattern is supported by an immediately preceding epenthetic high vowel:

(206) **Epenthesis** Ø → *i* / ____ C′ where C′ is an unsyllabified consonant

The backness of the inserted vowel (symbolized /i/ for brevity) is determined by harmony rules. Under (207) is an example of Epenthesis applying to resolve a triconsonantal cluster created by High Vowel Syncope. Periods are used to separate syllables from unsyllabified segments and from each other.

(207) a. *yík.tib* 'he writes' M 84

b. /yiktib-ɨ/ 'they write' M 84
yík.t.bɨ High Vowel Syncope (1)
yíkitbɨ Epenthesis (5)
yíkitbu *ɨ* → *u* / ____ #

Note that this derivation is quite similar to that in (203c). However, Epenthesis does not apply there, since it would split a geminate cluster: *gassmih* → **gasismih*. It is normal for epenthesis rules to be blocked in geminates (Kenstowicz and Pyle 1973); and I assume that the standard autosegmental account of this phenomenon (due to Jonathan Kaye; reviewed in Hayes 1986a, 326–28) would hold good for Cyrenaican.

Surface forms like [iktíb] are derived from underlying /kitib/ by High Vowel Syncope and Epenthesis. The underlying form is justified by the surface final stress and by forms derived from /kitib/ in other contexts; compare (217a).

(208) /kitib/ 'books' M 87
kitíb Stress Assignment (discussed below)
k.tíb High Vowel Syncope (202)
iktíb Epenthesis (206)

One final rule that plays a role here has a rather odd structural description but nonetheless is fairly widespread among Bedouin dialects (Irshied and Kenstowicz 1984, 114). It deletes a short vowel in an open syllable that is itself followed by an nonfinal open syllable; thus the rule requires a trisyllabic environment:

(209) **Trisyllabic Syncope** V → Ø / ____ CVCV

In effect, this rule can be illustrated only for underlying low vowels, since in the same context a high vowel would be expected to delete anyway by High Vowel Syncope (202). The derivations in (210a, b) justify the underlying form /katab/; (210c) illustrates Trisyllabic Syncope.

(210)	a. /katab/ → *kitáb*	} by Raising (204)	'he wrote'	M 84
	b. /katab-na/ → *kitábna*		'we wrote'	M 89
	c. /katab-at/		'she wrote'	M 84
	katábat	Stress Assignment (see below)		
	k.tábat	Trisyllabic Syncope (209)		
	k.tíbat	Raising (204)		
	iktíbat	Epenthesis (206)		

The four rules just given must be ordered in a way consistent with the following. (a) Both Syncope rules must precede Epenthesis, since Epenthesis repairs the consonant clusters they create. Moreover, high vowels in open syllables created by Epenthesis do not undergo High Vowel Syncope: /libs/ → [líbis] 'clothes' M 85. (b) As already noted, Raising must follow High Vowel Syncope, to avoid feeding it. (c) Raising cannot apply to vowels in open syllables created by Epenthesis: /katab-t/ → [kitábit] 'I/you (m. sg.) wrote' M 91. Therefore Epenthesis must follow Raising. As we will ultimately see, the actual order needed is Trisyllabic Syncope, High Vowel Syncope, Raising, Epenthesis.

6.3.7.2 *Earlier Accounts of Cyrenaican Stress*

Langendoen 1968 has provided a generative reinterpretation of Mitchell's original Firthian analysis. Since Langendoen's basic assumptions about phonology (e.g., that it involves ordered rules and derivations) are closer than Mitchell's to those employed here, I will review his account. Langendoen's basic analytical strategy is as follows: the two Syncope rules are assumed to be applicable only to stressless vowels, which indeed is a characteristic constraint on syncope. Since the stress rule must have access to underlying strings to apply correctly, it is ordered before Syncope.

The stress rule that results turns out to be rather complex. It assumes as a starting point the pattern of Cairene Arabic (§ 4.1.3), then adds the following modifications (Langendoen, p. 105): "Disyllables of the form *CVCVC* are stressed on the final syllable. . . . In words of three or more underlying syllables, if both the prefinal and the syllable preceding the prefinal are open, the prefinal syllable is accented if it contains a low vowel; otherwise the initial syllable is accented." By "initial" Langendoen appears here to mean "the first of the two candidate syllables," that is, the antepenult; compare his example /yingátilɨ/ → *yingítlu* 'they will be killed' (p. 107).

The complexity of this stress rule seems to argue against Langendoen's analysis; note that he does not attempt to formalize it (nor will I). Moreover, the unusual reference to vowel height in the rule recapitulates a distinction made by High Vowel Syncope (202); I show below that under a different analysis this duplication is unnecessary.

Langendoen points out (1968, 107) that his account also leads to an ordering paradox. Consider the derivation in (211a), with the underlying form justified in (211b). (For additional cases of this type see M 90, 98.)

(211) a. /yingatil-ɨ/ → [yingítlu] 'they (m.) can or will be killed' M 89
b. /yingatil/ → [yíngitil] 'he can or will be killed' M 89

The crucial question in (211a) is how to get the boldface /a/ to raise to /i/. In principle, this could be done by invoking the rule of Raising. The problem with this is that we already know that Raising follows High Vowel Syncope (see (205c)), and the application of High Vowel Syncope first would put the /a/ into a closed syllable (/yingatilɨ/ → /yin.gat.lɨ/), so that Raising would no longer be applicable. The two derivations (205c) and (211a) thus require conflicting orderings for Raising and High Vowel Syncope.

Langendoen's answer was to posit that the two rules apply simultaneously; that is, in /yingatilɨ/, the /a/ raises "while" the /i/ deletes. However, there is little independent support for such a mode of rule application. This, together with the complexity of Langendoen's stress rule, suggests it is worth considering alternative solutions.

Theoretical developments since Langendoen's study make such an alternative possible. We now have evidence that syncope rules sometimes do apply to stressed vowels, as well as a theory (§ 3.8.1) that predicts the outcome when this happens: stress does not disappear, but migrates to an adjacent syllable within the same metrical constituent. I argue that something along these lines occurs in Cyrenaican.

6.3.7.3 *Analysis*

Adopting the stress migration hypothesis, we can reconstruct the underlying stress pattern, which turns out to be a familiar one: that of Munsee/Unami (§ 6.3.3), Cayuga (§ 6.3.5), and, closer to home, Negev Bedouin Arabic (§ 6.3.6). The rules are as in (212):

(212)
a. **Foot Construction** — Form iambs from left to right. Degenerate feet are forbidden.
b. **Foot Extrametricality** — Foot → ⟨Foot⟩ / ___ $]_{word}$
c. **Word Layer Construction** — End Rule Right

In (213) are some simple examples, in which no syncope rule applies.

(213)

a. 'he wrote' M 84	b. 'they (f.) accompanied' M 91	c. 'office' M 84	
(×)	(×)	(×)	Foot Construction,
(. ×)	(. ×)⟨(. ×)⟩	(×)⟨(×)⟩	Word Layer
˘ –	˘ – ˘ –	– –	Construction
katáb	*tarɑ́:fagɑn*	*máktab*	
[kɨtáb]	[tɨrɑ́:fɨgɑn]	[máktab]	Raising (204)

d. 'you (f. pl.) wrote' M 89	e. 'her quarrel' M 91	
(×)	(×)	Foot Construction
(. ×)⟨(×)⟩	(×) (. ×)	Word Layer Construction
⏑ – –	– ⏑ – ⏑	
katábtan	*maʕrakɨtha*	
kitábtan	*maʕrɨkɨtha*	Raising
[kitábtan]	[maʕrɨkɨtta]	other rules

In (213a), foot extrametricality is blocked since it would exhaust the stress domain. Extrametricality is blocked in (213e) because the rightmost foot is not peripheral.

Cyrenaican allows superheavy syllables, which attract stress in final position. To analyze this, I follow the analysis employed here for other Arabic dialects (§ 5.2.1): the final consonant in such syllables remains unsyllabified until late in the derivation, and as a result keeps the rightmost foot from being peripheral. The rightmost foot therefore cannot be extrametrical, and attracts stress:

(214) (×) 'cups' M 90
(. ×) (×)
⏑ – –
fanaːǰiː.l → [fina˙ǰíːl]

In this example, the first /a/ undergoes Raising, and the underlying long /aː/ shortens to half-long when not under main stress (M 96 n. 30).

The next set of forms are cases where High Vowel Syncope or Trisyllabic Syncope deletes a stressless vowel. The results are as would be expected:

(215) **High Vowel Syncope**

a. 'he was killed' M 85	b. 'his bundles' M 85 (cf. *ilfáf* 'bundle' M 91)	c. 'they (m.) write' M 84	
(×)	(×)	(×)	Foot Construction,
(. ×)	(. ×)⟨(×)⟩	(×) ⟨(. ×)⟩	Word Layer
⏑ –	⏑ ⏑ –	– ⏑ ⏑	Construction
gitil	*lifaf-ih*	*yi-ktib-ɨ*	
g.tíl	*l.fáfih*	*yík.t.bɨ*	High Vowel Syncope (202)
—	*l.fífih*	—	Raising (204)
igtíl	*ilfífih*	*yíkitbɨ*	Epenthesis (206)
[igtíl]	[ilfífih]	[yíkitb**u**]	*ɨ* → *u* / ___ #

(216) **Trisyllabic Syncope**

a. 'trees SING.-3 F.' = 'her tree' M 90	b. 'write-FEM.-3 M. SG. = she wrote it' M 84	c. 'she was killed' M 87	
(×) (. ×)(×) ˘ ˘ – ˘ *šaǰar-it-ha*	(×) (: ×) (×)⟨(×)⟩ ˘ ˘ – – *katab-aːt-ih*	(×) (×) (. ×)⟨(×)⟩ – ˘ ˘ – *in-gatal-at*	Foot Construction, Word Layer Construction,
š.ǰarítha	*k.tabáːtih*	*in.g.tálat*	Trisyllabic Syncope (209)
š.ǰɨrítha	*k.tibáːtih*	*in.g.tɨ́lat*	Raising (204)
išǰɨrítha	*iktibáːtih*	*inigtɨ́lat*	Epenthesis (206)
[išǰɨrɨ́tta]	[iktibíɛtih]	[inigtɨ́lat]	other rules

Next, consider a case where High Vowel Syncope causes leftward migration of stress within the iambic foot. For purposes of the following derivation, I assume that when there are adjacent syllables in which High Vowel Syncope could apply, it applies from right to left. This assumption will ultimately prove unnecessary. The underlying suffixed form in (217a) can be justified by comparison with (208) /kitib/ → *iktɨ́b.*

(217)

a. /kitib-ih/ 'books-his' M 88	b. /faːkih-it-ih/ 'fruit-SING.-3 SG. = his fruit' M 88	
(×) (. ×) ⟨(×)⟩ ˘ ˘ – *k i t i b i h*	(×) (×) (. ×)⟨(×)⟩ – ˘ ˘ – *f a ː k i h i t i h*	Foot Construction Word Layer Construction
(×) (. ×) ⟨(×)⟩ – – *k i t Ø b i h*	(×) (×) (. ×)⟨(×)⟩ – – – *f a ː k i h Ø t i h*	High Vowel Syncope
(×) (×) ⟨(×)⟩ – – *k i t Ø b i h*	(×) (×) (×)⟨(×)⟩ – – – *f a ː k i h Ø t i h*	Migration of stress within the foot
[kítbih]	[faˑkíhtih]	surface form

It was forms like these that led Langendoen 1968 to posit a stress rule sensitive to vowel height. Similar forms like (210c), (216c) which have an underlying low vowel in the penultimate syllable retain stress on this vowel. In the present analysis, this difference has a simple explanation: low vowels do not delete in this position, so stress does not shift to the preceding syllable. The stress rule can be stated to assign stress uniformly to the penult in words with these syllable patterns, letting the independently motivated High Vowel Syncope rule account for the difference in surface stress.

Consider next case (211a), the one which leads to an ordering paradox in Langendoen's account. My assumption is that Trisyllabic Syncope is ordered before High Vowel Syncope; as a result, it is possible to delete both vowels of a foot. In the representation in (218), I include two forms for clarity: the normal layered representations, and more "official" representations, in which the brackets occur on the terminal string (§ 3.5).

(218) a. /yingatil–ɨ/ Underlying form

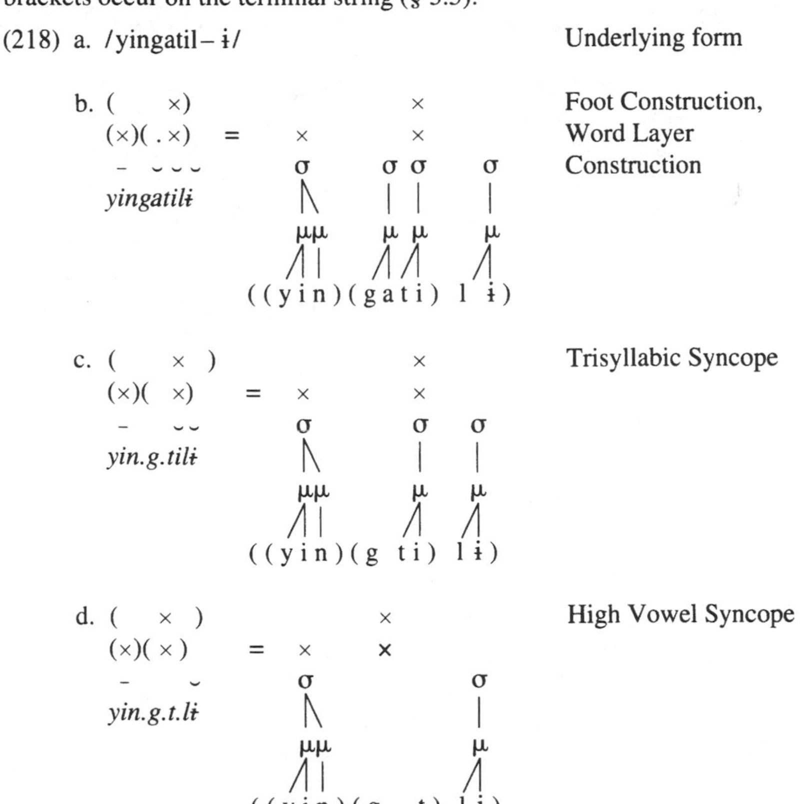

The form in (218d) involves essentially a "floating" grid column: all the syllables in the domain of which the boldface **×** is the head have been eliminated.

The floating grid column receives a resting place after Epenthesis and syllabification have applied, creating a new docking site:

(219)
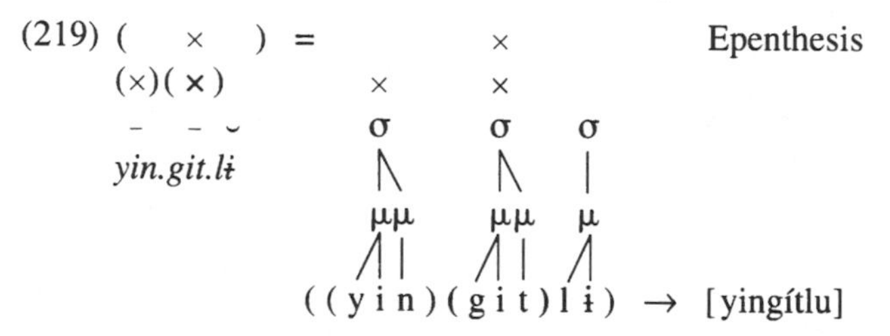

Note that the syllable /git/ must be selected for docking, since otherwise the lower / × / of the floating grid column would move outside its domain, violating the Faithfulness Condition (§ 3.8.1). The crucial / × / and the domain that constrains it are shown in boldface.

In this analysis, the high vowel in the penultimate syllable of *yingítlu* is not a raised /a/, as Langendoen suggested, but rather a stressed epenthetic vowel. This makes possible a consistent ordering of the rules: Trisyllabic Syncope precedes High Vowel Syncope (otherwise we would derive **yingátlu*), High Vowel Syncope precedes Raising (as shown in (205c)), and Epenthesis comes last.

This completes the basic analysis. There are two additional cases that lead to minor adjustments. Consider first a case in which Trisyllabic Syncope is applicable at more than one location in the string. The derivations in (220) justify the relevant underlying representations.

(220)

a. 'trees' M 90	b. 'trees-SING. = a tree' M 97	
/šɑǰɑr/	/šɑǰɑr-it/	underlying form
šɑǰɑ́r	*šɑǰɑ́rit*	Stress Assignment
—	*š.ǰɑ́rit*	Trisyllabic Syncope
šɨǰɑ́r	*š.ǰɨ́rit*	Raising
—	*išǰɨ́rit*	Epenthesis
[šɨǰɑ́r]	[išǰɨ́rah]	other rules (M n. 40)

The interesting case arises when we add the m. sg. suffix /-ih/ to /šɨ̆ǰɑr-it/, obtaining /šɑǰ*ɑr-it-ih/ 'his tree', surface form šɨ̆ǰɨrtih* M 90. In this form, both vowels of the stem are vulnerable to Trisyllabic Syncope. In fact, both must delete, so I will assume that this rule applies either left to right or simultaneously across the board. Moreover, High Vowel Syncope can also apply to the penultimate vowel, so there are three stranded consonants.

(221) a. (×) Foot Construction,
(. ×)⟪. ×⟫ Word Layer Construction
⏑ ⏑ ⏑ –
šaǰaritih

b. (×) Trisyllabic Syncope
(×) ⟪. ×⟫
⏑ –
š . ǰ.ritih

c. (×) High Vowel Syncope
(×) ⟪×⟫
–
š . ǰ.r.tih

At this point we must consider how Epenthesis applies to a string of three unsyllabified consonants. One pattern that has been noticed in other Arabic dialects is that epenthesis "economizes" on epenthetic vowels; that is, epenthetic vowels are placed so that the string of consonants is properly syllabified using the minimum possible number of vowels. Selkirk 1981 and J. Ito 1989 propose theoretical accounts of this phenomenon: epenthesis is construed not as a linear rule, but as a syllabification algorithm. Since syllabification is known to be maximal (i.e. to incorporate as many consonants as possible), this offers an explanation of the "economizing" effect.

I will not formalize such an account here, but simply assume that the Epenthesis rule (206) is guided by the principle of maximal syllabification. In addition, I assume left-to-right application.[16] Thus in (221c) Epenthesis begins by scanning the first two consonants, syllabifying them both with a single epenthetic vowel. The newly created syllable provides a docking site for the stranded main stress:

(222) (×) Epenthesis
(×) ⟪ ×⟫ (first iteration)
– –
šɨǰ.r.tih

The epenthetic vowel surfaces as [ɨ]: following Kenstowicz 1979, I assume that Syncope strands a floating backness autosegment, which redocks onto the epenthetic vowel, rendering it [+back].

The final output is derived by a second application of Epenthesis, which supports the stray /r/ with another epenthetic vowel. This triggers resyllabification of the preceding /ǰ/; and assuming feet do not break up syllables, also reaffiliation of the /ǰ/ to the following foot:

16. This is why I cannot follow Ito's proposal in its details: under her theory the left-to-right direction needed for Cyrenaican would place the epenthetic vowels in the wrong position with respect to the consonants.

(223)		
	(×) (×)((. ×)) ˘ – – [šɨǰɨrtih]	Epenthesis (second iteration)

The intermediate form /šǰrtih/ resembles representations posited by Owens (1980), who attempts to do without underlying high short vowels entirely, deriving them all by Epenthesis. A difficulty with this is that the position of high vowels is contrastive; as in /simiʕt/ → [ismíʕit] 'I/you (m. sg.) heard' M 86 versus /yiktib/ → [yíktib] 'he writes' M 84. For Owens, both forms are underlying /CCCC/, and special diacritics must be added to make Epenthesis work correctly. This strongly suggests that high vowels actually are present in the underlying forms, as the analysis here assumes.

Given the left-to-right, economizing behavior of Epenthesis, we may now loosen the assumptions made about High Vowel Syncope in (217): rather than applying from right to left, this rule could equally well be left to right, or across the board. The derivations for (217) would be as in (224):

(224)	a.	b.	
	(×) (. ×) ((×)) ˘ ˘ – *kitibih*	(×) (×) (. ×)((×)) – ˘ ˘ – *fa:kihitih*	Foot Construction, Word Layer Construction
	(×) (×)((×)) – *kØtØbih*	(×) (×) (×)((×)) – – *fa:kØhØtih*	High Vowel Syncope (left-to-right or across-the-board: grid column floats)
	(×) (×) ((×)) – – *kitbih* [kítbih]	(×) (×) (×)((×)) – – – *fa:kihtih* [faˑkíhtih]	Epenthesis (grid column docks) V: → Vˑ where not main-stressed

6.3.7.4 *Conclusion*

To summarize the analysis: (a) I basically adopt Langendoen's segmental rules, but permit Syncope to delete stressed vowels. (b) The stress system is treated as a commonplace iambic pattern. (c) The remainder of the analysis relies on general principles: migration or floating of stress under vowel deletion, and the treatment of Epenthesis as a maximizing form of syllabification. Under the revised account, the stress rule is rendered formally simple and typologically ordinary, and simultaneous rule application is not needed.

The analysis provides a further instance of migration of stress under vowel deletion, where the direction of migration is predicted by bracketing that is independently needed to derive stress. I noted in § 3.8.1 that the Cyrenaican case can be usefully compared with the similar case of Bedouin Hijazi Arabic (Al-Mozainy 1981; Al-Mozainy, Bley-Vroman, and McCarthy 1985). The Bedouin Hijazi dialect has almost exactly the same rule of Trisyllabic Syncope as Cyrenaican. But its stress pattern is based on moraic trochees, being ambiguous between the pattern of Latin (§ 5.1.4) and Palestinian Arabic (§ 6.1.1). It is striking, then, that in Bedouin Hijazi, stress stranded by Trisyllabic Syncope shifts to the right; whereas in Cyrenaican, stress shift is essentially to the left. Bani-Hassan Arabic (Irshied and Kenstowicz 1984; § 8.10) also fits into this picture: it, too, has Trisyllabic Syncope, and since its feet are moraic trochees, it shifts stress rightward.

Cyrenaican also bears on the issue of persistent versus non-persistent stressing, discussed in § 5.4.1. In Cyrenaican, metrical structure is quite regular in deep representations, but utterly disrupted on the surface. In particular, while Cyrenaican appears to respect an underlying ban on degenerate feet (there are no degenerate-size words), the result of the segmental rules is to create them on the surface: (216c), (223). Depending on assumptions about derived constituent structure, one also may find "impossible" /– ◡́/ feet in (215b, c), (216), and (220b), and /– –́ / feet in (215a) and (223); the absence of phonetic correlates of secondary stress in Cyrenaican (due perhaps to Conflation, § 5.5.2) prevents this from being fully determined. Cyrenaican thus appears to be a good example of a stress system in which foot parsing is NOT persistent, in the sense of § 5.4.1. If parsing were persistent, it would remove and reparse the ill-formed feet, resulting in a quite different observed stress pattern.

6.3.8 Stress and Syllabification in the Yupik Languages

The Yupik Eskimo languages all have left-to-right iambic stress, with numerous variations on the basic theme among individual languages and dialects. In the following discussion of various Yupik systems, three theoretical issues are at stake. First, certain Yupik dialects have been analyzed as requiring a kind of even iamb (§ 4.3.1). I argue that such an expansion of the universal foot inventory is not required, and that there are advantages to an alternative analysis that uses the normal iambs found in other languages. Second, Central Alaskan Yupik has been proposed as having (at least) a ternary syllable weight distinction, that is, CVV/CVC/CV. I suggest that the intermediate status of CVC results instead from assigning it either one or two moras, depending on the context (§ 5.6.1). Finally, I argue that the Yupik languages provide a clear example of "persistent" stressing (§ 5.4.1).

My analysis is highly indebted to the work of a group of scholars, among them M. Krauss, S. Jacobson, O. Miyaoka, J. Leer, and A. Woodbury, who have uncovered the structure of these remarkable languages over the last twenty-five years. Their research on Yupik prosody is presented in Krauss 1985a; Woodbury 1981, 1987; and other studies. The Yupik scholars have not only discovered very intricate patterns in the data, but also explicated them in cogent formal analyses, which my account follows to a large extent.

The Yupik family consists of several separate languages, of which the most widely spoken, Central Alaskan Yupik, has several dialects. Special attention will be paid to the varieties listed in boldface in (225).

(225) **Yupik Languages** (Krauss 1985b; Jacobson 1985)
  a. Naukanski (Siberia)
  b. **Pacific Yupik** (Southern Alaska) (§ 8.8)
  c. Central Siberian Yupik
    Dialects: i. Chaplinski (Siberia)
      ii. **St. Lawrence Island** (§ 6.3.8.1)
  d. Central Alaskan Yupik (Southwestern Alaska)
    Dialects: i. **Norton Sound** (§ 6.3.8.2)
        Unaliq subdialect (§ 6.3.8.6 (259))
        Kotlik subdialect
      ii. Hooper Bay/Chevak
        Hooper Bay subdialect
        Chevak subdialect (§ 6.3.8.4)
      iii. **General Central Yupik** (§ 6.3.8.3, 5–7)
      iv. Nunivak Island

I will follow an expository practice established by Yupik scholars, starting with the prosodically simplest language of St. Lawrence Island, then proceeding through languages and dialects of increasing complexity. Three crucial papers from which I have drawn data and insights are Jacobson 1985 (hereafter J), Miyaoka 1985 (hereafter M), and Woodbury 1987 (hereafter W). Phonetic transcriptions often reflect the application of segmental rules not relevant here.

6.3.8.1 *St. Lawrence Island*

In the St. Lawrence Island dialect of Central Siberian Yupik (Krauss 1975, 1985d; Jacobson 1985), stress follows a straightforward iambic pattern: it falls on nonfinal heavy (= CVV(C)) syllables and on the nonfinal even-numbered members of a light-syllable string, counting from left to right. For reasons given in § 6.3.8.8, I hypothesize that the perceived lack of stress on final syllables is due to the intonational system, and that eligible final syllables do receive grid marks. In the representations, I include these grid marks but do not mark their syllables with accents. Given this assumption about final position, St. Lawrence Island stress can be derived by parsing the word into iambs from left to right, with CVV(C) counted as heavy:

(226) a. 'he wants to make a big boat' J 26

(. ×) (. ×) (. ×)
˘ ˘ ˘ ˘ ˘ ˘
aŋyáχɬaχɬáŋyuxtuq

b. 'he wants to make a big ball' J 26

(×) (. ×) (. ×)
– ˘ ˘ ˘ ˘ ˘
á:ŋqaχɬáχɬaŋyúxtuq

Throughout Yupik, I assume that degenerate feet are forbidden, since light-syllable content words are apparently forbidden.[17] As the sources disagree on main stress, I do not attempt to analyze it here.

In St. Lawrence Island Yupik, there are segmental rules that increase the durational contrast of the foot, following the Iambic/Trochaic Law (§ 4.5.3). A rule of Iambic Lengthening creates canonical /⏑ –́/ iambs by lengthening non-final short stressed vowels in open syllables: /qayani/ → *qayá:ni* 'his own kayak' J 27. Iambic Lengthening creates only canonical /⏑ –́/ iambs, and never the /–́/ type, since a short vowel cannot be stressed unless it falls in the second syllable of a foot.

A more exotic lengthening phenomenon is found among older St. Lawrence Island speakers, who lengthen even long vowels in open syllables, creating overlong [V::]: /qaya:ni/ → *qayá::ni* 'in his (another's) kayak' J 27. Such overlength is also accompanied by a special falling intonation. Although overlong vowels are not common in the world's languages, it is easy to see why they might arise here: as Leer 1985b, 136–37, points out, they allow the phonemic distinction between an iambically lengthened short vowel and an underlying long vowel to be maintained on the surface, just as in Choctaw and Chickasaw (§ 6.3.2). The communicative value of maintaining such a distinction is emphasized by Leer; note the glosses given for the two preceding examples, as well as for (257a, b).

This avoid-neutralization factor helps explain a restriction on the overlength rule: it may only apply in disyllabic feet. This can be seen by comparing (227a), with surface overlength, and (227b), with none (weight depicted in moraic notation for clarity):

(227) a. (. ×) → (. ×) 'in his (another's) kayak' J 27
μ μμ μ → μ μμμ μ
/qaya:ni/ → *qayá::ni*

b. (. ×)(×) → (. ×)(×) 'in his (another's) drum' J 27
μ μ μμ μ → μ μμ μμ μ
/saɣuya:ni/ → *saγú:yá:ni*

In (227b), the syllable /ya:/ is stressed only by virtue of having an underlying long vowel; had its vowel been short, it would have been skipped over. Thus, unlike in (227a), overlengthening is not needed to cue underlying length.

To summarize, the rules I propose for St. Lawrence Island Yupik are as in (228):

17. In fact, even degenerate STEMS are forbidden: Jacobson 1984 documents phonological "conspiracies" that insure that the stem will contain at least one foot, both in St. Lawrence Island Yupik, where /–/ = (C)VV(C), and in Central Alaskan Yupik, where /–/ = CVC. See, however, Krauss 1985d for cases of degenerate feet in St. Lawrence Island.

(228) a. **Syllable Weight** /¯/ = CVV(C), /˘/ = CVC, CV

b. **Foot Construction** Form iambs from left to right. Degenerate feet are forbidden.

c. **Overlengthening**

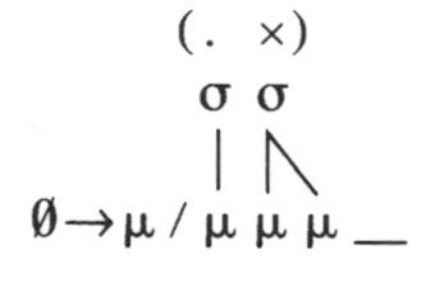

d. **Iambic Lengthening**

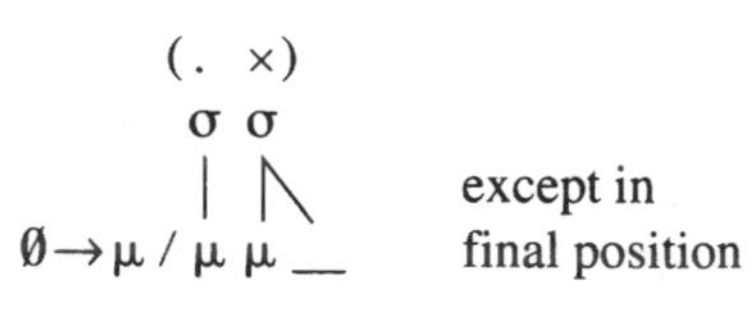

Rules (228c, d) plausibly might be collapsed, but are stated separately for clarity. They arguably need not specify that the inserted mora is filled by vowel lengthening, since only long vowels make weight in this language. In addition, the ban on crossed association lines would prevent the vowel from linking to the inserted mora in a closed syllable (see § 7.3.2, however, for an alternative account).

6.3.8.2 *Pre-Long Strengthening in Norton Sound*

I turn now to dialects of Central Alaskan Yupik (225d), treating them in order of increasing prosodic complexity. The simplest dialect is that of Norton Sound, whose prosodic system resembles that of St. Lawrence Island but with interesting differences.

First, CVC syllables in initial position (and only in this position) are counted as heavy: /aŋyaχpaka/ → *áŋyaχpá:ka* 'my big boat': compare St. Lawrence Island *aŋyáχpaka* (Krauss 1985c, 21). To account for this, I suggest (following § 5.6.1) that initial CVC is bimoraic in Norton Sound, but noninitial CVC is monomoraic. Put more formally, the rule of Weight by Position (§ 3.9.2) is restricted in Norton Sound to initial syllables. Assuming the same footing rules as in St. Lawrence Island, we derive stressings like (229):

(229)

```
μμ μ   μ μ        syllabification
aŋyaχpaka

(× )(.  ×)        Foot Construction (228b)
μμ μ   μ μ
aŋyaχpaka

(× )(.  ×)        Iambic Lengthening (228d)
μμ μ   μμ μ
áŋyaχpá:ka
```

The second area where Norton Sound prosody differs from St. Lawrence Island concerns the functional problem mentioned above: how are long vowels derived by Iambic Lengthening to be kept distinct from underlying long vowels? Recall that in St. Lawrence Island, the distinction is maintained by making underlying long vowels overlong. This is not the case for Norton Sound, which lacks the rule of Overlengthening (228c). Instead, a rather different strategy is pursued: in cases where ambiguity might result, the syllable preceding the long vowel is stressed. In addition, where this syllable is underlyingly light, it is made heavy by gemination of the following consonant. This is illustrated by the contrasting pairs in (230)–(231):

(230)	a. /qayapixkani/ →	*qayá***:pix***ká:ni*	'his own future authentic kayak'	J 31
	b. /qayapixka:ni/ →	*qayá***:píx***ká:ni*	'in his (another's) future authentic kayak'	J 31
(231)	a. /qayani/ →	***qa****yá:ni*	'his own kayak'	J 30
	b. /qaya:ni/ →	***qáy****yá:ni*	'in his (another's) kayak'	J 30

In (230), the underlying forms are kept apart by stress alone, whereas in (231) both stress and gemination are involved.

I will refer to the rule responsible for these effects as **Pre-Long Strengthening.** It is an interesting problem to formalize this rule. First, there is an ordering paradox: Pre-Long Strengthening applies only to odd-numbered members of a light-syllable string, with even-numbered members undergoing Iambic Lengthening instead, as in (232). Thus Pre-Long Strengthening apparently must follow Foot Construction, because it respects the alternating count of light syllables that feet formally represent.

```
(232)  (. ×)(× )              (. ×)(× )         'with their (other's) future   J 27
        μ  μ  μμ  μ            μ  μμ μμ  μ        steambath material'
       maqika:txun    →    maqí:ká:txun       (*maqíkká:txun, with Pre-Long
                                                 Strengthening)
```

As before, there is no functional motivation for Pre-Long Strengthening in even position: surface stress unambiguously shows that the following vowel is underlyingly long.

The ordering of Pre-Long Strengthening after Foot Construction seems paradoxical: ideally, the stressing effect of Pre-Long Strengthening ought to emerge directly from the stress rules. I will consider two approaches to this problem.

One approach, which I will ultimately argue against, is that developed in Woodbury 1985b, 1987 and Leer 1985c and adopted in modified form by Weeda 1989, 1990 and Halle 1990. Since Woodbury presents the most detailed version, I will refer to his analysis below. In this proposal, the foot template is a kind of even iamb: /⏑ ⏑́/ is allowed, but /⏑ –́ / feet are forbidden (for other even iambs, see § 4.3.1). Degenerate feet are allowed, and heavy syllables are defined as (C)VV(C). This proposal derives the examples of (230) and (231) as follows. In (230b) *qayapixka:ni,* two syllables *pix* and *ka:* must be footed separately, since /pixká:/ would be an ill-formed /⏑ –́/ foot:

(233) a. (. ×)(. ×)(×)
μ μ μ μμ μ
*qaya**pixka**:ni*
*[qayá:pixká:ni]

b. (. ×)(×) (×)(×)
μ μ μ μμ μ
qayapixka:ni
[qayá:píxká:ni] (actual form)

This accounts for the stress on pre-long *-píx-*. For pre-long CV syllables, Woodbury posits a rule whereby nonfinal feet of the form $[CV]_F$ are "bulked" by gemination. Thus in (234) (= (231b)), the light syllable *qa-* is bulked to *qay-* after it receives stress by foot parsing:

(234) (×)(×) → (×)(×)
μ μμ μμ μμ
/qaya:ni/ *qáyyá:ni*

This analysis correctly derives the facts. Moreover, since it only needs to assign stress once, it overcomes the ordering paradox mentioned above. But it is subject to a number of objections. First, it weakens the explanatory force of metrical theory, by expanding the inventory of foot templates to include /⏑ ⏑́/. The /⏑ ⏑́/ foot arguably is an unnatural one: as we have seen, iambic rhythm, both linguistic and otherwise, is characterized by a tendency toward unequal duration. Indeed, this tendency is manifested in Yupik itself: all Yupik languages have a rule of Iambic Lengthening like (228d), which converts /⏑ ⏑́/ feet to /⏑ –́/. The existence of this rule must count as a mystery in the even-iamb analysis, since it creates surface feet that are ill formed under the basic foot template.

Consider now an alternative analysis. I propose that Pre-Long Strengthening is a separate rule, ordered after Foot Construction, and that the feet are the ordinary /⏑ σ́/ iambs argued for elsewhere in this book. To make such an analysis work, two problems must be overcome.

First, if Pre-Long Strengthening follows Foot Construction, then it must do two things: stress the pre-long syllable; and if this syllable is CV, make it heavy by gemination. But a constrained phonological theory should not allow two such disparate actions to be carried out by a single rule.

Second, if we allow Pre-Long Strengthening directly to add a stress in syllables that precede long vowels, then we are greatly weakening metrical theory by allowing rules of the form "σ → σ́ / P ___Q"; that is, 'assign stress in context X'. The whole point of the research program is to constrain stress rules by limiting the set of foot parsing algorithms.

These two problems have the same solution: analyze Yupik footing as persistent, in the sense of § 5.4.1. In particular, the rule of Pre-Long Strengthening should directly affect only quantity: it makes the relevant syllable heavy by adding a mora, and the segmental details of how the heavy syllable is manifested are attributed to syllabification. The stressing action of the rule is attributed to persistent stressing, which would reparse the newly heavy syllable into a well-formed foot. Under this account, it is not necessary to posit a rule of the form "σ → σ́ / P ___ Q," nor a rule that changes both stress and quantity at once.

I formulate Pre-Long Strengthening as in (235):

(235) **Pre-Long Strengthening**

```
            (.    ×)
             σ    σ
             ⁞\   |\
Ø → μ / μ  __  μ  μ
                  \/
                [-cons]
```

The rule states that the first syllable of a foot is to be made heavy when the second syllable dominates a long nucleus. Note that rules of pretonic strengthening similar to (235) may be found in Tiberian Hebrew (McCarthy 1979b) and sporadically for Italian (Repetti 1989, 41). In a similar way, Maithili (§ 6.1.6) has retained geminates only in pretonic position.

Pre-Long Strengthening would apply to the forms of (230b) and (231b) as in (236):

(236)

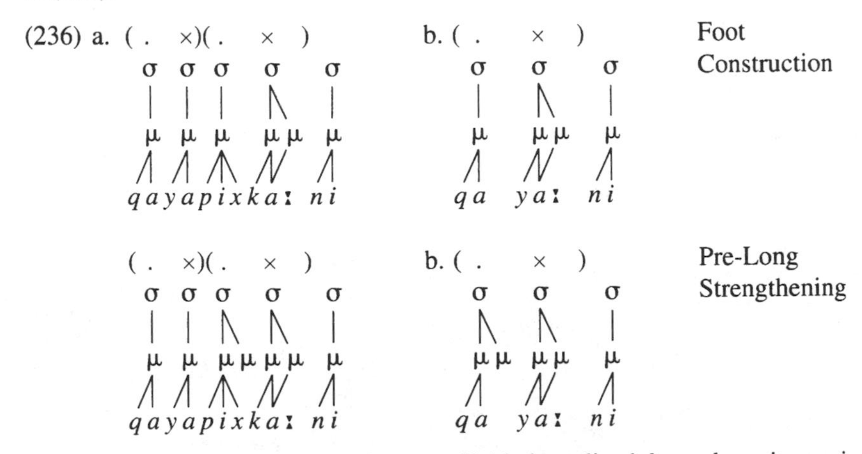

The way in which the newly heavy syllable is realized depends on its environment. I posit for Yupik the principles of syllabification in (237):

(237) **Stray Mora Association**

Stray moras associate with:
a. Non-moraic syllable-final consonants, if present; otherwise
b. Stressed vowels; otherwise
c. The onset of the following syllable.

I assume that (237) is automatic in languages that allow bimoraic CVC syllables. Part (b) accounts for Iambic Lengthening (228d). Part (c) can be independently motivated, in that it associates stray moras inserted by other rules of the language as well (see (259) below, as well as Jacobson 1985, 31–32). The moras inserted in (236) are affiliated by (237a, c) as in (238):

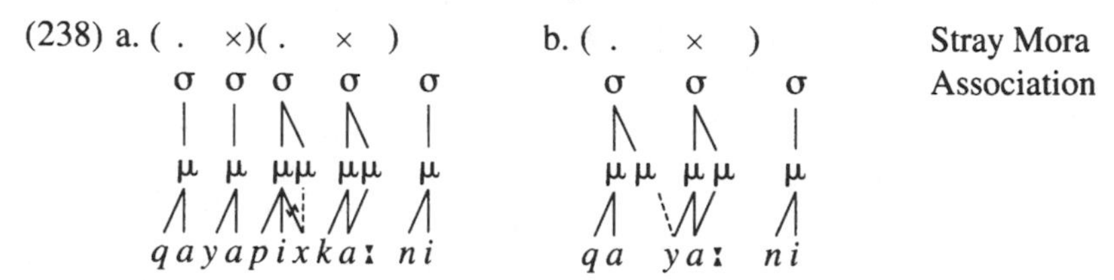

Because Yupik footing is persistent, Pre-Long Strengthening induces a shift in foot structure: since /μμ μ́μ/ is not a well-formed foot, it gets deleted, then refooted to /μ́μ/ + /μ́μ/.

(239) (. ×) → → (×)(×)
μμ μμ μμ μμ μμ μμ

Applied to (238a, b), this yields the correct results.

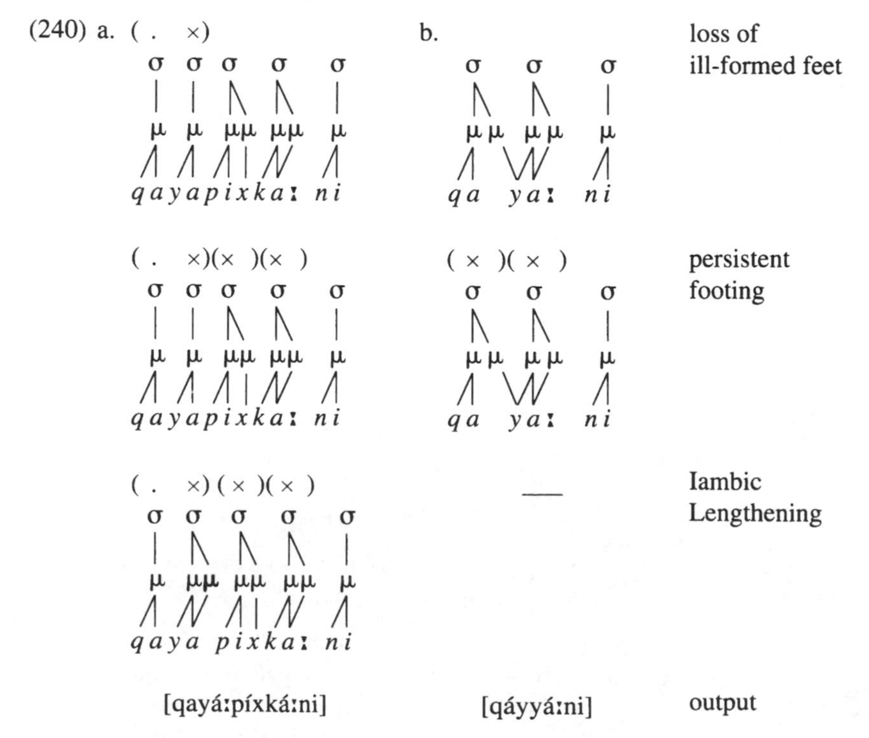

To summarize the proposed solution: first, Pre-Long Strengthening is a weighting rule, and not (directly) a stress rule; second, Foot Construction is persistent, and intervenes everywhere to restore the well-formedness of feet. Under these assumptions, we need not write a rule that alters both stress and weight at once. The stressing effects of Pre-Long Strengthening are simply the results of the regular iambic stress rule, applied persistently. Moreover, we need not weaken metrical theory by allowing stress to be assigned contextually; foot parsing continues to be the only mechanism of stress assignment.

A possible objection to this analysis is the following: the insertion of a mora is simply an indirect way of defining a basic /⏑ ⏑́/ foot template, since all basic /⏑ –́/ feet get broken up to /–́ –́/. As it turns out, however, this objection holds only for the dialect under discussion (Norton Sound). In other Central Alaskan Yupik dialects (§ 6.3.8.3), CVC is heavy throughout the word, and /⏑ –́/ feet ARE allowed when /–/ is constituted by CVC. My general view is that the function of Pre-Long Strengthening is not to avoid /⏑ –́/ feet, but to make underlying vowel length deducible from surface representations: Pre-Long Strengthening need not apply before CVC, because there is no phonemic contrast between heavy and light CVC (comparable to CVː vs. CV) that needs to be made audible on the surface.

Defenders of the basic /⏑ ⏑́/ template must consider another issue as well: in the only well-attested case of this template (Yupik), the unfootable /⏑ –/ sequences actually surface as /– –/, with bulking. Under the proposal made here, this is not an accident, since it is the bulking that makes /⏑ –/ unfootable, not the other way around. In the pure form that parses ⏑ – sequences as /⏑́/ –́ /, the /⏑ ⏑́/ template is unattested.

I thus claim two advantages for the account proposed here: it avoids the need for a poorly supported expansion of universal foot theory; and the rule of Iambic Lengthening is a natural consequence of the analysis rather than a mystery, since it creates canonical /⏑ –́/ iambs.

A final note concerning Pre-Long Strengthening: given that the mora inserted by this rule may dock onto a syllable-final consonant, one wonders if this might also be the case for Iambic Lengthening. Recall that in St. Lawrence Island Yupik, syllable-final consonants may not be moraic, so Iambic Lengthening has effects only for /CV/. But in Norton Sound, heavy CVC is permitted, so it is possible in principle for Iambic Lengthening to create heavy CVC:

(241)	(. ×) (. ×)		(. ×)(. ×)	'he wants to get a big kayak'	J 30
	μ μ μ μ μ	→	μ μμ μ μμ μ		
	qayaχpaŋyuxtuq		*qayáχpaŋyúxtuq*		

The change looks vacuous in phonemic transcription, but in the detailed phonetics, the light versus heavy CVC distinction actually does have real consequences. Woodbury notes (1981, 46, for Chevak; p.c. for the rest of Alaskan Yupik) that the class of syllable-final consonants that I analyze as having their own mora (derived either by Pre-Long Strengthening or by Iambic Lengthening) have greater phonetic length; this plausibly is the phonetic manifestation of their bearing their own mora.

6.3.8.3 *Destressing in General Central Yupik*

General Central Yupik is the most widely spoken dialect of Central Alaskan Yupik, of which Norton Sound, just covered, is the northernmost dialect (225d). The stress pattern of General Central Yupik basically resembles that of Norton Sound, and the reader should assume the same rules except as indicated. The crucial difference lies in the treatment of CVC syllables, summarized in (242):

(242) a. CVC preceding non-final CV is stressed: *á**ŋyáχ**paka* 'my big boat'; cf. Norton Sound *á**ŋyaχ**páka* (→ *áŋyaχpá:ka*) (Krauss 1985c, 21).

b. In other environments, CVC is treated as in Norton Sound; in particular,

i. Initial CVC is stressed: ***ák**ŋ̥iχtátŋ̥a* 'they hurt me (interr.)' M 59.

ii. Before CVC or finally, CVC is stressed just in case the preceding syllable is not. In sequences of CVC, this results in an alternating count: *ák**ŋ̥iχtát**ŋ̥a, qa**yáχpaŋyúxtuq*** 'he wants to get a big kayak' J 30.

iii. Before CVV(C), posttonic CVC is stressed by (235) Pre-Long Strengthening: (230b) /qayapixka:ni/ → *qayá:**píx**ká:ni*

iv. Before final CV in final position, posttonic CVC is not stressed: *á**ŋyax**ka* 'my two boats' M 68

I will first review Woodbury's (1987) account of the General Central Yupik pattern. His foot construction rule (p. 695) states that two short-voweled syllables σ_1 and σ_2 may form a binary foot under the following conditions: "if word-initial, σ_1 is open (CV); . . . if σ_1 is closed and σ_2 is open, σ_1 σ_2 is word-final." The first of these conditions insures that initial CVC will attract stress; compare (229). The second can be restated more clearly for our purposes as a filter:

(243) *(.   ×)   except in final position
CVC CV

The filter might be construed as a prohibition on feet having trochaic quantity and iambic stress. In (244), it forces construction of a monosyllabic CVC foot (244a) rather than disyllabic /CVCCV/ (244):

(244) a. (. ×)(×) (. ×) = (230a)   b. *(. ×)(. ×)
μ μ μ μ μ   μ μ μ μ μ
*qaya**pix**kani*   *qaya**pixka**ni*
→ [qayá:píxkani]   → *[qayá:pixká:ni]

Norton Sound, which lacks filter (243), derives (244b) as the actual output.

Woodbury 1985a, 1987 notes that these facts imply a three-way hierarchy of weight: CVV is always stressed, CVC often attracts stress, and CV is stressed only when the alternating binary count allows it. In Woodbury's analysis, the hierarchy is directly incorporated into the foot construction rules, because CVC is counted as heavy for the filter (243), but as light for the basic /⏑ ⏑́/ foot template; cf. (242b.ii) *qayáχpaŋyúxtuq,* where /paŋyúx/ is a /⏑ ⏑́/ foot. The triple distinction is a challenge to the theory assumed here, where it is assumed that foot construction may only make use of binary quantity distinctions (see § 7.1 for general discussion). In what follows, I propose an analysis under which this principle may be maintained.

Two ingredients of my account have already been presented: (a) stressing in Yupik is persistent, accommodating foot structure to changes made in syllable structure; (b) as argued above, CVC in Yupik may vary between one and two moras, depending on context. In addition, I will propose a Destressing rule.

I must first clarify how destressing would work in a system with persistent stressing and variable syllable weight. The crucial point is that any heavy syllable that was destressed but allowed to remain heavy would immediately have its stress restored, since /–́/ is a well-formed iamb. For this reason, it is plausible to treat destressing not as foot removal, but as the loss of heavy quantity. This change will often create degenerate feet, which (since Yupik bans them) will then be removed by persistent footing (§ 5.4.1).

For General Central Yupik, I propose that initial syllabification always gives CVC two moras; that is, the special treatment that Norton Sound extends to initial CVC (§ 6.3.8.2) is made general throughout the word. Thus under iambic parsing, CVC will always attract stress at an early stage of the derivation. The case that needs special pleading is where CVC LOSES stress, the opposite of Woodbury's account. In fact, the stress-loss environment is simple to state: it is **double clash,** the environment shown in (245):

(245) × × ×
σ CVC σ

This is a natural environment, given Hammond's (1984a) generalization that destressing rules universally resolve clash. The destressing rule can be expressed as in (246):

(246) **General Central Yupik Destressing in Clash**
Remove a mora from CVC in the context / × ___ ×

This rule applies iteratively, from left to right. The restriction to CVC might be explained on functional grounds: no phonemic contrast is lost by converting CVC from bimoraic to monomoraic status, but the phonemic vowel length contrast would be obliterated if the rule applied to CVV. Alternatively, we might seek an account within an enriched theory of moraic structure (see § 7.3), or in the theory of syllable prominence (§ 7.1). I leave this issue open.

What is crucial here is that under this analysis the two kinds of heavy syllables, CVC and CVV, are treated identically (as /–/) by Foot Construction. As I argue in chapter 7, it is quite generally the case that foot construction respects only mora count, whereas other rules such as destressing may make reference to additional properties of the syllable.

Destressing works as follows. In the sample forms in (247) (= (242b.ii), (242a)), the syllable types CV, CVC, and CVV count as /⏑/, /–/, and /–/ respectively. The examples also show the first pass of footing.

(247) a. (×)(×)(×) b. (×)(×)(. ×) Initial syllabification
μμ μμ μμ μ μμ μμ μ μ and footing
akŋ̥iχtatŋ̥a *aŋyaxpaka*

In these forms, the only syllable that is in the double clash environment is boldface /ŋ̥iχ/. Its quantity is reduced to short, as in (248):

(248) a. (×)(×)(×) b. — Destressing in Clash (246)
μμ **μø** μμ μ (= mora loss)
akŋ̥iχtatŋ̥a

Since the degenerate foot that results is ill-formed, it is deleted by persistent footing. Following the principles of persistent stressing laid out in § 5.4.1, the stray syllable is adjoined to the following foot, creating a canonical iamb:

(249) a. (×)(. ×) b. — deletion of ill-formed foot
μμ μ μμ μ
akŋ̥iχtatŋ̥a

This derives the correct outputs, *ákŋ̥iχtátŋ̥a* and *áŋyáxpaka.*

The derivations demonstrate the feasibility of an account of General Central Yupik based on destressing rather than a triple quantity distinction in the foot construction rules. I now consider a possible objection, plus two arguments in favor of this approach.

The potential objection is that the destressing rule, by applying iteratively, derives just the kind of alternating stress pattern that is normally attributed to footing.

(250) (×)(×)(×)(×)(×) (×)(. ×)(. ×) 'he wants to get
μμ μμ μμ μμ μμ → μμ μ μμ μ μμ a big boat' J 30
aŋyaχpaŋyuxtuq *áŋyaχpáŋyuxtuq*

However, there is a precedent for such destressing elsewhere, in Southern Sierra Miwok (Broadbent 1964; § 6.3.9).

In this language, primary stress is assigned by creating a single iamb at the left edge of the word, and secondary stress normally falls on later heavy syllables. But Broadbent notes (p. 17) that "in a long sequence of [heavy] syllables, the even-numbered ones tend to be less heavily stressed than the odd-numbered ones, counting from the beginning of the [heavy]-syllable sequence." We can attribute this pattern to (a) construction of unbounded, right-strong, quantity-sensitive feet following the initial iamb (see § 7.2 for unbounded feet); (b) left-to-right iterative destressing, with a left-side clash environment. Such an account is required because heavy syllables alternate in stress only when they appear in sequence. Note that other than the difference in the clash environment, the Sierra Miwok rule is the same as in Central Alaskan Yupik.

In the next two sections, I present arguments for the Destressing approach. § 6.3.8.4 shows how a simple generalization of Destressing can account for the stress pattern of Chevak Yupik. Following an excursus on destressing in penultimate position, § 6.3.8.6 presents a second argument, based on the interaction of Destressing in Clash with a rule of Schwa Deletion.

6.3.8.4 *Destressing in Chevak Yupik*

Chevak Yupik (Woodbury 1981), like General Central Yupik and Norton Sound, is a dialect of Central Alaskan Yupik (225d). It has basically the same stress pattern as General Central Yupik, except that in one class of words, we find fewer stresses in Chevak. An example is the word meaning 'they having sleds': *ikámʁíʁlutəŋ* in General Central Yupik vs. *ikámʁiʁlutəŋ* in Chevak (Woodbury 1981, 43, 100). The "missing stresses" of Chevak occur on post-tonic CVC syllables that precede a stressless syllable.

This difference can be accounted for with a trivial difference in the Destressing rules: whereas Destressing in General Central Yupik requires a two-sided clash (246), in Chevak Destressing requires only a clashing stress on the left. I formulate the Chevak version of the rule as in (251); it applies to *ikamʁiʁlutəŋ* as in (252):

(251) **Chevak Destressing in Clash**
Remove a mora from CVC in the context / × ___

(252) a. (. ×)(×)(. ×) initial footing
μ μμ μμ μ μμ
ikamʁiʁlutəŋ

b. (. ×) (. ×) Destressing in Clash
μ μμ μ μ μμ
ikámʁiʁlutəŋ

The only reparsing that takes place here is loss of the degenerate foot. Since persistent footing applies only to stray syllables (§ 5.4.1), we cannot extract the following syllable *lu* from its foot to form the disyllabic foot */ʁiʁlú/. The syllable /ʁiʁ/ must therefore remain stray. This accounts for the characteristic ternary stress intervals of Chevak. Note that to characterize Chevak stress without Destressing would require a complex ternary footing procedure, deviating even further from a constrained universal foot inventory.

Destressing for Chevak was in fact first proposed by Woodbury 1987, 699, for final syllables only. I have generalized Woodbury's rule to apply further within Chevak, and also formulated a similar rule for General Central Yupik, with a different clash environment. Destressing allows us to simplify the foot construction rules, maintaining the principle that footing respects only binary distinctions of syllable quantity.

6.3.8.5 *Destressing of Penultimate Syllables*

A Destressing phenomenon that I have not yet accounted for is that noted in (242b.iv): the lack of stress on posttonic CVC syllables that precede a word-final light syllable, as in *áŋyaxka.* The absence of stress in this position is expected in any event for Norton Sound (where non-initial CVC syllables are light), and for Chevak (where Destressing requires only a preceding stress). But *áŋyaxka* is also found in General Central Yupik, where normally CVC gets stressed when it precedes a CV syllable (cf. (242a)).

In Woodbury's analysis, *áŋyaxka* is handled by suspending in final position the filter (243) that forbids /CVC CV́/ feet. We could add a similar stipulation, allowing Destressing in Clash to apply with a one-sided environment when the target stress is the rightmost in its word. For General Central Yupik, the revised rule would look like (253):

(253) **General Central Yupik Destressing in Clash** (revised)
Remove a mora from CVC in the following contexts:
a. / × ____ ×
b. / × ____ . . . $]_{\text{word}}$

Following Steriade's (1987) approach to rule locality, I interpret (253b) as applying to the rightmost × in a word, which need not be on the final syllable. The rule would derive (242b.iv) *áŋyaxka* from *áŋyáxka* as in (254) (for absence of final stress, see § 6.3.8.8):

```
(254) a. (×  )(×  )         Foot Construction
          μμ  μμ  μ
          aŋyaxka

      b. (×  )(×  )         Destressing in Clash
          μμ  μØ  μ
          aŋyaxka

      c. (×  )              deletion of ill-formed foot
          μμ  μ   μ
          aŋyaxka

      d. (×  )(.   ×)       persistent footing
          μμ  μ   μ
          áŋyaxka
```

It is actually possible that neither account needs to stipulate an extra rule for such forms, owing to the following facts. Word-final light syllables in Central Alaskan Yupik often surface as heavy when another word follows. In particular, a vowel-initial word causes a preceding CV syllable to appear as CVV, and a consonant-initial word causes a preceding CV syllable to appear as CV**C**, where the boldface C constitutes the first half of a geminate. As Weeda 1989, 1990 points out, if this "tacit" weight is represented underlyingly, then the loss of stress on *yax* in *áŋyaxka* follows straightforwardly under either account: for Woodbury, filter (243) is inapplicable; and under my proposal the covertly heavy syllable *ka* would attract stress and trigger Destressing in Clash, with the two-sided environment. In (255) /Ø/ represents the phonologically empty position:

(255) (×)(×)(×) (×)(. ×)
μμ μμ μμ → μμ μ μμ
aŋyaxkaØ *áŋyaxkaØ*

Since little hinges on this, I will adopt the less interesting strategy here and simply assume the Destressing rule of (253b).

6.3.8.6 *Schwa Deletion*

With the background just presented, the Destressing analysis of § 6.3.8.3 can now be further supported by a rule of Schwa Deletion: a dialect split within Central Alaskan Yupik can be analyzed using different orderings of Schwa Deletion and Destressing in Clash.

Before presenting the Schwa Deletion facts, it will be useful to review the rules given so far for the Central Alaskan Yupik dialects, in their correct order:

(256) a. **Foot Construction** Form iambs from left to right.
Degenerate feet are prohibited.
/ – / = CVV, initial CVC in Norton Sound;
= CVV, all CVC elsewhere.

b. **Destressing in Clash** (251), (253)
(not found in Norton Sound, different environments in General Central vs. Chevak)

c. **Pre-Long Strengthening** (235)

d. **Iambic Lengthening** (228d)

Foot Construction must precede all other rules, since they refer to foot structure. Destressing in Clash must precede Pre-Long Strengthening, because the latter has the last say when both rules compete for the same syllable:

(257) a. 'they hurt me (indic.)' M 59 cf. b. 'they hurt me (interr.)' M 59

a.	b.	
ákŋ̥íχtáːtŋ̥a	*ákŋ̥íχtátŋ̥a*	Foot Construction
ákŋ̥iχtáːtŋ̥a	*ákŋ̥iχtátŋ̥a*	Destressing in Clash
ákŋ̥íχtáːtŋ̥a	—	Pre-Long Strengthening
[ákŋ̥íχtáːtŋ̥a]	[ákŋ̥iχtátŋ̥a]	Output

Moreover, in General Central Yupik, where Destressing in Clash requires a stressed syllable on the right, the stresses derived by Pre-Long Strengthening are created "too late" to trigger Destressing in Clash:

(258) 'there is nothing wrong with me' M 60

čaŋátə́nʁitúa	Foot Construction (*tua* is the one CVV syllable)
—	Destressing in Clash
*čaŋátə́n**ʁít**túa*	Pre-Long Strengthening
*ča**ŋáː**tə́nʁíttúa*	Iambic Lengthening

Lastly, Iambic Lengthening must follow Pre-Long Strengthening, because the long vowels derived by Iambic Lengthening do not trigger Pre-Long Strengthening (232); only underlying long vowels do ((230b), (231b)).

Consider now a rule of Schwa Deletion. Many schwas alternate with zero in Yupik dialects. We can understand the motivation for deletion if we consider the Yupik vowel inventory, which (diphthongs aside) is long /iː aː uː/, short /i a u ə/. The crucial asymmetry is that only schwa lacks a long counterpart. This restriction holds even for the output of Iambic Lengthening, which never produces a long schwa.

If Iambic Lengthening cannot produce *[əː], what happens to stressed open syllables containing schwa? In St. Lawrence Island Yupik, Iambic Lengthening simply does not occur for this vowel. For Central Alaskan Yupik, a heavy syllable (hence a canonical iamb) is sometimes achieved in this context by geminating the following consonant. This is true, for instance, in the Unaliq subdialect of Norton Sound (225d.i):

(259) (. ×) Foot Construction 'real name' J 30
μ μ μ
atəpik

(. ×) gemination after stressed schwa
μ μμ μ
atə́ppik

In other Central Alaskan Yupik dialects, there is a more remarkable solution to the problem: the offending schwa is deleted.

(260) **Schwa Deletion** $ə́ \rightarrow \emptyset$ / ____ $]_{syl}$

In other words, whenever we would expect the rules to derive a stressed schwa in an open syllable, the schwa instead is dropped. For example, from underlying /qanʁutəkaqa/ 'I speak about it' J 34, we would expect **qánʁutə́kaqa*. What actually surfaces is [qánʁútkaqa], with schwa deleted. The same underlying morphological sequence surfaces with the schwa preserved when the alternating count of light syllables leaves schwa stressless: /qanaːtəkaqa/ → [qánnáːtəkáːqa], also 'I speak about it' J 34.

Schwa Deletion is distributed among the Central Alaskan Yupik dialects (see (225)) as follows: Norton Sound–Unaliq has no Schwa Deletion, treating stressed schwa as in (259). Most Chevak speakers have Schwa Deletion across the board. For all others (General Central Yupik, Norton Sound–Kotlik, Hooper Bay, Nunivak Island, and optionally for some younger speakers of Chevak), the pattern of (259) is found when the flanking consonants are identical (the context of "antigemination"; McCarthy 1986); otherwise Schwa Deletion applies.

Under the theory assumed here, deletion of stressed vowels is predicted to result in a migration of stress to a neighboring syllable within the foot. Since Yupik dialects are iambic, stress shift should be to the left, which is indeed the case. For the Schwa Deletion dialects, the derivation corresponding to (259) is as in (261):

(261) (. ×)(×) Foot Construction
μ μ μμ
atəpik

(. ×)(×) Schwa Deletion
μ μμ
at pik

(×)(×) stress migration within the foot
μμ μμ
átpik

When Schwa Deletion applies to syllables later in the word, it can create stress clashes. We can ask: Are such clashes then resolved by Destressing in Clash ((251), (253))? Investigating the crucial facts, Jacobson 1985 and Miyaoka 1985 found a dialect division in Central Alaskan Yupik, which crosscuts the classification of (225). According to Jacobson 1985, 45, a particular set of dialects I will refer to loosely as "Inland" displays a pattern that can be described by ordering Schwa Deletion after Destressing in Clash. The other dialects, which I will call "Coastal," order Destressing in Clash after Schwa Deletion. The ordering difference is illustrated in (262).

(262) a. **Inland Dialects** b. **Coastal Dialects** ‘I talk about them [2]’ J 45

a. Inland Dialects	b. Coastal Dialects	‘I talk about them [2]’ J 45
(×)(. ×)(×) μμ μ μ μμ μ *qanʁutəkaxka*	(×) (. ×)(×) μμ μ μ μμ μ *qanʁutəkaxka*	Foot Construction
(×)(. ×)(. ×) μμ μ μ **μ** μ *qanʁutəkaxka*	(ordered later)	Destressing in Clash (253) (Inland dialects)
(×)(×)(. ×) μμ μμ μ μ *qanʁutkaxka*	(×) (×)(×) μμ μμ μμ μ *qanʁutkaxka*	Schwa Deletion (260), migration of stress
(ordered earlier)	(×) (. ×) μμ μ μμ μ *qanʁutkaxka*	Destressing in Clash (253) (Coastal dialects)
[qán**ʁút**kaxka]	[qánʁut**káx**ka]	output

In a similar fashion, underlying /quːyuʁnitəkaxka/ ‘I smile at them [2]’ J 45 becomes [qúːyúʁnítkaxka] in Inland and [qúːyuʁnítkaxka] in Coastal: in this word, the stress retracted from deleted schwa triggers rather than undergoes Destressing.

Observing the distribution of dialects, Jacobson 1985, 45, notes: “It appears that [Inland] is probably the original Yupik pattern and [Coastal] is a modification of that pattern.” This is supported by the speech of the Central Kuskokwim region, which lies between Coastal and Inland territory: here, Inland-style forms are characteristic of older speakers and Coastal-style forms of younger ones (W 698). Under my analysis, the innovation in the Coastal dialects can be treated as a classical case of rule reordering (Kiparsky 1982c), motivated by the opacity of Destressing in Clash in the older dialect.

Woodbury’s account of Yupik encounters an ordering paradox in the Coastal dialect: stress assignment must precede Schwa Deletion, to establish the locations where Schwa Deletion is applicable. But it must also follow Schwa Deletion, in order to assign stress to the correct positions to the right of the deleted schwa. Woodbury suggests that such an ordering paradox is not a defect of the analysis. His assumption is that rules like Foot Construction and Schwa Deletion form a single rule block, and that it is the entire block which iteratively scans the word from left to right. However, to my knowledge such a form of rule application is not attested in any other language. The account proposed here avoids an ordering paradox because Destressing in Clash is a different rule from Foot Construction.

To summarize so far: the last four sections have proposed an analysis of the relation of stress and quantity in the more complex Central Alaskan Yupik dialects south of Norton Sound. The analysis makes use of ordinary iambs, with a binary distinction of weight in which both CVC and CVV count as heavy. Crucial to this account is the rule of Destressing in Clash, which removes stress from CVC syllables in clashing environments, the definition of clash varying across dialects. The arguments for the analysis are that it generalizes trivially to account for the facts of Chevak Yupik, and that the dialect variation for words with deleted schwa emerges from variable rule orderings of Schwa Deletion and Destressing, with no ordering paradoxes.

6.3.8.7 *Clitics and Cyclic Stress*

In Central Alaskan Yupik, the same segmental string can receive a different prosodic pattern depending on whether it contains a clitic or not. For example, underlying /aŋyaxka#mi/ 'how about my two boats?' M 68, with clitic *#mi,* surfaces as *áŋyaxká:#mi;* whereas underlying /aŋyaxka-mi/ 'in the material for boats' M 68, with suffixal *-mi,* surfaces as *áŋyáxkami* (Chevak *áŋyaxkami*). In this section, I offer an analysis of this phenomenon.

It is reasonable to assume that the [a:] in *áŋyaxká:#mi* is derived by Iambic Lengthening (228d). This is because the lengthening respects an alternating count: cf. *ayúqɬi#kiq* 'I hope it will be like (. . .)' M 71, where the pre-clitic syllable immediately follows a stress and is not lengthened.

The crucial question is why Iambic Lengthening should apply in /aŋyaxka#mi/ but not in /aŋyaxka-mi/. I suggest that Foot Construction and Destressing in Clash are applied cyclically, with new cycles induced by clitics but not suffixes. Under these assumptions, the correct lengthening patterns follow from the rules already given.

(263)				
a.	(×) (×) (. ×) μμ μμ μ μ *anyaxka-mi*	(×)(×) μμ μμ μ *aŋyaxka*	(. ×) μ μμ μ *ayuqɬi*	First Cycle: Foot Construction
b.	—	(×) μμ μ μ *aŋyaxka*	—	Destressing in Clash (253b)
c.	—	(×)(. ×) μμ μ μ *aŋyaxka*	—	persistent footing
d.	—	(×)(. ×) μμ μ μ μ *aŋyaxka#mi*	(. ×) μ μμ μ μμ *ayuqɬi#kiq*	Second Cycle: addition of clitic
e.	—	—	(. ×)(. ×) μ μμ μ μμ *ayuqɬikiq*	Foot Construction
f.	—	(×)(. ×) μμ μ μμ μ *áŋyaxká:mi*	—	Iambic Lengthening
g.	[ányáxkami]	[áŋyaxká:mi]	[ayúqɬikiq]	output

The crucial part of the derivation is (263b, c), where during the first cycle, Destressing in Clash and persistent footing create a foot from the sequence /yaxka/ in /aŋyaxka#mi/. Since /aŋyaxka-mi/ has no inner cycle, it cannot undergo these changes, and surfaces with a quite different stress pattern. Similar examples of clitic stress (M 67–73) follow from the same analysis, though additional rules must be added for vowel-initial clitics (M 74).

6.3.8.8 *Phrase-Final Destressing*

The behavior of phrase-final syllables is problematic for my account. Although Yupik scholars hear these as unstressed, I have nonetheless assigned them grid marks in contexts where the rules would derive them. Here, I defend this decision, considering possible accounts of the phonetic outcome.

To begin, I follow Woodbury 1987 in asserting that WITHIN THE WORD PHONOLOGY, final syllables are stressed if they meet the structural description of the stress rules. Here are the arguments: (a) Stress on final syllables does surface whenever a word is not final in what Woodbury (p. 699) calls the Intonational Phrase. For example, the final syllables of /nunakaː/ and /qayamun/ are stressless when phrase final: [nunáː**kaː**], [qayáː**mun**]. But if another word follows in the same phrase, the stress surfaces phonetically: [nunáː**káː#**tamáːna] 'that (extended one) is his land' M 70, [qayáː**mún#**təkíːtuq] 'he came to the kayak' M 72. (b) Miyaoka 1985, 65, notes that Phrase-Final Destressing is blocked when a certain type of intonation is present, so the final stresses can sometimes surface even when they occur phrase-finally. (c) Derivation (258) indicates that Pre-Long Strengthening can apply before word-final CVV syllables. Since this rule applies to the weak syllable of a disyllabic foot, there must be a foot in final position to trigger it here. (d) We also need the word-final feet to trigger Destressing in Clash in General Central Yupik, where the rule only applies with a two-sided clash:

(264)	*maqíkátxún*	Foot Construction	'with the future	J 37
	*maqí**ka**txún*	Destressing in Clash	steambath materials'	
	maqíːkatxún	Iambic Lengthening		
	maqíːkatxun	(standard transcription)		

What is really at issue is whether the word-final stresses that are also phrase-final are later removed by a rule of Phrase-Final Destressing, as proposed by Woodbury (p. 699). A number of problems arise if we posit such a rule. (a) Culminativity (§ 3.1), a good candidate for a linguistic universal, is violated, because words that have just one foot must be regarded as completely stressless in isolation: St. Lawrence Island *nuna* 'land' (Krauss 1985d, 49), Central Alaskan Yupik *əna* 'house' J 31. (b) A low-level phonetic rule (Woodbury 1981, 46–49) that assigns surface degrees of stridency to fricatives acts as if final syllables are footed. Thus the putative Phrase-Final Destressing rule would have to be ordered among rules of the phonetics, even though it is a categorial rule of the phonology. (c) The view that phrase-final syllables are always unfooted would demolish the claim that footing in Yupik is persistent. For example, in Chevak *ikámʁiʁlutəŋ* (252), the surface representation would contain three consecutive unfooted syllables. (d) A Phrase-Final Destressing rule would violate the universal pointed out by Hammond 1984a: destressing rules always apply in the context of clash. For instance, this would not hold for the Chevak example just cited.

All these problems are avoided if we adopt an alternative account proposed originally by Woodbury 1985b: instead of setting up a Phrase-Final Destressing rule, treat the relevant phenomena as part of the rules that interpret metrical structure phonetically. Under this account, final syllables may be metrically strong on the surface but lack (some of) the phonetic properties assigned to strong syllables in other environments. The most salient such property is the ability to bear pitch accents, the intonational tones that link to stressed syllables. (For Yupik tonal association, see Woodbury 1989, 1990.) Under this account, the intonational rules would be modified (perhaps through extrametricality; Weeda 1989) so as not to assign pitch accents to syllables that are final in the Intonational Phrase, much as in Cayuga (§ 6.3.5). Note that this fits in with the observation of Miyaoka cited above: for those contours where the intonation system assigns the final syllable a pitch accent, it sounds stressed.

This solution is not without empirical difficulties. Woodbury notes (p.c.) that "the notion of [phrase-final] stress is undermined by a strong tendency to devoice [final] syllables when [a final intonational L boundary tone] is present." Such devoicing is characteristic of stressless syllables. Woodbury further observes that phrase-final syllables sound unstressed even when they bear high pitch as a result of high intonational boundary tones.[18]

The situation is thus unclear. The case for a late rule of Phrase-Final Destressing seems plausible, but the proposed rule is bizarre enough from a typological point of view (lack of clash environment, violation of culminativity, and ordering among phonetic rules) to make me reluctant to accept it, and prefer instead to attribute the phenomenon to the intonational system, as just proposed. Further research on Yupik intonation may help resolve this issue.

6.3.8.9 *Conclusions*

My proposed account of Yupik prosody (see also § 7.3.2, § 8.8. for further discussion) differs crucially from earlier ones in that it does not require us to expand the inventory of foot construction algorithms beyond those which are

18. To these arguments might be added the fact that Iambic Lengthening (228d) is blocked in phrase-final position. This would follow if the putative Phrase-Final Destressing rule precedes it. But there is an alternative account of the blockage: as noted in § 6.3.8.5, final CV often counts as heavy, suggesting that it contains an empty final prosodic or segmental position. This position would be expected to block Iambic Lengthening.

well attested in other languages. The crucial facts that appeared to require such algorithms followed instead from independent rules of Pre-Long Strengthening and Destressing in Clash. The revised approach greatly simplifies the metrical analysis of the Chevak dialect (§ 6.3.8.4) and overcomes the Coastal Yupik rule ordering paradox faced by other analyses (§ 6.3.8.6). In addition, under my approach Iambic Lengthening can be seen as a normal rule that creates canonical feet, rather than as a puzzling surface contradiction to the foot template.

The proposed analysis provides evidence for the idea that CVC syllables may vary in their quantity (§ 5.6.1): this claim was crucial to the analyses of initial stress in Norton Sound (§ 6.3.8.2), Pre-Long Strengthening (235), and Destressing in Clash ((251), (253)).

The analysis also supports the existence of persistent foot construction, as outlined in § 5.4.1. This device made it possible to write a simple rule of Pre-Long Strengthening, whose effects are purely quantitative, with the concomitant stress derived by persistent footing.

6.3.9 Other Iambic Systems

(a) **Kashaya** (Pomoan, Northern California; Oswalt 1961, 1988) possesses an iambic system of great theoretical interest. A substantial metrical analysis of Kashaya appears in Buckley 1991. Buckley suggests a kind of "chained" extrametricality of the sort I propose to prohibit (§ 5.2.1), though he also notes that Kashaya is a pitch accent language and that tonal alternatives to chained extrametricality are conceivable.

(b) **Yidiɲ** (Queensland, Australia; Dixon 1977) has an unusual metrical system, discussed from the viewpoint of the Iambic/Trochaic Law in Hayes 1985 and McCarthy and Prince 1986, as well as by HV and Kirchner 1990, 1992. Yidiɲ is apparently unique in having a kind of "labeling harmony": if at least one foot of a word constitutes a canonical /⏑ –́ / iamb, then all the feet of the word are made iambic; otherwise all feet are made trochaic. As Kager (1993) points out, Yidiɲ also seems to tolerate "even iambs" (§ 4.3.1) at intermediate stages of the phonological derivation, that is, feet of the form /– ⏑́/, /– –́/. It is significant, however, that in every such case, the initial heavy syllable of the foot is not underlying, but derived by a phonological rule. I conjecture, then, that (a) Yidiɲ is basically a normal iambic language; (b) the vowel lengthening rules are blocked when they would impede the creation of a canonical /⏑ –́ / iamb (i.e. by strengthening its weak syllable); and (c) in words that lack a canonical iamb, metrical structure takes the form of syllabic trochees instead.

In some additional iambic languages, we have no evidence that foot construction iterates across the word: only a single iamb, constructed at the left edge of the word to derive main stress, can be documented from the data in the literature.

(c) **Hopi** (Northern Uto-Aztecan, Arizona; Whorf 1946; Kalectaca 1978; Jeanne 1982; Seaman 1985; a metrical analysis appears in Hayes 1981, 77–79). The initial syllable is stressed if it is heavy (= CVː, CVC), otherwise the second. This suggests construction of an iamb at the left edge of the word, with a word layer created by End Rule Left. Content words of the form CV do not occur. Hopi also has final syllable extrametricality and thus exhibits the "unstressable word syndrome"; it is resolved by incorporation, as discussed in § 5.3. Whorf and Seaman report secondary stress, but its patterning is as yet poorly understood.

(d) **Ossetic** (Iranian, Georgia; Abaev 1964). The initial syllable of a phrase is stressed if it contains a long vowel; otherwise the second syllable is stressed. Degenerate-size words are permitted. The analysis would involve construction of an iamb at the left edge of a phrase (heavy = CVː(C)), with a "phrase layer" created by End Rule Left. Stressless short vowels in initial position tend to reduce or delete, following the Iambic/Trochaic Law.

(e) **Sierra Miwok** (Miwok-Costanoan, Northern California) comprises three closely related languages: Northern (Callaghan 1987), Central (Freeland 1951), and Southern (Broadbent 1964). In all three, the initial syllable is stressed if it is heavy (= CVː, CVC); otherwise the second syllable is stressed. This can be analyzed by constructing an iamb at the left edge of a word. Words of degenerate size are apparently forbidden.

In all three languages, any initial disyllabic foot necessarily takes the canonical form /⏑ –́/. Callaghan 1987, 19, attributes this to a specific rule of iambic vowel lengthening. The other authors do not posit such a rule; rather, the /⏑ –́/ target is achieved through an elaborate conspiracy involving prelengthening suffixes, -CCV suffixes, nonconcatenative morphology, and stem-structure constraints.

For Northern Sierra Miwok, Callaghan very briefly notes an alternating length pattern, which may indicate iterative iambic footing. Secondary stress in the Central language is difficult to make sense of, given the limited data (Freeland 1951, 8). For discussion of secondary stress in the Southern language, see § 6.3.8.3.

For Central Sierra Miwok, the "unstressable word syndrome" is in effect in words of the form /⏑ σ/; see § 5.3.

(f) **Maidu** (Penutian, Northern California; Shipley 1964; Robbins 1991). Stress falls on an initial heavy syllable (CVC only, as there is no vowel length distinction), otherwise on the second syllable. A large number of stems have lexicalized initial stress, argued by Robbins to involve an underlying degenerate foot. Stressed vowels become long ("about one and one-half to two morae," Shipley, p. 10) in open syllables; if such syllables are interpreted as phonologically bimoraic, then the rule can be interpreted as "repairing" all underlying degenerate-size words.

(g) **Cambodian** (Mon-Khmer; Huffman 1970; Nacaskul 1978; Griffith

1991). Here, the iamb serves not so much as an algorithm for stress assignment as a template for word structure. Cambodian distinguishes so-called **major** and **minor** syllables, which may plausibly be identified with /–/ and /⏑/ respectively. Major syllables are freely distributed, but minor syllables may only occur when directly followed by a major syllable. All major syllables are stressed, all minor syllables are stressless, and main stress is final.

This phonotactic pattern has a plausible interpretation in terms of iambs: all syllables of a word must form part of a foot, and all feet must consist of /⏑ –́/ or /–́/ iambs. The marked /⏑ ⏑́/ form of the iamb, which violates the Iambic/Trochaic Law, is excluded. These constraints account for the distribution of minor syllables. The segmental phonology of Cambodian is iambic in character, in that it includes rules that reduce and shorten vowels of minor syllables, thus reinforcing the durational contrast of the iambic foot. The Cambodian pattern appears to be widespread among Mon-Khmer languages.

Iambic systems discussed elsewhere in this book are Seminole/Creek (§ 4.1.2), Asheninca (§ 7.1.8), Pacific Yupik (§ 8.8), Winnebago (§ 8.9), and, possibly, the more controversial cases in the remainder of this chapter.

6.3.10 Right-to-Left Iambs?

All directional-iterative iambic systems discussed so far involve left-to-right stress assignment. In this and the following section, I consider the case for right-to-left iambs. In general, it seems that the existence of such systems is hard to prove.

(a) **Turkish.** A right-to-left iambic analysis is argued for by Barker 1989 (see Kaisse 1985; Hammond 1986; and HV for other analyses). In a subset of the Turkish vocabulary, words ending in /– ⏑ σ/ bear antepenultimate stress, while words ending in /⏑ ⏑ σ/ and /– σ/ are stressed on the penult. Barker accounts for this by marking the final syllable extrametrical, then creating an iamb at the right edge of the stress domain. A provision equivalent to the Priority Clause (§ 5.1.6) prevents a penultimate light syllable from forming a foot on its own when it is preceded by /–/ (265d):

(265) a. (×) / . . . ⏑ –́ ⟨σ⟩ b. (×) / . . . – –́ ⟨σ⟩ c. (. ×) / . . . ⏑ ⏑́ ⟨σ⟩ d. (×) / . . . –́ ⏑ ⟨σ⟩

An alternative is to invoke relatively powerful extrametricality rules so that all final syllables may be treated as light. In this case, the same pattern can be derived with right-to-left moraic trochees, with foot extrametricality in clash:

(266) a. (×) / . . . ⏑ –́ ⏑ b. (×) / . . . – –́ ⏑ c. (× .) / . . . ⏑ ⏑́ ⏑ d. (×)⟪(× .)⟫ / . . . –́ ⏑ ⏑

(b) **Sarangani Manobo** (Philippines, variety described by Meiklejohn and Meiklejohn 1958; see also § 6.1.9) and **Javanese** (Austronesian, Indonesia; Herrfurth 1964) are both Austronesian languages and have a similar stress pattern: main stress falls on the penult unless it is schwa, in which case main stress falls on the ultima. This can be analyzed by treating syllables with schwa as light, forming right-to-left iambs, letting the rightmost foot be extrametrical in clash (§ 5.2.2), and applying End Rule Right:

```
(267) a. (     ×)       b. (     ×)      c. (   ×   )
         (×)(. ×)          (×)(. ×)         (×)(×)
          —  ⏑ ⏑́             —  ⏑ –́           —  –́ ⏑

      d. (    ×)            e. ( ×)
         (×)(×)⟨(×)⟩           (. ×)
          —  –́  —               ⏑ –́
```

Extrametricality is blocked in (267a, b) because there is no clash, and in (267c) because the rightmost foot is not peripheral.

(c) **Malay** (Austronesian, Malaysia), in the variety described by Winstedt 1927, works just like Javanese and Sarangani Manobo, except that only open syllables with schwa count as light. In the variety described by Lewis 1947, Foot Extrametricality applies even when there is no clash, so forms analogous to (267b) receive antepenultimate stress.

(d) **(Tiberian) Biblical Hebrew** has been analyzed by McCarthy 1979b with right-to-left iambs. However, such an analysis faces difficulties in transferring information about main stress placement to the putative iambic feet responsible for vowel reduction. This has led Rappaport 1984 and HV to suggest that metrical structure can occupy separate and simultaneous planes, a proposal that considerably weakens the predictive power of the theory. Churchyard 1990, 1992 argues for an alternative analysis which involves (essentially) moraic trochees rather than iambs, and does not necessitate multiple planes.

(e) **Tübatulabal** (Voegelin 1935) is a Uto-Aztecan language of Southern California; for earlier metrical discussion see Wheeler 1979a, 1979b; Prince 1983a; Crowhurst 1991a, 1991b. Tübatulabal has the following stress pattern: (a) final syllables are stressed; (b) heavy syllables (CV:) are stressed; (c) every other light syllable before a heavy syllable is stressed. This forms a plausible case for right-to-left iambs, an analysis proposed by Crowhurst. Such an analysis is problematic for the theory from two points of view. First, the Priority Clause (§ 5.1.6), which requires that degenerate feet be avoided if further parsing could create a proper foot, runs afoul of the Tübatalabal facts: in words ending in /. . . – ⏑/, the final syllable would wrongly be skipped over to create a proper foot headed by the penult; compare Turkish, Sarangani Manobo, and Javanese (this section), where this skipping appears actually to occur. Second, light syllables occurring word-initially and to the right of a heavy syllable can be stressed under the alternating count. Under an iambic analysis, such syllables would form degenerate feet in weak position, which I have proposed to exclude entirely.

As Kager 1989 has pointed out, it is feasible to analyze Tübatulabal using moraic trochees instead. The crucial idea is to place the main stress in final position prior to foot construction, so that we have top-down stressing. Since the only degenerate foot is in strong position, this will circumvent the problems with weak degenerate feet faced by the iambic analysis.

(268) **Tübatulabal Stress**

a. **Word Layer Construction** — End Rule Right
b. **Foot Construction** — Form moraic trochees from right to left. Degenerate feet are allowed in strong position.

(269) a. 'he is looking out' V 57 — b. 'you may cross it' V 116

```
(     ×)              (              ×)      Word Layer
 ˘ – ˘                  ˘ –  ˘ ˘ ˘           Construction
ele:gɨt               hatda:wahabi

(     ×)              (              ×)      Foot Construction
     (×)                            (×)      (first iteration:
 ˘ – ˘                  ˘ –  ˘ ˘ ˘           satisfies CCC)
ele:gɨt               hatda:wahabi

(     ×)              (              ×)      Foot Construction
  (×)(×)                 (×) (× .)(×)        (remaining
 ˘ – ˘                  ˘ –  ˘ ˘ ˘           iterations)
elè:gɨ́t               hatdà:wàhabí
```

c. 'he is the one who was going along pretending to cry' V 75

```
(                        ×)      Word Layer
 ˘ ˘ – ˘ –  ˘ ˘ ˘ ˘              Construction
anaŋa:lilɔ:gɔpɨganan

(                        ×)      Foot Construction
                        (×)      (first iteration:
 ˘ ˘ – ˘ –  ˘ ˘ ˘ ˘              satisfies CCC)
anaŋa:lilɔ:gɔpɨganan

(                        ×)      Foot Construction
(× .)(×) (×)   (× .)(×)          (remaining
 ˘ ˘ – ˘ –  ˘ ˘ ˘ ˘              iterations)
ànaŋà:lilɔ̀:gɔpɨ̀ganán
```

In this analysis, the final main stress created in top-down mode forces a degenerate foot underneath it to satisfy the Continuous Column Constraint, and also licenses this foot by placing it in strong position.

Whether the final main stresses derived by this analysis are actually present is a difficult question. Voegelin is not entirely clear on this point, stating that the "theoretically" strongest stress is final, but elsewhere that stresses are equal. On a tape recording of Tübatulabal utterances in the UCLA Phonetics Laboratory archives (speaker anonymous), the final main stress is plainly audible. Megan Crowhurst (p.c.) has also listened to tapes from a native speaker, and also hears final stress. Munro 1977 also believes stress is final. I conjecture, therefore, that Voegelin heard final stress as well, but was reluctant to state this firmly in the absence of secure phonological diagnostics. However, Voegelin and Voegelin (1977) note that in the decades since Charles Voegelin did his original fieldwork, the (dying) language has changed its prosody greatly. Thus the question must remain open.

A final point worth making is that in its recent history, Tübatulabal suffered loss of final short vowels and shortening of final long vowels. As Wheeler 1979a, 1979b points out, the modern stress system may be the restructured reflex of a more straightforward earlier system, which would have involved right-to-left moraic trochees, plus End Rule Right.

(f) **Aklan** (Chai 1971; Hayes 1981) and **Cebuano** (Shryock 1993). These closely related Philippine languages have very similar stress systems, treated by Hayes and Shryock using right-to-left iambs. As with Tübatulabal, it is difficult to exclude firmly the alternative of right-to-left moraic trochees, and thus preserve the ban on weak degenerate feet. Shryock's arguments, which go into considerable detail, tend to favor the iambic account.

(g) **Weri** (Goilalan, New Guinea; Boxwell and Boxwell 1966) assigns final main stress, with alternating stresses on preceding syllables. Weri can be analyzed identically to Tübatulabal, as in (268). However, since Weri has no syllable weight distinction, the feet could equally well be syllabic or moraic trochees.

Considering all these cases together, my opinion is that the evidence to decide the issue of right-to-left iambs conclusively appears to be lacking. All the cases either are subject to trochaic reanalysis or are based on quite sketchy data ((b, c) above).

No matter how the issue is decided, the theory will face unresolved puzzles. If there are no true right-to-left iambic systems, why should this gap occur, given that syllabic trochees and moraic trochees can be assigned in either direction? Alternatively, if we do permit right-to-left iambs, we must deal with the question of why in a number of cases, the Priority Clause is violated, and degenerate feet are freely tolerated in weak position.

Concerning the scarcity of right-to-left iambs, one possibility is that the gap is accidental. For syllabic trochees, the most widely found foot type, the number of documented left-to-right systems is quite a bit higher than the number of right-to-left systems (§ 6.2). (For moraic trochees, the number of clearly iterative systems is too small to establish a secure answer.) Moreover, a preference for left-to-right footing seems plausible a priori, since it requires less phonological pre-planning in speaking. A different account of the left/right asymmetry in iambs, based on very different fundamental assumptions, is offered by Kager (1993).

6.3.11 Even Iambs?

I turn finally to cases in which the appropriate foot template appears to be the "even iamb" (§ 4.3.1), taking the form in (270):

(270) (. ×)
σ σ

In § 6.3.5, I discussed the case for even iambs in Seneca, showing how the evidence disappears under the deeper understanding of long vowels reached by Michelson 1984. Below, I consider some further languages that look like they require even iambs:

(A) SOUTHERN PAIUTE (Uto-Aztecan, Utah; Sapir 1930, 1949; Hayes 1981; HV; Jacobs 1990). In this language, even syllables are stressed, counting from left to right. The word layer rule is End Rule Left, with main stress thus falling on the second syllable.

An additional complication occurs in words with an even number of syllables: the penult is stressed, and the final is not, thus going against the alternating count. I assume that (a) final syllables are extrametrical and (b) the degenerate foot created on the penult of even-syllabled words is repaired by incorporation (§ 5.3), as in (271):

(271) (. ×)(. ×)(. ×)(×) → (. ×)(. ×)(. ×)(× .)
σ σ σ σ σ σ σ ⟨σ⟩ → σ σ σ σ σ σ σ σ

For discussion of syllable merger in Southern Paiute, see § 5.6.2.

(B) ARAUCANIAN (Chile; Echeverría and Contreras 1965). Even syllables are stressed, counting from left to right. The word layer is assigned by End Rule Left, thus creating second-syllable main stress. I assume that the prominence heard on final CVC syllables by Echeverría and Contreras is a perceptual effect; they do not present phonological diagnostics to support their transcription.

(C) ONONDAGA (Chafe 1970, 1977; Michelson 1988; § 6.3.5). Main stress is normally penultimate, but can be final or antepenultimate under special circumstances. It can be analyzed with a syllabic trochee, with stressed suffixes and extrametrical suffixes. To account for certain lengthening rules (§ 6.3.5), it is assumed that all even-numbered syllables within the sequence counting from the beginning of the word up to the main stress are metrically strong. This suggests left-to-right even iambs.

(D) DAKOTA (Siouan, Great Plains; Chambers 1978; Shaw 1985a, 1985b). Stress falls on the second syllable of a word, and on monosyllables. This suggests that a single even iamb is assigned at the left edge of a word.[19]

At this point we can assess the extent to which these examples counterexemplify the theory proposed here. The apology for them will consist of two claims. First, in strictly formal terms all these languages are fully compatible with the theory, though they go against the claims of the Iambic/Trochaic Law. Second, for some of these languages, the violations of the Iambic/Trochaic Law are repaired in certain contexts by segmental rules.

The reason all the languages discussed above can be derived under the theory is that they all lack a syllable weight distribution, at the relevant level of the derivation. (For Onondaga, this restriction holds for pretonic syllables, which form the domain of the secondary stress rule.) Within moraic theory, this restriction is represented by making all syllables light, as in (272):

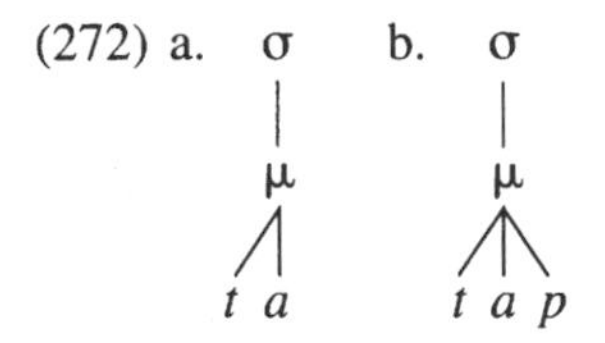

Based on this, I propose to analyze the languages above simply as iambic. The iambic foot template includes as an option /◡ ◡́/, which accommodates the "even iamb" cases, provided they contain no heavy syllables. The languages listed above can thus be thought of as defective iambic languages, in which there are no heavy syllables for the stress rule to encounter.

I am claiming, then, that there is no foot template of the form (273a), in that all cases that appear to require (273a) are equally amenable to analysis in terms of (273b), the "standard" iamb.

(273) a. **Even Iamb** b. **Standard Iamb**

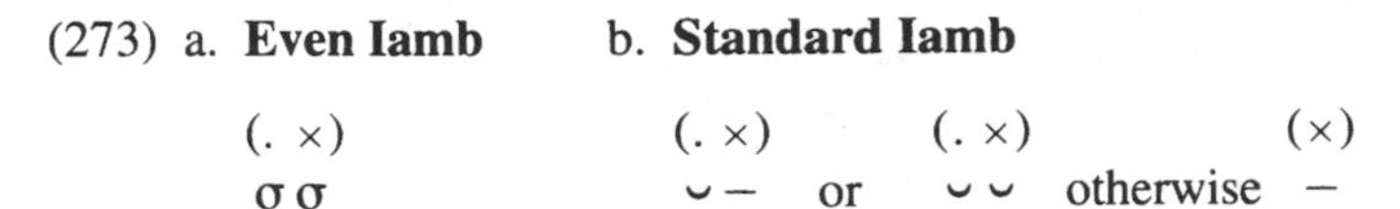

19. Shaw's detailed analysis of the Stoney dialect also includes a second stress rule, claimed to be sensitive to syllable quantity: that is, stress the final when it is heavy and the penult when the final is light. Since there is no phonemic vowel length in Dakota, and the claimed quantity distinction is one of final CVC vs. CVCC, there is an alternative: assume final C in CVCC to be temporarily extrasyllabic, as in Arabic (§ 5.2.1), and let penultimate stress be assigned by syllable extrametricality and End Rule Right. The stray consonant would block extrametricality, so that final CVCC would attract stress.

This claim is empirically falsifiable. To see why, consider the mirror image of the even iamb, the syllable trochee. Here, we find cases where a quantity distinction exists, but stress is nevertheless assigned to every other syllable, irrespective of quantity; for discussion and examples, see § 5.1.9. In contrast, there appear to be no cases of this sort among iambic systems. This is predicted by the theory proposed here, which has no mechanisms to derive truly quantity-insensitive iambic stress.

In the case of Dakota, my account apparently recapitulates history: according to Miner 1979, Common Mississippi Valley Siouan, the ancestor of Dakota, had an ordinary iambic pattern. Dakota reached its present state by losing its long vowels; words that originally had long vowels in their initial syllables have synchronically exceptional initial stress, at least in some dialects. For further discussion, see § 8.9.4.

Nevertheless, the stress rules noted above to some degree go against the predictions of the theory, since the canonical iamb, /˘ –́/, is not created by the stress rule. That is, although these languages conform to the letter of the theory, they do not conform to its spirit, in that they go against the Iambic/Trochaic Law.

In this connection, it is noteworthy that /˘ –́/ feet do occur in surface representations in some of these languages.

In Onondaga and Seneca, rules apply to lengthen the vowel of a /˘ ˘́/ foot, creating surface /˘ –́/; for discussion, see § 6.3.5.

In the analysis of Harms 1966, 1985, Southern Paiute creates canonical /˘ –́/ iambs on the surface by gemination. Harms argues that voiceless vowels are underlying. When they occur stressed before a voiceless stop or affricate, they are voiced, and the consonant is geminated, creating a canonical /˘ –́/ foot:

(274) (×) → (×) 'to fall' Harms 1966, 233

(. ×) → (. ×)

˘ ˘ ˘ → ˘ – ˘

ḁčḁkɨ → ***ḁčakkɨ***

Even in alternative accounts, where gemination is underlying, the gemination alternations conform to the Iambic/Trochaic Law, when considered under the two-layered theory of syllable quantity proposed in § 7.3.

Araucanian has reduction and (in certain cases) optional deletion of weak syllables in the foot when the vowel is /ɨ/, as in /kɨθáw/ → [kəθáw, kθáw] 'work' EC 134.

It can be seen, then, that of the cases above, Southern Paiute, Onondaga, Seneca, and Araucanian manage to conform in some ways to the Iambic/Trochaic Law, even though they lack the underlying heavy syllables that would be needed to conform to it fully.

6.3.12 Conclusion

I conclude this section with a number of remarks concerning iambic languages in general.

A theme I have emphasized is the tendency of segmental phonology in iambic languages to exaggerate the quantitative contrast within the foot, through either lengthening of the final stressed syllable or reduction of the stressless one. Obviously, not all iambic languages do this, but the tendency seems fairly widespread.

A characteristic that is shared by many rules of iambic lengthening is the avoidance of neutralization; a lengthened short vowel tends not to be merged with an underlying long vowel. For example, Choctaw, Chickasaw, and St. Lawrence Island Yupik avoid such neutralization by having three surface degrees of length. Central Alaskan Yupik avoids neutralization with Pre-Long Strengthening (§ 6.3.8.2). Hixkaryana, Maidu, and Lake Iroquoian (at the stage where the rule was introduced) have no vowel length contrast in the first place; and Iambic Lengthening by gemination is similarly non-neutralizing in Menomini and Yupik. However, Iambic Lengthening of vowels does appear to be neutralizing in Menomini, Potawatomi, Kashaya, Yidiɲ, and Sierra Miwok.

A mysterious property of iambic lengthening rules is their tendency not to apply to syllables in word-final position, as in Hixkaryana, Choctaw and Chickasaw, Lake Iroquoian, Yupik, Sierra Miwok, Yidiɲ, and Kashaya. I have incorporated this restriction in various ways into the individual analyses, but find this unsatisfactory. If the avoidance of final iambic lengthening is truly general, it deserves general explanation.

Another property that seems to be unusually common in iambic languages is the lack of a higher layer of metrical structure, as in Choctaw, Chickasaw, Eastern Ojibwa (dialect described by Bloomfield), Seneca, Macushi (§ 6.3.1), and perhaps Hixkaryana, Yupik, and Yidiɲ. It is plausible that iambic rhythm, being based on durational contrast, would be less closely linked to the hierarchical rhythmic patterns characteristic of intensity contrast, as found in trochaic languages.

Note finally the curious generalization that most iambic languages are (or were) spoken in the Americas. Given that Dryer 1989 has argued that typological similarities can extend over linguistic areas that comprise entire continents (including, for example, the Americas), this geographical asymmetry may not be a coincidence. On the other hand, throughout the Americas the iambic languages are interspersed with great numbers of non-iambic languages, so the idea that iambic stress is an areal phenomenon of the Americas should not be taken as a certainty.

• 7 •

Syllable Weight

This chapter addresses specific issues in the theory of syllable weight. My central proposal is that what has previously been treated as a single phenomenon of weight is in fact best regarded as the domain of two quite distinct theories, a theory of syllable quantity and a theory of syllable prominence. By recognizing two distinct phenomena, we can impose substantial restrictions on the behavior of each. Combined, the theories form a general account of how stress rules may refer to properties of the syllable.

The later portions of the chapter discuss additional problems in the theory of weight: § 7.2 applies the theory of prominence to unbounded stress systems; § 7.3 discusses cases in which different prosodic processes in a single language access different criteria of quantity; and § 7.4 discusses miscellaneous problems.

My starting point is moraic theory, outlined in § 3.9.2. To review the crucial assumptions: the weight of a syllable is indicated by its mora count. In most languages with a weight distinction this can vary from one to two, though some languages permit heavier syllables. Short and long vowels contrast in underlying representations by bearing one versus two moras respectively. Coda consonants license an additional mora in some languages, via a rule of Weight by Position. This rule can vary in its structural description, sometimes only assigning a mora to a subclass of coda consonants, or to coda consonants in particular contexts (§ 5.6.1). Onset consonants never license a mora, accounting for the characteristic pattern that the presence or length of an onset makes no difference to weight.

In § 3.9.2, I reviewed two fundamental arguments in favor of moraic theory: that it predicts the phenomenon of moraic conservation, and that it represents only those prosodic elements that may be counted by phonological rules.

7.1 QUANTITY VERSUS PROMINENCE

Consider now some distinctions of syllable weight that cannot be analyzed under moraic theory. The problematic cases fall into two types.

First, syllable weight can involve more than two degrees. For a number of such cases, the theory as already proposed can provide an account, by factoring out a part of the three-way hierarchy using a separate rule. Such strategies include consonant extrametricality (Arabic, § 4.1.3, § 6.1.1); treating super-heavy syllables as the equivalent of a heavy + light sequence (Hindi, § 6.1.7; Estonian, § 8.5); or assigning variable weight to CVC (Yupik, § 6.3.8). But there are more difficult cases, where there is a multiple weight hierarchy, and the stress rule must consider three or more weights SIMULTANEOUSLY in order to locate stress on the heaviest syllable. Examples of this sort, discussed below, include Pirahã (§ 7.1.7), Asheninca (§ 7.1.8), and certain dialects of Hindi (§ 7.1.2).

In addition, syllable weight occasionally involves structural properties that cannot be represented with moraic structure. For example, in the weight hierarchy for Pirahã, the presence or absence, as well as the voicing, of the syllable onset makes a difference in weight. In Serbo-Croatian, Lithuanian, and Golin (§ 7.1.3), high-toned syllables attract stress in preference to low-toned ones, and the high-toned syllables cannot be equated with bimoraic syllables.

The situation, then, is that the theory of syllable weight is faced with a residue of cases that do not follow from moraic theory. Moreover, syllabic constituency theory (cf. § 3.9.2) does no better. A pessimistic conclusion is that there are no linguistic universals of interest in this area, and that theories of syllable weight at best express a commonplace pattern among endless possibilities.

In my opinion this is too pessimistic, and is the result of an erroneous assumption: that what we pretheoretically consider as syllable weight should necessarily be treated as a single phenomenon within the theory. Suppose to the contrary that two phenomena are involved, as follows.

First, weight can be thought of as a property of the **time** dimension: a syllable is heavy because it is long. This is the viewpoint of moraic theory: the moras form an abstract characterization of a syllable's phonological duration. Under the Iambic/Trochaic Law, a canonical iambic foot is canonical because of its uneven durational form, that is, /μ $\acute{\mu}\mu$/; and a canonical moraic trochee is likewise canonical because of its even durations; /$\acute{\mu}$ μ/. To emphasize the idea that weight as depicted by moras represents duration, I will employ the term **quantity** for such weight distinctions.

There is a second sense in which syllable weight can be thought of: as raw prominence or perceptual salience. Heavy syllables, or syllables with high tone, or syllables with low vowels, and so on, tend to sound louder than other syllables. Normally, this variation is phonologically irrelevant, but it appears that some languages take differences in prominence and phonologize them, making them the basis of true phonological stress rules.

The proposal here is to formalize this distinction, accounting for weight with a theory of quantity, based on moraic structure; and a theory of prominence, based on a different representation, which encompasses the whole set of phonetic properties (weight included) that make syllables sound louder.

A formalism for representing prominence has been proposed by Everett and Everett 1984; Everett 1988; and Davis 1989: a kind of grid is employed in which more prominent syllables have higher columns. For example, the hierarchy of prominence employed in Pirahã is as depicted in (1):

(1)

a.	b.	c.	d.	e.
×				
×	×			
×	×	×		
×	×	×	×	
×	×	×	×	×
KVV	GVV	VV	KV	GV

where $K = \begin{bmatrix} C \\ -voice \end{bmatrix}$

$G = \begin{bmatrix} C \\ +voice \end{bmatrix}$

Such prominence grids can be referred to by metrical rules.

A crucial property of prominence, I claim, is that it is **irrelevant to foot construction.** In the theory proposed here, the templates for weight-sensitive feet (iambs and moraic trochees) are defined in terms of the theory of quantity; that is, in terms of moraic content. Foot construction accordingly has no access to prominence. Moreover, this is not an arbitrary stipulation: the foot templates are based on the Iambic/Trochaic Law, which in turn is related to phonological duration and therefore to mora count.

This is what sets up the crucial empirical prediction of this approach; namely, that stress rules that refer to prominence are limited to mechanisms of the theory other than footing, namely the End Rule, destressing, and extrametricality. This restriction rules out a wide variety of logically conceivable stress rules. In (2) are some hypothetical examples:

(2) a. **Tonal Iambs** Stress all high-toned syllables and the even-numbered members of any sequence of low-toned syllables (= left-to-right iambs, with high tone counted as heavy).

b. **Vowel-Sonority Cairene** The rule of Cairene Arabic (§ 4.1.3), modified as follows: the vowel system contains only /i, a, u/; /a/ is heavy, and /i, u/ are light.

c. **Multiweight Iambs** Consider the syllables of a word by pairs, going from left to right. If on the hierarchy of (1), σ_n is lighter than σ_{n+1}, form /σ_n σ_{n+1}/ into a right-headed foot and move on to σ_{n+2} σ_{n+3}. Otherwise, make σ_n into a foot and move on to σ_{n+1} σ_{n+2}. Repeat until the word is fully parsed.

In effect, I am claiming that stress rules have a choice: one possibility is to refer to a simple criterion of syllable weight (i.e. quantity, under moraic theory) and have access to the rich inventory of stress assignment processes that foot construction permits. The other possibility is to employ a rather unconstrained criterion of syllable weight (i.e. prominence, represented with grid columns) and have access only to a more impoverished inventory of stress assignment devices.

I emphasize that the dichotomy above applies at the level of RULES, not of stress systems. The more interesting cases below involve a mixture of footing rules that respect quantity (or ignore weight entirely) and non-footing rules (destressing, end rules, extrametricality) that respect prominence.

The first to suggest that a subset of metrical rules might be restricted in how they refer to syllable structure was Davis 1988, 4. My proposal divides the pie somewhat differently than Davis's, in that I include only foot construction in the set of rules that must reference pure quantity.

As Davis points out, the theory of HV is in a sense entirely a theory of syllable prominence: weight differences are ALWAYS represented by projecting grid columns, so there is no quantity/prominence distinction. To the extent that the distinction is valid, it constitutes evidence for the approach taken here, where foot construction refers directly to moraic structure rather than preassigned grid columns.

7.1.1 How Rules Refer to Prominence

I now consider the theory of prominence in more detail. The proposals of Everett and Everett 1984 and Davis 1989 that prominence be represented with a grid is supported by the kinds of prominence-sensitive rules we find: typically stress falls on the most prominent syllable of a word, with a default direction (rightmost or leftmost) in the event of ties. This can be derived by projecting grid columns according to prominence and labeling them with End Rule Left or (as in (3)) End Rule Right:

```
(3)                            (                    x        )
             x        x                    x        x
       x     x     x  x     x  →     x     x     x  x     x
    x  x  x  x  x  x  x  x  x     x  x  x  x  x  x  x  x  x
    σ  σ  σ  σ  σ  σ  σ  σ  σ     σ  σ  σ  σ  σ  σ  σ  σ  σ
```

Another property that suggests grid representation is found in the examples of § 7.1.2: a rule of syllable extrametricality is blocked if it would mark as extrametrical the most prominent syllable of the word. This can be characterized as a constraint on grid rows: it is not permitted to make all the marks on a row extrametrical.

A number of problems arise in using prominence-based grids in languages that also have foot structure. In Asheninca (§ 7.1.8), prominence (based on a combination of weight and vowel quality) determines which of the last two feet in the word will receive main stress. It sometimes happens that a foot earlier in the word will have a more prominent head than either of the last two feet, as in (4), taken from (47a) below:

```
(4)    ×   ×
       ×   ×   ×             } grid layers projecting prominence
       ×   ×   ×   ×
     ( ×) ( ×)(.  ×)(. ×)       foot layer
       –    –  ⌣  ⌣  ⌣ ⌣⟨⌣⟩
     nʲàːwʲàːtawákariri
```

Nevertheless, one of the last two feet (in this case, the penultimate one) still gets the main stress. It is not immediately clear how this can be done while respecting the Continuous Column Constraint:

```
(5) * (            ×    )       word layer
       ×   ×
       ×   ×   ×             } grid layers representing prominence
       ×   ×   ×   ×
     ( ×) ( ×)(.  ×)(. ×)       foot layer
       –    –  ⌣  ⌣  ⌣ ⌣⟨⌣⟩
     nʲàːwʲàːtawákariri
```

A similar problem from Asheninca is that a foot head always wins out in stressing over a syllable that is not a foot head, even when the latter has more prominence. In (6), the non-head syllables of the rightmost two feet contain /a/, which counts as more prominent in Asheninca than /i/, the vowel of the foot heads. Nevertheless, it is a foot head that receives main stress.

```
(6) (          ×)        'he said it without thinking'       Payne 1990, 198
    (. ×)(. ×)(. ×)
     ⌣ ⌣  ⌣ ⌣ ⌣ ⌣⟨⌣⟩
    ikàntasìtaríra
```

In general, among the stress systems that include both footing and prominence-based rules, there appear to be no clear cases in which syllables that are not foot heads win out over foot heads for main stress.

A reasonable way to resolve these conflicts between prominence grids and foot structure is to suppose that the prominence grid is not a true metrical grid, forming a permanent, inherent property of representations, but rather a temporary computational device employed by individual rules. There are two observations that support this view. First, in Asheninca different rules refer to different hierarchies of prominence; see § 7.1.8.3. Second, prominence grids frequently involve severe and pervasive clashes, with little if any pressure to resolve them; see for example (36h) below. This goes completely against what we have seen among true metrical grids.

In light of this, I will posit a rather modest prominence theory, which regards the prominence grid as a mere temporary computational device. I will depict prominence grids with asterisks on a separate inverted plane, below the syllable string. The crucial notion is this: constituent construction creates a constituent on the true metrical plane, but it must satisfy the structural requirements of the End Rule on the prominence plane.

The idea of using a separate plane as a computational device is a central one in HV, which permits such planes to have full metrical structure, in order to account for cyclic stress and other phenomena. The separate planes assumed here are considerably less ambitious, intended to serve solely as a depiction of prominence available for rule application within the true metrical representation. Prominence planes may appear in structural descriptions, but not in structural changes.

Here is a schematic example. Suppose that syllables containing /a/ are more prominent than other syllables; that there are no feet; and that the rule creating the word layer is End Rule Right. This would create a representation for a schematic word as in (7):

```
(7) (          ×     )      metrical plane
    pipupapipápupi
     * * * * * * *          prominence plane
         *   *
```

Next, consider what happens if the metrical plane contains feet. As already noted, it appears that a non-foot head, no matter how prominent, never receives main stress over a foot head. For this reason, I propose that only those syllables that are eligible for promotion in the metrical layer being created may have their prominence projected on the prominence plane. In (8) are two hypothetical examples, which also assume End Rule Right and prominence for /a/:

```
(8) a. (              ×  )  b. (          ×        )  metrical plane
       ( . ×)(. ×)(. ×)        ( . ×)(. ×)(. ×)
        ˘ ˘ ˘ ˘ ˘ ˘ ˘            ˘ ˘ ˘ ˘ ˘ ˘ ˘
       papupapipupípu          pipapapápupipu
          *   *   *               *   *   *      prominence plane
                                  *   *
```

The crucial point is that the /a/'s in (8a) cannot be stressed, because they are not projected on the prominence plane.

The scheme just outlined is stated somewhat more formally in (9), as a proposal for a parametrically specified version of the End Rule.

(9) **Prominence-governed End Rule (Right/Left)**

a. Construct a specified metrical constituent $\mathcal{C}$.
b. Establish a hierarchy of prominence among syllables.
c. On a separate prominence plane, project the prominence values of all potential docking sites within $\mathcal{C}$. Docking sites are foot heads where feet have been constructed, otherwise all syllables in $\mathcal{C}$.

d. Assign a grid mark to 𝒞, aligned with the right/leftmost of the syllables having the highest grid columns on the prominence plane.
e. Delete the prominence plane.

I conjecture that rules of destressing and extrametricality may also construct temporary prominence planes, though in the absence of enough cases to study I will not pursue this here.

As to what factors can render a syllable more prominent, the following apparently must be included: heavy syllable quantity, lowness in vowels, high tone, the presence of syllable-final /ʔ/, and the presence or voicing of syllable-initial consonants. Theories of speech production and perception are relevant to this question, since many prominence factors seem to have a natural phonetic basis: low vowels are typically acoustically louder, high pitch is correlated with greater loudness, longer stimuli tend to sound louder, and so on (Denes and Pinson 1963; Lehiste 1970).

A minimal claim that can be made at this point is that any given factor influences prominence in only one direction; for instance we do not expect to find a language in which greater prominence is associated with light syllables, high-voweled syllables, low-toned syllables, and so on.

The following are examples of stress assigned by prominence grids.

7.1.2 Hindi Dialects

In § 6.1.7 I discussed a moraic trochee stress pattern for Hindi supported by evidence from metrics, noting that the literature contains a large number of differing descriptions. I suggested that the metrically supported pattern may be archaic, and that the creation of superheavy syllables in Hindi through vowel loss has led to innovating stress rules. Here, I discuss one such rule. The evidence consists only of subjective judgments, but since three sources agree to a fair extent on the facts, it seems plausible that the pattern corresponds to phonological reality for some speakers.

In the pattern described by Kelkar 1968, Hindi stress is based on three phonological weights: superheavy (CV:C, CVCC), heavy CV:, CVC), and light (CV). The data presented may be summarized as follows: stress falls on the heaviest available syllable, and in the event of a tie, the rightmost nonfinal candidate wins. Kelkar also presents secondary stress patterns, which I will not attempt to analyze. The main stress rule can be expressed as in (10):

(10) a. **Prominence Projection** Project syllable prominence according to the following hierarchy:

***: CV:C, CVCC
**: CV:, CVC
*: CV

b. **Syllable Extrametricality** σ → ⟨σ⟩ / ____ $]_{word}$
Condition: do not make the only grid mark on a layer extrametrical.

c. **Word Layer Construction** End Rule Right

Rule (10a) might be generalized to something like "translate weight into prominence," assuming CV:C, CVCC are trimoraic in Hindi. This follows Davis (1989), who applies the idea to the analysis of the rather different Hindi stress pattern reported in Gupta 1987. Rule (10b) prohibits exhausting a prominence grid layer with extrametricality (cf. the boldface prominence grid marks in (13)). This constitutes the analogue of the ordinary prohibition on exhausting a domain in true metrical grids (§ 3.11). I conjecture that it is a general principle applying to prominence grids rather than a language-particular rule.

The following are examples from Kelkar 1968, 27–29, with sample representations derived by the rules above.

(11) **Prominence Hierarchy:** /=/ > /–/ > /⌣/

a. /– –́ ⌣ –/ *ka:rí:gari:* 'craftsmanship'

b. /=́ ⌣ – –/ *šó:xǰaba:ni:* 'talkative'
/=́ – –/ *ré:zga:ri:* 'small change'

```
( .  ×  .)          ( ×    .  .)
  –  –  ⌣⟨–⟩          =    ⌣  – ⟨–⟩
ka:rí:gari:         šó:xǰaba:ni:
  *  *  * *           *    *  *  *
  *  *    *           *       *  *
                      *
```

(12) **Ties Resolved: Rightmost Non-Final**

a. /⌣ ⌣́ ⌣/ *samíti* 'committee'

b. /–́ –/ *qísmat* 'fortune'
/– –́ –/ *ro:zá:na:* 'daily'
/⌣ –́ –/ *ruká:ya:* 'stopped (trans.)'

c. /=́ =/ *ró:zga:r* 'employment'
/= =́ =/ *a:smá:nǰa:h* 'highly placed'
/=́ – =/ *á:smã:ǰa:h* 'highly placed' (variant)

```
( .  ×)       ( .  ×)        (.    ×)
  ⌣  ⌣⟨⌣⟩       –  – ⟨–⟩       =    =  ⟨=⟩
samíti        ro:zá:na:      a:smá:nǰa:h
  *  **         *  *  *        *    *   *
                *  *  *        *    *   *
                               *    *   *
```

(13) **Extrametricality Blocked When It Would Exhaust a Layer**

a. /⏑ –́/ *kidʰár* 'which way'
   /⏑ ⏑ –́/ *rupiá:* 'rupee'

b. /⏑ =́/ *ǰaná:b* 'sir'
   /– =́/ *asbá:b* 'goods'
   /⏑ – =́/ *musalmá:n* 'Muslim'
   /– ⏑ =́/ *inqilá:b* 'revolution'
   /– – ⏑ =́/ *parvardagá:r* 'God as a protector'

```
( .  . ×)     ( .   .     ×)
  ⏑  ⏑ –        –   –   ⏑ =
r u p i á :   parvardagá:r
  *   * *       *   *   * *
        *       *   *     *
                          *
```

With slight differences, the same algorithm of stress assignment holds for the varieties of Hindi described by Sharma 1969 and Jones 1971. Sharma deals with only two syllable weights, treating /=/ syllables like /– ⏑/ sequences; this leads to different stress in forms like (11b) and (12c). Jones recognizes only an ordinary light/heavy distinction, again resulting in different outcomes for some forms. For Sindhi, a neighboring Indo-Aryan language, Khubchandani 1969 presents the same rule, again with just two syllable weights.

The algorithm for Hindi that Kelkar provides in his article actually differs slightly from what is presented above, in that he resolves ties by giving main stress to the second-to-last heaviest syllable, rather than to the rightmost non-final one. No forms that could resolve this question (e.g. /– – ⏑/, /= = –/) are presented by Kelkar or by Jones. Sharma and Khubchandani stress such words according to the rule given here.

A neighboring Indo-Aryan system, Bhojpuri (Tiwari 1960; Shukla 1981), can also be analyzed with the rules of (10). For Bhojpuri, the pattern of vowel reduction suggests that moraic trochees, assigned as in Maithili (§ 6.1.6), are also present. These feet are compatible with the prominence system, since only foot heads ever receive main stress.

7.1.3 Stress Based on Tone

In Serbo-Croatian (Inkelas and Zec 1988), Lithuanian (Halle and Kiparsky 1981; Blevins 1991), and Golin (Chimbu, New Guinea; Bunn and Bunn 1970), stress appears to be dependent on tone. In Golin, the last high-toned syllable in the word is stressed, with final stress in words having all low tones. For Serbo-Croatian and Lithuanian, stress falls on the first syllable linked to a high tone, and there are no low-toned words.[1] Lithuanian has been treated by HV as a case of mora-bearing stress rather than a tone-based system; for discussion of the disagreement, see § 3.9.1 and Blevins 1991.

I analyze the Golin pattern as in (14); (15) gives schematic examples illustrating the outcome:

(14) a. **Prominence Projection** Construct a prominence grid according to the following hierarchy:
  ** high-toned syllables
  * low-toned syllables

b. **Metrical Structure** End Rule Right

(15)
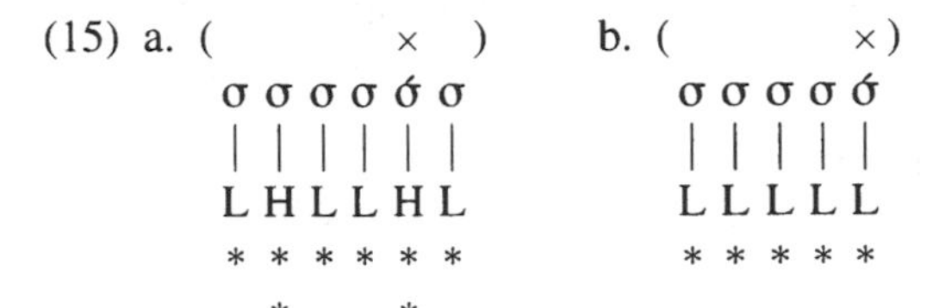

7.1.4 Klamath

The stress pattern of Klamath (Oregon; Barker 1964) has been analyzed metrically by Levin 1985a; Hammond 1986; and HV. Klamath stress shows an interesting interaction between foot construction, based on quantity, and an End Rule, based on prominence. The basic stress pattern resembles Classical Latin (§ 5.1.4; § 5.3), as shown below. It should be noted that the stress rules as stated by Barker (pp. 35–37) fail to assign a stress to certain strings, such as CVC(C)VCVC(C) or CVCVC(C). Such strings include one form that Barker actually lists, namely *Glégatk* 'dead' B 38. A quasi-Latin pattern seems to be the best reconstruction of Barker's intent.

(16) a. A heavy penult is stressed (/ – / = CVː, CVC):

/⏑ –́ ⏑/	*naq'áːqbli*	'puts a flat object back on one's lap'	B 35
/– –́ ⏑/	*gaːmóːla*	'finishes grinding'	B 35
/– –́ ⏑/	*gatbámbli*	'returns home'	B 36

b. If the penult is light, stress falls on the antepenult:

/⏑́ ⏑ ⏑/	*č'áw'iga*	'is crazy'	B 36
/–́ ⏑ ⏑/	*ʔápʔota*	'promises'	B 36

c. Disyllables bear initial stress:

/⏑́ ⏑/	*bóč'o*	'wild celery'	B 37
/–́ ⏑/	*gépgi*	'come!'	B 37

1. Similar claims made for Fore (Nicholson and Nicholson 1962, cited in Hayes 1981) were not supported in later work (Pike and Scott 1963; Scott 1978).

Klamath differs from Latin in the following way: main stress must fall on the rightmost long-voweled syllable, no matter what position of the word it occupies:

(17)	a. /⏑ V́ː – ⏑/	*čat'**á**ːwipga*	'is sitting in the sun'	B 35
	b. /⏑ V́ː – – ⏑ /	*gaw'**í**ːnapgabli*	'is going among them'	B 37
	c. / – V́ː/	*n'isq'**á**ːk*	'little girl'	B 37

Barker supports this stress pattern by showing how the position of main stress determines the alignment of intonation contours. He states that in cases like (17a, b) the penult bears secondary stress (as indeed it does in the proposed analysis) but expresses doubt about secondary stress elsewhere. Following other accounts, I will not attempt a general treatment of secondary stress here.[2]

My analysis of Klamath involves the construction of ordinary moraic trochees, with the word layer based on prominence structure. The rules for footing, given in (18), are a modified version of the Latin rules in § 5.1.4. The differences are: (a) footing is iterative; (b) degenerate feet are allowed (since Klamath tolerates degenerate-size words: *pse* 'daytime', *tmo* 'grouse'; Barker 1963); (c) extrametricality holds only for short-voweled syllables.

(18) a. **Syllable Extrametricality** σ → ⟨σ⟩ / ____ $]_{word}$ where σ does not dominate Vː

b. **Foot Construction** Form moraic trochees from right to left. Degenerate feet are allowed in strong position.

	a.	b.	c.	d.	e.
(19)	(×)(×)	(×) (×)	(×)(×)	(× .)	(×)
	⏑ – ⟨⏑⟩	– – ⟨⏑⟩	– – ⟨⏑⟩	⏑ ⏑⟨⏑⟩	⏑ ⟨⏑⟩
	naq'aːqbli	*gaːmoːla*	*gatbambli*	*č'aw'iga*	*bóč'o*

e.	f.	g.	h.
(×)	(×)(×) (×)	(×) (×)(×) (×)	(×) (×)
– ⏑⟨⏑⟩	⏑ – – ⟨⏑⟩	⏑ – – – ⟨⏑⟩	– –
ʔapʔota	*čat'aːwipga*	*gaw'iːnapgabli*	*n'isq'aːk*

For creation of the word layer, we project the prominence of every foot head, since foot heads are the only candidates for promotion at the word layer. The prominence hierarchy used is as in (20):

(20) **: VV syllables
 *: other syllables

By forming a word layer and labeling it with End Rule Right based on the prominence grids, we obtain the correct outcomes:

2. Hammond 1986 derives the posttonic secondary stresses, but not the other instances of secondary stress described by Barker.

(21)
a. (×) / (×) / ˘ – ⟨˘⟩ / *naq'á:qbli* / * / *
b. (×) / (×) (×) / – – ⟨˘⟩ / *ga:mó:la* / * * / * *
c. (×) / (×)(×) / – – ⟨˘⟩ / *gatbámbli* / * *
d. (×) / (× .) / ˘ ˘⟨˘⟩ / *č'áw'iga* / *
e. (×) / (×) / ˘ ⟨˘⟩ / *bóč'o* / *

e. (×) / (×) / – ˘⟨˘⟩ / *ʔápʔota* / *
f. (×) / (×) (×) / ˘ – – ⟨˘⟩ / *čat'á:wipga* / * * / *
g. (×) / (×) (×) (×) / ˘ – – – ⟨˘⟩ / *gaw'í:napgabli* / * * * / *
h. (×) / (×) (×) / – – / *n'isq'á:k* / * * / *

Following this stage, the prominence plane is deleted; the output representations also show the loss of degenerate feet in weak position. It can be seen that the analysis captures the basic pattern of Barker's description: stress roughly follows the Latin type at the footing stage, but the prominence of CV: syllables asserts itself at the prominence-grid stage.

An inelegant aspect of the analysis is the need to refer to long vowels twice: they project as more prominent, and syllables containing them cannot be extrametrical. The latter condition is necessary in order to derive final stress when /V:/ occurs in the final syllable. This redundancy is shared (in different guises) by Hammond's and HV's analyses. It might be thought that the problem would disappear if we allow syllables that are not foot heads (e.g. final long-voweled syllables) to project prominence grid columns. But such a solution would wrongly assign final stress to other final syllables, such as **ɢlegátk* for *ɢlégatk* 'dead' B 38.

It is worth noting that in the Klamath intonational system, long vowels invariably attract high tone (Barker 1964, 35–38). It is thus conceivable that the prominence system is based on tone rather than length (§ 7.1.3).

7.1.5 Mam

The phonology of Mam (Mayan; Guatemala) is analyzed by Nora England in a grammar (1983) and in her introduction to the dictionary of Maldonado, Ordonez, and Ortiz 1986. The stress pattern is summarized below; forms with stress inferred from her description have parenthesized page references.

(22) a. If a word contains a long vowel, it is stressed:

ʔaɗú:ntl	'work'	E 37
waɗná:ya	'I worked'	E 37
ɗá:nɓil	'medicine'	E 28
tʈ͡ʂ'á:qan	'old (3 sg.)'	(E 185)
nčó:koša	'my fine thread'	(E 50)

b. If a word contains no long vowel, stress is placed on the last syllable closed by a glottal stop:

nʂ̪úʔxala	'my wife'	(E 68)
t͡ʂ̪óʔwɓax	'blanket'	(E 69)
puʔláʔ	'dipper'	E 37

c. If the preceding two rules cannot apply, stress will fall on a final closed syllable, and otherwise on the penult:

ʂ̪pičáʄ	'raccoon'	E 38
spík'ʲa	'clear'	E 38

England supports this stress rule with segmental evidence: certain unstressed vowels are subject to rules of reduction and syncope (pp. 43–45), whose effects are omitted above for clarity. These rules can be observed in the word for 'sugar cane', phonemically /pat͡s'on/ (p. 44). In isolation, it is realized [pt͡s'ón], that is, the final CVC attracts stress, and the first vowel is syncopated. The possessed form 'his/her sugar cane' undergoes morphologically conditioned lengthening of the first vowel (p. 67), which then attracts stress: /tpat͡s'on/ → /tpaːt͡s'on/ → /tpáːt͡s'on/. The stressless second vowel then reduces to schwa: [tpáːt͡s'ən].

England actually gives two versions of the stress rule. In the one given above (from her introduction to Maldonado, Ordonez, and Ortiz 1986), only syllable-final glottal stops attract stress. In England 1983, any vowel that precedes a glottal stop is stress-attracting. The evidence of vowel reduction (*ɓuʔux, ɓəʔux* 'a lot' E 47) suggests that the syllable-based description is correct.

The description of (22a, b) implies a three-way hierarchy of weight. Actually, the hierarchy is only two-way, since morphophonemic processes (England 1983, 53–54) ensure that CVː and CVʔ syllables do not occur in the same word. Other rules (pp. 50–51) eliminate any cases with two CVː syllables, so ties are resolved by the stress rule only in the case of CVʔ . . . CVʔ.

I account for the pattern of (22c) using right-to-left moraic trochees. The ban imposed on degenerate feet reflects the apparent absence of CV content words.

(23) **Foot Construction** Form moraic trochees from right to left.
Degenerate feet are forbidden absolutely.

(×) (×)	(×) (× .)	(×) (×)	(×) (× .)	(×)(×)	(×)	(× .)
– –	– ᴗ ᴗ	– –	– ᴗ ᴗ	– –	ᴗ –	ᴗ ᴗ
tt͡ʂ̪'aːqan	*nčoːkoša*	*t͡ʂ̪oʔwɓax*	*nʂ̪uʔxala*	*puʔlaʔ*	*ʂ̪pičaʄ*	*spik'ʲa*

The projection of prominence is defined by the rule in (24):

(24) **Prominence Projection** Project a prominence grid according to the following hicrarchy:

**:	VV, Vʔ
*:	other

Main stress can then be assigned by End Rule Right, whose structural description must be satisfied on the prominence plane:

(25) **End Rule Right**

(×)	(×)	(×)	(×)	(×)	(×)	(×)
(×) (×)	(×)(× .)	(×) (×)	(×)(× .)	(×)(×)	(×)	(× .)
– –	– ᴗ ᴗ	– –	– ᴗ ᴗ	– –	ᴗ –	ᴗ ᴗ
tt͡ʂ̩'áːqan	*nčóːkoša*	*t͡ʂ̩óʔwɓax*	*nʂ̩úʔxala*	*puʔláʔ*	*ʂ̩pičáʠ*	*spík'ʲa*
* *	* *	* *	* *	* *	*	*
*	*	*	*	* *		

The prominence plane is then deleted.

England points out that syllable-final /ʔ/ in Mam behaves almost like a prosodic feature rather than a segment: it is invisible to rules of vowel merger (p. 36), occurs as a "floating segment" in certain affixation processes (p. 54), and is realized phonetically as low tone after a long vowel (p. 35). She plausibly suggests that the stress behavior of /CVʔ/ is related to its semi-prosodic status.

7.1.6 Negev Bedouin Arabic (continued)

I left a promissory note in the discussion (§ 6.3.6) of Negev Bedouin Arabic. To review, most of the facts follow from an analysis in which iambs are assigned from left to right, the rightmost foot is made extrametrical, and main stress is assigned by End Rule Right. The one case where this does not derive the correct results is in words of the form /– ᴗ ᴗ σ/. Here, in addition to the predicted penultimate stress, initial stress is also possible.

In principle, this is not a serious problem, since the initial syllable is also the head of a foot, so that one might plausibly posit an optional rule to shift the main stress leftward onto it. This is shown by the representation in (26) (cf. § 6.3.6 (200)):

```
(26) (       ×)
     (×) (. ×)
     –   ᴗ ᴗ⟨–⟩
     ankitalaw
```

However, a general rule of optional leftward shift will not suffice, since there are a fair number of other forms where my source, Blanc 1970, does not point out any alternative stressing. (It is conceivable that this is simply an omission, but I will assume here that the gap is genuine.) The problem is to trim back the set of cases in which a leftward stress shift will overgenerate.

As it turns out, the only case where the main stress may fall on a foot that is not the rightmost one (modulo foot extrametricality) is where the rightmost metrical foot is headed by a light syllable and the preceding foot is headed by a heavy, as in (26). In all other cases, both foot heads are heavy, or the rightmost foot has a heavy head and its left neighbor a light. (In principle, the competing feet could both have light heads, but this case does not arise.)

We can account for this by adding a prominence component to the Negev analysis, as in (27):

(27) **Prominence Projection** Project a prominence grid as follows:
**: Heavy syllables
*: Light syllables

We then apply End Rule Right, deriving the correct results. Representations for (26), along with other sample cases from § 6.3.6, are given in (28):

(28)
```
a. (×)  (. ×)⟨(×)⟩   ( .  ×)(×) ⟨(×)⟩   (×) (×) (×) ⟨(×)⟩   Foot
    –    ˘ ˘ –        ˘  ˘ –   –         –   –   –   –      Construction,
   ankitalaw         taħata:niy         baššibriyyih        Foot
                                                            Extrametricality

b. (×         )      (        ×)        (          ×)       Projection of
   (×)  (. ×)⟨(×)⟩   ( .  ×)(×) ⟨(×)⟩   (×) (×) (×) ⟨(×)⟩   prominence grid,
    –    ˘ ˘⟨–⟩       ˘  ˘ –   ⟨–⟩       –   –   –  ⟨–⟩     End Rule
   ánkitalaw         taħatá:niy         baššibríyyih        Right
   *     *                *  *           *   *   *
   *                         *           *   *   *
```

The analysis derives the retracted version of *ankitalaw*, and also retains the correct stressings for the other forms. To drive the penultimate-stress version of *ankitalaw*, I propose to revise the prominence hierarchy so that it assigns greater prominence to heavy syllables only optionally:

(29) **Prominence Projection (Revised)**
Project a prominence grid as follows:

Heavy syllables }
Light syllables } optionally heavier than

If the indicated option is not taken, then the prominence hierarchy becomes degenerate, reducing the application of the End Rule to the normal type:

```
(30) (         ×)         (          ×)          (            ×)        End Rule
     (×)  (.  ×)《(×)》   ( .   ×)(×)  《(×)》  (  ×)  (×)  (×)  《(×)》  Right
      –    ˘   ˘⟨–⟩        ˘    ˘   –   ⟨–⟩       –    –    –   ⟨–⟩
     ankitálaw            taħatá:niy            baššibríyyih
      *        *                *  *               *    *    *
```

7.1.7 Pirahã

Pirahã is a Mura language of Brazil, whose prosodic system is analyzed in detail by Everett and Everett 1984 and Everett 1988. Metrical accounts other than the Everetts' include Davis 1985; HV; Prince 1990; and Hammond 1990b. The basic stress pattern is as follows: of the last three syllables in a word, main stress falls on the strongest according to the hierarchy in (31):

(31) KVV > GVV > VV > KV > GV   where K is [– voice], G is [+ voice]

(Syllables of the form V do not occur.) In the event of ties, the rightmost candidate wins.

Everett 1988 argues that this pattern is not simply an artifact of listeners' auditory judgments, but reflects a genuine rule of Pirahã phonology. In particular, posttonic syllables may optionally devoice or delete, alternations of stress within paradigms trigger tonal alternations, and stress errors in Everett's own pronunciation were corrected by native speakers.

I summarize an analysis by Everett 1988, 235–38. In this account, conventional metrical devices create the three-syllable window. A stress shift rule, governed by prominence, then establishes the surface position of stress.

The trisyllabic limit is derived using the normal procedure assumed here for antepenultimate stress languages (§ 3.11):

(32) a. **Syllable Extrametricality** $\sigma \rightarrow \langle\sigma\rangle$ / ____ $]_{\text{word}}$

b. **Foot Construction**   Form a syllabic trochee from the right.

The next rule is a troublesome one from the present point of view:

(33) **Extrametricality Annulment**
Annul extrametricality, adjoining the extrametrical syllable to the foot.

This is needed to permit stress on final syllables. Although extrametricality annulment is a standard procedure in a number of theories (e.g. Hayes 1981, where it forms the universal convention of Stray Syllable Adjunction), we have seen reasons not to adopt it here (§ 5.2.3). Pirahã is to my knowledge the only case that supports Stray Syllable Adjunction in the present framework.

An alternative to Stray Adjunction is suggested by Prince 1990: we retain the rules of (32), but apply End Rule Right within the Minimal Word, under the theory of Prosodic Circumscription developed in McCarthy and Prince 1990. The Minimal Word would be defined here as a single canonical foot plus an extrametrical syllable. This is an appealing solution, though its general implications have yet to be fully explored. As Lombardi and McCarthy (1991, 70) note, the ability of Prosodic Circumscription to apply recursively gives it considerably more power than the original theory of extrametricality that it subsumes, since only the latter is constrained by the Peripherality Condition.

A further alternative is simply to posit a ternary foot as a basic element of the system. Some evidence that this is not correct for Pirahã is given below.

For now, I will simply assume the existence of some means of establishing a trisyllabic domain in final position, and continue the summary of Everett's analysis. The labeling of the trisyllabic domain is carried out by the End Rule, making reference to a prominence projection:

(34) a. **Prominence Projection** Project prominence grids as follows:

```
*****: KVV   where K = voiceless consonant
 ****: GVV         G = voiced consonant
  ***:  VV
   **:  KV
    *:  GV
```

b. Apply End Rule Right within the final trisyllabic domain.

Here, the End Rule might better be thought of as an instance of Move X (§ 3.4.2), since the procedure for creating a final trisyllabic domain necessarily places a grid mark in domain-initial position.

The effects of the rules are shown below, with examples taken from Everett and Everett 1984, 107–9. I use IPA /ˈ/ for stress, since /´/ is reserved for high tone. The rules creating trisyllabic domains form the structures in (35):

(35) a. (× .) *kaː gai* 'word'
b. (× . .) *soi oá ga hai* 'thread'
c. (× . .) *poː gáí hi aí* 'banana'
d. (× . .) *ʔa pa baː si* 'square'
e. (× . .) *ʔí bo gí* 'milk'
f. (× . .) *ʔa ba pa* (proper name)
g. (× .) *ko po* 'cup'
h. (× . .) *paó hoa hai* 'anaconda/rainbow'

Projecting prominence (34a) and applying End Rule Right (34b), we derive the correct outcomes:

```
(36) a. ( ×   . )     b.  ( .   .  × )     c.     (×   . . )
        'kaː gai        soi oá ga 'hai         poː 'gáí hi aí
          *   *             *  *   *                 *   * *
          *   *             *      *                 *   * *
          *   *             *      *                 *     *
          *   *                    *                 *
          *                        *
```

```
d.   (.  ×  .)       e. ( ×  .  .)      f. (.  .  ×)
     ʔa pa 'baː si      'ʔí bo gí          ʔa ba 'pa
        *  *  *          *  *  *            *  *  *
        *  *  *          *                  *     *
           *
           *

g. ( .  ×)           h. ( .    .    × )
   ko 'po               paó hoa 'hai
    *  *                  *    *    *
    *  *                  *    *    *
                          *    *    *
                          *    *    *
                          *    *    *
```

Note in particular example (36c), where the most prominent syllable /poː/ is outside the foot and thus fails to receive a prominence grid column (cf. (9c)) and surfaces as stressless.

We can now consider some evidence against a simple ternary foot. This involves the Pirahã nominalizing suffix *-sai,* which according to Everett is extrametrical: it cannot receive the stress, even when it is more prominent than the two preceding syllables. In Everett's account, *-sai* is lexically marked as an exception to Extrametricality Annulment (33). It therefore cannot be adjoined to the preceding binary foot; so that in words containing *-sai,* the window for stressing consists of the antepenult and the penult:

```
(37)  ( .  ×)            'remover'                 Everett 1988, 227
     ʔíːto'pi⟨sai⟩
         *  *
         *  *
```

As Everett points out (1988, 233), the existence of an extrametrical suffix argues against the possibility mentioned above of allowing primitive ternary feet. In conjunction with extrametricality, such a stress window would consist of the fourth, third, and second syllables from the end, which is incorrect.

To sum up: Pirahã stress is troubling from the viewpoint of the general theory advocated here, in that it apparently requires us to appeal either to Extrametricality Annulment or to the direct use of (a particular kind of) Minimal Word in stress assignment. However, from the narrower viewpoint of the theory of prominence versus quantity, Pirahã is an encouraging example. Its unusual weight criterion has nothing to do with the footing rule, which ignores weight and simply counts syllables. The part of the Pirahã phonology that does refer to a rich prominence hierarchy is the End Rule. In my theory, this is a rule type to which prominence (as opposed to quantity) may be relevant.

7.1.8 Asheninca

A remarkably intricate prominence-based system is found in Asheninca, an Arawakan language of Peru. The data and insights, and most of the formal analysis below, are taken from Judith Payne's (1990, hereafter P) work on the Pichis dialect. Further information on this dialect may be found in D. Payne 1983. Additional data and analysis for the Apurucayali dialect are presented by Payne 1981; Payne, Payne, and Sanchez 1982; and Spring 1989b, 1990.

Though I will suggest minor revisions to Payne's analysis, my intent is not so much to improve on it as to show how the Asheninca pattern supports the general theory of syllable weight proposed here. Payne shows clearly that Asheninca must refer to a hierarchy of four degrees of syllable weight. I propose to resolve this hierarchy into a number of independent factors: there are two degrees of QUANTITY (the only possibility, under the theory); there are three degrees of syllable PROMINENCE; and for the fourth degree of weight I will appeal to the action of segmental rules. Crucially, where Asheninca stress involves foot construction, it is a model of a proper quantity-respecting language, invoking a binary, purely moraic weight criterion. Where complexity sets in is precisely in the other parts of the system: the assignment of main stress based on prominence, the destressing rules, and the stress shifts under vowel deletion.

Following Payne, I first describe the basic distribution of stresses, then perturbations of this basic pattern, then the selection of main stress.

7.1.8.1 *Foot Construction*

The fundamental division of Asheninca syllables is into heavy CVV(C) versus light CV(C). I assume that the former are bimoraic, the latter monomoraic; that is, there is no rule of Weight by Position. Stress is assigned in a familiar iambic pattern: to heavy syllables, and to the even members of light-syllable strings. An overriding requirement is that final syllables are never stressed, except by a late stress shift (P 190).

As a first step toward analyzing this pattern, I state the foot construction rules as in (38):

(38) a. **Syllable Extrametricality** $\sigma \rightarrow \langle\sigma\rangle$ / ____ $]_{\text{word}}$

b. **Foot Construction** Form iambs from left to right.

The rules at this stage must permit degenerate feet, whose ultimate disposition is discussed below. In the examples in (39), the stress marks give the pattern that is ultimately derived on the surface.

(39)

a.	b.	c.
'take care of her' P 188	'he bought it for her' P 188	'you threw it out' P 190
(. ×)(. ×)(. ×)(×)	(. ×)(. ×)(. ×)	(×)(. ×)(×)
˘ ˘ ˘ ˘ ˘ ˘ ˘⟨˘⟩	˘ ˘ ˘ ˘ ˘ ˘⟨˘⟩	– ˘ ˘ ˘⟨˘⟩
pamènakòwentákero	*hamànantàkenéro*	*pò:kanákero*

d. (. ×)(. ×) (×)(. ×) 'I rested a while' P 190
˘ ˘ ˘ – – ˘ – ⟨˘⟩
nomàkorʲà:wàitapá:ke

e. (×) 'here' P 188
˘ ⟨˘⟩
háka

In the Apurucayali dialect (Payne, Payne, and Sanchez 1982, 186–87), the same foot structure is found, but CVC syllables (only CVN, N = [+nasal], occur underlyingly) count as heavy.

I next discuss two perturbations of the basic iambic pattern.

7.1.8.2 *Perturbation I: /i/ Drop*

The very lightest syllables in Payne's hierarchy consist of a strident consonant plus /i/: /sʲi, t͡si/. These have a special status in Asheninca phonology. Phonemically, they are ambiguous between /si, t͡si/ and /sʲi, tsʲi/, there being no contrast of palatalization in this context. Phonetically, they normally are [šɨ, tsɨ], with a centralized vowel. When they occur before a voiceless consonant, their vowel is deleted by a rule of /i/ Drop (40a). Where /i/ is stressed, the stress behaves in the expected way (§ 3.8.1), shifting leftward onto the preceding syllable within the same foot (40b, c).

(40) a. /i/ **Drop** $i \rightarrow \emptyset \, / \begin{bmatrix} \text{C} \\ +\text{strident} \end{bmatrix} \underline{\quad\quad} \begin{bmatrix} \text{C} \\ -\text{voice} \end{bmatrix}$

b. (. ×) → (×) 'dog' P 195
˘ ˘⟨˘⟩ → – ⟨˘⟩
otsiti → *ótsti*

c. (. ×)(. ×)(×) → (×) (. ×)(×) 'he has intestinal parasites' P 195
˘ ˘˘ ˘˘ ⟨˘⟩ → – ˘ ˘˘ ⟨˘⟩
isʲitanetatʲa → *ìsʲtanétatʲa*

I assume that the derived closed syllable is heavy, since it may survive a stress clash with a following syllable; see data in P 195.

Asheninca also shows RIGHTWARD stress shift in trochaic feet, as in (41). These feet arise by a process of incorporation, described in (45b).

(41) (×) → (×) → (×) 'intestinal worm' P 190
(× .) /i/ Drop (× .) Stress Shift (×)
˘ ˘ → ˘ → ˘
sʲitsa → *sʲtsa* → *sʲt͡sá*

There are other cases of stress shift involving /sʲi, t͡si/ that are harder to analyze. In particular, we get leftward shift even when the vowel of /sʲi, t͡si/ precedes a voiced consonant and therefore does not delete:

(42) a. /ko**tsí**ro/ → [**kó**tsɨro] 'machete' P 195

b. /no**sʲì**yapìt͡satàntanàkaróri/
→ [**nò**sʲɨyapìt͡satàntanàkaróri] 'that I escaped from her' P 196

Payne (P 196) attributes this to an explicit rule of Stress Transfer, which accomplishes as a language-specific rule exactly what is done by universal convention in the case of deletion.

One is tempted to reanalyze the retained [ɨ] vowel here as a kind of nonsyllabic consonant release, so that the shifts in (42) could also be analyzed as stress shift under vowel deletion. This is a fairly radical interpretation of Payne's phonetic observations (P 196), however. Trochaic stress shift does not occur before voiced consonants (*sʲíma* 'fish' P 189).

/i/ Drop is a very late rule (P 196): it appears to be ordered after the Destressing rules discussed below (P 195); moreover, it creates surface forms (e.g. (41)) that violate an otherwise general ban on degenerate feet. Native speakers apparently are unconscious of the deletion, since they spell the deletion site with letter *i* (P 208).

7.1.8.3 *Perturbation II: Destressing*

The other perturbation of the basic iambic pattern results from a destressing rule: stressed light syllables lose their stress when they precede another stress, with certain segmental conditions attached. The formulation in (43) is preliminary, to be amended in (49b).

(43) **Destressing** × → Ø / _____ × Obligatory if ᴗ = C*i*,
ᴗ otherwise optional

In (44) are examples of how Destressing applies:

(44) a. (. ×)(×)(. ×)(×) → (×)(. ×)(×) → (. ×)(. ×)(×)
ᴗ ᴗ – ᴗ ᴗ ᴗ ⟨ᴗ⟩ → ᴗ ᴗ – ᴗ ᴗ ᴗ ⟨ᴗ⟩ → ᴗ ᴗ – ᴗ ᴗ ᴗ ⟨ᴗ⟩
oŋkitaitamanake *oŋkitaitamanake* *oŋkitàitamánake*
'in the morning' P 191

b. (. ×)(. ×)(×) → (. ×) (×) → (. ×) (. ×)
ᴗ ᴗ ᴗ ᴗ – ⟨ᴗ⟩ *optional* ᴗ ᴗ ᴗ ᴗ – ⟨ᴗ⟩ ᴗ ᴗ ᴗ ᴗ – ⟨ᴗ⟩
atiripayeːni *atiripayeːni* *atìripayéːni*
or [atìripàyéːni] 'people' P 191

The derivations assume that footing is persistent in Asheninca; the destressed light syllable is adjoined to the following heavy syllable to form a /ᴗ –́/ foot, following the principles laid out in § 5.4.1. This assumption turns out to be crucial in the account of main stress below.

The condition on Destressing reflects the status of /i/ as the least sonorous of the Asheninca vowels. This also shows up in the prominence-based rules for main stress discussed below. As Payne points out, a somewhat different hierarchy of prominence is used for Destressing than for main stress assignment. This supports the view that prominence is a computational device used by individual rules, and not a permanent property of representations.

7.1.8.4 *Main Stress*

The selection of a particular foot to bear main stress follows a quite complicated pattern, based both on syllable prominence and on whether a foot is degenerate or not.

First, if a word contains only one foot, naturally that foot takes the main stress:

(45) a. (×) 'my canoe' P 193 b. (×) (×) 'water' P 189
(. ×) (×) (× .)
⌣ ⌣⟨⌣⟩ ⌣⟨–⟩ → ⌣ –
nopíto *hínʲa:* *hínʲa:*

Short words like (45b) (similarly (39e)) form an instance of the "unstressable word syndrome" (§ 5.3). I posit that they are repaired by Incorporation (§ 5.3), as shown. This is what creates the trochaic feet discussed in (41).

For the remaining cases, let F_n be the rightmost foot in the word and F_{n-1} be the immediately preceding foot. The simpler examples are those where F_n is not degenerate. As Payne notes, these involve a clear hierarchy of prominence, which I will notate as in (46):

(46) ***: CVV
**: C*a*, C*o*, C*e*, C*i*N, where N = a nasal consonant
*: C*i*

Main stress goes on the more prominent of F_{n-1} and F_n, and on F_n if there is a tie. This is shown in the examples of (47):

(47) a. /CiN, Ca, Ce, Co/ are stronger than /Ci/:
(×) (×)(. ×)(. ×) 'what he saw in a vision' P 198
– – ⌣ ⌣ ⌣ ⌣⟨⌣⟩
nʲa:wʲa:tawakariri → [nʲà:wʲà:tawákariri]

b. /CV:/ is stronger than any short-voweled syllable:
(×)(. ×) 'type of bee' P 198
– ⌣ ⌣⟨⌣⟩
ma:kiriti → [má:kiriti]

c. The rightmost foot wins in a tie:
(×)(×) 'type of partridge' P 198
– – ⟨⌣⟩
sa:sa:ti → [sà:sá:ti]

(. ×) (. ×)(. ×)(. ×)(. ×) 'he thought about it for a while' P 198
⌣ ⌣ ⌣ ⌣ ⌣ ⌣ ⌣ ⌣ ⌣ ⌣⟨⌣⟩
iŋkiŋkisʲiretakotawakeri → [iŋkìŋkisʲiretàkotàwakéri]

(. ×)(. ×)(. ×) ‘I went in vain’ P 198
˘ ˘ ˘ ˘ ˘ ˘ ⟨˘⟩
*nawisawetan**a**ka* → [nawìsaw**è**tan**á**ka]

d. As we would expect, where a stronger syllable follows a weaker one, the stronger one wins:

(. ×)(. ×) ‘I returned’ P 201
˘ ˘ ˘ ˘ ⟨˘⟩
*nop**i**yan**a**ka* → [nop**ì**yan**á**ka]

(. ×)(. ×)(×) ‘people’ P 191
˘ ˘ ˘ ˘ – ⟨˘⟩
*atirip**a**y**e:**ni* → [atìrip**à**y**é:**ni],
or [atìrip**a**y**é:**ni] by Destressing (43)

The cases where F_n is degenerate follow a more complicated pattern, described in (48):

(48) a. In case of a tie, the LEFTMOST foot wins:

(. ×) (. ×)(×) ‘I wanted (it) in vain’ P 188
˘ ˘ ˘ ˘ ˘ ⟨˘⟩
*n o k o w a w **e** t **a** k a* → [nokòwaw**é**t**a**ka]

(×)(. ×)(×) ‘you stepped on him’ P 190
– ˘ ˘ ˘⟨˘⟩
*pa:tik**a**keri* → [pà:tik**á**keri]

b. As before, /Ca/ is stronger than /Ci/:

(. ×)(×) ‘it costs’ P 193
˘ ˘ ˘⟨˘⟩
*op**i**n**a**ta* → [op**i**n**á**ta]

c. But /Co, Ce/ count either as stronger than /Ci/, or else the same strength, leading to free variation:

(. ×)(×) ‘he turned around’ P 193
˘ ˘ ˘ ⟨˘⟩
*ip**i**ts**o**ka* → [ip**i**ts**ó**ka], [ip**í**ts**o**ka]

d. Where instances of /Ci/ compete, either may win:

(. ×)(×) ‘cinnamon’ P 193
˘ ˘ ˘⟨˘⟩
*kaw**i**n**i**ri* → [kaw**í**n**i**ri], [kaw**i**n**í**ri]

e. As expected, where a stronger syllable precedes a weaker one, the stronger one wins:

(. ×)(×)(×) ‘however’ P 191
˘ ˘ – ˘ ⟨˘⟩
*kantim**ai**t**a**tsʲa* → [kantim**ái**t**a**tsʲa]

In general, a degenerate foot has less ability to take stress over the preceding foot than a proper foot: compare the different outcomes for /Ci . . . Co/, /Ci . . . Ce/, /Ci . . . Ci/ and for other ties.

To account for this complex situation, Payne posits a sequence of destressing rules that can trim away the rightmost foot in a word. After destressing has applied, End Rule Right assigns main stress to the rightmost surviving foot. Her rules are stated (in the notation of this book, but with equivalent effect) as in (49):

(49) a. **Long-Distance Destressing** (= P ex. (34))

$$\times \rightarrow \emptyset \,/\, \underset{\alpha}{\times} \ldots \underset{\beta}{\underline{\quad\quad}}\,]$$

where (a) . . . contains no ×
(b) syllable α is stronger on the hierarchy of (46) than syllable β

b. **Prestress Destressing** (= P ex. (25))

$$\times \rightarrow \emptyset \,/\, \underset{\beta}{\underline{\quad\quad}}\, \underset{\alpha}{\times}$$

where (a) β = C*i*, α = C*a* or CVV (applies obligatorily)
or (b) β = CV, α = CVV (applies optionally)
or (c) β = C*i*, α = C*i*, C*o*, C*e* (applies optionally)

c. **Poststress Destressing** (= P ex. (15d))

$$\times \rightarrow \emptyset \,/\, \times\, \underset{\smile}{\underline{\quad\quad}}$$

d. **Word Layer Construction** (= P (32)): End Rule Right

Note that Prestress Destressing is needed independently (albeit in a simpler version) to trim back clashing stresses earlier in the word; compare (43). Payne's destressing rules must be applied exactly in the order stated. Otherwise, more than one foot will be removed in certain cases, and stress will fall too far to the left.

This is clearly a reasonably elegant account of a very complex situation. I offer an alternative for two reasons. First, Long Distance Destressing must incorporate a variable, and thus goes against the general typological observation (Hammond 1984a) that all destressing rules remove clash. I propose to replace Long Distance Destressing with a prominence-based End Rule. Second, Payne's rule of Poststress Destressing (49c) is set up so that it will apply only to degenerate feet; this is the joint effect of the clash restriction and the limitation to light target syllables. Elsewhere (§ 5.1) we have seen that languages tend to avoid degenerate feet entirely. Therefore, in the analysis below I reformulate Payne's Poststress Destressing as a language-particular implementation of this ban. There is additional evidence for a ban on degenerate feet in Asheninca, at least for the Apurucayali dialect, in that words of degenerate length are forbidden (Spring 1990), and an Apocope rule (Payne 1981, 120) is blocked when it would create a degenerate word. (The ban must exempt forms derived by the very late rule of /i/ Drop, as noted above.)

The rules implied by the above discussion are as in (50):

(50) a. **Colon Construction**
Form a binary colon (§ 5.5.1) over the rightmost two feet. Labeling: End Rule Right, based on the following prominence hierarchy (= (46)):

```
***:  CVV
 **:  Ca, Co, Ce, CiN
  *:  Ci
```

b. **Prestress Destressing** (repeated from (49b))

$\times \rightarrow \emptyset \, / \, \underset{\beta}{\underline{\quad}} \; \underset{\alpha}{\times}$ where (a) β = C*i*, α = C*a* or CVV (applies obligatorily)
or (b) β = CV, α = CVV (applies obligatorily)
or (c) β = C*i*, α = C*i*N, C*o*, C*e* (applies optionally)

c. **Degenerate Foot Removal**
Eliminate a degenerate foot where this would not violate culminativity (§ 3.1); i.e., where it is not the only foot in the word.

Below, I show construction of the final colon, along with the prominence grids that determine its labeling. In the first row of examples, the rightmost foot is degenerate; in the second row it is a proper foot.

```
(51) a. (     ×)      b. (     ×)      c.   (     ×)      d.     (×   )
        (. ×)(×)         (. ×)(×)         (×)(. ×)(×)         (.  ×)(×)(×)
         ˘ ˘ ˘⟨˘⟩          ˘ ˘  ˘⟨˘⟩         – ˘ ˘ ˘⟨˘⟩           ˘  ˘ – ˘ ⟨˘⟩
        opinata          ipit͡soka         pa:tikakeri        kantimaitat͡sʲa
          * *              *  *               * *                  * *
          * *                 *               * *                  * *
                                                                   *

   e.          (   ×    )   f. (×     )   g. (    ×)    h.    (          ×)
     (×) (×)(.  ×)(. ×)       (×)(. ×)      (×)(×)         (.  ×)(.  ×)(.  ×)
      –    –  ˘  ˘ ˘ ˘⟨˘⟩      –  ˘ ˘⟨˘⟩       –  – ⟨˘⟩        ˘   ˘ ˘   ˘ ˘  ˘ ⟨˘⟩
     nʲa:wʲa:tawákariri       má:kiriti      sa:sá:ti       nawisawetanáka
              *    *          *    *         *  *                  *    *
              *               *              *  *                  *    *
                              *              *  *
```

Following Colon Construction, prominence grids are deleted. For the examples of (51e–h), this procedure derives the final outputs, and no further rules are applicable.

For the remaining examples (51a–d), the rule of Prestress Destressing applies next, much as it does in earlier stretches of the word. Since footing is persistent in Asheninca, any degenerate foot that follows the destressed syllable gets a light syllable adjoined to it and thus becomes well-formed:

```
(52) a. (    ×)    b. (    ×)   (    ×)    c. ——   d.   ( ×  )
        (. ×)         (. ×)(×)     (. ×)                 (. ×)(×)
        ⌣ ⌣ ⌣⟨⌣⟩       ⌣ ⌣  ⌣⟨⌣⟩  ⌣ ⌣  ⌣⟨⌣⟩               ⌣ ⌣ – ⌣ ⟨⌣⟩
        opinata       ipit͡soka,  ipit͡soka              kantimaitat͡sʲa
```

In principle, Prestress Destressing could be formulated to refer to prominence grids, but we will not pursue this possibility here. As noted above, there are differences between the prominence patterns required for colon labeling versus destressing: for example, /Co/ is optionally heavier than /Ci/ for Prestress Destressing (50b), but not for colon labeling (50a).

The last step of the derivation is to remove the temporary license on degenerate feet.[3] This causes the grid mark of the colon layer to become stranded, and it migrates leftward to a legal docking site within the colon, satisfying the Faithfulness Condition (§ 3.8.1). Applying these processes in (51b, c, d), we derive the remaining surface forms.

```
(53) a. (    ×)    b. ( ×  )   (    ×)    c.   ( ×  )     d.   ( ×  )
        (. ×)         (. ×) ø    (.  ×)        (×)(. ×)ø        (. ×)ø
        ⌣ ⌣ ⌣⟨⌣⟩       ⌣ ⌣  ⌣⟨⌣⟩  ⌣ ⌣  ⌣⟨⌣⟩      – ⌣ ⌣ ⌣⟨⌣⟩        ⌣ ⌣ – ⌣ ⟨⌣⟩
        opináta       ipít͡soka,  ipit͡sóka       pàːtikákeri      kantimáitat͡sʲa
```

Note that in (53a) and the second output for (53b), the reparsing induced by Destressing creates proper feet, and thus blocks foot removal in these words. (The process of Incorporation illustrated in (45b) must be ordered after destressing, since it does not rescue degenerate feet.)

My analysis predicts one difference in the data: posttonic secondary stresses not in clash, as in (51e, f), should survive to the surface, whereas Payne's analysis removes them. Since the cues for posttonic secondaries are often subtle, the discrepancy is a small one. In the Apurucayali dialect (Payne, Payne, and Sanchez 1982), secondary stress is transcribed in such words.

Summing up, the rules under this analysis (in order) are Syllable Extrametricality (38a), Foot Construction (38b), Colon Construction (50a), Prestress Destressing (50b), Degenerate Foot Removal (50c), and /i/ Drop (40). Of these, all but (50a, c) follow Payne. The advantages of this reanalysis are that it adheres to more restrictive assumptions than Payne's in disallowing long-distance

3. Note that there is no need to suppose that degenerate feet survive to this point in the derivation unless they are in strong position; in this sense Asheninca does not counterexemplify the theory of degenerate feet laid out in § 5.1.

destressing, and that it draws a closer connection between the stress pattern and the ban on degenerate-size words.

7.1.8.5 *Conclusion*

This does not exhaust the stress phenomena of Asheninca. Payne notes the existence of exceptional suffixes that always take main stress; I follow her view in assuming they bear underlying stress on highest grid layer. She also notes patterns of destressing and stress shift that occur in fast speech; I propose that these are non-phonological, and reflect auditory judgments of prominence. If this position is not correct, then the facts would require substantial revision of the theory proposed here.

The main theoretical result follows either under my own analysis or from Payne's. I have drawn here a distinction between syllable **quantity,** depicted moraically, and syllable **prominence,** depicted with asterisk columns in an auxiliary grid. Quantity alone is relevant to foot construction, with prominence limited to other metrical operations, such as the End Rule and destressing. In both analyses of Asheninca considered, this claim holds up: the rule of foot construction refers to a narrow, moraically definable distinction between heavy and light syllables. It is only when we study the rules of destressing and main stress location (the End Rule, in my analysis), that the unconstrained criteria of syllable prominence (e.g. /i/ vs. other vowels; /a/ vs. other vowels) become apparent. Asheninca is thus an excellent example of the quantity/prominence distinction.

7.2 Unbounded Stress Systems

The theory of syllable prominence is an ingredient in the account I will propose in this section for unbounded stress systems (§ 3.3); that is, systems that are sensitive to syllable weight, but place no limits on the distance between stresses or between stress and word boundary. In (54) and (55) is a list of such cases; the headings of (54) and (55) are from Prince 1985.

(54) **Defaults to Opposite Side**

a. **Rightmost Heavy Syllable, Otherwise Leftmost Syllable**

Classical Arabic (McCarthy 1979a) / – / = CVV, CVC.
Kuuku-Yaʔu (Pama-Nyungan, Cape York, Australia; Thompson 1976) / – / = CVV. Secondary stress on initial syllables, heavy syllables. Thompson reports secondary stress on posttonic syllables, perhaps attributable to a pitch effect.
Huasteco (Mayan, Mexico; Larsen and Pike 1949) / – / = CVː.
Chuvash (Turkic, Russia; Krueger 1961) / – / = full vowel, treated here as bimoraic.
Eastern Cheremis (Finno-Ugric, Russia; Sebeok and Ingemann 1961) / – / = full vowel; treated here as bimoraic. Another option is simple final stress, by End Rule Right.

b. **Leftmost Heavy Syllable, Otherwise Rightmost Syllable**

Komi (Yaz'va dialect; Finno-Ugric, Russia; Itkonen 1955; Lytkin 1961)
Weight is complex: all non-high vowels pattern as heavy, while most high vowels are treated as light in some instances and heavy in others. I conjecture a (possibly abstract) length distinction.
Kwakw'ala (Kwakiutl) (Wakashan, British Columbia; Boas 1947; Bach 1975; Zec 1988) heavy = VV or V[+son].

(55) **Defaults to Same Side**

a. **Rightmost Heavy Syllable, Otherwise Rightmost Syllable**

Aguacatec (Mayan, Guatemala; McArthur and McArthur 1956) /–/ = CVː.
Golin (§ 7.1.3)
Western Cheremis (Finno-Ugric, Russia; Itkonen 1955) /–/ = full vowel, treated here as bimoraic; final syllables are extrametrical.

b. **Leftmost Heavy Syllable, Otherwise Leftmost Syllable**

Amele (Gum, Papua New Guinea; Roberts 1987) /–/ = CVC, no phonemic CVV. Morphology influences stress.
Au (Torricelli, New Guinea; Scorza 1985) /–/ = contains [i, a, aː, o, u]; /◡/ = contains [ɨ, ʌ]. Vowel length apparently irrelevant.
Indo-European accent: Sanskrit, Russian, Lithuanian (Halle and Kiparsky 1977, 1981; HV) /–/ = high tone (alternatively: lexically accented syllables).
Khalkha Mongolian (Altaic; Zebek and Schubert 1961; Austin, Hangin, and Onon 1963; Street 1963) /–/ = CVV.
Lhasa Tibetan (Odden 1979) /–/ = CVV.
Lushootseed (Salishan, Washington; Hess 1976; Odden 1979) "heavy" = non-schwa vowel; vowel length is phonemic, but short non-schwa vowels attract stress.
Murik (Lower Sepik, New Guinea; Abbott 1985) /–/ = CVV.
Yana (Hokan, Northern California; Sapir and Swadesh 1960) /–/ = CVV, CVC.

For reasons given in § 3.3, it is difficult to find evidence to favor one theory for these cases over another; the proposal I provide here, based on syllable prominence, is mostly for concreteness' sake.

For the "default to same side" cases, my analysis is essentially that of Prince 1983a, 76–77, except with a distinct prominence grid:

(56) a. Project the prominence distinction.
b. Form a word layer, labeled with End Rule (Left/Right).

In (57) I give an example for the pattern "leftmost heavy, otherwise leftmost," which employs End Rule Left; End Rule Right would yield the mirror image.

```
(57) a. ( . . ×  . . . . . )      b. ( × . . . . . . . )
          ◡ ◡ –́ ◡ ◡ – ◡ ◡              ◡́ ◡ ◡ ◡ ◡ ◡ ◡ ◡
          * * * * * * * *              * * * * * * * *
              *     *
```

For the "default to opposite" cases, I adopt the analysis proposed in Prince 1976a and assumed in much subsequent work. This analysis assumes an unbounded, quantity-sensitive foot template: such feet may be either right-headed or left-headed, and require that their weak syllables be light. In (58) is the analysis for the "rightmost heavy, otherwise leftmost" case:

(58) a. **Foot Construction** Form left-headed, quantity-sensitive unbounded feet.

b. **Word Layer Construction** End Rule Right

```
i. (                 ×     )      ii. ( ×                 )
   ( ×  . )( ×  . . )( ×  . . )        ( × . . . . . . . )
     ◡ ◡    – ◡ ◡    –́ ◡ ◡              ◡́ ◡ ◡ ◡ ◡ ◡ ◡ ◡
```

The "leftmost heavy, otherwise rightmost" case would be the mirror image, with right-headed feet and End Rule Left.

The "default-to-same" cases in principle could also involve foot construction. For example, the pattern "leftmost heavy, otherwise leftmost" could be analyzed as in (59):

(59) a. **Foot Construction** Form left-headed, quantity-sensitive unbounded feet.

b. **Word Layer Construction** Form a prominent grid, with /–/ > /◡/. End Rule Left

```
(          ×              )      ( ×                 )
( ×  . )( ×  . . )( ×  . . )     ( × . . . . . . . )
  ◡ ◡    –́ ◡ ◡    – ◡ ◡            ◡́ ◡ ◡ ◡ ◡ ◡ ◡ ◡
  * *    * * *    * * *            * * * * * * * *
         *        *
```

The use of feet in deriving unbounded stress predicts that phenomena that result from the ban on degenerate feet could arise in unbounded stress systems. Such appears to be the case in Malayalam, as noted in § 5.1.5. Further, since the default-to-opposite systems must employ feet, I predict that they may refer only to syllable quantity, not to prominence. This appears to be true: with the possible exception of Komi, the only cases above that refer to non-moraic prominence distinctions (Au, Golin, Indo-European, Lushootseed), fall in the default-to-same class.

The analysis of "default-to-same" systems proposed here makes no use of "obligatory branching" feet (Halle and Vergnaud 1978; Hayes 1981), in which the head is required to be a heavy syllable. My proposal thus forms part of the argument that the Obligatory Branching Parameter can be eliminated from metrical theory (§ 4.3.1).

Unbounded feet of any sort face a difficulty: the apparent absence of unbounded-foot extrametricality, as pointed out in a different framework by Wheeler 1979a. See Prince 1985 for a proposal to eliminate unbounded feet as a primitive notion of the theory.

7.3 DUAL CRITERIA OF WEIGHT IN MORAIC THEORY

I turn in this section to a remaining problem within the theory of quantity, here taken to be moraic theory.

As Hyman 1985 points out, syllables in moraic theory are defined as heavy or light for all phonological and morphological processes of a language. In this respect the theory contrasts with a theory of syllabic constituency positing both Nucleus and Rhyme (§ 3.9.2), where rules have the choice of which constituent they refer to. Recent research suggests that the predictions of the weaker syllabic constituency theory are correct for some cases. Crowhurst 1991a argues that in Tübatulabal (§ 6.3.10), closed syllables count as light for stress but as heavy for purposes of reduplication. Steriade 1991 argues for dual criteria of weight in Ancient Greek and many other languages, and proposes an explicit theory of the phenomenon, discussed below in § 7.3.3.

One class of problematic examples Steriade points out was foreseen earlier by Selkirk 1988 and is also discussed by Tranel 1992: there are languages that have geminates, but do not count syllables closed by geminates as heavy (see § 7.3.2 for a list). Assuming a theory in which geminates must bear a mora (as opposed to being merely ambisyllabic), these languages also show that a single count of moras is insufficient for describing all prosodic processes.[4]

The use of a single criterion of weight arguably is a tendency across languages. But for the cases given, the simplest version of moraic theory is inadequate. What is needed is a theory that allows such dual distinctions of weight to be made, but which retains the advantages in predictive power that moraic theory holds over alternatives.

As a proposal in this direction, I suggest that languages may make distinc-

4. A caveat is that languages often convert gemination contrasts into featural contrasts, e.g. of fortition or voicing. Some plausible examples of this are Korean, Eastern Ojibwa (§ 6.3.4), and Malayalam (Mohanan 1989, 620). In many cases it is difficult to determine which analysis (gemination vs. feature) is correct. In a featural account, lack of weight follows straightforwardly.

tions among moras. Adapting an idea of Prince 1983a, let us suppose that moras form a kind of grid within the syllable, where the height of a column depends on the sonority of the segment it is associated with. Since these syllable-internal grids arguably are distinct from stress grids (e.g., they tolerate clash), I use μ's instead of ×'s in the notation to make the difference clear. I assume that the syllable-internal grid obeys the Continuous Column Constraint, so that any segment that bears weight on the higher layer must also bear weight on the lower layer.

In (60) are sample representations, given for a language that treats CVC as heavy for some distinctions but light for others:

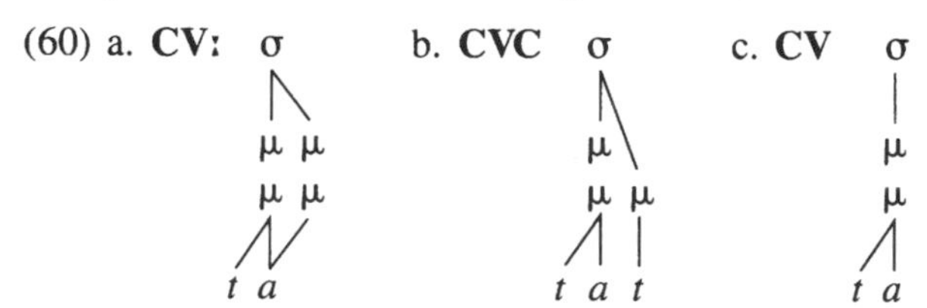

In such a language, processes that treat CVC as heavy may be expressed as referring to the lower layer of the syllable-internal grid, while processes that treat CVC as light would refer to the higher layer.

Representations like (60) plausibly are employed only as the exceptional case. In many languages, CVC behaves consistently as light or heavy for all phonological processes. For such languages, single-layer moraic structures suffice, and I assume tentatively that they do not employ a second layer.

I assume further that in two-layer systems, the heights of moraic columns are persistently adjusted to conform to the sonority of the segments that license them. Thus in a case of ordinary compensatory lengthening, the second mora of the derived long vowel is assigned a mora on the higher layer:

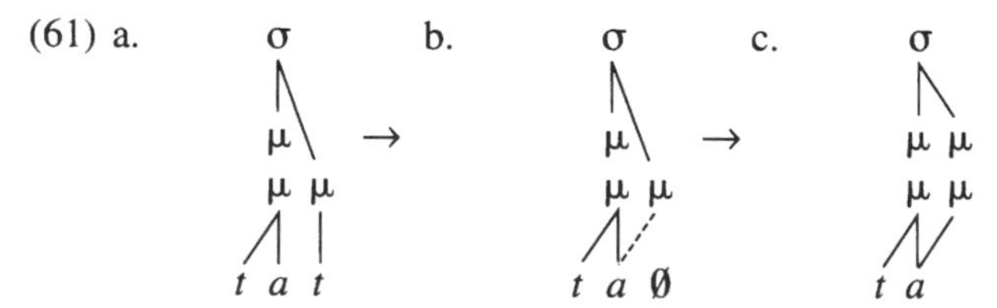

This avoids the possibility of long vowels being represented as light as a result of their derivational history, a possibility argued against by Levin 1985a.

I further conjecture, adapting a claim of Steriade 1991, that the requirements of syllable-external prosody (e.g. footing, word minima, tonal docking) are characteristically enforced on the higher moraic layer, while syllable-internal requirements (e.g. mora population limits) are characteristically enforced on the lower layer. However, neither principle is exceptionless, and we will see that there appears to be pressure for well-formedness requirements to be met on both layers at once.

The crucial point of two-layer moraic representations is that they permit us to represent dual criteria of weight, while retaining the main virtues of moraic theory (§ 3.9.2), namely, they still predict the conservation of moraic count in compensatory processes (in this case, lower-level moras), and they still include in the representation only elements that actually are counted by rules (here, moras on a given layer).

Two-layer representations can account for the cases presented by Crowhurst 1991a and Steriade 1991. I will not demonstrate this here, since the reanalyses are straightforward. Instead, I mention some further cases in which the two-layer theory can appropriately describe variable weight patterns.

7.3.1 Laryngeal Metathesis in Cayuga

Cayuga (§ 6.3.5) has left-to-right iambic stress, with CVː but not CVC counted as heavy. There is in Cayuga a segmental rule called Laryngeal Metathesis (Foster 1982, 69–71), which works approximately as in (62):

(62) **Laryngeal Metathesis**
In weak position of a metrical foot: CV*ʔ* → C*ʔ*V
CV*h* → CV̥ / ____ C
CV*h* → C*h*V / ____ V

That is, on the weak side of an iambic foot, /ʔ/ and prevocalic /h/ metathesize with the preceding vowel, and preconsonantal /h/ merges with the preceding vowel to create a voiceless vowel. As the rest of the phonology of Cayuga shows (Foster 1982; Michelson 1988), /VʔV/ and /VhV/ sequences are syllabified /Vʔ.V/ and /Vh.V/, so in all cases of Laryngeal Metathesis, the laryngeal segment in the input is syllable-final.

What all these changes have in common is that they convert syllables ending in VC into syllables ending in V. This is an expected outcome for the weak syllable of an iambic foot: its phonological duration is reduced, following the Iambic/Trochaic Law. But under a single-layered moraic theory, this characterization cannot be formally expressed, since a CVC syllable is represented as monomoraic.

In a two-layered theory, we suppose that CVC syllables in Cayuga are bimoraic on the lower layer only. Laryngeal Metathesis converts syllables that are monomoraic only on the higher moraic layer (the one on which the stress pattern is computed) to syllables that are monomoraic on all layers:

(63)
```
a.   σ           σ      b.   σ           σ      c.   σ           σ
     |\          |           |\          |           |\          |
     μ \   →     μ           μ \   →     μ           μ \   →     μ
     μ  μ        μ           μ  μ        μ           μ  μ        μ
    /\  |      / /|         /\  |       /|          /\  |       //|
   C V  ʔ     C ʔ V        C V  h      C V̥         C V  h      C h V
```

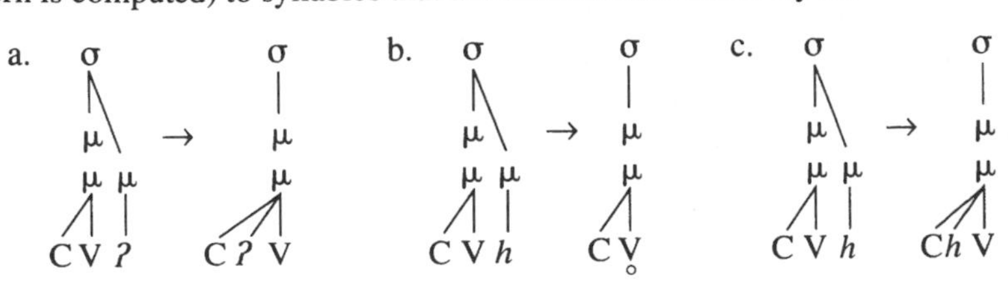

The metrical feet in which Laryngeal Metathesis has applied thus are brought into conformity with the Iambic/Trochaic Law at all moraic layers.

7.3.2 Yupik Quantity

Central Alaskan Yupik, discussed in § 6.3.8, appears not to require use of two-layer moraic structures. However, in principle there would be no difficulty in restating the rules of Yupik phonology using two-layered representations: alternations between heavy and light CVC can be expressed as alternations between CVC with two strong moras versus CVC with a strong and a weak (i.e. (64a) vs. (64b)). Syllable-final consonants with a strong mora are assumed to be phonetically longer than those with a weak mora (§ 6.3.8.2).

(64) a. **Heavy CVC** b. **Light CVC**

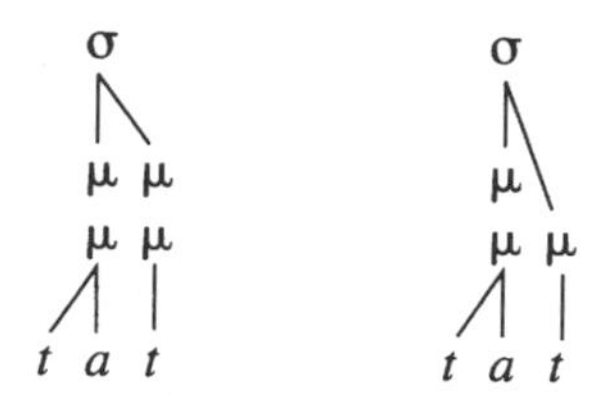

This is important, because two related and prosodically similar languages, Pacific Yupik (Leer 1985a, 87) and Seward Peninsula Inupiaq Eskimo (Kaplan 1985, 194), do appear to require two-layer representations. In particular, for these languages syllables that are closed by a geminate count as light in the stress rule. Such syllables lose their gemination when they are not stressed, so that on the surface the heavy/light distinction works consistently across both layers of the moraic grid. The two-layer theory would also suffice to describe other cases where geminates fail to make their syllable heavy: Chuvash (§ 7.2), Ossetic (§ 6.3.9), and various Algonquian languages (§ 6.3.4).

There are lengthening and shortening rules in Yupik that also appear to require the two-layer theory. To make sense of these, it will help first to endorse a traditional view concerning the characteristic patterns of vowel length rules: lengthening is commonly blocked in, and shortening is commonly restricted to, closed syllables. The common-sense view of this is that both phenomena are based on overall weight limits: in moraic terms (e.g. Trubetzkoy 1969, 180), closed syllable lengthening is avoided because it creates trimoraic syllables, and closed syllable shortening is favored because it repairs them. It can be observed in support of this that in languages where coda consonants make no weight at all, long vowels typically are tolerated in closed syllables; whereas they are often forbidden in languages where coda consonants make weight. This is illustrated in (65):

(65) a. **Weightless Coda Consonants, CVVC Allowed**

Asheninca (Pichis dialect, § 7.1.8)	Malayalam (§ 5.1.5)
Cahuilla (§ 6.1.3)	Menomini (§ 6.3.4)
Cayuga (§ 6.3.5)	Nyawaygi (§ 6.1.9)
Chuvash (§ 7.2)	Ossetic (§ 6.3.9)
Eastern Ojibwa (§ 6.3.4)	Potawatomi (§ 6.3.4)
Huasteco (§ 7.2)	Sarangani Manobo (§ 6.1.9, § 6.3.10)
Javanese (§ 6.3.10)	Tümpisa Shoshone (§ 6.1.9)
Komi (§ 7.2)	Votic (§ 6.2.3)
Kuuku-Yaʔu (§ 7.2)	Wargamay (§ 6.1.4)
Kwakw'ala (§ 7.2)	Western Cheremis (§ 7.2)
Lenakel (§ 6.1.8)	Yidiɲ (§ 6.3.9)

b. **Weighted Coda Consonants, CVVC Forbidden**

Arabic (§ 4.1.3, § 6.1.1, § 6.1.2, § 6.3.6–7)	Hopi (§ 6.3.9)
Choctaw/Chickasaw (§ 6.3.2)	Japanese (Vance 1987)
Hausa (Newman 1987)	Kashaya (§ 6.3.9)
Tiberian Hebrew (§ 6.3.10)	Maithili (§ 6.1.6)
Hindi (§ 6.1.7)	Sierra Miwok (§ 6.3.9)
Hixkaryana (§ 6.3.1)	Turkish (§ 6.3.10)
	Yana (§ 7.2)

Note that (65b) includes cases where the prohibition is only in nonfinal syllables or at some stage of the derivation. For (65b), see also the discussion of Osthoff's Law (Ancient Greek) in Steriade 1991.

The typology just given seems correct as a tendency, but there are exceptions. In one direction, we find languages like Latin (§ 5.1.4), Seminole/Creek (§ 4.1.2), General Central Yupik (§ 6.3.8), Klamath (§ 7.1.4), or Estonian (§ 8.5), in which CVC is heavy but CVVC is allowed—these are languages that exceptionally allow trimoraic syllables, or special crowding of segments onto moras. More crucial for present purposes are cases in which CVC is light but CVVC is avoided. St. Lawrence Island Yupik (§ 6.3.8.1) and Pacific Yupik (§ 8.8) both block the rule of iambic vowel lengthening in closed syllables, even though syllable-final consonants fail to make weight for purposes of stress. (They do make weight in initial syllables in Pacific Yupik, just as in Norton Sound (§ 6.3.8.2), but this does not affect the argument, since Iambic Lengthening can only apply to non-initial syllables.)

We can resolve the contradiction by using two-layered representations: for purposes of stress, weight is computed on the upper layer of the mora grid, while the two-mora population limit that constrains Iambic Lengthening is computed on the lower layer:

(66) a. **Lengthening Allowed** b. **Lengthening Blocked**

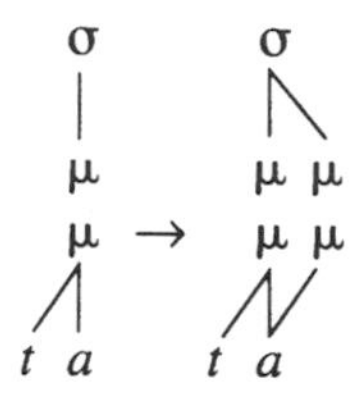

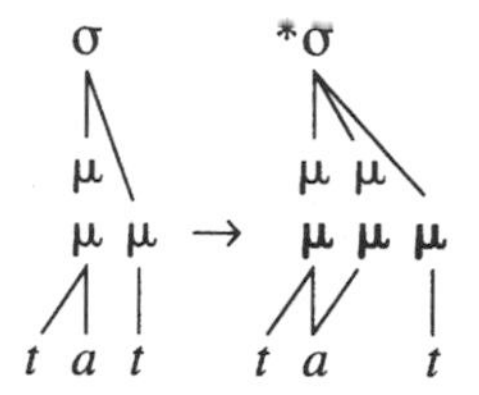

Iambic Lengthening is blocked in closed syllables because it would create a trimoraic sequence on the lower moraic layer (boldface).

Pacific Yupik also has the opposite case: a rule of Compression (Jacobson 1985, 36–38; Leer 1985a, 89; Woodbury 1987, 701) shortens a long vowel or diphthong in a closed syllable, despite the fact that coda consonants do not in general make their syllable heavy. Again, we can express quantity correctly for purposes of stress on the upper layer of the mora grid, while expressing Compression by imposing a two-mora population limit on the lower layer.

(67) a. b.

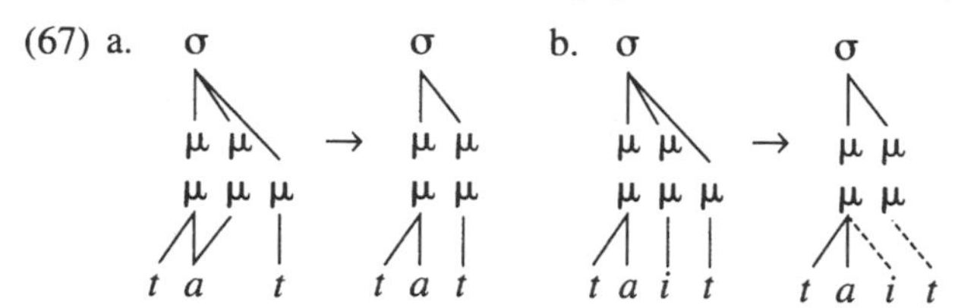

7.3.3 Summary and Comparison

The two-layered theory of the mora forms a means of accounting for the (probably unusual) cases in which weight is counted differently for purposes of different phonological processes. A virtue of the theory is that it retains the restrictiveness of the original moraic theory with regard to compensatory processes and phonological counting (§ 3.9.2).

Steriade 1991 also addresses the question of dual weight criteria, making a rather different proposal: there is just one representation for weight, involving a single layer of moras. Individual rules may refer to this representation in different ways, depending on the sonority level of the segments the moras dominate. Steriade's proposal is appealingly parsimonious, and it is not easy to distinguish it empirically from the proposal made here. A possible argument can be made on the basis of the analysis of Pacific Yupik in § 8.8: there, it is proposed to make a structural distinction between heavy and light CVC, where the second C forms half of a geminate. Since the relevant moras have the same sonority in each case, Steriade's single-layer account could not distinguish between the two.

The strategy followed by Crowhurst 1991a, stated in general terms, is to add or remove moras at the appropriate stage in the derivation, as individual rules require (for similar cases, see Mohanan 1989 and § 6.3.5). Aside from the unhappy history of this strategy in other domains (Kenstowicz 1970), this account largely removes the possibility of regarding metrical structure as a persistent organizing principle in phonology. In particular, the heavy CVC syllables that would have to be created late in the derivation in Cayuga and Yupik often occur in weak position in the foot, which would make numerous feet ill-formed at the surface.

Selkirk (1988, 1990), who foresaw the inconsistent quantity problem, proposes an elegant and economical solution: to use the root nodes of segments as the depiction of segment count. In this view, geminates and long vowels have two root nodes, but only one set of feature values, all of which are doubly linked. Selkirk's account severs many of the connections between segmental length and syllable weight, allowing among other things the possibility noted above of geminates that do not make their syllable heavy.

Although the two-root theory of length is supported with arguments from a wide range of phonological phenomena, it appears to be insufficient as an account of the particular problem of dual weight criteria. The difficulty is this: the tier of root nodes includes onset root nodes. But the phenomena at issue ignore onsets; that is, they are genuinely weight-based. Thus, for example, Cayuga Laryngeal Metathesis reduces syllable weight, but it does not reduce the number of root nodes. Similarly, the more liberal criterion of quantity employed in Ancient Greek (Steriade 1991) counted VC as heavy, but not CV, even though the two have the same number of root nodes. The upshot is that Selkirk's proposal can describe languages that employ both a weight criterion and a segment count criterion, but what we actually need to account for are languages that have two distinct weight criteria.

7.4 MISCELLANEA

The account advanced here for how syllable structure influences stress is based on a central proposal, moraic theory, and two amplifications: the two-layered theory of moras (§ 7.3), to handle cases of "inconsistent" syllable quantity, and the theory of prominence (§ 7.1). Below, I discuss some miscellaneous problems.

In Aklan (§ 6.3.10), two prefixes *ka-* and *ga-* are claimed by Hayes 1981 to have purely diacritic weight. This claim has been adopted by Halle and Vergnaud (HV, p. 23), who take it as an argument for their approach to syllable weight (see discussion in § 8.5.4). However, Hayes (1981) was not aware that Aklan apparently has phonemic vowel length (Isidore Dyen, p.c.), which is not marked in his source material, Chai 1971. It seems highly plausible that *ka-* and *ga-* actually are long-voweled prefixes.

Lahiri and Koreman 1988 and Kager 1989 argue that for purposes of stress assignment in Dutch, CVC and CV_iV_j (i.e. diphthongal syllables) count as heavy, whereas CV: counts as light. Since Dutch stress would be treated in the theory assumed here using moraic trochees (Lahiri and Koreman), this odd weight distinction must be considered one of quantity, not prominence.

Addressing this problem, van der Hulst and van Lit 1988 and Smith et al. 1989 argue that in underlying forms, Dutch does not have a normal distinction of vowel length; rather, we have something akin to Trubetzkoy's (1969) feature of **close contact** (*Silbenschnittkorrelation*). The vowels termed "short" or "lax" are required to occur in closed syllables, that is, in close contact with a following consonant. Since closed syllables are heavy, "short" vowels attract stress (Kager 1989, 229). The vowels called "long" or "tense," which are basically monomoraic, impose no closed-syllable requirement. Thus they can occupy light syllables and be skipped over by the stress rule. When "long" vowels get stressed, they are phonetically lengthened, but when they are stressless, they can be quite short. In particular, "long" vowels in penultimate position that have been skipped over by the stress rule can be even shorter than "short" vowels in the same context (Nooteboom 1972). The phonetic facts thus support the view that the Dutch "long" vowels are not a counterexample to the view of weight adopted here.

A useful survey of some further problematic cases is provided by Davis 1988. Of his examples, Pirahã is treated above within the theory of prominence, and (as Davis notes) Western Aranda (Strehlow 1945) is dealt with by Archangeli 1986 and by HV using initial vowel extrametricality, amplified by a percolation convention. In the case of Madimadi (Hercus 1986), the syllable weight criterion is mostly of the ordinary type, where CVC and CVV are heavy. But certain syllable onsets cause the preceding syllable to count as heavy, attracting (probably iambic) stress. Inspection of Hercus 1986 shows a substantial overlap between the onsets that attract stress and those that induce lengthening of the preceding vowel; this perhaps might form the basis of at least a diachronic account of the Madimadi pattern, which in any event is fairly irregular. Davis also presents onset-sensitive cases from Italian and from English.

All the onset-sensitive analyses Davis proposes involve types of metrical rules other than foot construction. Thus none impinge on my central claim: that foot construction, being governed by the Iambic/Trochaic Law, has a special status in referring only to the "orthodox" syllable weight patterns depictable within moraic theory.

• 8 •

Ternary Alternation and Weak Local Parsing

8.1 OUTLINE OF THE THEORY

Much recent theorizing in generative grammar focuses on the idea of locality: we obtain interesting and valid predictions by constraining rules to apply within bounded domains. In phonology, the principle of locality often takes the form of limiting what can be counted: a reasonable conjecture is that phonological rules can count only to two.

Theories of universal foot inventories have formed part of the locality research program. Apart from early proposals by Halle and Vergnaud 1978 and McCarthy 1979b, foot inventories have usually excluded feet that require any counting higher than two.

Hayes 1981 attempted to show how locality in this sense can be maintained in the face of apparent counterexamples. Many languages place stresses at a distance of three syllables from the word edge, as in English *Pámela,* or from another stress, as in ***Wìnnepesáukee***. Hayes argued that such effects could be best explained not by including ternary feet in the universal inventory, but by other devices that are independently motivated, namely extrametricality (§ 3.11) and destressing (§ 5.1.7).

However, there are stress systems that allow ternary alternation freely across the entire word. Such patterns cannot be described under a theory positing just binary feet, extrametricality, and destressing rules, and thus constitute a challenge to the principle of locality in phonology.

This chapter addresses the question of how such languages can be incorporated into a local theory of stress. Before presenting my proposal, I note some criteria that a theory of ternary alternation ideally should meet.

(a) Ideally, most of the formal apparatus needed should already be present in the theory.

(b) The theory should provide a formal means of characterizing ternary alternation as marked, since it seems fairly certain (pace Levin 1988a) that the phenomenon is quite unusual, especially in comparison to binary alternation. The literature I have examined describes dozens of languages with alternating binary stress, but only a few with ternary patterns.

(c) An adequate theory should be restrictive in allowing for the attested ternary cases, but not expanding the power of the framework to the point where it can describe anything. For example, stress alternation at four-syllable intervals appears to be completely unattested, and an adequate theory should exclude it.

The following is a basic outline of the analysis to be proposed, which is similar in many respects to the independent work of Hammond 1990a, 1990b.

Consider first the view adopted here (§ 4.1.1, § 5.1, § 5.2.3) that foot construction does not always exhaust the string of syllables: in cases where degenerate feet are forbidden, parsing often leaves syllables stray, typically in odd-numbered strings. Destressing can also create unfooted syllables.

Suppose we let some syllables go unparsed "deliberately." Assuming that only single syllables can do this, we obtain ternary alternation:

(1) (× .) (× .) (× .) (× .)
  ◡◡ ◡ ◡◡ ◡ ◡◡ ◡ ◡◡ ◡ . . .

The sequences of stressless syllables consist of the weak syllable of a foot plus an unfooted syllable.

Now, the normal scheme I have assumed for footing excludes (1), since feet must be constructed adjacently whenever the syllable string permits this. Thus to create forms like (1) we must add a new parsing algorithm that permits nonadjacent footing. We next ask: if feet are not adjacent, how close do they have to be? If the general approach of seeking locality principles is valid, then this distance must be small. Here, I assume that it must be the smallest definable prosodic distance, namely a single mora. Since foot construction cannot split up syllables (§ 3.9.1, § 5.6.2), this is equivalent to a single light syllable.

Pursuing this scheme, I propose the parameter in (2). Assuming that foot parsing examines a finite "window" (syllable sequence) at each iteration, the parameter governs how the window is advanced once a foot has been constructed:

(2) **Foot Parsing Locality Parameter**

a. **Strong Local Parsing** When a foot has been constructed, align the window for further parsing at the next unfooted syllable. (unmarked value of the parameter)

b. **Weak Local Parsing** When a foot has been constructed, align the window for further parsing by skipping over /◡/, where possible. (marked value of the parameter)

Ternary alternation is the result of weak local parsing, which creates forms like (1), while binary alternation results from strong local parsing.

Under this scheme, the general theory still posits no ternary feet. In fact, we will see that the known ternary patterns simply reflect the three kinds of feet (moraic trochee, syllabic trochee, iamb) already adopted here.

A way of thinking of this scheme is that strong local parsing attempts to make the feet adjacent, while weak local parsing attempts to separate them by /◡/. Such attempts do not always succeed. For example, assuming parsing into moraic trochees in either direction, (3a) must be parsed as shown even under weak local parsing, since there is no /◡/ to skip. Similarly, (3b) parses as shown even under strong local parsing, because of the Priority Clause (§ 5.1.6).

(3) a. (×)(×)   b. (×)   (×)
   – –   – ◡ –

It is only when the parsing window includes at least two light syllables that differences will be seen.

8.2 CAYUVAVA

Consider now a concrete example of weak local parsing. The stress system of Cayuvava (Bolivia; Key 1961, hereafter K; Key 1967, hereafter M) was pointed out in Hayes 1981, 107, as a problem for any theory limited to binary foot templates. Previous metrical analyses of Cayuvava include Levin 1985b, 1988a; HV; and Spring 1989a.

The Cayuvava stress pattern is summarized in (4), which gives the normal stressing for words of two to eleven syllables (monosyllables are apparently lacking). There is no vowel length distinction, so /V_iV_i/ represents a disyllabic sequence rather than a long vowel; in other words, Cayuvava appears to have only light syllables.

(4)

a.	/σ́ σ/	*éɲe*	'tail'	K 144
b.	/σ́ σ σ/	*šákahe*	'stomach'	K 144
c.	/σ σ́ σ σ/	*kihíbere*	'I ran'	K 144
d.	/σ σ σ́ σ σ/	*ariúuča*	'he came already'	K 149
e.	/σ̀ σ σ σ́ σ σ/	*ǰìhiraríama*	'I must do'	M 71
f.	/σ σ̀ σ σ σ́ σ σ/	*maràhahaéiki*	'their blankets'	K 150
g.	/σ σ σ̀ σ σ σ́ σ σ/	*ikitàparerépeha*	'the water is clean'	K 149
h.	/σ̀ σ σ σ̀ σ σ σ́ σ σ/	*čàadiròboβurúruče*	'ninety-nine (first digit)'	M 60
i.	/σ σ̀ σ σ σ̀ σ σ σ́ σ σ/	*medàručečèirohíiɲe*	'fifteen each (second digit)'	M 61
j.	/σ σ σ̀ σ σ σ̀ σ σ σ́ σ σ/	*čaadàirobòirohíiɲe*	'ninety-nine (second digit)'	M 60

As can be seen, stress is initial in disyllables, otherwise antepenultimate. If enough syllables are available, stress also falls on the sixth and ninth syllables from the end, following a clear ternary pattern. Levin 1988a, 105–8, provides arguments that the rightmost stress is the strongest, as well as discussion of irregular penultimate and final stress in certain morphological contexts.

8.2.1 Analysis

The analysis I propose for Cayuvava is as in (5):

(5) a. **Syllable Extrametricality** $\sigma \rightarrow \langle\sigma\rangle$ / ____ $]_{word}$

b. **Foot Construction** Form syllabic trochees from right to left:
   i. Employ weak local parsing.
   ii. Degenerate feet are allowed in strong position.
   iii. Footing is non-persistent.

c. **Word Layer Construction** End Rule Right

We could also assume moraic trochees, since in a language without heavy syllables, these are equivalent to syllabic trochees. The weak rather than the strong prohibition on degenerate feet (§ 5.1.1) must be invoked, even though monosyllables are lacking, since there are exceptional forms with final (main) stress, such as (13). Cross-linguistically, this seems to be unusual: languages usually tolerate degenerate feet in exceptionally stressed words only if they tolerate them in monosyllables as well.

Under these rules, the first iteration of foot construction produces the output for (4h):

```
(6)                     (×  .)
      ◡◡ ◡◡ ◡  ◡ ◡ ◡⟨◡⟩
      čaadiroboβururuče
```

Under weak local parsing, the second foot to be created is placed not adjacent to the first one, but separated from it by a single light syllable. Continuing the process, we stress every third syllable:

```
(7)           (×  .)  (×  .)          (×.) (×  .)  (×  .)
   →    ◡◡ ◡◡ ◡  ◡ ◡ ◡⟨◡⟩  →    ◡◡ ◡◡ ◡  ◡ ◡ ◡⟨◡⟩
        čaadiroboβururuče       čaadiroboβururuče
```

The selection of the antepenult for main stress is determined by End Rule Right:

```
(8) (                ×  )
    (×.) (×  .)  (×  .)
     ◡◡ ◡◡ ◡  ◡ ◡ ◡⟨◡⟩
    čàadiròboβurúruče
```

This accounts for the basic pattern, found in its purest form with words having $3n$ syllables, n an integer. The other cases work as follows.

When a word has $3n + 1$ syllables, the initial syllable is skipped by weak local parsing, and no further foot construction may take place:

```
(9)  (          ×  )
        (×  . ) (× .)
     ˘  ˘  ˘  ˘˘˘⟨˘⟩
     maràhahaé iki
```

In disyllables, foot construction creates a degenerate foot, which survives because Word Layer Construction places it in strong position:

```
(10) (×)
     (×)
      ˘ ⟨˘⟩
      eɲe
```

One might in addition suppose that extrametricality is revoked, as in § 5.3, though this has no apparent empirical consequences.

The most interesting cases in Cayuvava are those with $3n + 2$ syllables, $n > 0$ (i.e. five, eight, and eleven). Here, at least one foot can be constructed in the normal way, satisfying culminativity. At the left edge of the word, however, we get a **double upbeat;** the first two syllables remain stressless, as in (4d, g, j).

As HV points out, the double upbeat is anomalous. Among bounded iterative stress systems, the overwhelmingly more common outcome is that a foot is constructed if there is space available for one. Most versions of metrical theory account for this with the principle that if there is sufficient space available for a foot or a grid mark, then one must be assigned. Some versions of the theory (e.g. Prince 1983a, or the present theory) block stress assignment when only one syllable is available. But no theory blocks stress assignment when there are two syllables. Given the outcome in most bounded stress systems, this seems the correct prediction for a theory to make.

As Levin (1985b, 1988a) points out, the claim of stresslessness for initial syllables in $3n + 2$ words is empirically confirmed by segmental phonology. Rules of vowel deletion that fail to apply to stressed vowels do apply to the first syllable of a $3n + 2$ word, for example *piriβéβere* ~ *priβéβere* 'you are running' K 144.

The anomaly of $3n + 2$ words in Cayuvava cannot be attributed to the ternarity of the system. Pacific Yupik (§ 8.8) and Winnebago (§ 8.9) have ternary stress systems that are close to the mirror image of Cayuvava. When there are two syllables available for stress assignment in these languages, then a stress is assigned.

Under the theory proposed here, the "double upbeats" of Cayuvava can be derived if we assume that foot construction is not persistent. The relevant cases will work as follows. Initial footing yields the outcomes in (11):

```
(11) a. (×) (× .)        b. (×) (×  .) (×  .)        c. (×) (×.)   (×.)   (×.)
        ˘ ˘˘˘ ⟨˘⟩           ˘ ˘˘  ˘ ˘ ˘  ˘⟨˘⟩           ˘˘  ˘˘  ˘  ˘˘  ˘  ˘˘ ⟨˘⟩
        ariuuča             ikitaparerepeha             čaadairoboirohiiɲe
```

In all of these, the right-to-left scan skips over the peninitial syllable and places a degenerate foot on the initial. When the word layer is constructed, this degenerate foot falls in weak position and is therefore removed:

```
(12) a. (   × )          b. (         ×  )           c. (                 × )
        (×) (× .)           (×) (×  .) (×  .)           (×) (×.)   (×.)   (×.)
        ˘ ˘˘˘ ⟨˘⟩           ˘ ˘˘  ˘ ˘ ˘  ˘⟨˘⟩           ˘˘  ˘˘  ˘  ˘˘  ˘  ˘˘ ⟨˘⟩
        ariuuča             ikitaparerepeha             čaadairoboirohiiɲe

        (   × )             (         ×  )              (                 × )
            (× .)               (×  .) (×  .)                 (×.)   (×.)   (×.)
        ˘ ˘˘˘ ⟨˘⟩           ˘ ˘˘  ˘ ˘ ˘  ˘⟨˘⟩           ˘˘  ˘˘  ˘  ˘˘  ˘  ˘˘ ⟨˘⟩
        ariúuča             ikitàparerépeha             čaadàirobòirohíiɲe
```

Under the assumption that Cayuvava footing is not persistent, the initial disyllabic sequence cannot be footed further, and the derivation ends, yielding the correct surface forms. In contrast, Pacific Yupik and Winnebago have persistent footing, and they create feet out of the stray syllable sequences. Compare also the case of disyllables in Cayuvava (10): here the degenerate foot is in strong position, and thus can survive to the surface.

As mentioned above, main stress follows different patterns in various morphological contexts. For example, certain imperative verb forms have final stress, with secondary stress computed on the usual ternary basis. To account for such cases, I assume top-down stressing (§ 5.4.3), triggered by an early morphologically governed rule of End Rule Right. Assuming that the main stressed syllable created by this rule cannot become extrametrical, this will force the creation of a final degenerate foot, yielding the correct result:

```
(13) (      ×)        (      ×)       (      ×)                'give it to me!'
      ˘ ˘ ˘ ˘   →           (×)   →   (× .) (×)                           K 149
     borejere         ˘ ˘ ˘ ˘         ˘ ˘ ˘ ˘
                      borejere        bòrejeré
```

There are also monosyllabic pre-accenting suffixes, which I propose to analyze, following Levin 1988a, as exceptions to the extrametricality rule (5a).

8.2.2 Earlier Accounts of Cayuvava

It is useful to compare the ability of differing theories to account for the double upbeats of Cayuvava. Suppose first that we try to account for Cayuvava using ternary feet. The approach of Levin 1985b is to form dactyls; that is, ternary initially stressed feet:

```
              (× . .)                        (× .)
(14) Form     σ σ σ   if possible; otherwise form   σ σ.
```

The elsewhere case is needed to stress disyllables. Consider the footing that this analysis provides for representative words of two, five, and six syllables:

```
(15) a. (×    )    b. *(     ×        )    c. (           ×         )
        (×   .)       (×   )(× .    .)       (×   .   .) (× .     .)
         σ   σ         σ   σ σ σ    σ         σ   σ   σ   σ σ     σ
         é  ɲ e        à r i ú u č a          ǰ ì h i r a r í a m a
```

Representations (15a) and (15c) are correct, but (15b) must have its initial secondary stress removed, presumably by a rule of destressing. It is here that the dactylic analysis leads to a problem. It appears to be a general principle (Hammond 1984a; § 3.4.2(d)) that rules of destressing always apply in an environment of stress clash. Any rule that removed the initial stress of (15b) would necessarily be a counterexample to this generalization.

This point is made by Levin 1985b, who suggests that as a consequence the clash generalization will have to be abandoned, and that the correct general characterization of destressing is somewhat different: it occurs in feet that are of non-maximal size for the particular stress rule.

This proposal is subject to two objections. First, the clash generalization is widely supported across languages and should not be abandoned cavalierly. Second, Levin's alternative of destressing non-maximal feet would have exceptions of its own: there are clear cases in which maximal feet are removed when they are involved in a clash, in Icelandic (§ 6.2.2.4), French (Dell 1984), and English (Hayes 1982b, 257–60).

Other ternary-foot analyses of Cayuvava have been proposed by HV and by Levin 1988a. These make use of amphibrachic feet, that is, ternary feet with stress in the middle. Amplified with syllable extrametricality, they can derive the basic ternary pattern. For $3n + 2$ words, Levin removes initial degenerate feet by a language-specific rule, as in (16):

```
(16) (×)(. × .)             (. × .)
      σ  σ σ σ ⟨σ⟩  →    σ  σ σ σ ⟨σ⟩
      a r i u u č a      a r i ú u č a
```

This seems an improvement over the dactylic analysis, since languages seem to avoid degenerate feet even when there is no clash (§ 5.1.8.2). For further discussion of amphibrachs, see § 8.4(a, c, e).

HV's analysis is similar to that of Levin 1988a but has a more ambitious account of $3n + 2$ words. This relies on a proposed universal Recoverability Condition (p. 10), which requires that the location of constituent boundaries must always be recoverable from the stress pattern. Five-syllable words cannot have initial stress, since with the amphibrachic feet HV assumes, this would lead to two possible bracketings (17a, b). Only by leaving out the initial stress can the bracketing be made unambiguous (17c):

(17) a. *(×)(. × .)
σ σσσ ⟨σ⟩
àriúuča

b. *(× .)(× .)
σ σ σ σ ⟨σ⟩
à r i ú u č a

c. (. × .)
σ σσσ ⟨σ⟩
a r i ú u č a

A possible problem with this account is that the Recoverability Condition apparently has no purpose in the HV theory other than to account for the Cayuvava facts.[1] In addition, it appears that the Recoverability Condition is empirically falsified by the facts of Pacific Yupik and Winnebago. As noted above, for analogous forms these languages behave in the opposite way to Cayuvava. For further discussion of the Recoverability Condition, see Elenbaas 1992.

8.3 AN ALTERNATIVE VERSION OF WEAK LOCAL PARSING

One aspect of the analysis I have proposed that may seem puzzling is why it invoked extrametricality. A plausible alternative would be to attribute antepenultimate stress in Cayuvava to weak local parsing itself: that is, define weak local parsing so as to place the feet the minimal prosodic distance not just from each other, but from word edge as well.

There are two reasons not to adopt this approach. First, Cayuvava turns out to be exceptional among ternary languages: for all the other cases, the analysis is simpler if we assume that weak local parsing governs only the separation of the feet from each other, and not from word edge. Since final syllable extrametricality is not uncommon, we should not be surprised if at least one ternary language incorporates it.

The second reason not to revise weak local parsing in this way is that we wish to avoid excess descriptive power. If weak local parsing involves distance to the edge of the word as well as distance between feet, then we could combine it with extrametricality to derive pre-antepenultimate stress, as in (18):

(18) (× .)
. . . σ σ σ ⟨σ⟩ #
skipped by extrametricality
skipped by weak local parsing

Since the rule "stress the fourth syllable from the end" appears to be completely unattested, it seems better to limit weak local parsing to governing only the distance between feet.

1. It also excludes unbounded constituents in which the head may occur on any syllable. However, such feet are plausibly excluded in any event, because they cannot be used to define stress rules.

8.4 SUMMARY OF THE ARGUMENTS

Looking ahead to the rest of this chapter, here are the arguments that support the proposed approach.

(A) ECONOMY. My proposal does not require the addition of new foot templates to metrical theory: with the three templates employed here, weak local parsing suffices. This keeps the theory from having excess power; for example, it is inherently incapable of describing a system of quadrisyllabic alternation.

(B) FULL INSTANTIATION. The possibilities predicted by the theory are fully attested. In particular, all three foot types occur in weak local parsing systems:

(19) a. **Syllabic Trochees** Cayuvava (§ 8.2), Estonian (§ 8.5), other Finno-Ugric languages (§ 8.6), Auca (§ 8.10), Mantjiltjara (§ 8.10)
b. **Moraic Trochees** Sentani (§ 8.7), Bani-Hassan Arabic (§ 8.10)
c. **Iambs** Pacific Yupik (§ 8.8), Winnebago (§ 8.9)

(C) RELATING BINARY AND TERNARY SYSTEMS. The foot parsing locality parameter (2) implies a close connection between weak local systems and their minimal equivalents with strong local parsing. This connection turns out to be a characteristic property of weak local systems: they typically occur in close "proximity" to the corresponding strong local system. Either weak and strong local parsing occur as free or conditioned variants within a single language, or else closely related languages or dialects have the same stress system, differing only in whether locality of parsing is weak or strong. In the case of related languages, the argument is at the diachronic level: it appears that the parsing locality parameter can shift its value as varieties diverge from one another in the course of language change.

Cases of language-internal variation of the sort described occur in Estonian (§ 8.5), Finnish (§ 8.6), Karelian (§ 8.6), Hungarian (§ 8.6), Sentani (Elenbaas 1992; § 8.7), Pacific Yupik (§ 8.8), Winnebago (§ 8.9), Mantjiltjara (§ 8.10), and Auca (§ 6.2.1, § 8.10). Cross-linguistic variation of this type is found in Finnish, Hungarian, and the Yupik languages.

The connections just observed would not follow from the amphibrach theory of ternary alternation, where one would have to claim that variant forms within a language differed in a very fundamental way, namely the foot template involved. This predicts that we could have free variation involving essentially any feet, such as iambs versus trochees, which appears to be false.

(D) PRESERVING LOCALITY. The theory is compatible with the general effort to impose locality restrictions on phonological systems. The apparent non-locality of ternary systems is treated as the combined result of two local principles: binary feet and a minimal distance requirement between feet.

(E) SCOPE. The ternary system found in Estonian and related Finno-Ugric languages has a straightforward account under weak local parsing, as we will see, but not under previously proposed theories of ternary alternation. This system is treated in the next section.

There is an alternative approach rather similar to what is pursued here, which would also capture most of the above generalizations: namely, add a single parameter that doubles the set of basic foot templates, by adjoining to each one a single light syllable position. The resulting expanded foot templates could then be used to derive ternary stress directly, without weak local parsing. The difficulty with this is that we cannot predict the SIDE of the basic foot to which the additional /◡/ position should be added: for example, is the expanded syllabic trochee to be constructed as /σ́ σ/ + ◡ = /σ́ σ ◡/, or as ◡ + /σ́ σ/ = /◡ σ́ σ/? Inspection of the examples in this chapter suggests that there is no general principle that will predict where the extra /◡/ position should be added. The correct generalization—that the extra skipped light syllables always lie in the FORWARD DIRECTION of the parse—follows from weak local parsing.

8.5 ESTONIAN

The account of Estonian stress below draws heavily on the intensive analysis of Prince 1980. Prince's work is based on Hint 1973 and others, and focuses on the nature of the third degree of quantity, or **overlength.** I show here that Prince's treatment can be naturally adapted into a theory that posits weak local parsing. The Estonian pattern also has important consequences for metrical stress theory in general, discussed in § 8.5.4.

8.5.1 Data

In Estonian, main stress normally falls on the initial syllable of a word, except in some borrowings. Secondary stress is largely predictable, except for a restricted set of stems and affixes that carry or trigger idiosyncratic secondary stress.

Hint 1973 groups together the non-initial primary stresses and the non-predictable secondary stresses, referring to them collectively as **morphologically bound** stress. Hint argues that every morphologically bound stress, whether primary or secondary, begins its own phonological word, which may be smaller than the grammatical word. With Prince, I follow Hint's proposal, and focus on the predictable patterning of stress within the phonological word. All the data to be considered are forms that lack morphologically bound stresses, and thus consist of exactly one phonological word and one grammatical word.

Within the phonological word, the strongest stress always falls on the initial syllable. In words without overlength, secondary stresses are assigned iteratively from left to right at intervals of two or three syllables. Provided appropriate conditions are met, there can be free variation in the placement of secondary stress. The basic pattern is this: if a syllable σ_n in the word bears stress, then one option is to place another stress on σ_{n+2}. However, just in case σ_{n+2} is light (CV), σ_{n+3} may be stressed instead. Thus the word *teravamaltt* (21h) /◡ ◡ ◡ – / has two options: *téravàmaltt* or *téravamàltt.* That is, if we take σ_n to be the initial syllable *te-*, then σ_{n+2} (*va*) is light, so stress can fall on either σ_{n+2} (*va*) or σ_{n+3} (*maltt*). However, the word *sóoyemàttel* (21i) / – ◡ – ◡/ has only one option, because σ_{n+2}, *-mat-*, is heavy. This algorithm applies iteratively in longer words. In sum, stress always has the option of being based on a simple binary count, but in many forms there is the additional option of ternary alternation, the crucial condition being that the third syllable in the ternary interval be light.

There is a complication: at the end of a word, only a heavy syllable may bear stress. "Heavy" in this context includes CVV, CVVC, CVCC, but not CVC. This is due, I assume, to a rule of Consonant Extrametricality (cf. § 3.11):

(20) **Consonant Extrametricality** $C \rightarrow \langle C \rangle \;\_\_\_\_ \;]_{\text{phonological word}}$

Because of this rule, word-final CVC is treated as CV and thus counts as light.

Here is an example of how the ban on word-final stressed light syllables affects the stress pattern. The third syllable of (21g) *osavama* is light, so that in principle the stress contour **ósavamà* would be possible. This is forbidden, however, by the ban on word-final stressed light syllables, so only *ósavàma* is allowed. In contrast, in *téravàmaltt* ~ *téravamàltt,* the final syllable is heavy, and both options are permitted. Note that this pattern essentially excludes the possibility that final secondary stress is just an auditory effect, as suggested for parallel cases in § 5.1.8.2: if *téravamàltt* had no true final secondary stress, it would be parallel to **ósavama,* which is ill-formed. Therefore, in *téravamàltt,* absence of secondary stress on the penult argues for the presence of stress on the final syllable.

At this point we can consider the full data, taken from Hint 1973. For longer words, the charts below take advantage of the fact that the quantity of the first two syllables of a word without overlength makes no difference to the stress pattern of the word; thus for four-syllable words, there are really only four logical possibilities to consider rather than sixteen.

(21) **Estonian Stress in Words without Overlength**

a.	/◡́ ◡/	*pálat*	H 203
b.	/◡́ –/	*pálatt*	H 203
c.	/–́ ◡/	*páttu*	H 156
d.	/–́ –/	*nóorikk*	H 156

e. /σ́ σ ◡/	*ósava*	H 157
f. /σ́ σ –̀/	*ósavàtt*	H 157
g. /σ́ σ ◡̀ ◡/	*ósavàma*	H 159
h. /σ́ σ (◡̀) (–̀)/	*téravàmaltt, téravamàltt*	H 159
i. /σ́ σ –̀ ◡/	*sóoyemàttel*	H 159
j. /σ́ σ –̀ –/	*párimàtteltt*	H 159
k. /σ́ σ (◡̀) (◡̀) ◡/	*pímestàvale, pímestavàle*	H 161
l. /σ́ σ (◡̀) (◡̀) (–̀)/	*ǘlistàvamàit, ǘlistavàmait*	H 162
m. /σ́ σ (◡̀) (–̀) ◡/	*pímestàvasse, pimestavàsse*	H 161
n. /σ́ σ (◡̀) (–̀) (–̀)/	*ÿ́ppettàyattèks, ÿ́ppettayàtteks*	H 162
o. /σ́ σ –̀ ◡ ◡/	*válusàttele*	H 161
p. /σ́ σ –̀ – ◡/	*pímestàttutte*	H 161
q. /σ́ σ (◡̀) (◡̀) (◡̀) ◡/	*ósavàmalèki, ósavamàleki*	H 163
r. /σ́ σ (◡̀) (–̀) (◡̀) ◡/	*vállutàyattèka, vállutayàtteka*	H 163
s. /σ́ σ (◡̀) (–̀) (–̀) ◡/	*érinèvattèsse, érinevàttesse*	H 163
t. /σ́ σ –̀ ◡ ◡̀ ◡/	*ÿ́pettùstelèki*	H 163
u. /σ́ σ –̀ – ◡̀ ◡/	*kárastàttuimàle*	H 163
v. /σ́ σ –̀ ◡ (◡̀) (–̀) (–̀)/	*úsaltàttavàmattèks, úsaltàttavamàtteks*	H 163

The rest of the data involve words which contain overlength. Drawing on earlier work, Prince characterizes overlength as a property of syllables, not segments: an overlong syllable receives a dose of extra duration, distributed by rule to various segments at or near the end of the syllable. Because of this extra syllabic length, it is possible to find three-way surface contrasts of segmental length in Estonian: short segments, long (i.e. geminate) segments, and long segments that have received the effects of overlength. The idea that overlength is a syllabic property and not a segmental one accords with the intuitions of some native speakers, who hear overlength as an "extra heavy stress" on the relevant syllable. Overlong syllables also have special pitch patterns as well.

I follow Prince's transcription system for overlength, which combines an overlength mark /ː/ with orthographic gemination. The symbol /=/ is used to indicate an overlong syllable.

Historically, the overlong syllables of Estonian usually derive from disyllabic sequences (Tauli 1954; Mürk 1981), and they behave much like disyllabic sequences in the synchronic stress pattern—compare Hindi (§ 6.1.7) for a similar phenomenon, with the same historical origin. For example, an overlong syllable in Estonian may be directly followed by another stress, whereas a stressed syllable without overlength must have at least one stressless syllable intervening between it and the next stress. Moreover, in a disyllabic word of the form /= –/, the final heavy syllable must bear stress, just as the final heavy syllable of a word of the form /σ σ –/ (without overlength) must also be stressed. Thus as a rough approximation, the possibilities for secondary stress in words beginning with /=/ are the same as the possibilities in words beginning with /σ σ/.

There are exceptions to this general rule, however. The sequence /=́ – σ̀/ is possible, although its putative equivalent /σ́ σ – σ̀/ is excluded. Moreover, according to Eek 1975, words of the form /= ⌣ ⌣/ may have the stress pattern /=́ ⌣ ⌣/ as well as the expected /=́ ⌣̀ ⌣/, but the putative equivalent /σ́ σ ⌣ ⌣/ is impossible.[2]

Bearing this limited equivalence in mind, consider the basic data involving overlength. In the examples in (22), the overlong syllable is always initial in the domain. This is necessarily the case, since I assume, following earlier work, that any non-initial overlong syllable must begin a new phonological word.

(22) **Estonian Stress in Words with Overlength**

a.	/=́ ⌣/	*árs:ti*	H 157
b.	/=́ –̀/	*vánkk:rìtt*	H 157
c.	/=́ (⌣̀) ⌣/	*káu:kèle, káu:kele*	H 158; Eek 1975, 15
d.	/=́ (⌣̀) (–̀)/	*yǽl:kètest, yǽl:ketèst*	H 158
e.	/=́ –̀ ⌣/	*yúl:kètte*	H 203
f.	/=́ (–̀) (–̀)/	*hái:kùstest, hái:kustèst*	H 158
g.	/=́ (⌣̀) (⌣̀) ⌣/	*tö́ös:tùsele, tö́ös:tusèle*	H 160
h.	/=́ ⌣ –̀ ⌣/	*káh:tlevàile*	H 159
i.	/=́ ⌣ –̀ –/	*kɤ́h:kleyàtteks*	H 159
j.	/=́ (–̀) (⌣̀) ⌣/	*tö́ös:tùstele, tö́ös:tustèle*	H 160
k.	/=́ (–̀) (–̀) ⌣/	*tö́ös:tùstesse, tö́ös:tustèsse*	H 160
l.	/=́ (–̀) (–̀) (–̀)/	*áu:sàimattèks, áu:saimàtteks*[3]	H 161
m.	/=́ (⌣̀) (⌣̀) (⌣̀) ⌣/	*trúu:tùselèki, trúu:tusèleki*	H 162
n.	/=́ (⌣̀) (–̀) (⌣̀) ⌣/	*tɤ́es:tùsettàki, tɤ́es:tusèttaki*	H 163
o.	/=́ ⌣ –̀ – ⌣/	*káu:kemàttesse*	H 162
p.	/=́ (–̀) (⌣̀) (⌣̀) ⌣/	*kín:tlùstelèki, kín:tlustèleki*	H 162
q.	/=́ (–̀) (–̀) (⌣̀) ⌣/	*káu:kèttessèki, káu:kettèsseki*	H 162

8.5.2 Prince's Account

The analysis proposed below is a modified version of that presented by Prince 1980, which I summarize here. For Prince, the Estonian foot is a syllabic trochee, which will normally contain two syllables. However, when just one heavy

2. I should not cite Eek's observation without also mentioning the existence of Eek 1982, an article which characterizes Hint's description of Estonian stress (and Prince's analysis based on it) as almost entirely in error. My own discussion assumes tentatively that Hint's thoughtful and detailed presentation must have some basis in the prosody of Estonian, at least for the variety Hint describes. Clearly further research on the phonetic correlates of Estonian stress would be helpful in resolving the controversy.

3. The latter variant is marginal, occurring under certain conditions of sentence stress or emphasis (H 161).

syllable is left over when a word is parsed into feet, that heavy syllable may form a foot. Prince derives this case by generalizing the rule that defines the syllabic trochee:

(23) Foot → u u where u = σ or μ (Prince 1980, ex. (27a.i))

When u is taken to be /μ/, this defines /–́/ as a foot. Since feet are normally constructed maximally, /–́/ will serve as a default syllabic trochee when (and only when) just one heavy syllable is left in the parse. In § 5.1.9, I suggested that Prince's idea of /–́/ as default is applicable to syllabic trochees in general.

The syllabic trochees alone define the stress options where counting is always binary. To account for the ternary intervals, Prince permits composite ternary feet, by the rule in (24) (his (27a.ii)):

(24) A foot in Estonian can be a foot followed by a light syllable.

This yields structures like ((σ́ σ) ⏑). The limitation to light syllables ensures that ternary alternation will only occur when the third syllable in the ternary interval is light.

To account for the behavior of overlong syllables, Prince assumes (25):

(25) An overlong syllable counts as a single foot.

This statement conceals its significance. Rather than applying a rule to make overlong syllables into feet, Prince claims that it is the other way around: monosyllabic foot status is itself the defining property of overlong syllables. In his view, overlength is not a segmental property, but is simply the result of how the phonetic length rules of the language manifest foot structure. (A slight complication is that heavy word-final syllables that bear secondary stress are not overlong.)

Recall now that overlong syllables may also participate in ternary intervals; compare (22c). To allow for this, Prince states that rule (24), which licenses composite feet, may be invoked recursively, yielding a structure like (((=́) ⏑) ⏑). However, since this recursive definition would also allow ill-formed ((((=́) ⏑) ⏑) ⏑), (((((=́) ⏑) ⏑) ⏑) ⏑), and so on, Prince adds the qualification in (26):

(26) No foot may end in more than two weak syllables. Prince 1980, ex. (27a.iv)

Provision (26) restricts rule (24) to two applications after a minimal (=) foot, giving (((=́) ⏑) ⏑), and to just one application after a (σ́ σ) foot, giving ((σ́ σ) ⏑). These are the maxima that are observed. Note that by stating the restriction linearly (with syllable count) rather than hierarchically, rule (26) accounts for one of the non-parallelisms between minimal (σ́ σ) feet and minimal (=́) feet: the fact that the latter can have two light syllables grafted onto them, whereas the former can have only one.

The other non-parallelism between (=́) and (σ́ σ) is that when another stress immediately follows, ((=́) –) counts as a possible foot, but *((σ́ σ) –) does not. To describe this difference, Prince adds one additional rule:

(27) Foot + σ can be a foot / _____ foot Prince 1980, ex. (27a.iii)

This allows / – / to be stressless after /=/, just in case the next syllable is stressed. In other positions, feet of the form (=́ –) are ruled out.

Main stress is assigned to the first foot in the word, by mechanisms equivalent in effect to End Rule Left.

These rules largely suffice to generate the complex pattern of Estonian stress. In (28) are some examples, given in the notation of this book:

```
(28) a. ( ×           )   (×          )          b.   (×          )
        ((× . ) . )(× )   (× .)(×   .)                (×   .)(×  .)
          ◡  ◡  ◡   –       ◡  ◡  ◡   –               –    ◡   –  ◡
        téravamàlt⟨t⟩, téravàmalt⟨t⟩ = (21h)          sóoyemàtte⟨l⟩ = (21i)

     c.  (×        )  ( ×        )              d. ( ×        ) (×        )
         (× )(×  . )  ((× )  .)(×)                 (((× )  .).) (×  )(× .)
          =    ◡ –      =     ◡ –                    =     ◡ ◡    =    ◡ ◡
        yǽl:kètes⟨t⟩, yǽl:ketès⟨t⟩ = (22d)          káu:kele, káu:kèle = (22c)

     e. ( ×             )        (×             )
        ((× )   . )(×   .)       (× )((×   .) . )
          =     –   –   ◡          =     –  –   ◡
        tő̋ös:tustèsse    or    tő̋ös:tùstesse = (22k)
```

Case (28a) shows the optionality in stressing allowed by the /◡/-adjunction rule (24); such optionality is absent in (28b), where the third syllable is heavy, and thus cannot be adjoined into the initial foot. Example (28a) also shows the possibility of a word-final heavy syllable forming a foot; final light syllables can never be stressed (cf. (21)–(22)). Rule (25), which equates overlong syllables and monosyllabic feet, is illustrated in (28c); the second variant of this form also shows that light syllables can be adjoined to overlong syllables to form a foot. Recursive adjunction of two light syllables is shown in (28d). Finally, rule (27), which allows /=́ – / to form a foot provided it immediately precedes another stress, is documented by (28e). Observe that in none of the relevant cases above (e.g. (22j, p)) is /= – / allowed to form a foot in other circumstances.

It should be emphasized that while Prince's analysis (like the revised version of it below) captures the basic patterning of stress, there are puzzling residual questions it does not address. For example, the rules generate variants not listed in Hint 1973: /=́ ◡ ◡/ for (22c), /=́ ◡ ◡ ◡̀ ◡/ for (22m), /=́ ◡̀ – ◡/ for (22h), /=́ ◡̀ – –̀/ for (22i), and /=́ ◡̀ – –̀ ◡/ for (22o). Eek (1975), whose description diverges from Hint's in a number of areas, allows /≐ ⌣ ⌣/ (p. 15) but forbids /≐ ⌣̀ – –̀/ (p. 13). Regarding the last three forms, Hint apparently forbids them (p. 159) but does not in general exclude the possibility of a /⌣́ –/ foot following /=/; compare (22n). (There seems to be evidence (Harms 1964) that related Finnish (§ 8.6) avoids /⌣́ –/ syllabic trochees in some contexts; in such feet, quantity is directly opposite to stress.) Hint also lists cases in which diphthongs more freely tolerate adjacent stresses, especially in emphatic speech; perhaps these can be treated as optionally disyllabic CV.V sequences. I am unable to resolve these questions, and must leave them to future research, adopting Prince's rules as an elegant if tentative account of the system.

Below, I explore how Prince's analysis might be modified to take into account matters of general stress theory that have arisen since it was formulated. In particular, rule (26), which limits feet to three syllables, is something one would like to dispense with, given the overall goal of limiting the ability of rules to count. Rule (24), which adjoins light syllables to feet, also creates structures (e.g. /–́ – ⌣/) that go beyond the minimum of three basic foot types that I attempt to maintain here.

8.5.3 A Revised Analysis

For the cases without overlength, there is a straightforward translation of Prince's analysis into the theory of weak local parsing:

(29) a. **Consonant Extrametricality** C → ⟨C⟩ / ___ $]_{\text{phonological word}}$
b. **Foot Construction** Form syllabic trochees left to right:
(i) degenerate feet are banned entirely;
(ii) weak local parsing is optionally invoked;
(iii) footing is persistent.
c. **Word Layer Construction** End Rule Left

In (29b), the optional invocation of weak local parsing (§ 8.1) accounts for most of the cases of free variation in the data; it has the same function in my analysis as Prince's foot definition rule (24). Persistent footing is assumed, because when two stray syllables are left over at the end of the initial parse, they must be regrouped into a syllabic trochee. In this respect, Estonian contrasts with Cayuvava (§ 8.2), where parsing is non-persistent, and strings of two stray syllables may be left unfooted. The ban on degenerate feet reflects the lack of light-syllable words in Estonian.

Consider now an example of how this system applies to the string /⌣ ⌣ ⌣ ⌣ ⌣ ⌣/: (21q) *ósavàmalèki* ~ *ósavamàleki*. Scanning from left to right, we first construct a syllabic trochee:

(30) (× .)
⌣ ⌣ ⌣ ⌣ ⌣ ⌣
osavamaleki

At the next step we have two options. Adopting weak local parsing for this iteration, we obtain the form in (31a), with feet separated by a single /◡/. At this point no further parsing is possible, since degenerate feet are forbidden. Following Word Layer Construction we obtain (31b) *ósavamàleki*, a possible output:

```
(31) a. (× .)  (× .)           b. (×            )
        ◡ ◡ ◡  ◡ ◡ ◡  →           (× .)  (× .)
        osavamaleki                ◡ ◡ ◡  ◡ ◡ ◡
                                   ósavamàleki
```

If on the other hand at stage (30) we had selected strong local parsing, we would have obtained (32):

```
(32) (× .)(×  .)
     ◡ ◡ ◡  ◡ ◡ ◡
     osavamaleki
```

At this stage, the ultimate outcome is fixed, though there are two ways of getting there. If the next iteration selects strong local parsing, the derivation terminates as in (33), yielding *ósavàmalèki:*

```
(33)                           (×              )
    →  (× .)(×  .)(× .)   →    (× .)(×  .)(× .)
        ◡ ◡ ◡  ◡ ◡ ◡            ◡ ◡ ◡  ◡ ◡ ◡
        osavamaleki             ósavàmalèki
```

But if at stage (32) we had selected weak local parsing, the string available for footing would consist of only a light syllable:

```
(34) (× .)(×  .)
     ◡ ◡ ◡  ◡ ◡ ◡
     osavamaleki
               ||
               | available for parsing
               skipped under weak local parsing
```

As degenerate feet are forbidden, the first pass of foot construction must end here. However, footing is persistent, so the last two syllables can still be parsed after the initial left-to-right scan. The rest of the derivation will therefore look exactly like (33), and we get the same end result.

In (35) are further examples of forms from (21), showing how the analysis works. Any word whose derivation may involve persistent footing is marked P; for all of these, there is an alternative derivation in which the same disyllabic sequence is parsed directly by strong local parsing. Where there are two possibilities, the one from strong local parsing appears first. The word layer, assigning main stress to the initial foot, is omitted.

(35) **Words without Overlength**

a. (× .)
– –
nóorikk

b. (× .)
˘ ˘ ˘
ósava

c. (× .)(×)
˘ ˘ –
ósavàt⟨t⟩

d. (× .)(× .)
˘ ˘ ˘ ˘ P
ósavàma

e. (× .)(× .) (× .) (×)
˘ ˘ ˘ – ˘ ˘ ˘ –
téravàmalt⟨t⟩, téravamàlt⟨t⟩

f. (× .)(× .)
– ˘ – ˘
sóoyemàtte⟨l⟩

g. (× .)(× .)
˘ ˘ – –
párimàttelt⟨t⟩

h. (× .)(× .) (× .) (× .)
˘ – ˘ ˘ ˘ ˘ – ˘ ˘ ˘
pímestàvale, pímestavàle

i. (× .)(× .)(×) (× .) (× .)
˘ – ˘ ˘ – ˘ – ˘ ˘ –
ű̋listàvamài⟨t⟩, ű̋listavàmai⟨t⟩

j. (× .)(× .) (× .) (× .)
˘ – ˘ – ˘ ˘ – ˘ – ˘
pímestàvasse, pímestavàsse

k. (× .)(× .)(×) (× .) (× .)
– – ˘ – – – – ˘ – –
ýppettàyattèk⟨s⟩, ýppettayàttek⟨s⟩

l. (× .)(× .)
˘ ˘ – ˘ ˘
válusàttele

m. (× .)(× .)
˘ – – – ˘
pímestàttutte

n. (× .)(× .)(× .) (× .) (× .)
– – ˘ – ˘ ˘ – – ˘ – ˘ ˘
vállluttàyattèka P, *vállluttayàtteka*

o. (× .)(× .)(× .) (× .) (× .)
˘ ˘ ˘ – – ˘ ˘ ˘ ˘ – – ˘
érinèvattèsse, érinevàttesse

p. (× .)(× .)(× .)
– – – ˘ ˘ ˘
ýppettùstelèki P

q. (× .)(× .) (× .)
˘ – – – ˘ ˘
kárastàttuimàle P

r. (× .)(× .)(× .)(×) (× .) (× .) (× .)
˘ – – ˘ ˘ – – ˘ – – ˘ ˘ – –
úsaltàttavàmattèk⟨s⟩, úsaltàttavamàttek⟨s⟩

Consider now the treatment of overlong syllables. The analysis I will propose does not rely on metrical structure to represent overlength, but follows instead the proposal of Hayes 1989b, 293–97, whereby overlong syllables are represented as trimoraic. Hayes argues that such representations considerably simplify a crucial morphophonemic rule associated with overlength. Since it does not use metrical structure to depict overlength, the analysis is free to include overlong syllables within simple disyllabic feet, and in fact this option is crucial below.

The central analytical problem in the overlong stress pattern is that overlong syllables behave like disyllabic sequences in some contexts but not in others. The chart in (36) reviews how this works.

(36)		**Overlong Syllables**	**/σ́ σ/ Sequences**
a. **Same**	i.	Can be directly followed by stress.	(same)
	ii.	Can be followed by stressless /⏑/	(same)
	iii.	Cannot be followed by stressless / – / (except as noted below)	Cannot be followed by stressless / – /
b. **Different**	i.	/=́ ⏑ ⏑/ is a possible sequence.	*/σ́ σ ⏑ ⏑/ is not possible.
	ii.	/=́ – / can occur before another stress.	*/σ́ σ – / cannot occur at all.

Prince's account sorted out this pattern by assuming that overlong syllables are formally equivalent to disyllabic sequences (both may form a minimal foot). This automatically covers the parallels of (36a). Prince then added the extra provisions (26) and (27) to cover the non-parallels of (36b).

An alternative is to suppose that while an overlong syllable MAY be taken as the equivalent of a disyllable, this is not obligatory. Under this view, any disyllabic sequence, even if it includes an overlong syllable, may form a metrical foot. This option is available, since I represent overlength moraically and thus do not have to claim that overlong syllables are always monosyllabic feet. Thus I propose (37):

(37) An overlong (i.e. trimoraic) syllable may be treated as disyllabic.

The effects of this are less dramatic than it might seem, owing to the requirement (§ 3.9.1, § 5.6.2) that foot boundaries not occur within syllable boundaries. Because of this, the only overt consequence of (37) is that overlong syllables may occur as single feet.

As Prince points out, there is another sense in which overlong syllables act like disyllabic sequences in Estonian: the minimal word can be either two ordinary syllables or one overlong syllable. I posit that this constraint should be stated as follows: the minimal word is a maximal syllabic trochee. This is extended to overlong syllables by (37).

We observed in § 5.1.9 that the "syllabic" trochee in fact appears to be a more general structure, which may dominate a number of different terminal elements (e.g. moras, syllables, feet). One might speculate that overlong syllables in Estonian, and perhaps generally, have a kind of binary branching structure, along the lines of (38):

(38) $[_\sigma[\ \mu\ \mu\ ]\ [\ \mu\ ]\ ]_\sigma$

Rule (37) is intended to give the Estonian "syllabic" trochee the option here of treating the constituents [μ μ] and [μ] in (38) as its terminal nodes, for purposes of both foot construction and the minimal word constraint.

Overlong syllables must be excluded from weak position within a foot. Under the assumption adopted above in § 8.5.1, this will follow from the requirement that overlong syllables must occur initially within the phonological word.

We can now return to the stress pattern of overlong syllables. Where /=/ is treated as a disyllable, the analysis is just like Prince's; and thus derives the fact that /=/ (and only /=/) may be directly followed by stress (36a.i). Where /=/ is treated as a single syllable, the case of (36b.i) is derived: /≐ ⌣ ⌣/ consists of a disyllabic trochee, plus the stray light syllable permitted under weak local parsing. Treating /=/ as monosyllabic also permits /≐ –/ to be a foot, as in (36b.ii); and case (36a.ii) follows under either treatment of /=/.

What remains is case (36a.iii): /≐ –/ feet are excluded except when they occur immediately before another stress. The sequences in (39) are the ill-formed cases that must be excluded:

```
(39) a. * (×  . )        b. *(×  . )
        # =  –  #             =  –  ⌣
```

In its present state the analysis generates both.

My proposal for trimming back this overgeneration is based on the observation that /= –/ can also be parsed as two well-formed feet, namely /=/ + /–/. I propose that such parsing is enforced by rule, as in (40):

(40) **Reparsing** Reparse the contents of a foot, provided no stress clash is created.

The Reparsing rule would apply to (39a) as in (41). I assume that this form involves no clash, since /=/ is being counted as disyllabic.

```
(41) * (×  . )        (×)(×)
      # =  –  #      # =  –  #
```

Reparsing would also apply to (39b), creating the intermediate representation (42b). Since footing is persistent, the following stray light syllable is parsed into the preceding foot (§ 5.4.1), yielding (42c).

```
(42) a. *(×  . )         b. (×)(×)          c. (×)(×  . )
          =  –  ⌣    →       =  –  ⌣    →        =  –  ⌣
```

When the sequence /= – / occurs before a stress, Reparsing would create a clash and thus is blocked.

(43) (× .)(× . . . → *(×)(×)(× . . .
= – σ   = – σ

The input form is therefore retained. Note that / – / following /=/ CAN be stressed on the surface, but this would reflect a different initial parse that treated /=/ as disyllabic: /≐/ – σ/.

Although the Reparsing rule does not mention any syllable quantities in its structural description, this does not lead to overgeneration elsewhere in the system. For example, Reparsing cannot affect / –́ – /, since this would create clashing stresses. Illegal reparsing of initial /˘́ – / ˘ to ˘ / –́ ˘/ can be ruled out if we order Reparsing after End Rule Left: the main stress on the initial light syllable would then be protected by the Continuous Column Constraint (§ 3.4.2). Elsewhere, Reparsing creates only foot sequences that can be generated independently.

To summarize, the proposed analysis adds to the basic principles of metrical structure creation (29) the following: optional disyllabic interpretation of /=/ (37), and Reparsing (40). This suffices to generate the overlength data. In the representations in (44), R marks the forms to which Reparsing is applicable. Note that in all such cases, a different derivation can obtain the same result without application of this rule.

(44) **Words with Overlength**

a. (× .)(×)
= ˘ = ˘
árs:ti, árs:ti

b. (×)(×)
= –
vánkk:rìt⟨t⟩ R

c. (×)(× .) (× .)
= ˘ ˘ = ˘ ˘
káu:kèle, káu:kele

d. (×)(× .) (× .)(×) (×) (×)
= ˘ – = ˘ – = ˘ –
yǽl:kètes⟨t⟩, yǽl:ketès⟨t⟩, yǽl:ketès⟨t⟩

e. (×) (× .)
= – ˘
yúl:kètte R

f. (×)(× .) (× .)(×)
= – – = – –
hái:kùstes⟨t⟩, hái:kustès⟨t⟩

g. (×)(× .) (× .)(× .)(×) (× .)
= ˘ ˘ ˘ = ˘ ˘ ˘ = ˘ ˘ ˘
tő̈ös:tùsele, tő̈ös:tusèle, tő̈ös:tusèle

h. (× .)(× .) (×) (× .)
= ˘ – ˘ = ˘ – ˘
káh:tlevàile, káh:tlevàile

i. (×)(× .) (×) (× .)
= ˘ – – = ˘ – –
kɤ́h:kleyàttek⟨s⟩, kɤ́h:kleyàttek⟨s⟩

j. (×)(× .) (× .)(× .)
= – ˘ ˘ = – ˘ ˘
tő̈ös:tùstele, tő̈ös:tustèle

k. (×)(× .) (× .)(× .)
= – – ˘ = – – ˘
tő̈ös:tùstesse, tő̈ös:tustèsse

l. (×)(× .)(×) (× .) (× .)
= – – – = – – –
áu:sàimattèks, áu:saimàtteks

o. (× .)(× .) (×) (× .)
= ˘ – – ˘ = ˘ – – ˘
káu:kemàttesse, káu:kemàttesse

m. (×)(× .)(× .)(× .)(× .) (×) (× .)
= ⌣ ⌣ ⌣ ⌣ = ⌣ ⌣ ⌣ ⌣ = ⌣ ⌣ ⌣ ⌣

trúu:tùselèki, trúu:tusèleki, trúu:tusèleki

n. (×)(× .)(× .)(× .)(× .) (×) (× .)
= ⌣ – ⌣ ⌣ = ⌣ – ⌣ ⌣ = ⌣ – ⌣ ⌣

tɣ̌es:tùsettàki, tɣ̌es:tusèttaki, tɣ̌es:tusèttaki

p. (×)(× .)(× .) (× .)(× .)
= – ⌣ ⌣ ⌣ = – ⌣ ⌣ ⌣

kínt:lùstelèki R, *kínt:lustèleki*

q. (×)(× .)(× .) (× .)(× .)
= – – ⌣ ⌣ = – – ⌣ ⌣

káu:kèttessèki, káu:kettèsseki

To summarize the analysis, my basic foot construction rules simply translate Prince's account into the theory of weak local parsing. My account of stress in overlong syllables diverges somewhat more, being based on a moraic representation of overlength, variable treatment of overlong syllables, and Reparsing. This makes it possible to avoid the use of a syllable-counting rule like (26), which would go beyond the limits of rule locality assumed here. Where the analysis most strains the edges of the framework is in the rule of Reparsing (40). I see no better alternative to this rule. In defense of it one might note that it does not add any new structures to the basic inventory of metrical theory, but rather rearranges the string into different structures that exist independently.

I believe that the Estonian facts provide particularly strong support for the theory of weak local parsing. This arises from the natural way in which the theory distinguishes obligatory spacing of stresses from free variation. The minimum interstress distance in Estonian is associated with the trochaic foot template, whereas the variation in interstress distance is due to optional invocation of weak local parsing. This is a language-particular instantiation of the general argument given in § 8.4(c): stress assigned by weak local parsing commonly occurs in free variation with its strong local counterpart.

8.5.4 General Issues Related to Estonian Stress

The stress pattern of Estonian is useful for distinguishing between different theories of foot structure and of ternary alternation.

First, it appears that the ternary theory of HV, based on the amphibrach (§ 8.2.2), does not generalize to Estonian. Because the quantity-sensitive version of the HV amphibrach requires both of its weak positions to dominate light syllables, it cannot skip over the heavy syllables of Estonian that immediately follow a stress. To derive the Estonian facts, we would need a different kind of amphibrach, in which the first weak position must be a light syllable, but the second is quantitatively free:

(45) (. × .)
⌣ σ σ

Moreover, this amphibrach would have to be placed in free variation with an ordinary left-headed foot, to derive the optional cases of binary alternating stress.[4]

The issue is actually broader than this, because the hypothetical revised amphibrach of (45) would actually not be a possible foot template in HV's theory. To explain why this is so requires some background.

An influential idea of Prince 1976a (adopted e.g. in Halle and Vergnaud 1978; Hayes 1981; and here) is that syllable quantity affects stress indirectly: the templates for metrical feet specify which positions within the foot may be heavy or light. Given the templates that exist, this tends to place stress on heavy syllables. A contrary view, adopted by HV, is that the influence of quantity is direct, based on a rule that assigns a grid mark to all heavy syllables (HV 21–22); thus heavy syllables always get stressed. This is what makes (45) an impossible foot template for HV.

Estonian bears crucially on the weight-to-stress problem: although its stress rule is quantity-sensitive, it often skips over heavy syllables. This appears to be difficult to handle in HV's approach: heavy syllables must be projected on the grid, but in many cases must not be stressed. As far as I have been able to determine, appeals to a later rule of destressing would not solve the problem.

The mixed behavior of heavy syllables can be treated in the approach taken here and in Prince's work: the foot template of Estonian simply does not care about the quantity of the syllables it dominates, whereas the weak local parsing algorithm can skip over only light syllables. The indirect method of referring to syllable weight provides the flexibility needed to describe the Estonian pattern.

8.6 OTHER FINNO-UGRIC LANGUAGES

The Estonian stress system, minus the complications of overlength, is repeated in three related languages. Of these, Finnish and Karelian are closely related to Estonian, being neighboring languages or dialects of the Balto-Finnic group. Hungarian is genetically more distant, related at the level of Proto-Finno-Ugric (Collinder 1960).

The Finnish pattern (Sovijärvi 1956; Harms 1964; Carlson 1978) resembles that of Estonian, but with two differences. First, Finnish has no overlong syllables, and therefore lacks the rules pertinent to them. Second, in at least some varieties of Finnish, weak local parsing appears to be obligatory, so that the ternary option is always taken where this is possible. However, in Sovijärvi's

4. In addition, something must be said to avoid skipping over word-initial /˘/. Since this would follow straightforwardly from top-down stressing (§ 5.4.3), however, this should not be counted against the theory.

description, parsing appears to be optionally strong local. There are many complications in the Finnish system, related in part to morphology, and it is clear that further work needs to be done to develop a complete metrical analysis of Finnish stress.

For Karelian, Leskinen 1984 describes a system parallel to Finnish, with weak local parsing in free variation with strong local parsing. Final stress is avoided, which plausibly reflects persistent footing.

For Hungarian, the literature shows two patterns. The one described by Szinnyei 1912 appears to involve weak local parsing: "In independent words the main stress falls on the initial syllable. In addition, in longer words a secondary stress falls on the third and fifth syllables or (if the third syllable is light), the fourth and sixth, but never on the last" (p. 12, translated). This appears to be like the pattern of Finnic languages, with left-to-right construction of syllabic trochees and weak local parsing. The avoidance of final stress implies that /σ σ ⏑ σ/ words would be stressed /σ́ σ ⏑̀ σ/, with persistent parsing. Szinnyei's pattern is also described by Lotz 1939 (cited in Kerek 1971).

The pattern for Hungarian described by Balassa 1890 (cited in Kerek 1971); Hall 1938; and Sovijärvi 1956 is quantity-insensitive: main stress is initial, with secondary stresses on every other syllable thereafter. This would involve syllabic trochees, assigned with strong local parsing from left to right. (For secondary stress on final syllables, see § 5.1.8.2.) Hammond 1987 describes the same pattern, amplified with a colon layer (§ 5.5.1), assigning greater prominence to odd-numbered feet; the word layer again assigns initial main stress. Arany 1898 (cited in Kerek 1971) claims stress on the first and fifth syllables; plausibly he was recording only those stresses that are the strongest of their colon.

Assuming this variation represents a genuine dialect split, Hungarian supports my claim (§ 8.4) that related weak and strong local parsing systems are frequently found in close proximity.

Kerek's (1971) own account of Hungarian involves a kind of right-to-left Latin stress rule, with clash avoidance. If his analysis is correct it is a problem for the theory proposed here. Kerek does note data from other speakers that do not conform to his account, however.

8.7 SENTANI

Sentani is a language of the Sentani family, spoken in New Guinea. It is described in Cowan 1965. Other references on Sentani cited in Foley 1986, 240, are less detailed but basically corroborate Cowan's description. Sentani was noticed as a case of ternary stress by René Kager, and the analysis that follows is essentially the one provided by him in personal communication. Data from another dialect, and analysis in considerably greater depth, appear in Elenbaas 1992 and Elenbaas and Kager, forthcoming.

Main stress in Sentani falls on the final syllable of a word if it is heavy (/ – / = CVC; there is no phonemic vowel length). Otherwise, it falls on the penult (all data from Cowan, pp. 9–10, except as noted):

(46) a. /⏑ –́/ *falə́m* 'head'
b. /⏑́ ⏑/ *yóku* 'dog'
c. /–́ ⏑/ *kámbi* 'neck'
d. /⏑ ⏑́ ⏑/ *hokólo* 'young'
e. /⏑ –́ ⏑/ *ukə́wnə* 'he told him'

As in the cases of § 6.1.9 (129), this can be derived by right-to-left construction of moraic trochees, followed by End Rule Right. There are some cases of irregular final stress on a light syllable, e.g. *ifá* 'small canoe for men' C 10, which can be treated with pre-listed metrical structure.

Secondary stress follows a ternary pattern, which can be derived with leftward iterative footing under weak local parsing:

(47) a. /⏑̀ ⏑ ⏑ ⏑́ ⏑/ *ə̀dəkawále* 'I saw thee'
b. /⏑ ⏑̀ ⏑ ⏑ ⏑́ ⏑/ *adìləmihíbe* 'you two will collect them'
c. /⏑ –̀ ⏑ ⏑ ⏑́ ⏑/ *habə̀wnokokále* 'I struck him (aor.)'

As in Cayuvava (§ 8.2), the ternary alternation is accompanied by double upbeats in certain word shapes:

(48) a. /⏑ ⏑ –́/ *habakáy* 'tobacco'
b. /⏑ ⏑ ⏑̀ ⏑ ⏑ –́/ *adilə̀dəmihím* 'let me collect them'

The rules stated below account for this just as in Cayuvava, by making parsing non-persistent.

As expected in a quantity-sensitive system, ternary alternation is replaced by binary where ternary stressing would skip across a heavy syllable:

(49) a. /⏑ –̀ ⏑ ⏑́ ⏑/ *habə̀wdokóke* 'he hit me (aor.)'
b. /⏑ –̀ ⏑ –́ ⏑/ *ənàynewə́nde* 'they (pl.) will go tell him'

A further important generalization is that / – / in pretonic position is always stressless:

(50) a. /– –́/ *ankɛ́y* 'ear'
b. /– ⏑́ ⏑/ *howbóke* 'he killed (something)'
c. /– –́ ⏑/ *honkə́wnə* 'he burnt him'
d. /⏑ – –́ ⏑/ *honəmbónde* 'he will kill (something) for him'
e. /⏑ –̀ ⏑ – –́ ⏑/ *ənàynəkɛnsínde* 'they (pl.) will throw it away'

To account for this, the analysis will let such syllables receive stress at an intermediate stage, then trim this back with a pretonic destressing rule.

The full analysis, then, is as in (51):

(51) a. **Foot Construction** Form moraic trochees from right to left.
Parsing is weak local and non-persistent.
Degenerate feet are allowed in strong position.

b. **Word Layer Construction** End Rule Right

c. **Destressing** × → Ø / ____ ×

The weak ban on degenerate feet is adopted, since Sentani tolerates degenerate-size words: *fa* 'child', *i* 'fire' C 5.

Applied to the forms in (46)–(49), Foot Construction and Word Layer Construction yield the outputs in (52)–(55):

```
(52) a. (  ×)      b. (×   )     c. (×    )     d. (  ×  )     e. (  ×     )
        (×)           (× .)         (×)              (× .)           (×)
        ⏑ –           ⏑ ⏑           –   ⏑          ⏑ ⏑ ⏑          ⏑ –   ⏑
        falə́m         yóku          kámbi          hokólo         ukə́wnə

(53) a. (        × )     b. (          × )     c. (              × )
        (× .)  (× .)          (× .)  (× .)           (×)       (× .)
        ⏑ ⏑ ⏑  ⏑ ⏑          ⏑ ⏑⏑  ⏑ ⏑ ⏑            ⏑ –    ⏑ ⏑ ⏑ ⏑
        ə̀dəkawále            adìləmihíbe            habə́wnokokále

(54) a. (     × )     b. (           ×)
           (× )           (×  .) (×)
        ⏑ ⏑ –           ⏑ ⏑⏑ ⏑  ⏑ –
        habakáy         adilə̀dəmihím

(55) a. (        × )     b. (          ×    )
          (×)   (× .)          (×)   (×)
        ⏑ –   ⏑ ⏑ ⏑          ⏑ –  ⏑ –  ⏑
        habə̀wdokóke          ənàynewə́nde
```

The double-upbeat forms in (54) undergo derivations analogous to § 8.2 (12). The non-persistence of foot parsing is illustrated here not only by the double-upbeat forms, but also by (53c): here, the syllable *ko* is skipped over by weak local parsing, while *no* is skipped over by the Priority Clause (§ 5.1.6). Since parsing is non-persistent, the stressless /⏑ ⏑/ sequence remains on the surface.

The derivations in (56), for the forms of (50), invoke Destressing (51c):

```
(56) a. (   ×)      b. (   × )      c. (   ×    )
        (×) (×)        (×) (× .)       (×) (×)
        –   –          –   ⏑ ⏑         –   –   ⏑
        ankɛy          howboke         honkəwnə

        (   ×)         (   × )         (   ×    )
            (×)            (× .)           (×)
        –   –          –   ⏑ ⏑         –   –   ⏑
        ankɛ́y          howbóke         honkə́wnə
```

```
d. (      ×    )      e. (            ×    )
      (×) (×)              (×)    (×) (×)
   ᴗ  –   –  ᴗ            ᴗ  –  ᴗ  –  –  ᴗ
   honəmbonde              ənaynəkɛnsinde

   (      ×    )         (            ×    )
        (×)                 (×)       (×)
   ᴗ  –   –  ᴗ            ᴗ  –  ᴗ  –  –  ᴗ
   honəmbónde              ənàynəkɛnsínde
```

Since stressing is not persistent, pretonic heavy syllables remain stressless, even though they could in principle constitute moraic trochees.

A single additional form eludes the analysis: *handəbókə* 'we (pl.) killed (something)' C 9, which should bear an initial secondary stress. In light of parallel forms like (47a) and (49a), it is plausible that Cowan simply omitted secondary stress here; if not, then clearly the theory and analysis would need serious revision. In the data of Elenbaas 1992, 48, parallel forms do receive initial secondary stress.

Cowan notes that /ə/ in an open penult tends to lose stress to a preceding heavy syllable: *wə́wnəle* 'thou wilt say to him' C 9. For these cases, we might posit an optional destressing rule, sensitive to prominence (§ 7.1), and ordered before Word Layer Construction. Elenbaas notes further phenomena involving schwa, which appear problematic for the theory of weight presented in § 7.1.

Summing up, for purposes of stress Sentani is strikingly like Cayuvava (§ 8.2), having both ternary intervals (47) and double upbeats (48). In the theory adopted here, the two languages share weak local parsing, trochaic feet, and non-persistent stressing.

8.8 PACIFIC YUPIK

Pacific Yupik (also called Alutiiq) is the southernmost of the Yupik languages. Its dauntingly complex prosody is described and analyzed by Jeff Leer (1985a, 1985b, 1985c, 1989). Earlier metrical analyses other than Leer's include Rice 1988, 1989, 1990, 1992; Halle 1990; Hammond 1990b; Hewitt 1991; and Kager (1993).

Pacific Yupik has two major dialects, Koniag and Chugach. They share much of the prosody of other Yupik languages (§ 6.3.8): iambs assigned from left to right, Iambic Lengthening, and special treatment of syllables that precede CVV. Pacific Yupik differs in having ternary stress, which takes two forms: a relatively simple pattern in the Chugach dialect, and a pattern in Koniag which is found only under particular morphological circumstances, laid out in Leer 1985a, 118–28. I focus on Chugach here.

Syllable quantity: CVV is heavy, CV is light, and CVC is treated as in Norton Sound (§ 6.3.8.2), with initial CVC heavy and noninitial CVC light. I adopt the same analysis, in which Weight by Position applies only in initial syllables.

Pacific Yupik has geminates that fail to make weight. For these, I assume the analysis of § 7.3.2, in which moras occur on two layers. However, since this issue is relatively independent of the other problems of Pacific Yupik prosody, I will use representations below with just one layer. This corresponds to the higher moraic layer in a full analysis.

8.8.1 Words with All Light Syllables

The stress pattern of words with all light syllables is as follows: the second syllable is stressed, and in words of sufficient length, additional stresses fall every third syllable thereafter. Nonfinal stressed open syllables undergo iambic vowel lengthening, shown below in the phonetic forms. Page references below are to Leer 1985a, 1985b, and 1985c.

(57)

a. /⏑ ⏑́ ⏑/	*atáka* [atá:ka]	
	'my father'	L 116
b. /⏑ ⏑́ ⏑ ⏑́/	*akútamə́k* [akú:tamə́k]	
	'(kind of food) (abl. sg.)'	L 84
c. /⏑ ⏑́ ⏑ ⏑ ⏑́/	*atúquniki* [atú:quniki]	
	'if he (refl.) uses them'	L 113
d. /⏑ ⏑́ ⏑ ⏑ ⏑́ ⏑/	*pisúqutaqúni* [pisú:qutaqú:ni]	
	'if he (refl.) is going to hunt'	L 113
e. /⏑ ⏑́ ⏑ ⏑ ⏑́ ⏑ ⏑́/	*maŋáχsuqutáquní* [maŋáχsuqutá:quní]	
	'if he (refl.) is going to hunt porpoise'	L 113

This pattern has a straightforward account under the theory assumed here:

(58) **Foot Construction** Form iambs from left to right, under weak local parsing; footing is persistent.

(59) **Iambic Lengthening**

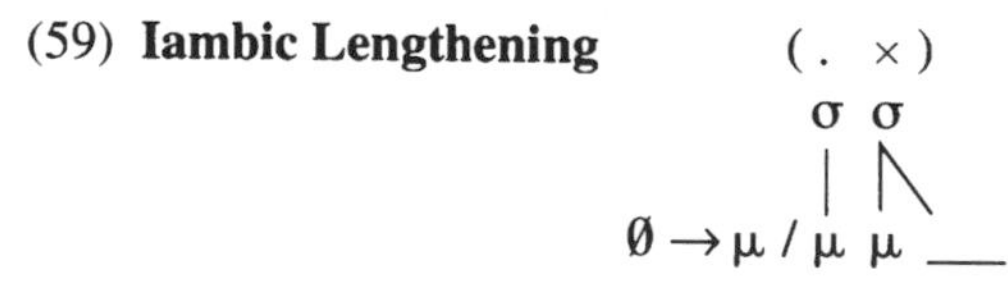

The analysis does not state which form of the ban on degenerate feet (§ 5.1.1), if any, is in effect; evidence is lacking to decide this issue (§ 8.8.5(b)). Iambic Lengthening is stated as a fully general rule; its effects are often trimmed back, however, by two shortening rules, which apply after stress:

(60) a. **Compression** (L 88–89) VV → V / ____ C $]_\sigma$ (§ 7.3.2)

b. **Final Shortening** (L 88–89) VV → V / ____ #

These rules give rise to the representations in (61):

(61) a. (. ×) b. (. ×)(. ×) c. (. ×) (. ×)

˘ ˘ ˘   ˘ ˘ ˘ ˘   ˘ ˘ ˘ ˘ ˘

atáka *akútamə́k* *atúqunikí*

[áː] [úː] [úː]

d. (. ×) (. ×) e. (. ×) (. ×)(. ×)

˘ ˘ ˘ ˘ ˘ ˘   ˘ ˘ ˘ ˘ ˘ ˘ ˘

pisúqutaqúni *maŋáχsuqutáquní*

[úː] [úː] [áː]

Note that where the ternary count would leave a string of two stressless syllables at the end, the final syllable is stressed (61b, e), which I account for by making footing persistent (§ 8.2.1). This assumption turns out to be somewhat problematic, however; see § 8.8.4. As Leer 1989 and Rice 1989, 1990 note, the presence of a stress on the final syllable of forms like (61b, e) is a counterexample to the Recoverability Condition of HV (§ 8.2.2), since the grid pattern is compatible with two foot bracketings: /akúta/mə́k/, /akú/tamə́k/.

8.8.2 Heavy Syllables

In (62)–(66) are representative words containing heavy syllables; recall that these consist of either CVC initially or CVV anywhere. In the phonetic forms, some segments are transcribed as half-long (Cˑ, Vˑ)); these are discussed below. Stress is not marked in the quantity schemata except in straightforward cases.

(62) **Initial Heavy Syllables**

a. /–́ ˘/	*pínka* 'mine (pl.)'	L 110
b. /–́ ˘ ˘́/	*ánŋaqá* 'my older brother'	L 110
c. /–́ ˘ ˘ ˘́/	*ánčiqukút* 'we'll go out'	L 84
d. /–́ ˘ ˘ ˘́ ˘/	*náːqumalúku* [náːqumalúːku] 'apparently reading it'	L 89
e. /–́ ˘ ˘ ˘́ ˘ ˘́/	*átmakutáχtutə́n* 'you're going to backpack'	L 116
f. /–́ ˘ ˘ ˘́ ˘ ˘ ˘́/	*átsaχsuqútaquní* [átsaχsuqúːtaquní] 'if he (refl.) is going to get berries'	L 113
g. /–́ ˘ ˘ ˘́ ˘ ˘ ˘́ ˘/	*tánnəʁliχsúqutaqúni* [tánnəʁliχsúːqutaqúːni] 'if he (refl.) is going to hunt bear'	L 113
h. /–́ ˘ ˘ ˘́ ˘ ˘ ˘́ ˘ ˘ ˘́ ˘/	*kúmlačiwíliyaqútaquníki* [kúmlačiwíːliyaqúːtaquníːki] 'if he (refl.) is going to undertake constructing a freezer for them'	L 113

(63) **Posttonic Heavy Syllables**

a. /–́ –́/ *ná:qá:* [ná:q'á] 'she's reading it' L 115
b. /–́ –́ ⏑/ *kálmá:nuq* [kálm'á:nuq] 'pocket' L 103

(64) **Heavy Syllables After One Light**

a. /⏑ –/ *ati:* [áttí] 'his father' L 110
nuyai [núyyái] 'her hair' L 102
b. /⏑ – ⏑/ *časa:ʁi* [čássá:i] 'his clock' L 96
(see L 96 for [ʁ] deletion)
c. /–́ ⏑ –́/ *ánčiquá* [ánčiq'uá] 'I'll go out' L 84
d. /–́ ⏑ –́ ⏑/ *íqɬukí:ŋa* [íqɬuk'í:ŋa] 'she lied to me' L 112
e. /⏑ ⏑́ ⏑ –́/ *uxáčimá:n* [uxá:čim'án] 'you must be good at it' L 112

(65) **Heavy Syllables After Two Lights**

a. /– ⏑ ⏑ –/ *atmaxčiqua* [átmaxčí'q'uá] 'I will backpack' L 115
na:mačiqua [ná:mačí'q'uá] 'I will suffice' L 84
ankutaχtua [ánkutáχt'uá] 'I'm going to go out' L 116
b. /⏑ ⏑ –/ *muluku:t* [mulú'k'út] 'if you take a long time L 87

(66) **Heavy Syllables After Three Lights**

/⏑ ⏑ ⏑ ⏑ ⏑ –/ *ulutəkutaʁa:* [ulú:təkutá'ʁ'á] 'he is going to watch her' L 104

Below, I go through the cases, amplifying the analysis along the way.

(A) Initial Heavy Syllables (62). These follow the rules given already: initial /–/ attracts stress, and a ternary pattern is derived for the sequence of following lights.

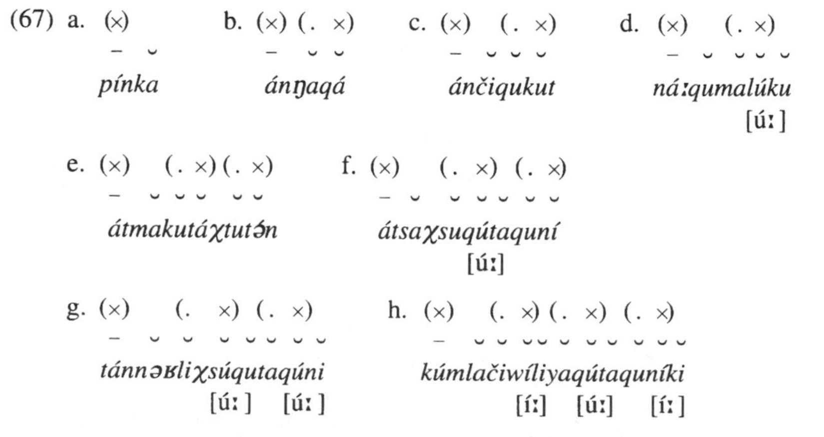

(B) Posttonic Heavy Syllables (63). These receive stress as predicted: /–/ cannot be skipped over in weak local parsing, nor can it form the weak syllable of an iamb. Thus we get consecutive stresses:

(68) (×) (×)
– – ˘
kálm'á:nuq

For discussion of half-length on /m/, see (75a).

(C) HEAVY SYLLABLES IN SECOND POSITION (64a, b). Here, Pacific Yupik behaves just like Central Alaskan Yupik (§ 6.3.8.2): the syllable that precedes CVV is stressed and made heavy by gemination; /nuyai/ → [núyyái]. The rule responsible for this is formulated in (72).

(D) HEAVY SYLLABLES /σ́ ˘ ˘ ___ (65a). Our default expectation for such words is shown in (69):

(69) (×) (. ×)
– ˘ ˘ –
átmaxčiquá → *[átmaxčiquá]

That is, weak local parsing would be expected to skip over the light syllable *max,* then form a canonical iamb on *čiqua.* The actual outcome, however, is [átmaxčí'q'uá].

My analysis deviates from earlier proposals in supposing that the third syllable in such examples actually is heavy at the stage where the stress rule applies. That is, the quantitative structure is as in (70):

(70) (×) (. ×) (×)
– ˘ – –
átmaxčí'q'uá

Under such a quantitative pattern, surface stress is as expected.

The segmental content of the heavy syllable in third position is quite interesting: according to Leer, we get a half-long consonant, preceded by a half-long vowel (transcribed (V:), but noted in L 87 as shorter). Together, they form a heavy syllable. This can be derived by inserting an empty mora to the left of CVV syllables, prior to foot construction, and linking it to both neighboring segments:

(71) 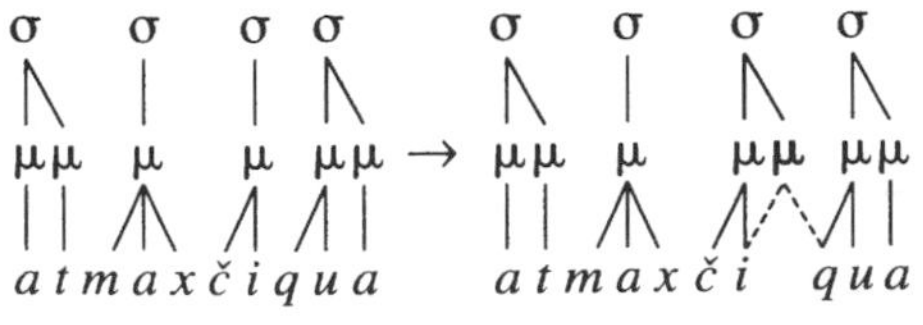 σ σ σ σ → σ σ σ σ
μμ μ μ μμ → μμ μ μμ μμ
a t m a x č i q u a → *a t m a x č i q u a*

In this view, semilong [i'] bears "1 1/2" moras, while semilong /q/ occupies half a mora plus the onset position. In contrast, a fully long vowel would occupy two moras, and a full geminate would occupy a full mora plus the onset.

The notion of shared moras has been argued for in the case of Sukuma by Maddieson and Ladefoged (1993).

The rule that inserts the mora in (71) is a more complex version of Pre-Long Strengthening, a rule we have seen before in Central Alaskan Yupik (§ 6.3.8.2 (235)).

(72) **Pre-Long Strengthening (Pacific Yupik)**

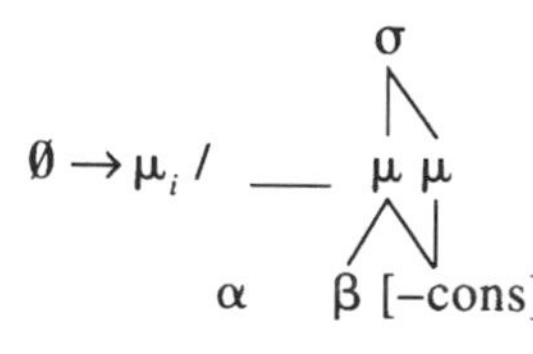

Conditions: (a) μ_i links to β where basic syllabification permits
(b) Otherwise, μ_i links to both α and β.

Pre-Long Strengthening is formulated to cover both the case of (71) and that of (64a) /nuyai/ → [núyyái]. When the mora is inserted after an INITIAL CV syllable, the syllabification conventions permit it to be filled entirely by consonant spreading, since CVC is a legitimate heavy syllable in initial position:

(73)
```
σ  σ          σ    σ
|  |\         |\   |\
μ  μμ   →    μ μ  μμ
/\ /\|        /\ \/\|
n u y a i    n ú  y a i   [núyyái]
```

Subsequent footing yields the correct stress pattern. Otherwise, Pre-Long Strengthening takes a non–structure-preserving course, dividing the inserted mora between consonant and vowel, as in (71). The derivation begun there is completed in (74):

(74) a.
```
(×)       (× )(× )
σ    σ    σ    σ
μμ   μ    μμ   μμ
a t max č i  q u a
```
Foot Construction (58)
(/max/ skipped by weak local parsing)

b.
```
(×) (.    ×) (× )
σ    σ    σ    σ
μμ   μ    μμ   μμ
a t max č i  q u a   →  [átmaxčí·q·uá]
```
persistent footing
(incorporation of /max/ into canonical iamb)

In § 6.3.8.2, I noted that Pre-Long Strengthening in Central Alaskan Yupik has a strong functional motivation: it distinguishes underlying long vowels from underlying short vowels that have undergone Iambic Lengthening. In Pacific Yupik, the functional motivation is even stronger, since this language obscures underlying vowel length not only with Iambic Lengthening, but also with Compression and Final Shortening (60a, b).

Pre-Long Strengthening would also be expected to apply after closed and long-voweled syllables, as shown in (75) for (65a) and (63a):

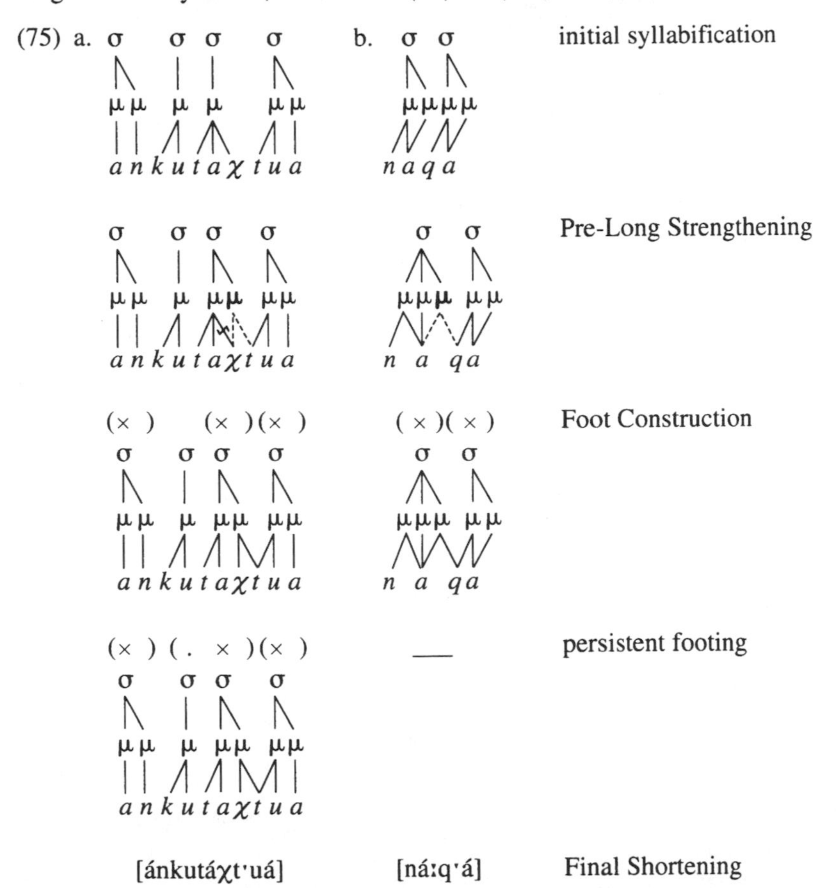

[ánkutáχt·uá] [náːq·á] Final Shortening

In (75a), the inserted mora is shared between two consonants, rather than between a consonant and a vowel. This has implications for the phonetic duration of [χ] which should be checked, but data are unavailable. In cases like (75b), where CVV follows a heavy syllable, the rules would derive a trimoraic syllable; from Leer's description (1985a, 87), it seems that such syllables are most likely adjusted to bimoraic, with loss of the medial mora: [ná·q·á].

(E) Heavy Syllables /# ˘ ˘ ___ (65b). These forms are derived in the same way as in (d) above:

(76)

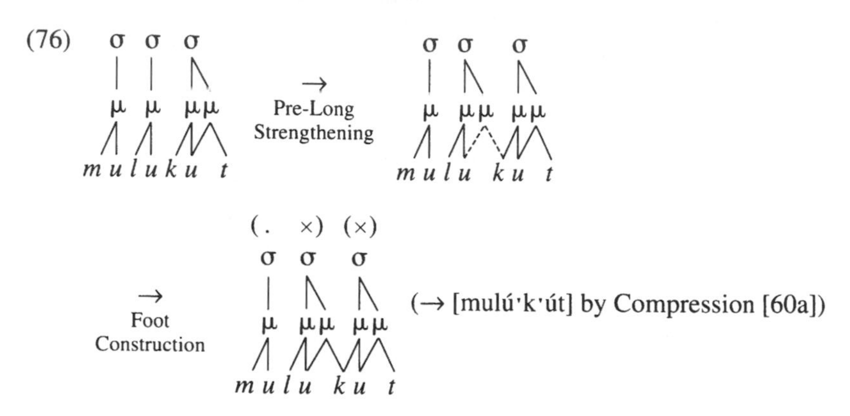

In these cases, the second syllable would be stressed in any event, since weak local parsing does not skip over an initial light syllable.

(F) Heavy Syllables /◡ ◡ ◡ ___ (66). These also follow from the rules given so far: the pre-CVV syllable is made heavy by Pre-Long Strengthening, and weak local parsing derives the correct output:

(77)

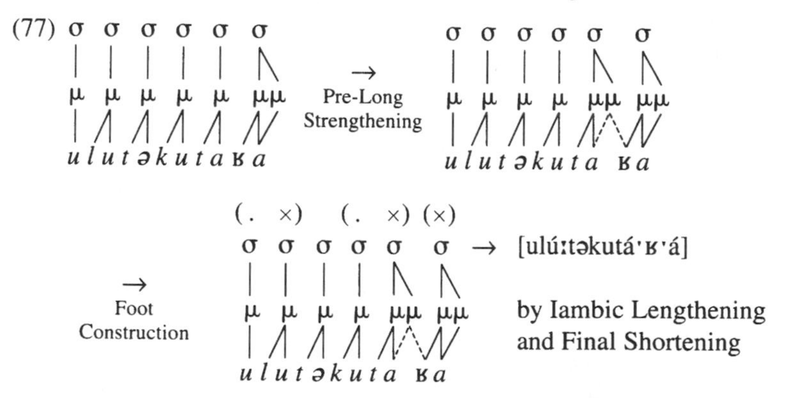

(G) Heavy Syllables / σ́ ◡ ___ (64c–e). The rules developed so far would apply as in (78):

(78)

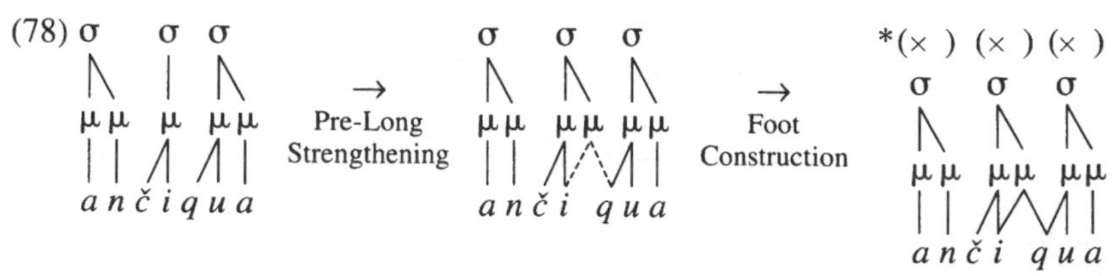

The correct form in fact bears stress only on the initial and final syllables. Recall that in General Central Yupik, CVC syllables that occur in the context

/× ___ × undergo a rule of Destressing in Clash (§ 6.3.8.3 (246)). I posit essentially the same thing for Pacific Yupik:

(79) **Destressing in Clash**
Remove the first mora from CVC in the context / × ___ ×

Applied to the output of (78), this derives (80a). Since /⏑/ is a degenerate foot in weak position, it is removed, yielding (80b). For the moment I assume that the stray syllable is adjoined to the following foot by persistent footing (80c); see, however, further discussion in § 8.8.4.

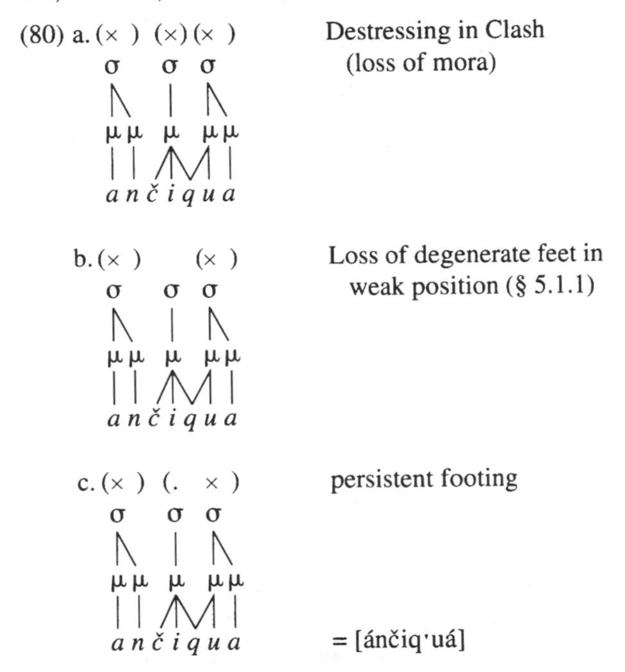

Destressing in Clash is formulated to remove the first mora of the syllable, so that the syllable-final consonant will remain ambisyllabic, hence half-long. A less stipulative alternative would be to rely on richer representations of constituency (e.g. Hayes 1990), whereby the consonant belongs to both syllable and mora. I will not explore this possibility here.

This completes the inventory of stress patterns. To summarize, the analysis involves essentially the same rules as General Central Yupik (§ 6.3.8): left-to-right iambs, Pre-Long Strengthening, Destressing in Clash, and Iambic Lengthening. Crucial differences are that parsing is weak local, and that Pre-Long Strengthening applies before stress and distributes its moras differently.

8.8.3 Phonetic Fortition

Leer's intensive description actually isolates FOUR surface degrees of consonant length (1985a, 86). In addition to short, half-long, and geminate consonants, Leer finds that there are consonants that are longer than short, but shorter than half-long; these he transcribes with the same [⨥C] notation he uses for half-long consonants. To keep things straight, (81) gives the descriptive (N.B. not analytical) labels to be used, placed under Leer's:

(81)				
Leer 1985a	lenis [C]	fortis [⨥C]	fortis [⨥C]	geminate [Cː]
This Book	[1 long] [C]	[2 long] [C]	[3 long] [Cˑ]	[4 long] [Cː]

Such fine distinctions are not always audible, but the four-way hierarchy can be inferred fairly clearly from phonological diagnostics. Of the four categories, only the distinction between surface [4 long] and all others is phonemic, expressed underlying as gemination. The distribution of [1–3 long] is as follows: [3 long] consonants occur / ___ VV (i.e. the environment of Pre-Long Strengthening (72)), [2 long] consonants occur foot-initially where the foot is not /CVV/, and [1 long] consonants occur elsewhere. Differences of length are accompanied by differences in voicing and fortition.

I follow Leer in assuming that [2 long] consonants are derived by a phonetic rule, stated in rough form in (82):

(82) **Phonetic Foot-Initial Lengthening**
Realize a non-moraic foot-initial consonant as slightly longer and more fortis.

As we will see below, in Pacific Yupik every CVV syllable begins a foot. Because of this, Leer treats [3 long] consonants as derived by a version of (82); presumably, the rule he has in mind would have an extra proviso to assign more duration to foot-initial consonants that precede VV. Since Leer treats [2 long] and [3 long] consonants in parallel fashion, he transcribes them the same.

The analysis proposed here divides up the pie somewhat differently: I follow Leer in the case of [2 long] consonants, but treat [3 long] consonants as semimoraic, having a different syllable structure as a result of Pre-Long Strengthening (72). Kager (1993) suggests that this may be a defect of my analysis, in that it splits up a single phenomenon into two rules. Below, I attempt to answer this objection.

Here are the ways in which [2 long] and [3 long] consonants are treated differently in Pacific Yupik: (a) Fricative Dropping: /ɣ/ and /ʁ/ are deleted in certain dialects when they are [1 long] or [2 long], but not when they are [3 long] or [4 long] (Leer 1985a, 93–94). (b) Fricative Devoicing: Various dialects devoice fricatives that are [3 long] or [4 long], but not [1 long] or [2 long] (Leer 1985a, 101). (c) Glide Hardening: /y, w/ are realized as [$ɣ^j$, $ɣ^w$] when they are [3 long] or [4 long], but not when they are [1 long] or [2 long] (Leer 1985a, 86). The general picture seems to be that the third degree of length behaves like the fourth degree, and not like the second. In my analysis this makes sense, since the [3 long] and [4 long] consonants are the ones that have a similar structural representation, in which the consonant is linked to a mora. The [2 long] consonants pattern differently because their length is not structurally represented, but derived by phonetic rule.

This argument can be taken further: for the three rules mentioned above, it appears from Leer's description that SYLLABLE-FINAL consonants are treated just like [3 long] and [4 long] consonants; that is, they resist Fricative Dropping and undergo Fricative Devoicing and Glide Hardening. The overall generalization, then, is that the language treats WEIGHT-BEARING consonants differently from consonants that are strictly in the onset (Steriade 1988a; for coda weight in stressless CVC, see § 7.3.2). This generalization can be made only if we represent [3 long] consonants moraically, as proposed here.

8.8.4 Fortition, Foot Boundaries, and Persistent Stressing

The existence of Phonetic Foot-Initial Lengthening (82) provides a diagnostic for foot boundaries. This gives rise to a problem with the analysis, pointed out by Leer (p.c.) and by Kager (1993). In (80c), repeated as (83a), the /č/ of *ánčiq'uá* is [1 long] and not [2 long], even though in my analysis it is foot-initial and should undergo Phonetic Foot-Initial Lengthening. The correct structure is (83b), with the medial syllable left stray.

```
(83) a. *(×  ) (.   × )    b.(×  )      (×  )
          σ     σ  σ         σ     σ  σ
          |\    |  |\        |\    |  |\
          μ μ   μ  μ μ       μ μ   μ  μ μ
          | |  /|\/  |       | |  /|\/  |
          a n č i q u a      a n č i q u a
```

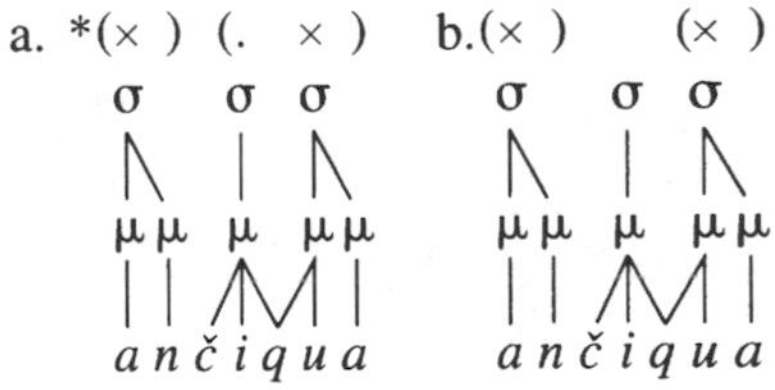

The structure of (83b) is also supported by the pitch patterns of these forms, discussed in § 8.8.5(f).

It is plausible that the derivation of (80) goes wrong in its third stage, where the medial syllable is adjoined to the following foot. To remedy this, I propose that footing in Pacific Yupik is not fully persistent: while at the initial stage of footing it may affect stray syllables left behind by weak local parsing (see (61b) and (75a)), it does not intervene to adjust metrical structure to the outcome of segmental rules. To allow this additional form of persistence is clearly a weakening of the theory, but appears to be empirically necessary.

8.8.5 Further Phenomena in Pacific Yupik

The above covers the central aspects of Pacific Yupik stress. Some further phenomena are dealt with below.

(a) Pacific Yupik includes patterns that evoke the issue of "Phrase-Final Destressing" in Central Alaskan Yupik (§ 6.3.8.8). The evidence that this is an intonational phenomenon, rather than being due to an actual destressing rule, seems somewhat stronger for Pacific Yupik; see Leer 1985a, 91–92.

(b) The rule of Final Shortening (60b) creates apparent degenerate feet, as in (77) (cf. Leer 1985a, 117). Whether this can be rationalized under the theory (e.g. by positing a right-strong word layer) is not clear. Information on the realization of such feet in non–phrase-final position might clarify this issue, since in phrase-final position one could argue that the degenerate foot has been lost.

(c) Pacific Yupik appears to show a more limited form of the phenomenon discussed for Central Alaskan Yupik in § 6.3.8.6, whereby stressed schwa is deleted, with leftward migration of stress within the foot (Leer 1985a, 109). There is also a complex set of rules that devoice schwa. When stressed schwa is devoiced, the stress migrates leftward in the same way as for deleted schwa (Leer 1985a, 105, 115).

Leer takes a quite different approach to schwa deletion and devoicing, in which the stress rules are greatly complicated precisely to avoid stressing deletable or devoiceable schwa in the first place. In favor of Leer's approach, however, is the existence of a small number of cases that are puzzling for the stress shift analysis. In particular, after a syllable that undergoes devoicing in a noninitial foot, we get strong instead of weak local parsing: /xłí:palisqə́xku-náku/ → [xłí:palísqə̥xkuná:ku] 'telling him not to make bread' L 115; also Leer (1989, exx. (26), (27), (29)). Assuming an analysis in which devoicing follows foot construction, foot parsing must somehow be able to look ahead to the subsequent devoicing of schwa. A similar interaction is described in Leer 1985a, 132.

(d) Binary rather than ternary alternation is required if a foot would cross a clitic boundary (Leer 1985a, 112–13). To analyze this, I suggest that as in Central Alaskan Yupik (§ 6.3.8.7), stress applies cyclically to the pre-clitic domain. This forces binary foot construction on the two syllables preceding the clitic:

(84) /atuquni-mi/ 'what if he (refl.) sings?' L 113

(. ×)(. ×) ˘ ˘ ˘ ˘ *atúquní*	First cycle: Foot Construction (persistent)	
(. ×)(. ×) ˘ ˘ ˘ ˘ ˘ *atuquni-mi*	Second cycle:	affixation of *-mi* Foot Construction inapplicable
[atú:quní:mi]	Iambic Lengthening	

(e) A few borrowed words involve stress assigned by lexically idiosyncratic strong local parsing, such as /ukkulutaq/ → [úkkulú:taq] 'garden', from Russian *ogoród* L 116. In the Koniag dialect, variation between weak and strong local parsing is pervasive, governed by morphological factors.

(f) Foot structure has interesting pitch correlates, discussed in Leer 1985a,

90–93; Rice 1988, 1989; and Hewitt 1991. Roughly, in the Chugach dialect discussed here, H tone is assigned to foot heads, L tone to nonfinal unfooted syllables, and the initial syllables of disyllabic iambs are unspecified, receiving interpolative pitch in the sense of Pierrehumbert 1980. A H adjacent on the tonal tier to a preceding H is dissimilated to H^+ (extra H); this rule applies iteratively from left to right. The Koniag dialect has different rules: it lacks H Dissimilation, and the tone it assigns to unfooted syllables is H, sometimes even H^+. These tones, together with fortition of the initial consonant of the following foot, arguably account for the quasi-stressed auditory impression created by stray syllables in Koniag (Leer 1985a, 91–92).

8.8.6 Other Approaches; Summary

The central difficulty in analyzing Pacific Yupik metrically is that of ensuring that there will be a foot boundary before all CVV syllables. Leer's (1985a, 1985c) approach is to forbid /⏑ –́/ feet at the stage of the initial parse. This means that when a /⏑ ⏑ –/ sequence is encountered, the first /⏑/ cannot be skipped across, since this would leave an illegal /⏑ –/ sequence left over to form a foot. Thus the initial /⏑ ⏑/ sequence is footed instead. Halle 1990 and Hewitt 1991 also follow this strategy, implemented in different ways.

I have already stated an objection to this in § 6.3.8.2: if foot structure is an organizing principle of phonology, it is odd that Pacific Yupik should have Iambic Lengthening, given that it creates putatively illegal /⏑ –́/ feet. Moreover, introducing the /⏑ ⏑́/ type of even iamb arguably expands the universal inventory of metrical structures beyond what is necessary.

These objections have less force when applied to the analysis of Kager (1993), which is based on a quite different general theory. Kager's analysis (inspired in part by Rice 1988, 1990) explicitly distinguishes between parsing feet and surface feet, where the latter are created only by later processes. In the case of iambs, the only quantity-sensitive parsing foot that is permitted under the theory is /⏑ ⏑́/, whereas /⏑ –́/ may occur as a surface foot. Thus for Kager the limitation to an /⏑ ⏑́/ template in the initial parse is a principled one.

It is not clear whether Kager's theory of iambic stress is sufficiently general. Winnebago, discussed in the following section, is, like Pacific Yupik, a ternary language that assigns iambs from left to right. In Winnebago, /⏑ ⏑ –/ sequences must work in the opposite way to Pacific Yupik, with a foot formed on the /⏑ –/ substring (§ 8.9.2 (96k, m)). Thus for Winnebago, a general prohibition on parsing feet of the form /⏑ –́/ is a handicap. In the theory assumed here, Winnebago is a straightforward case; it differs from Pacific Yupik only in that it lacks a rule of Pre-Long Strengthening.

To sum up, I have analyzed the Chugach dialect of Pacific Yupik as an example of weak local parsing. In the one set of cases where the pattern deviates from expectations (§ 8.8.2(d)), I provide an alternative account, in which [3 long] segments are represented as semi-moraic. This account is justified by the fact that [3 long] segments pattern like other weight-bearing segments in their phonological behavior (§ 8.8.3).

My claim that corresponding weak and strong local parsing systems characteristically occur in close proximity (§ 8.4) is abundantly confirmed by the data, both within individual Pacific Yupik dialects and in comparing across Yupik languages. Pacific Yupik also appears to require a weakening of the theory, in allowing foot parsing to be specified as persistent only at the initial stage of footing.

8.9 WINNEBAGO

Winnebago is a Siouan language described and analyzed by Susman 1943; Miner 1979, 1989; Hale and White Eagle 1980; Hale 1985; and HV. The language has long been of interest to metrical theory. Hale and White Eagle analyzed Winnebago in an early metrical framework, proposing a general convention that governs the interaction of epenthesis and metrical structure. In HV's version of their analysis, this convention is adopted and extended as the Domino Condition. Halle 1990 returns to the Winnebago issues, addressing whether single syllables may be split between metrical feet.

By following the views of Miner 1979, 1989, it appears possible to formulate an account which both improves on the empirical predictions of earlier metrical analyses, and invokes only the theoretical apparatus assumed in this book. This account views Winnebago as a pitch accent language, having tonal as well as metrical rules. As we will see, Winnebago also forms a (problematic) case of iambic ternary alternation.

8.9.1 Basic Data

The data below are taken from Susman 1943 (hereafter S); Miner 1979 (hereafter M), 1981, 1989; Hale and White Eagle 1980 (hereafter HWE); White Eagle 1982; HV; and Halle 1990. They are organized by syllable count and weight; in Winnebago, /⏑/ = CV(C) and / – / = (C)VV(C). Nasalized vowels are transcribed /Ṿ/. The first data set consists of forms that involve no applications of the epenthesis rule known as Dorsey's Law, discussed below.

(85) **Monosyllables**

/ –́ /	*wą́:k*	'man'	M 27
	čí:	'house'	HWE 130

(86) **All Lights**

a. /⏑ ⏑́/	*hižą́*	'one'	M 28
	wajé	'dress'	HWE 118
b. /⏑ ⏑ ⏑́/	*hotaxí*	'expose to smoke'	M 28
	waγįγį́	'ball'	HWE 118

c. /˘ ˘ ˘́ ˘/	*haračábra*	'the taste'	M 28
	hočįčį́nįk	'boy'	HWE 118
d. /˘ ˘ ˘́ ˘ ˘́/	*hokiwároké*	'swing (n.)'	M 28
e. /˘ ˘ ˘́ ˘ ˘ ˘́/	*hokiwároroké*	'swing (v.intr.)'	Miner 1981, 342

(87) **Initial Heavy Syllables**

a. /– ˘́/	*ho:čą́k*	'Winnebago'	M 27
	mą:táč	'promise (1 sg.)'	HWE 127
b. /– ˘́ ˘/	*ho:čą́gra*	'the Winnebago'	M 27
	wa:kítʔe	'speak to (1 sg.)'	HWE 121
c. /– ˘́ ˘ ˘́/	*waipéresgá*	'linen'	M 28
	ha:kítuǰìk	'I pull it taut (plain)'	HWE 118
d. /– ˘ ˘ ˘ ˘/	*wi:rágųšgerá*	'the stars'	M 28
(two outcomes)	*wi:rágųšgèra*	'the stars'	HWE 117
e. /– ˘́ ˘ ˘ ˘́ ˘/	*hi:žúgokirúsge*	'double-barreled shotgun'	M 28
f. /– –́/	*nąwą́:k*	'he (moving) was singing'	S 10
g. /– –́ ˘/	*mą:čáire*	'they cut a piece off'	M 29
	bo:tá:ną	'he hit him'	Halle 1990, 149
h. /– –́ ˘ ˘ ˘́/	*yu:kí:hinąngkì*	'if I could mix them . . .'	Halle 1990, 149

(88) **Noninitial Heavy Syllables**

a. /˘ –́/	*haǰá:k*	'he (moving) saw'	S 10
b. /˘ –́ ˘/	*kirí:ną*	'returned'	Halle 1990, 149
c. /˘ ˘ –́/	*hitʔatʔá:k*	'he (moving) talked'	S 10
d. /˘ ˘ –́ ˘/	*hitʔetʔéire*	'they speak'	M 29

e. /⏑ ⏑ –́ ⏑ ⏑ ⏑́/ *hižąkí:čąšgunį́*
'nine' M 25

f. /⏑ ⏑ –́ ⏑ ⏑ –́ ⏑ ⏑́/ *hižąkí:čąšgunį́ą́nągá*
'nine and' M 25

g. /⏑ ⏑ ⏑́ –́ ⏑/ *haragínąį́če*
'you will suffer for it' S 49, 139

h. /⏑ ⏑ ⏑́ ⏑ ⏑ –́ ⏑/ *nįkšikšínįkǰąné:ną*
'it will not be weak for you' S 41

i. /⏑ ⏑ ⏑́ ⏑ ⏑ –́ ⏑ ⏑́/ *waγįγį́gišgapʔų́įžeré*
'baseball player' M 25

j. /⏑ ⏑ ⏑́ ⏑ ⏑ –́ ⏑ –́ ⏑ ⏑́/ *waγįγį́gišgapʔų́įžereánągá*
'baseball player and' M 25

Monosyllables, which must be heavy, are accented (85). In words consisting of all lights (86), an accent will fall on the third syllable if there are at least three syllables, otherwise accent falls on the final syllable. In some forms ((86e), (87e, h), (88e, f, h–j)), ternary accent intervals also occur later in the word. But other forms involve a binary count; compare (87d) above and (96o) below. If in the left-to-right sweep of accentuation, there are only two light syllables left in the word, the final syllable receives an accent, as in (86d), (87c), and (88f, i, j). A final posttonic light syllable is not accented: (86c), (87b, e, g), (88b, d, g, h).

Consider next the behavior of heavy syllables. When / – / occurs initially, it acts like /⏑ ⏑/, so accent falls on the immediately following syllable (87). But according to Miner, non-initial heavy syllables pattern differently: they attract the accent irrespective of the alternating count (88g), and the count is started over immediately to their right; compare (88e) *hižakí:čąšgunį́* and similarly (88f). Miner thus characterizes the accent rule as "mora-counting, but syllable-accenting" (1979, 28).

Another sense in which Winnebago is "syllable-accenting" is that there is no contrast between long nuclei with the first mora stronger versus those with the second mora stronger. Rather, the mora that is heard as more prominent is determined by the sonority of the vowels involved, according to the hierarchy /a > o > u > e > i/. This is shown in (89), adapted from Susman 1943, 27; /V:/ is transcribed /VV/ here for clarity.

(89) a. **First Mora More Prominent** b. **Second Mora More Prominent**

ái				*áa*			
ói	*óe*		*óo*				*oá*
úi	*úe*	*úu*				*uó*	*uá*
éi	*ée*					*eó*	*eá*
íi					*ié*	*ió*	*iá*

Susman notes that the weak member of a diphthong is phonetically shorter than the strong member. Her account of long nuclei is corroborated by Miner (1979, 28–29; p.c.). Because of this patterning, it is possible to assume that the metrical derivation is limited to assigning accent to a particular syllable, and that any apparent distinctions between V́V and VV́ merely reflect the syllable's internal sonority profile.[5]

Consider now some differences among the sources.

(A) Stress Subordination versus Downstep. Hale and White Eagle (and Miner 1989) transcribe the first accent in a word with /´/ and subsequent accents with /`/, whereas Susman 1943 and Miner 1979 use /´/ for all accents. This difference is largely a matter of interpretation. Hale and White Eagle treat Winnebago as a stress system, with primary and secondary stresses, while Miner 1979 analyses the system as pitch accent: "An accented syllable in Winnebago has noticeably high pitch relative to unaccented syllables in the same word. Although pitch seems to be the chief acoustic correlate of accent in this language, an accented syllable may have relatively greater intensity as well" (p. 25). Under this view, what might otherwise be analyzed as a primary–secondary stress difference can be thought of as tonal downdrift: "When more than one syllable in a word or stretch of utterance is accented, there is a downstep or terracing effect, each successive accented syllable having a slightly lower pitch and intensity than the last preceding" (p. 25). Miner 1979, 26, and Susman 1943, 9, note that downstep is suspended at certain speaking rates. They also agree that downstep is phrasal, not word-bounded (but cf. Hale and White Eagle 1980, n. 3, arguing for word-bounded downstep).

Based on these observations, I will assume that Winnebago has a tonal prosody; that is, it is a pitch accent system with optional downdrift. For this reason, I will treat all accents as equal, and will not use the symbol /`/ or word layers in metrical representations.

(B) Quantity. A second difference among the sources is that unlike Miner, Hale and White Eagle allot no role to syllable quantity, treating all long vowels as if they were disyllabic. In principle this means that certain long vowels could have stress on their second mora. But in the case of monosyllables, Hale and White Eagle add a special rule (pp. 130–31) to ensure that long vowels will bear greater prominence on their first mora, which accords with Susman's and Miner's descriptions. In the absence of any indication to the contrary, I assume that the facts of syllable-internal prominence in Hale and White Eagle's data are the same: that at least on the surface the accent rule treats syllables as units.

5. The discussion in Miner 1989, 169 suggests that the rising diphthongs of (89b) count as light syllables, which seems to be the normal cross-linguistic pattern for rising diphthongs (Donegan 1978). I have not been able to locate the data needed to evaluate Miner's contention.

(C) BINARY VERSUS TERNARY ALTERNATION. Hale and White Eagle (1980) mostly cite data in which binary alternation predominates after the first accent, whereas the forms in Miner 1979 are always ternary where the rules allow. Indeed, Miner hears (87d) as *wiːrágųšgerá,* whereas Hale and White Eagle have *wiːrágųšgèra.* However, according to Miner (p.c.), there is actually no dialect difference involved. Miner's more recent work (1989, 152) includes binary forms and adopts the binary rule, and Hale and White Eagle include one form (96k) that might be analyzed as ternary. Susman 1943 lists binary forms amidst mostly ternary data. It appears that different morphological contexts give rise to binary or ternary patterns, in a way that is not well understood.

A possibly relevant phenomenon is pointed out by Susman: Winnebago has "stress-neutral" morphology of the type found in English (*SPE* 84–87), as in the compounds in (90):

(90)

a. *ʔáː#ǰiré*	(not *ʔaːǰíre)	'say-start; it started to say'	S 48
b. *waší#kirigí*	(not *wašikírigí)	'dance-return; to come back dancing'	S 58

It is easy to imagine that stress-neutral morphology could serve as an explanation for ternary intervals: for example, hypothetical /◡ ◡ ◡ # ◡ ◡ ◡/ would be accented as two words, yielding /◡ ◡ ◡́ # ◡ ◡ ◡́/. But this cannot be the full explanation, since there are many examples of ternary alternation involving affixes that are not stress-neutral. Such examples include (91a, b); (91c, d) are included to demonstrate the non–stress-neutral status of *-nį* 'not' and *-kǰąne* 'FUTURE'.

(91) a. /š-rušʔák-nį-kǰąné/ → *šurušʔágnįkǰąné* S 52, 140
2.-unable-not-FUT. 'it will not be impossible for you'
b. *roː-rá-gų-nį-kǰą́ne* 'you need not' S 48
$want_1$-INFIXED 2-$want_2$-not-FUT.
c. *gigi-**nį́**-ną-gaǰą́* 'he should not have done it to him' S 53, 136
cause-not-should-then
d. *tʔeː-**kǰą́ne*** 'he would die' S 48, 137
die-FUT.

The rules given by Susman 1943, 55, for accenting close-knit compounds also derive ternary alternation.

In what follows I will assume that the difference between binary and ternary alternation results from a difference in rule application: as elsewhere (§ 8.1), ternary alternation results from weak local parsing, binary alternation from strong. However, it should be borne in mind that the factors determining which one takes place are at present obscure. It is to be hoped that future research by specialists can clarify this issue.

8.9.2 Dorsey's Law

Winnebago accent interacts in interesting ways with an epenthesis process known as Dorsey's Law. This process, argued to be a synchronic phonological rule by Miner 1979, 1989, breaks up certain underlying consonant clusters by inserting a copy of the following vowel. The rule can be stated roughly as follows; see Miner 1979, 26, for a detailed listing of the clusters involved.

(92) **Dorsey's Law** Miner 1981, 342

$$\begin{bmatrix} -\text{sonorant} \\ -\text{voice} \end{bmatrix} \begin{bmatrix} -\text{syllabic} \\ +\text{sonorant} \end{bmatrix} \text{V}$$

$$1 \qquad 2 \qquad 3 \quad \rightarrow \quad 1\,3\,2\,3$$

Depending on where the insertion takes place, the inserted vowel can either take the accent itself, or else shift it from its expected location.

It is useful to divide the examples into those where Dorsey's Law applies early in the word (interacting with the initial accent), and those where it applies later (interacting with later accents). Cases of the first type are listed in (93)–(95). /δ/ stands for the syllable created by Dorsey's Law, and the inserted vowels are cited in boldface.

(93) **Dorsey's Law in Initial Position**

a. /δ ⏑́/	*p**a**rás*	'flat'	M 27
b. /δ ⏑ ⏑́/	*š**u**rušgé*	'you (sg.) untie it'	M 30
	*š**a**wažók*	'mash (as potatoes) (2 sg.)'	HWE 124
c. /δ ⏑ ⏑́ ⏑/	*k**e**rejų́sep*	'Black Hawk'	M 30

(94) **Dorsey's Law in Second Position**

a. /⏑ δ ⏑́/	*hok**e**wé*	'enter'	M 30
b. /⏑ δ ⏑ ⏑́/	*hik**o**rohó*	'prepare, dress (3 sg.)'	M 30, HWE 128

(95) **Dorsey's Law in Third Position**

a. /⏑ ⏑ δ́ ⏑/	*hanįpš**ą́**ną*	'I swam (declar.)'	White Eagle 1982, 314
b. /⏑ ⏑ δ́ ⏑ ⏑́/	*hirak**ó**rohò*	'prepare , dress (2 sg.)'	HWE 128
c. /⏑ ⏑ δ́ ⏑ ⏑́ ⏑/	*hirak**ó**rohònį*	'you don't prepare, dress'	HV 31
d. /⏑ ⏑ δ́ ⏑ ⏑́ ⏑ ⏑́/	*hirak**ó**rohònįrà*	'the fact that you do not dress'	Halle 1990, 149
e. /– δ́ ⏑/	*ro:k**é**we*	'dress, paint face'	M 30
	*mą:š**á**rač*	'promise (2 sg.)'	HWE 127
f. /– δ́ ⏑ ⏑́/	*wa:p**ó**rohí*	'snowball making'	M 30
	*ya:k**ó**rohò*	'prepare, dress (1 sg.)'	HWE 128

The pattern is as follows. If the cluster broken up by Dorsey's Law is initial, so that the epenthetic vowel is the first vowel of the word, we get normal third syllable accent (93). If the epenthetic vowel ends up in second position, then accent surfaces on the fourth rather than the expected third syllable (94). Finally, if the epenthetic vowel ends up in third position, it receives the accent (95).

As Miner notes, this looks contradictory. To get fourth-syllable accent in (94), we would want to order Dorsey's Law after the accent rule, but to get accented epenthetic vowels in (95), we must order accent after Dorsey's Law. Resolving this contradiction is a central challenge of Winnebago accentology.

The remaining Dorsey's Law cases, where epenthesis takes place later in the word, are not so easily sorted into categories. In (96) is a list of relevant cases.

(96) **Additional Dorsey's Law Cases**

a.	/δ –́ ⏑/	*k**a**ráire*	'they departed returning'	M 29
b.	/δ ⏑ δ́ ⏑/	*k**e**rek**é**reš*	'colorful'	M 30
c.	/⏑ δ ⏑ δ́ ⏑/	*wik**i**rip**á**ras*	'cockroach'	M 30
		*wak**i**rip**á**ras*	'flat bug'	HWE 131
d.	/⏑ δ ⏑ δ́ ⏑ δ́ ⏑/	*wak**i**rip**ó**rop**ò**ro*	'spherical bug'	HWE 131
		*gik**ą**nąk**ą́**nąpš**ą̀**ną*	'it is shiny'	M 1989, 170
e.	/– δ́ ⏑ δ́ ⏑/	*wa:p**ó**rop**ó**ro*	'snowball'	M 30
f.	/– ⏑́ δ ⏑́/	*ro:rák**e**wé*	'you dressed him'	S 13
g.	/– ⏑́ δ ⏑ ⏑́/	*ra:gák**ą**nąšgé*	'ant'	M 1981, 342
h.	/⏑ ⏑ ⏑́ δ ⏑ ⏑́/	*hirokíyap**o**rokšé*	'he rolled them up together'	S 29
i.	/⏑ ⏑ ⏑́ δ ⏑ ⏑́ δ ⏑́/	*harakíš**u**rujìkš**ą**nạ̀*	'pull taut (2 sg. declar.)'	HWE 126
j.	/⏑ ⏑ ⏑́ ⏑ δ́ ⏑/	*hakirúǰikš**ą̀**ną*	'pull taut (3 sg.-declar.)'	HWE 126
k.	/⏑ ⏑ ⏑́ ⏑ δ ⏑ δ́ ⏑/	*hiratʔátʔaš**ą**nąkš**ą̀**ną*	'you are talking'	HWE 130
l.	/⏑ ⏑ –́ δ ⏑́/	*hakeweákš**ą**ną́*	'he is entering (moving)'	M 29
m.	/– ⏑́ ⏑ δ ⏑́/	*wi:pámąk**e**ré*	'rainbow'	M 1981, 342
n.	/– –́ δ ⏑ ⏑́/	*ču:giáš**ą**nąpké*	'kingbird'	M 1981, 342
o.	/– ⏑́ ⏑ ⏑́ δ ⏑́/	*ha:kítuǰìkš**ą**nạ̀*	'pull taut (1 sg. declar.)'	HWE 118

8.9.3 The Proposals of Hale/White Eagle and HV

The first metrical analysis of Winnebago was proposed by Hale and White Eagle. For brevity, I will review below only the revised version of their analysis given by HV (pp. 30–34), of which the central rules are as in (97):

(97)

a.	**Extrametricality**	The first mora is extrametrical.
b.	**Foot Construction**	Going from left to right, organize the remaining moras into right-headed binary feet.
c.	**Destressing**	A mora to the right of a stressed mora is destressed.

For (97b), it should be recalled that HV describe only Hale and White Eagle's data, in which non-initial accents are assigned on a binary count. I will not explore here how their analysis might be extended to account for the ternary examples. HV address the issue of quantity indirectly, by having the rules count moras rather than syllables; note that contrary to assumptions I have made (§ 5.6.2), HV assume that single syllables may fall into two metrical feet.

The rules of (97) apply as in (98), which is adapted to my notation.

(98)
(. ×)(. ×) = (86d)
μ μ μ μ μ → μ μ μ μ μ
⟨ho⟩kiwaroke → *⟨ho⟩kiwároké*

The metrical structures serve as the basis for explicating the accent shifts induced by Dorsey's Law. The core of the account is a general principle called the Domino Condition, which is a revised version of a convention originally proposed by Hale and White Eagle.

(99) **Domino Condition** (HV 33)
The introduction of an additional position inside a bounded constituent destroys that constituent and all constituents to its right [for cases of rightward construction] and all constituents to its left [for cases of leftward construction]. Constituent structure is reimposed on the affected substring by a subsequent reapplication of [constituent construction].

Here are examples of how the Domino Condition works. In (94b) *hikorohó,* from /hikroho/, the accent rules will produce the configuration in (100a). Dorsey's Law then introduces the syllable *ko* (100b). Since the epenthetic syllable is not inside the foot, nothing happens, and the correct surface pattern is derived:

(100) a. (. ×) b. (. ×)
μ μ μ → μ μ μ μ
⟨hi⟩kroho → *⟨hi⟩**ko**rohó*

This should be compared with (95b) /hirakroho/ → *hirakórohó,* where the epenthetic vowel gets the accent. Initial footing would yield (101) (HV's framework freely permits degenerate feet):

(101) (. ×)(×)
μ μ μ μ
⟨hi⟩rakroho

When Dorsey's Law applies, the inserted syllable is spanned by the foot *rakro.* The Domino Condition therefore removes the structure dominating the spanned syllable, as well as all feet on the right (102b). The Domino Condition also requires that the affected string be reparsed, deriving the correct output, (102c):

(102) a. (. ×)(×) b. c. (. ×)(. ×)

μ μ μ μ μ → μ μ μ μ μ → μ μ μ μ μ

⟨hi⟩rakoroho → *⟨hi⟩rakoroho* → *⟨hi⟩rakórohó*

The derivation in (103) shows why the Domino Condition must delete not just the foot split by epenthesis, but all further feet to the right as well:

(103) (. ×) (×) Extrametricality, = (96d)
μ μ μ μ Foot Construction
⟨wa⟩kripropro

(. ×) (×) Dorsey's Law
μ μμ μ μ μ μ
⟨wa⟩kiriporoporo

μ μμ μ μ μ μ
⟨wa⟩kiriporoporo Domino Condition (deletion)

(. ×)(. ×)(×) Domino Condition (reparsing)
μ μμ μ μ μ μ
⟨wa⟩kiriporoporo

(. ×)(. ×) Destressing (97c)
μ μμ μ μ μ μ
⟨wa⟩kiripóropóro

Were we to delete and reparse only the foot split by Dorsey's Law, we would derive the incorrect **wakiripóroporó.*

I now mention two potential problems with this analysis, one concerning the data, the other the theory.

First, as Miner 1989 points out, HV's analysis fails empirically in a broad class of forms: those with at least three underlying syllables and an initial Dorsey's Law cluster:

(104) (. ×) Foot Construction = (93c)
μ μ μ
⟨kre⟩ǰųsep

(. ×) Dorsey's Law (Domino Condition inapplicable)
μ μ μ μ
**kereǰųsép*

The correct accentuation is *kereǰų́sep,* and similarly for parallel forms (Miner 1989, 154).

In addition, there is reason to doubt the generality of the Domino Condition on which the analysis rests. In Mohawk (Michelson 1988, 139), a rule of epenthesis much like Dorsey's Law (Michelson's "*e*-epenthesis II") creates surface exceptions to the general pattern of penultimate stress:

(105) a. /té-k-rik-s/ → *tékeriks* 'I put them next to each other' M 133
b. /ʌ́-k-r-ʌʔ/ → *ʌ́kerʌʔ* 'I'll put it into a container' M 134

If penultimate stress is the result of a final trochaic foot, the Domino Condition will wrongly create *[tekériks] and *[ʌkérʌʔ]. In Lenakel (§ 6.1.8), an optional epenthesis rule (Lynch 1974, 82–83) likewise splits a foot without restructuring, in examples like (106):

```
(106) (    ×     )          (    ×      )
      (× .)(×   .)          (× .)(× .  .)
       ˘  ˘  ˘   ˘           ˘  ˘  ˘ ˘  ˘
      neluyaŋyaŋ   →   nèluyáŋɨyaŋ      (*nelùyaŋɨ́yaŋ)      'twig'
```

As far as I can determine, the only data that actually support the Domino Condition are the Winnebago data.[6]

The Domino Condition could be supported on purely theoretical grounds if it could be made to fall out from fundamental principles of foot construction, and HV 132–38 provide just such an account. But that formulation derives incorrect results: **wakiríparás* for (96c) and **wakiríporóporó* for (96d). These stressings are listed in HV 31–32; the correct stressings are *wakiripáras* M 30, HWE 131, and *wakiripóropóro* HWE 131. HV elsewhere give correct derivations (from which (103) above is taken), but these are based on a preliminary, unformalized version of the Domino Condition.

Summing up, it seems worth developing an alternative to the HWE/HV analysis, both because of the empirical problem it faces in forms like (104), and because of the doubtful status of the Domino Condition on which it rests.

8.9.4 Miner's Analysis

Miner's (1979) discussion makes two crucial points. First, as noted earlier, Winnebago is plausibly analyzed as a pitch accent system, the dominant cue for syllable prominence being high pitch. This suggests that prosodic derivations in Winnebago might involve tonal rules as well as rules creating metrical structure. Second, Miner shows that, at least historically, Winnebago has undergone an **accent shift:** accent in Winnebago falls one syllable to the right of where it occurred in the ancestor language. Miner compares Winnebago with Chiwere, which preserves the Common Mississippi Valley Siouan accent:

6. Martin 1992 suggests a possible case in Seminole/Creek; however, the Domino Condition he invokes appears to be radically different from HV's.

(107)	**Chiwere**	**Winnebago**	
a.	*was**ó**se*	*wašoš**é***	'brave'
	*wan**á**xe*	*wanąγ**í***	'spirit'
b.	*k**í**ða*	*ki:z**á***	'fight'
	*l**ó**hą*	*ro:h**ą́***	'much; many'

From such evidence the accent rule for the protolanguage can be reconstructed:

(108) **Common Mississippi Valley Siouan Accent**
Accent the second syllable if the initial syllable is light. Otherwise accent the initial syllable.

From this original pattern (/CVCV́CV/ for (107a), /CV́:CV/ for (107b)), Winnebago has preserved vowel length and shifted accent; Chiwere has lost vowel length but preserved accent, now rendered phonemic.

Recall now the basic contradiction found in Dorsey's Law examples: in some positions, the syllable inserted by Dorsey's Law seems to be ignored for accentual purposes, whereas in others the inserted syllable actually takes the accent. Miner's elegant contribution is to show that this results from the historical ordering of Dorsey's Law: it followed basic accent assignment but preceded Accent Shift. This resolves the contradiction as in (109):

(109)	a.		b.		
	hikroho	= (94b)	*hirakroho*	= (95b)	proto-forms
	hikróho		*hirákroho*		early accent (108)
	*hik**o**róho*		*hirák**o**roho*		Dorsey's Law
	*hikoroh**ó***		*hirak**ó**roho*		Accent Shift

The other cases work similarly, with one exception. To account for cases where Dorsey's Law applied in initial position (93c), Miner assumes that the inserted vowel counted here (and only here) as an extra syllable. Miner 1989, 167, suggests that the sound change was more advanced in initial position.

(110)		
	krejǔsep	proto-form
	*k**e**rejǔsep*	Early Dorsey's Law (initial position)
	*ker**é**jǔsep*	proto-accent
	*kerej**ǘ**sep*	Accent Shift

That something of the sort did happen is shown by the initial accent in Chiwere words that began with Dorsey's Law clusters, such as *gléblą* 'ten' M 27.

Since modern Winnebago accent is the historical product of the earlier rule plus accent shift, any analysis that attempts to predict modern accent directly is claiming that restructuring has taken place; that is, that the accent rule (108) and Accent Shift have been replaced synchronically by a single, quite different accent rule. Here, I reject restructuring, following Miner, and assume that the historical changes still function as synchronic rules.

I restate Miner's account with the rules in (111). Because the accent pattern of (108) is a normal iambic one, the metrical part of the analysis is quite simple:

(111) a. **Foot Construction** Form iambs from left to right.

b. **Tonal Rules**

i. **H Tone Assignment** Assign H tone to foot heads.

ii. **Tone Shift** σ σ (i.e., shift H tone one syllable to the right)
H

In this (preliminary) analysis, accent is conceived as tonal, though located on the basis of metrical structure. This allows Tone Shift to be expressed as an autosegmental rule, which avoids a violation of the Faithfulness Condition (§ 3.7). My assumption is that the secondary phonetic correlate of intensity is a phonetic concomitant of high tone, rather than being represented metrically. Note that similar rules that shift all tones rightward by one syllable have been observed in Bantu languages: Kikuyu (Clements and Ford 1979) and Chaga (McHugh 1990). Rightward tone shift is also observed as a phonetic tendency in Japanese (Hata and Hasegawa 1989).

In (112) are derivations illustrating the rules:

(112) a.

(. ×) *hikroho*	(. ×)(. ×) *hirakroho*	Foot Construction
(. ×) *hikroho* H	(. ×)(. ×) *hirakroho* H H	H Tone Assignment
(. ×) *hikoroho* H	(. ×) (. ×) *hirakoroho* H H	Dorsey's Law
(. ×) *hikorohó* H	(. ×) (. ×) *hirakórohó* H H	Tone Shift

In the output, tonal and metrical structures disagree; I assume that the H tone is the perceptually prominent element. An observation of Susman's (1943, 9) suggests that the second syllable may in fact retain its metrical strength: "in some words, [a] secondary stress on the second syllable seems about as strong as the normal stress on the third."

Under this account, it is not necessary to split heavy syllables into different feet, as in (113a) (cf. Halle 1990 and, for discussion, § 5.6.2). Heavy initial syllables simply attract their own foot, then transmit H tone to the next syllable (113b).

(113) a. (. ×) ⟨*ho*⟩*očąk*

b. (×) – ◡ *ho:čąk* | H → (×) – ◡ *ho:čą́k* | H = (87a)

The same holds for medial heavy syllables; compare (115b) below.

8.9.5 Extending Miner's Account

As the previous section showed, Miner's account offers an alternative to the Domino Condition in describing the Dorsey's Law facts. Below, I extend his approach to a broader range of phenomena.

(A) TERNARY ALTERNATION. It can be shown that even under a Tone Shift analysis, the ternary forms above require weak local parsing (§ 8.1) in the metrical part of their derivations. In the derivation for (86e), the step shown in (114a) shows the construction of iambs from left to right using weak local parsing. H tones are then shifted one syllable to the right (114b).

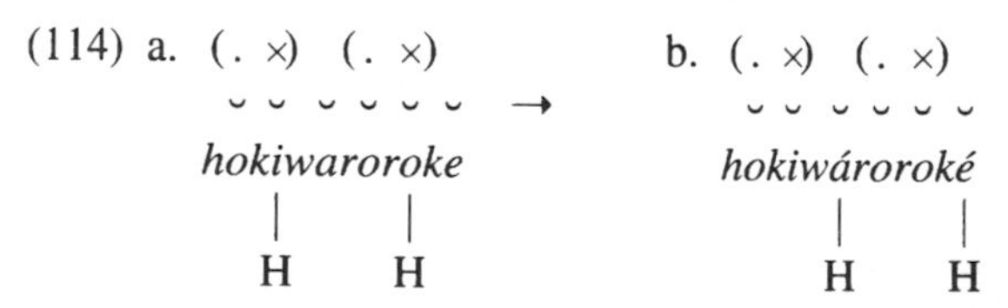

As noted above, I will leave untreated the morphological and possibly lexical factors that determine whether weak local or strong local parsing is selected in a particular word.

(B) HEAVY SYLLABLES IN LONGER WORDS. The rule of Tone Shift (111b.ii) requires an additional complication: other than in isolated cases (e.g. Susman 1943, 50–51), it apparently cannot shift H tone off non-initial heavy syllables. This is shown in (115) for (88b) and (87g); other examples are (87h), (88d–j), and (96a). In these derivations and others that follow, I omit metrical structure at stages of the derivation where it is not relevant.

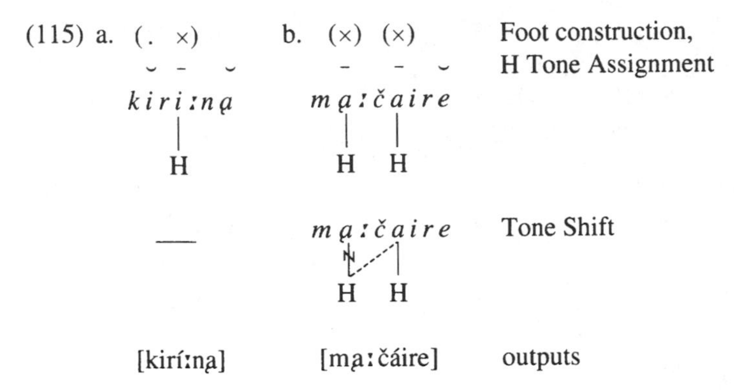

In (115a), Tone Shift off the non-initial heavy syllable *-ri:-* is blocked. In (115b), Tone Shift cannot apply off *-čai-*, but it can off *mą:-*, since it is initial. The effect of this shift is observable only in the removal of H tone from *mą:-*, since *-čai-* retains its own H. I assume either that the Twin Sister Convention (Clements and Keyser 1983, 95) deletes one of the H tones, or else two H tones linked to a single syllable have the same phonetic effect as one.

(c) Persistent Footing. Footing in Winnebago must be persistent, in the sense of § 5.4.1. With non-persistent footing, we could get "double off-beats" under weak local parsing, analogous to the double upbeats of Cayuvava (§ 8.2). Persistent footing ensures the presence of a final accent:

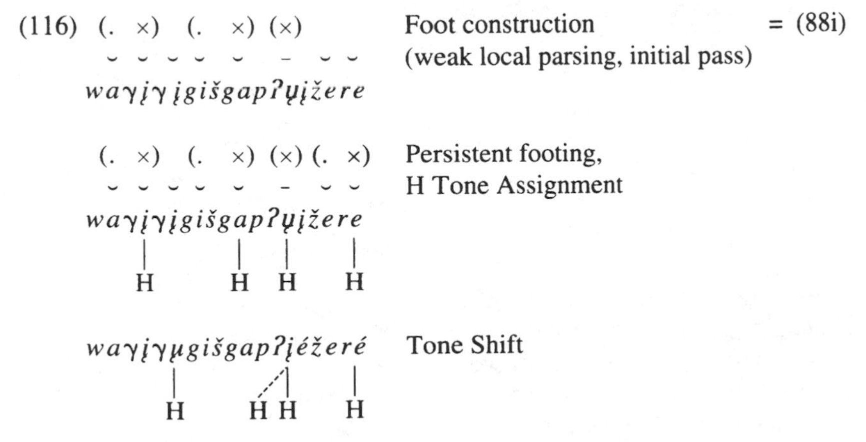

As noted in § 8.2.2, the ternary forms in Winnebago are problematic for the Recoverability Condition of HV: they are a mirror image of the Cayuvava forms that have a double upbeat, but they lack the double offbeat that the Recoverability Condition predicts.

(D) RESOLUTION OF HH SEQUENCES. Because of Tone Shift, consecutive H tones sometimes pile up on the penultimate and final syllables. Such sequences are resolved on the surface by deleting the second H. I state the rule of H Tone Deletion in (117), giving examples in (118).

(117) **H Tone Deletion**

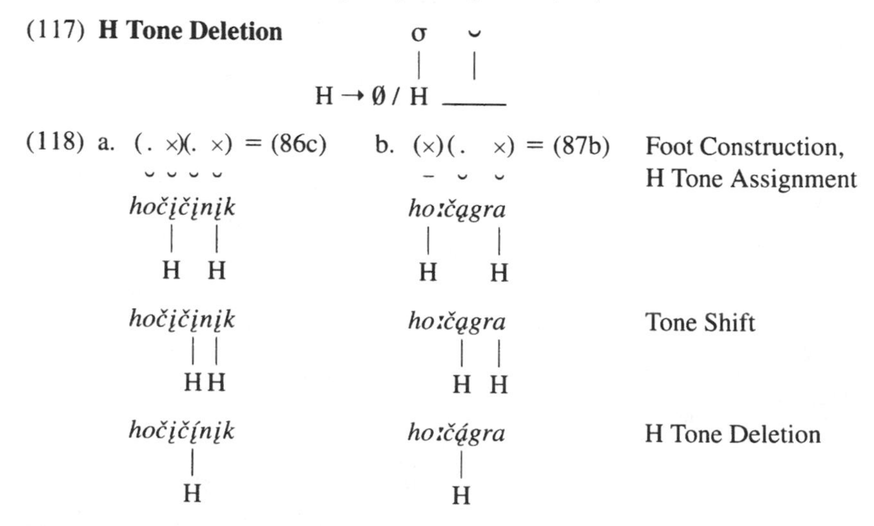

The limitation to light syllables is motivated by (88g). Note that (117) is a version of Meeussen's Law, a tonal rule that occurs commonly in Bantu languages (Clements and Goldsmith 1984, 7).

(E) ACCENTS THAT FOLLOW DORSEY'S LAW SEQUENCES. Dorsey's Law sequences have the curious effect of making later accents in the word come earlier than one would expect. I have already noted one such case, involving Dorsey's Law sequences in initial position. For such examples we would expect a derivation like /kreǰų́sep/ → /kreǰųsép/ → *[kereǰųsép], with surface accent on the fourth syllable. In fact, we get *kereǰų́sep* (93c), with accent coming unexpectedly early on the third syllable. I noted above that HV's analysis derives such forms incorrectly, and reviewed Miner's historical account of the facts. But I have not yet provided a synchronic account.

As it turns out, Dorsey's Law sequences are also followed by unexpectedly early accent in other environments. Consider the forms (95c), (96e), and (96d), repeated below. Naive expectation is as in (119):

(119)

a. /hirakrohoni̜/	b. /waːpropro/	c. /wakripropro/	underlying forms
hirákrohóni̜	*wáːpropró*	*wakrípropró*	accent
hirákorohóni̜	*wáːporoporó*	*wakiríporoporó*	Dorsey's Law
**hirakórohoní̜*	**waːpóroporó*	**wakiripóroporó*	Tone Shift

The correct forms are *hirakórohónį, wa:póropóro,* and *wakiripóropóro.* (For the first form, assuming weak rather than strong local parsing would still derive the wrong result.) In all three cases, the boldface accent in the actual forms comes one syllable earlier than expected.

Both the initial Dorsey's Law cases and the cases just given obey the following generalization: THE SYLLABLE TO WHICH DORSEY'S LAW APPLIES ACTS AS IF IT WERE HEAVY (cf. Susman 1943, 31, n. 21). Under this assumption, the relevant derivations work as in (120):

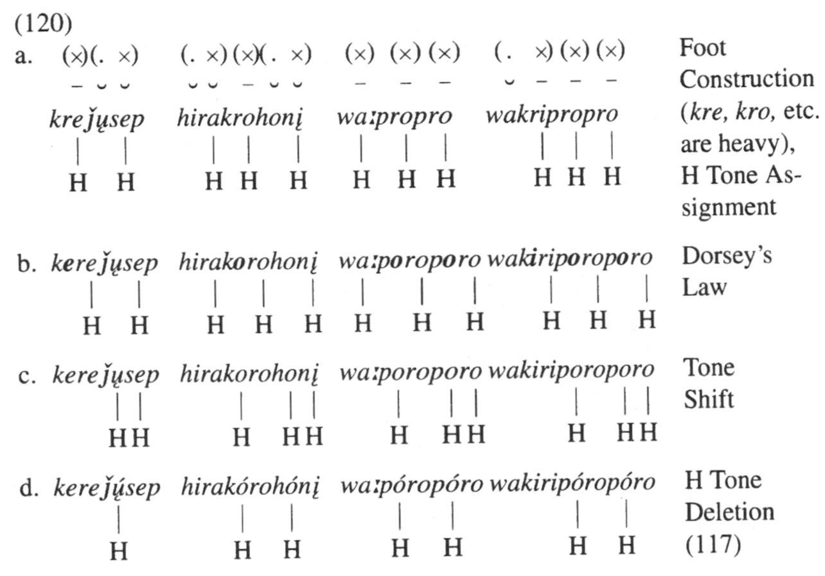

The assumption that Dorsey's Law syllables are everywhere counted as heavy is supported by additional evidence. As Miner 1989, 52, notes, the minimal word in Winnebago is at least two moras; that is, a word cannot consist of a degenerate foot (CV(C)). But a Dorsey's Law syllable can be a word, as in (93a) /pras/ [parás]. If Dorsey's Law syllables are underlyingly heavy, this follows without further stipulation; but if such syllables are light, then the underlying stem structure condition must be complicated, taking the form "a stem must either be minimally bimoraic or eligible for Dorsey's Law."

As a way of formally representing the underlying heavy quantity of Dorsey's Law syllables, I propose the representations in (121), in which the sonorant consonant bears a mora:

(121)

a. σ	b. σ	c. σ
μ μ	μ μ	μ μ
k r e	*k n ų*	*p r o*

These representations might be thought of as an intermediate phase between simple monomoraic /kre, knų, pro/ and the fully disyllabic (hence bimoraic) surface forms. The latter are derived by converting the weight-bearing sonorant consonant into a full syllable, using an epenthetic copy vowel:

(122)
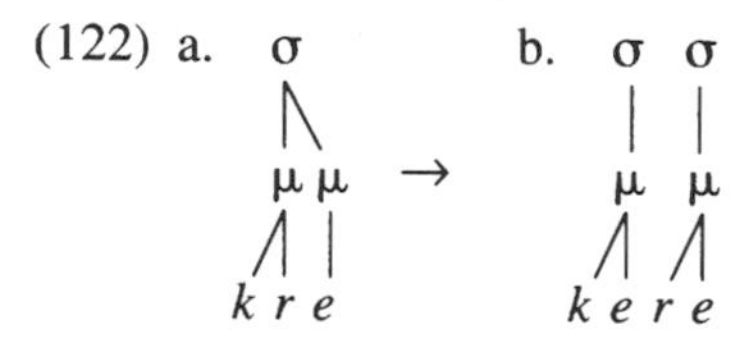

Stage (122a) is relevant to the minimal word constraint and to foot construction, both of which treat Dorsey's Law sequences as heavy monosyllables. Stage (122b) is relevant to the late rule of Tone Shift, where the sequence must count as two syllables.[7]

Syllables like (122a) have a rising sonority profile, just like the rising diphthongs mentioned in (89). Thus H tone docks onto their second mora, from which it may undergo Tone Shift if the latter is applicable (120).

The derivations in (123)–(126) illustrate heavy weight for Dorsey's Law sequences. For the forms (93a), (93b), (93c), and (96e), first feet are constructed, then H tones are assigned to the strongest mora of each foot head; in Dorsey's Law syllables these are the vowels. Note that in a strict representation, tones would dock onto moras; I dock them onto segments here for typographical convenience.

(123)
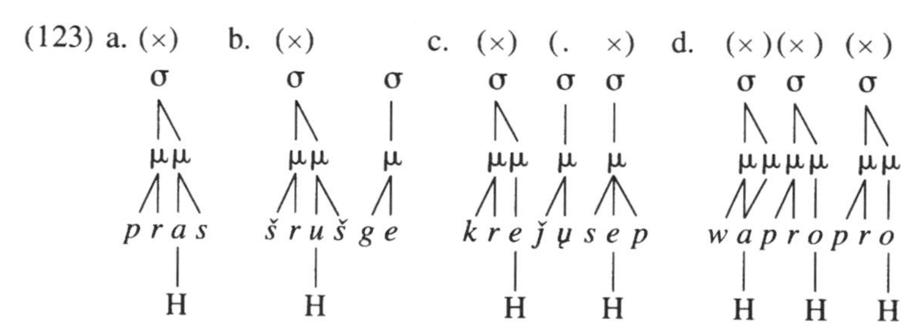

The application of Dorsey's Law is shown in (124). I omit metrical structure, since only the tones are phonetically relevant; under this circumstance it is impossible to obtain evidence for how metrical structure is accommodated to Dorsey's Law.

7. Clements 1991 makes the striking proposal that Dorsey's Law sequences are monosyllabic on the surface, noting among other things that they have shorter duration than ordinary CVCV sequences (Miner 1979, 26). With suitable adjustments made in the tonal rules, my account is compatible with this view.

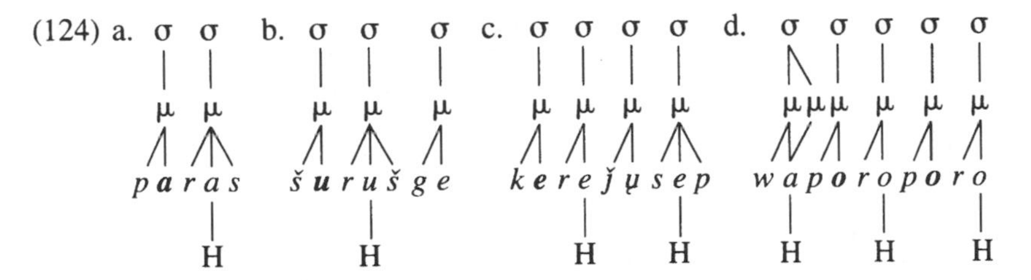

Tone Shift shifts all tones one syllable rightward; recall that H can shift off the heavy syllable of (124d) because it is initial.

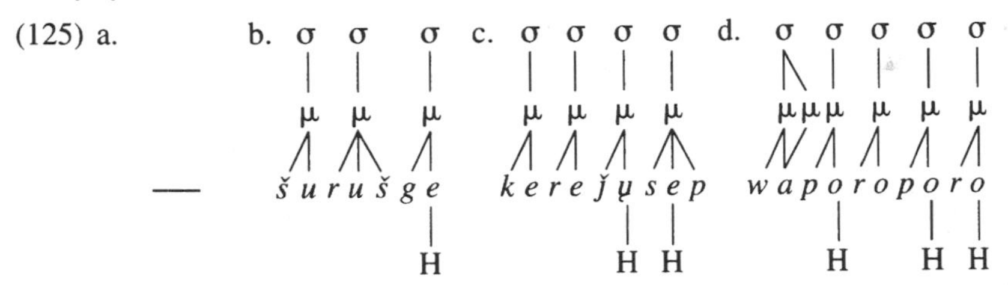

The surface forms derive from H Tone Deletion:

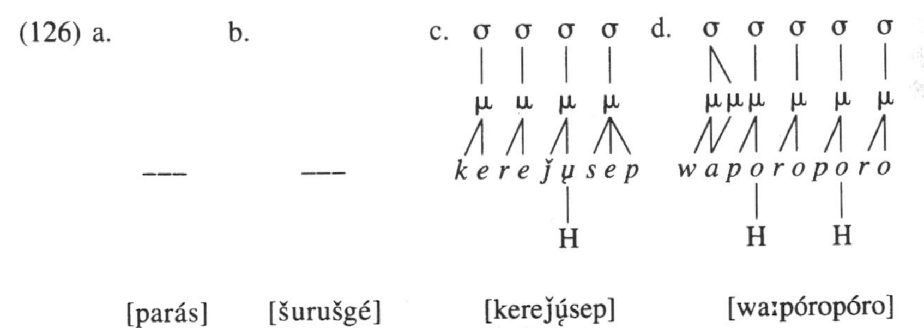

[parás] [šurušgé] [kereǰų́sep] [wa:póropóro]

To sum up, in the account proposed here the third-syllable accent on (93c) *kereǰų́sep* is made part of a more general phenomenon, namely the heavy quantity of Dorsey's Law syllables. This also accounts for the accentual pattern of other forms, as well as the minimal word constraint.

Here is the full analysis, with the rules stated in proper order:

(127)

a. **Foot Construction**	Form iambs from left to right. Varies between weak local and strong local parsing, under morphological conditioning. Degenerate feet are prohibited. Parsing is persistent.
b. **H Tone Assignment**	Associate H tone with the sonority peak of a syllable that is the head of its foot.

c. **Dorsey's Law**

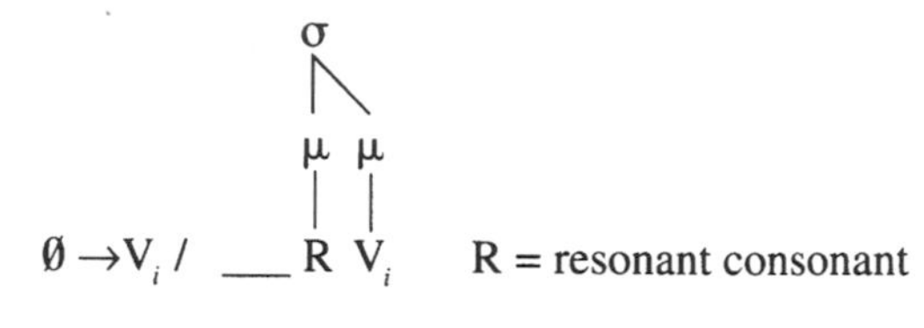

d. **Tone Shift**

σ_i σ Condition: σ_i is light if non-initial.

H

e. **Tone Deletion**

σ ˘

H → Ø / H ___

The rules presuppose that the syllabification conventions of Winnebago treat Dorsey's Law sequences as in (121).

Susman and Hale/White Eagle note additional facts that an adequate analysis should account for. In a number of cases, accent is assigned to the third vowel, whereupon a three-vowel sequence is resolved by syllable merger. The syllable created by merger then inherits the original H tone: /nąą-ha-ʔą/ 'weigh (1)' HWE 121–22 → /nąąháʔą/ (by accent assignment) → /nąąáʔą/ (by intervocalic /h/ deletion) → [ną́ːʔą] (by syllable merger and tonal inheritance). Such tonal stability under syllable merger is a characteristic pattern of tonal systems (Goldsmith 1976). Other instances of initial accented heavy syllables apparently cannot be explained in this way, and might be analyzed as lexical exceptions to Tone Shift.

My rules provide derivations for all the data above. Weak local parsing must be assumed for (86e), (87e), (88h–j), (91a–c), (96k, m), and the first variant in (87d). Strong local parsing must be assumed for (96j, o) and the second variant in (87d). In all other forms, the mode of parsing does not matter. For an alternative account of (96k), see Hale 1985.

8.9.6 Conclusions

Following Miner's original insight, the analysis here splits Winnebago derivations in two: an accentual derivation precedes Dorsey's Law, while a tonal derivation follows it. A disadvantage of this approach is that the tonal rules are non-trivial. The rule of Tone Shift (127d) in particular includes a fairly complex condition based on syllable weight. In addition, it is perturbing to what extent the tonal derivation looks like a stress derivation. While it is true that both Tone Shift and (127e) H Tone Deletion have counterparts in African tone languages, the rules nonetheless have properties that look metrical in character: the retention of H tone by all but initial heavy syllables, and the resolution of "clashing" H tones.

An avenue to explore would be whether Tone Shift might be construed as a wholesale Prosody Shift, moving both stress and tone. Such an analysis is proposed by McHugh 1990 for Chaga, a Bantu tone language with accentual characteristics. The implications this would have for the present theory, especially the Faithfulness Condition, go beyond the scope of this book.

The following are what I take to be the advantages of the proposal. (a) Unlike early analyses, it accounts for the full array of forms, notably the cases (§ 8.9.5(e)) where a Dorsey's Law sequence affects the position of accents later in the word. (b) Under my account, the metrical part of Winnebago phonology becomes typologically ordinary. All the mechanisms invoked in the analysis (left-to-right iambs, weak local parsing, and persistent footing) are attested in other languages. The analysis does not invoke initial extrametricality, which is quite rare cross-linguistically, and it need not let syllables be split up into metrical feet. (c) The analysis does not rely on the problematic Domino Condition.

My analysis is an attempt to include Winnebago in a general pattern established by Prince 1983a, 82–86, on Seneca, and Sauzet 1989 and Golston 1989 on Ancient Greek. These authors recognize that the study of pitch accent systems has a special status within metrical theory: since they are partly tonal systems, their derivations can include purely tonal rules. The surface result can be puzzling if one is attempting to cover the whole system with a purely metrical account. A judicious appeal to tonal rules in such systems allows us to simplify the metrical analysis greatly, as well as establish a simpler and more tightly constrained general metrical theory.

8.10 OTHER CASES OF TERNARY ALTERNATION

(a) **Auca,** discussed in § 6.2.1, shows weak local parsing in the "suffix train." The crucial case is that of a five-syllable suffix sequence: with weak local parsing into syllabic trochees from right to left, we derive the correct stressing:

```
(128) (×)                         'we two would have gone'    Pike 1964, 426
      (×) (×  .)  (×  .)
       ˘   ˘  ˘   ˘  ˘   ˘
      gó ] kæ̀dǫmǫnà͡į ba
```

Footing must be persistent, in order to assign two stresses to quadrisyllabic sequences; cf. *gó] támǫnápa* 'we two went' (Pike 1964, 426).

(b) **Mantjiltjara** (Pama-Nyungan, Western Australia; Marsh 1969) may also involve ternary stress. The feet are syllabic trochees, assigned from left to right, with main stress falling on the initial foot. Strong local parsing also occurs in some forms, and as in Winnebago the conditions that determine which is found are not understood. Parsing is not persistent, as shown by (129):

```
(129) (×          )   'fly!'                      Marsh 1969, 146
      (×   .)
       σ   σ σ σ
      párpakala
```

Here, the syllable /ka/ is skipped over by weak local parsing, and /la/ cannot be footed owing to a ban on degenerate feet. Under nonpersistent footing, the representation given survives as a surface form.

(c) **Bani-Hassan Arabic** is a Bedouin dialect of northern Jordan, whose prosody has been described by Kenstowicz 1983; Irshied and Kenstowicz 1984; and Kenstowicz 1986. Irshied and Kenstowicz provide a detailed metrical analysis, explicating some complex interactions between stress, morphological structure, and segmental phonology. Their analysis invokes left-to-right construction of uneven trochees, a metrical constituent I have proposed to dispense with (§ 4.3.1). I suggest that the crucial examples (all of the shape /–̀ ˘ ˘́ ˘/ can be treated instead with moraic trochees and weak local parsing, as shown in (130):

```
(130) a. (     ×  )    not:    b. (     ×  )      'she taught him'
         (×)  (× .)               (× .)(× .)      IK 117
          –  ˘  ˘ ˘                –  ˘  ˘ ˘
         ʕàllamátu⟨h⟩             ʕàllamátu⟨h⟩
```

Since long strings of light syllables are not found in Bani-Hassan, it unfortunately is not possible to determine whether the stress rule really would create ternary intervals if given the chance.

In other respects, Bani-Hassan appears to be much like Palestinian Arabic (§ 6.1.1). One further difference is that final feet are extrametrical only in a clash environment. Thus while (131a) is stressed as in Palestinian, (131b) is stressed on the final rather than the penultimate foot.

```
(131) a. (×)             'they taught'   b. (     ×  )     'she pulled you
         (×)⟨(×    .)⟩   IK 128             (× .)(× .)     (m. sg.)'
          –    ˘    ˘                        ˘  ˘  ˘ ˘     IK 130
         ʕállamu                            sàħabáta⟨k⟩
                                           (→ sħàbátak)
```

Case (131b) also shows that footing in Bani-Hassan must be persistent.

The surface form for (131b) illustrates another interesting phenomenon, first pointed out in Kenstowicz 1983, namely rightward shift of stress under syncope. The syncope rule in question is similar to Trisyllabic Syncope in Cyrenaican Bedouin Arabic, discussed in § 6.3.7.1.

A question arises of whether the secondarily stressed sequence [sħà] in (131b) forms a degenerate foot in weak position, contrary to the claims of § 5.1.1. Kenstowicz (p.c.) suggests that the vowel [à] (and similarly for parallel forms) may in fact be phonetically lengthened, and thus perhaps count as a heavy syllable. The lengthening can be regarded as a repair of the degenerate foot (§ 5.1.7) that permits it to survive in weak position.

•9•

Phrasal Stress

9.1 BACKGROUND

Assignment of stress at the phrasal level shows marked differences from assignment of word stress. In the case of English, at the phrasal stage of the derivation metrical structure has already been assigned to all syllables up to the word level. Thus phrasal stress usually carries out only the two operations in (1) (Selkirk 1984, 54–57):

(1) a. Assignment of relative prominence contours to strings of words, based on morphosyntactic structure, focus, and other factors

b. Adjustment via movement or deletion of the resulting contours in accord with rhythmic principles: e.g. avoidance of clash, even spacing of stresses

The foot-based operations of word stress assignment discussed in previous chapters are not well documented as phrasal processes. The word stress rule of Macedonian generalizes to the phrasal level in a few cases (Lunt 1952, 24–25), and similarly for Piro (Matteson 1965, 22). Bruce 1984 argues that Swedish alternating stress patterns are assigned without respect to word boundaries (in my terms, with top-down syllabic trochees, § 5.4.3), and similar cases are found in Cayuvava (§ 8.2), Italian (Nespor and Vogel 1989), and Modern Greek (§ 6.2.3) as described by Malikouti-Drachmann and Drachmann 1981, 287. It is perhaps more common for foot construction to extend its domain to include adjacent clitic-like words; this is discussed for Modern Hebrew by Bolozky 1982, for Fijian by Schütz 1985, for Lenakel by Lynch 1974, and for Yawelmani by Newman 1944, 29. The set of languages with phrasal foot construction may well be larger than previously suspected, since studies of alternating stress typically use only single words as data.

A substantial body of literature exists on phrasal stress assignment in English and a few other well-studied languages. Important work on English prior to the invention of metrical stress theory includes Newman 1946; Trager and Smith 1951; Chomsky, Halle, and Lukoff 1956; and especially *SPE*. Metrical studies include Liberman and Prince 1977; Prince 1983a; Selkirk 1984; Hayes 1984; HV; Kager and Visch 1988; Visch 1989; Nespor and Vogel 1989; Gussenhoven 1991; Churchyard 1992; and others. For less well documented languages, there appears to be rather little published research, so the typological basis of our knowledge is rather narrow.

Here, I propose an outline account of phrasal stress rules within the general framework adopted here, focusing on English. The theoretical discussion considers how the End Rule is to be formulated at the phrasal level. My proposal is based on the Continuous Column Constraint (§ 3.4.2) and the Faithfulness Condition (§ 3.7). The claim to be made is that by invoking these principles, we can eliminate two stipulative conditions placed on phrasal stress rules in earlier work: the Maximality Condition of Hayes 1984 and the Strong Domain Principle of Kager and Visch 1988. I also discuss the rule of Beat Addition (Selkirk 1984; Hayes 1984), providing an explicit formulation of it, and arguing against the view that the rule's effects can be attributed to general conventions.

9.2 PHRASAL STRESS OPERATIONS

To start, here is a brief overview of the types of phrasal stress rules that are commonly encountered, following earlier literature. Leaving aside footing at the phrasal level, noted above, these may be classified as follows: (a) End Rules; (b) Move X; (c) Destressing; (d) Beat Addition. These are discussed in § 9.2.1–4. The latter three operations appear to obey a principle of "eurhythmy," discussed in § 9.2.5.

9.2.1 The End Rule

At the phrasal level, the End Rule functions to establish relative prominence relations between members of a phrase. A common but not invariant pattern across languages is for syntactic phrases to receive final prominence, with compounds receiving initial stress. As an initial approximation, we can say that the End Rule at the phrasal level interprets morphosyntactic constituents as metrical constituents and assigns a peripheral grid mark to them. In (2) are two examples of the End Rule applying at the phrasal level:

(2) a. **Nuclear Stress Rule (English)**
The rightmost member of a phrase is strongest.

```
                                  (        × )
        (× .)      (× )          (× .) (× )
e.g.   Jesus  +  wept    →    Jesus wept
```

b. **Compound Stress Rule (Dutch)**
The leftmost member of a compound is strongest.

```
                               (×      )
        (× )      (×)          (× )(×)
e.g.   voet  +  bal    →    voetbal        'football'
```

When applied to more complex syntactic structures, End Rules can derive multiple degrees of stress, as in (3). (In this and many examples to follow, the foot layer is omitted for brevity. Nothing in the arguments is affected by this.)

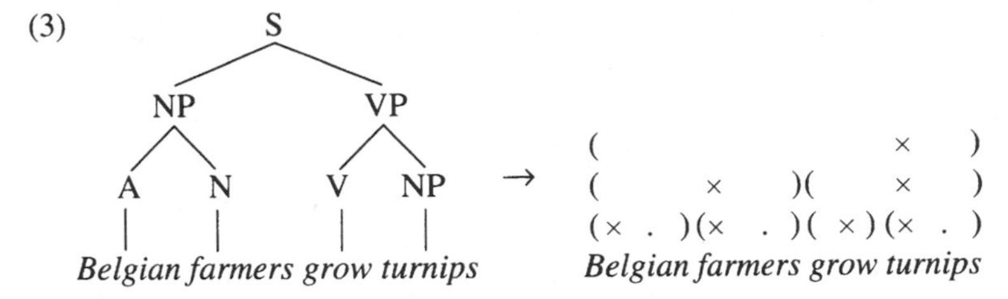

Such derivations show an important property: where rhythmic factors do not intervene (see below), relations of relative prominence are typically preserved under embedding (Liberman and Prince 1977). Thus the rising stress contour we find on the isolation form *Belgian farmers* is maintained when *Belgian farmers* is made part of a larger structure, as in (3). As Liberman and Prince point out, this is a natural consequence of a metrical theory, in which larger rhythmic structures are simply the structural combination of smaller ones.

There are a number of important issues surrounding the phrasal End Rule which I will only touch on here:

(a) As is well known, the Nuclear Stress Rule (2a) represents only the phonological default, and may be overridden by many factors, including focus marking and predicate–argument structure. Non-default phrasal stress assignment is an enormously complex area; for recent discussion see Gussenhoven 1984 and Gussenhoven, Bolinger, and Keijsper 1987. Compound stress is likewise a very complicated area in English, involving word extrametricality (Prince 1983a, 30) as well as large classes of often systematic exceptions. For expository purposes I give the simpler Dutch version of the rule (Visch 1989, 84).

(b) It is not clear whether the End Rule applies to morphosyntactic structure or to phonological domains of the sort proposed in the theory of the Prosodic Hierarchy (Selkirk 1980a; Nespor and Vogel 1986; Hayes 1989a). The latter option is argued for by Nespor and Vogel 1989. Since English is a language where it appears particularly difficult to determine how phonological domains are arranged, I leave this issue open. Given a sufficiently flexible phrasing system, I believe my results would hold irrespective of how this issue is decided.

(c) The analysis to follow treats stress apart from intonation, and in this respect differs from theories like Selkirk 1984 and Gussenhoven 1991, where the assignment of intonational pitch accents is integrated into the phrasal stress assignment system (cf. § 2.4). Gussenhoven's theory is an extremely streamlined one in which there are very few levels of stress, and most of the action of the phrasal stress phonology consists of manipulations of pitch accents. My view instead follows that of Pierrehumbert 1980, in which phrasal stress is an independent domain, and pitch accents are constrained to attach to the strongest available stresses. The main reason for this is that, in my judgment, phenomena such as the Rhythm Rule or Beat Addition apply irrespective of whether the relevant stresses are realized by pitch accents or not (see e.g. (66) below).

9.2.2 Move X

Move X is the formal representation of the Rhythm Rule in bracketed grid theory (§ 3.4.2). The Rhythm Rule in English retracts stress leftward when a stronger stress follows; thus *thirtéen,* but when a stronger stress follows, *thìrteen mén.* I state the general schema of Move X as in (4), following Prince 1983a, 33, 37, and § 3.4.2.

(4) **Move X** Move one grid mark at a time along its row. Where Move X resolves a stress clash, movement must take place along the row where the clash occurs.

In (5), the grid mark representing the main stress of *Tennessee* is moved leftward, under influence of the stronger stress on *Ernie:*

```
(5) (          ×  )       (         ×  )
    (        ×)(×  )  →   (×        )(×  )
    (×   .)(×)(×  .)      (×   .)(×)(×  .)
    Tennessee Ernie       Tennessee Ernie
```

As Prince 1983a showed (see review in § 3.4.2), the formulation in (4) correctly predicts two important universal patterns found in rhythm rules. First, since Move X must conform to the Continuous Column Constraint, the moved × seeks out the strongest available target; compare § 3.4.2 (11), /Sunset Pàrk Zóo/ → *Sùnset Park Zóo* (not **Sunsèt Park Zóo*). Second, the Move X schema predicts the universal pattern that Rhythm Rules move weak stresses away from stronger, never vice versa; compare the hypothetical, ill-formed example /kangaróo ìmitators/ → **kángaroo ìmitators,* from § 3.4.2 (12).

Move X is subject to language-particular restrictions governing direction of movement. In English, movement may only be leftward, so that for example the derivation in (6) (adapted from Liberman and Prince 1977) is excluded:

```
(6) ( ×           )      *( ×            )
    ( × ) (×      )  →    ( × ) (    × )
    ( × ) (×)(× )         ( × ) (×)(× )
    sports contest        sports contest
```

For similar reasons, *kangaróo ìmitators* must retain its input stress contour, and cannot become **kangaróo imitàtors.* In contrast, rightward stress shift is reported for German (Kiparsky 1966), Danish (Rischel 1972), Dutch (Kager and

Visch 1988), and Finnish (Hayes 1981, 122), as in the Danish example (7), from Rischel 1972, 221.

(7) (×   )   (×   )
(×)(×   ) → (×)(×)
(×)(×) (×   )   (×)(×) (×   )
vand luft pumpe   *vand luft pumpe*   'water-jet air pump'

Other restrictions are also found. For example, in Spanish (Solan 1979), Move X may apply only when it resolves clashing stresses on adjacent syllables (cf. *fundamentàl* + *ménte* → *fùndamentalménte* 'fundamentally' vs. *exàcta-ménte* 'exactly'), while in English this restriction does not hold: *Mississìppi múd* → *Mìssissippi múd.* In Lenakel (§ 6.1.8.3), Move X may move a grid mark only one syllable away.

9.2.3 Destressing

Destressing appears to occur more often as a word-internal rule than as a phrasal one. Cases of phrasal destressing occur in French (Dell 1984) and Italian (Nespor and Vogel 1989). The account of Destressing assumed here is given in § 3.4.2(d).

9.2.4 Beat Addition

Beat Addition, discussed in Liberman and Prince 1977; Selkirk 1984; Hayes 1984; and elsewhere, has the effect of increasing the degree of rhythmic alternation in a phrase by increasing the level of stress on particular syllables. As an example, consider the name *Farrah Fawcett-Majors.* The normal pattern of stress in English names requires rightward prominence. Thus *Majors* bears more stress than *Fawcett;* and *Fawcett-Majors* bears more stress than *Farrah.* This minimal set of prominence relations, along with word-internal prominence, is depicted in the schematic grid in (8):

(8)   ×
× . × . × .
Farrah Fawcett-Majors

However, this is not the usual way such a phrase would be pronounced; rather, one tends to place more stress on *Farrah* than on *Fawcett,* while keeping *Farrah* subordinated to the main stress on *Majors:*

(9)   ×
×   ×
× . × . × .
Fàrrah Fawcett-Májors

The same can be seen in *bìg black cát, Jòhnny's three bóoks, Bìll saw Súe,* and parallel examples.

Just what kind of rule should carry out this change has been a matter of debate in the literature. In one view (e.g. *SPE;* HV), such alternating patterns are the automatic consequence of general principles. The other view (e.g. Liberman and Prince 1977; Selkirk 1984; Hayes 1984; and others) is that an actual Beat Addition rule must be applied. I review the arguments for the latter view in § 9.3 and propose a particular version of Beat Addition in § 9.4.2 and § 9.6.

9.2.5 Eurhythmy

Phrasal stress rules typically conspire to achieve a particular rhythmic target. In general terms, the rules tend to create output configurations in which stresses are spaced not too closely and not too far apart. A grid having these properties is said to be **eurhythmic;** one can also speak of **degrees** of eurhythmy. It has been conjectured (Dell 1984, 116; Selkirk 1984, 36–37; Hayes 1984, 59) that the principles of eurhythmy are invariant across languages, and that they may extend beyond language into other cognitive domains.

Formulating precisely what is meant by eurhythmy is a difficult problem. Liberman and Prince 1977 and others have relied on the notion of **stress clash,** defined as the grid configuration of (10a). At the lowest layer, this prevents stresses on adjacent syllables; at higher layers, it can also rule out strong stresses that are not separated by a weaker stress (10b, c):

```
(10) a. *x x      b. *x       x      c. x       x
         x x           x       x         x   x   x
                       x x x x x         x x x x x
```

Hayes 1984 argues that this conception is inadequate; in particular that there is essentially no difference in the linguistic treatment of patterns like (10b) and (10c). What seems to work better is a kind of gradient principle: adjacent stresses are strongly avoided; stresses that are close but not adjacent are less strictly avoided; and at a certain distance (perhaps four syllables) the spacing becomes fully acceptable. Beyond the ideal distance, we find that stresses are too far apart, so that rhythmic phonology tends to interpolate stresses to fill the gap.

Nespor and Vogel (1989) note languages in which the main effect of eurhythmy is to resist stresses on adjacent syllables (considered by Hayes to be merely the extreme point on a continuum of dysrhythmy). On the basis of their data Nespor and Vogel argue that the definition of eurhythmy may vary parametrically across languages.

The principles of eurhythmy have been based on the well-formedness judgments of listeners. Interestingly, the work of Beckman et al. 1990, which tested whether the eurhythmy principles apply on-line in speaking, yielded negative results. I suggest that the process of planning speech on-line does not evaluate the relative eurhythmy of competing potential outputs when "deciding" whether to apply rhythmic adjustment rules, but that given time to reflect, speakers can judge the rhythmic well-formedness of the result.

In what follows, we will not need to decide between the various conceptions of eurhythmy. The crucial point is that other than End Rules, most rules of phrasal stress are optional; and their tendency to apply (or at least, a speaker's judgment of whether they should apply) appears to be based on the degree of improvement they induce in the eurhythmy of the string.

In the sections that follow, I develop an explicit theory of phrasal stress rules permitting the four rulc types covered above (End Rules, Move X, Destressing, and Beat Addition) to be expressed.

9.3. THE NEED FOR BEAT ADDITION

Where two constituents are sisters, there exists a domain in which an End Rule can apply, assigning them a prominence relation. The question here is what happens when two consecutive constituents are not sisters, as in the first two words of (8)–(9) *Farrah Fawcett-Majors.* As we saw above, it is often the case that such constituents do not receive equal stress; rather, one dominates the other, despite the lack of a domain in which the End Rule could apply.

Ideally, this pattern would not require us to add rules to the grammar, but would follow directly from how we set up our representations and formulate the End Rule. I consider two theories that have this property.

Suppose that when two constituents are metrically concatenated, we line them up so that the TOP of each grid occupies the same layer. Once this is done, we create a new top-layer constituent and apply the End Rule:

```
(11) a.               (          ×  )      output of first cycles
           (×  .)     (×   . ) (× . )
           Farrah  +  Fawcett-Majors

     b.    (×  .) (          ×   )         top-aligned concatenation
           Farrah (×   . ) (× . )
                  Fawcett-Majors

     c.    (                 ×   )         End Rule Right
           (×  .) (          ×   )
           Farrah (×   . ) (× . )
                  Fawcett-Majors
```

The alternating pattern that emerges is *Fàrrah Fawcett-Májors,* which is what we want. As René Kager (p.c.) has pointed out, the top-alignment algorithm is essentially a metrical replication of the conventions of *SPE:* in that pre-metrical theory, the same results follow from the use of a stress subordination convention (pp. 16–17), provided the rules are applied cyclically.

Another theory in which Beat Addition falls out automatically is that of HV. Their account is based on a proposed Stress Equalization Convention:

(12) **Stress Equalization Convention** (HV, p. 265)
When two or more constituents are conjoined into a single higher-level constituent, the [grid] columns of the heads of the constituents are equalized by adding [grid marks] to the lesser column(s).

Once the Stress Equalization Convention has applied, it is straightforward to create a new metrical layer by the End Rule, thus generating a prominence relation between the two sister constituents.

```
(13) a.              (        ×  )      outputs of first cycles
        (×  .)       (×   . ) (× . )
        Farrah  +  Fawcett-Majors

     b. (×   ) (            ×   )      Stress Equalization Convention
        (×  .) (×   . )  (× . )
        Farrah Fawcett-Majors

     c. (                   ×   )      End Rule Right
        (×   ) (            ×   )
        (×  .) (×   . )  (× . )
        Farrah Fawcett-Majors
```

As can be seen, the falling prominence contour on the sequence *Farrah* + *Fawcett* is an automatic consequence of the Stress Equalization Convention.

The Stress Equalization Convention approach does not always generate the same outputs as the top-row-alignment algorithm, as (14) shows:

```
(14) a. Top-Layer Alignment         b. Stress Equalization Convention
        (          ×       )            (             ×        )
        (×      ) (  ×     )            (×        ) (  ×       )
        (×)(×  ) (   × ) (× )           (×        ) (   × ) (× )
         Elroy's (×)(× ) sets           (×)(×  ) (×)(× ) (× )
                 canteen                Elroy's canteen sets
```

Since there seems to be no greater stress on *-roy's* than on *can-*, this example favors HV's account. However, I argue below that neither approach is correct in any event. The theory I will eventually propose generates the output of (14b).

The important thing to bear in mind about both solutions is this: the relative prominence of non-sister weak constituents is determined by degree of embedding; that is, *Farrah* is stronger than its non-sister *Fawcett* in (11) and (13) because it is less deeply embedded in the morphosyntactic tree.

Liberman and Prince 1977, 323–28, and Selkirk 1984, 163, have argued against this claimed correlation of stress and depth of embedding. Among their examples are long, left-branching compounds in English, which receive initial stress when the compound stress rules are applied to them. This initial stress often seems somewhat unnatural, but the reading can be encouraged by building up the compound one word at a time, as in the following example: a lamp that burns *whále oil* is a *whále oil lamp;* a stand for such a lamp is a *whále oil lamp stand;* a person who sells such stands is a *whále oil lamp stand dealer.* Once one has such a compound, one can inspect its contour of secondary stresses, for which both depth-of-embedding theories predict the same result:

(15)

a. **Top-Layer Alignment**

```
( ×                          )
( ×                  ) (×    )
( ×            ) ( × ) dealer
( ×     )(×  ) stand
( × ) (×) lamp
whale oil
```

b. **Stress Equalization Convention**

```
( ×                          )
( ×                  ) (×    )
( ×            ) ( × ) (×    )
( ×     )(×  ) ( × ) (×    )
( × ) (×)(×  ) ( × ) (×   .)
whale oil lamp stand dealer
```

This prediction seems incorrect. Perhaps there is more stress on *dealer* than on *oil, lamp,* or *stand* (though this perception might also be attributed to phrase-final lengthening; § 5.1.8.2); but there seems to be no rising prominence contour on the sequence *oil, lamp, stand.*

Other examples also argue against the depth-of-embedding hypothesis. For instance, in a four-word right-branching structure with right-strong stress, if we arrange the stresses properly an alternating rhythm becomes natural:

(16)

a. **Alternating Pattern**   b. **Prediction of Depth-of-Embedding Theories**

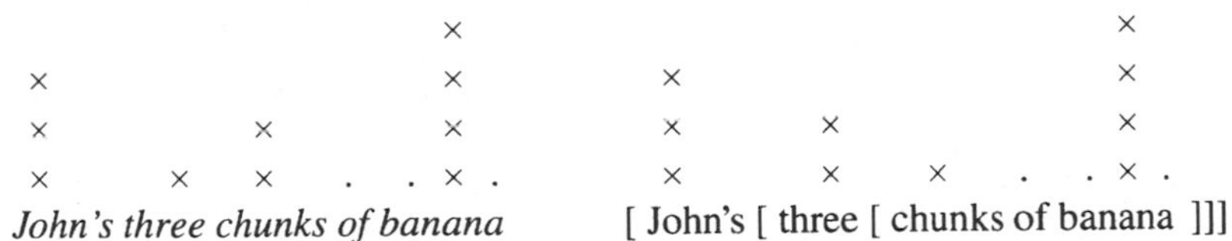

The depth-of-embedding theories generate only (16b), which seems less natural. Other crucial examples involve verbs, which, as various authors have pointed out (Schmerling 1976; Bing 1979; Ladd 1980), tend to resist receiving phrasal stress. If one arranges the words so that a verb can fall in weak position, alternating rhythm becomes dominant (example from Giegerich 1981):

(17)

a. **Alternating Pattern**   b. **Prediction of Depth-of-Embedding Hypothesis**

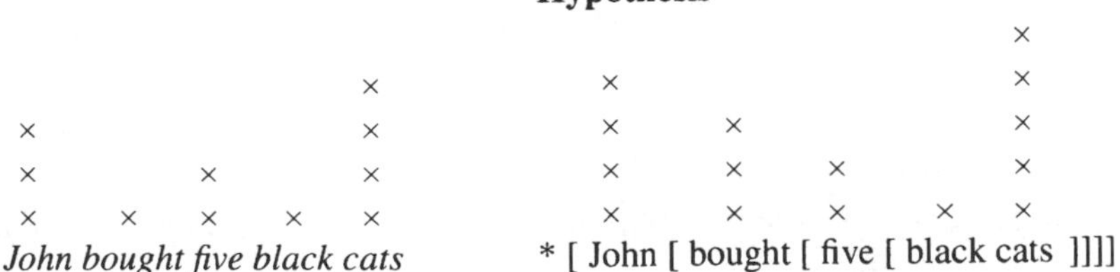

Here, the output of any depth-of-embedding theory seems quite awkward.

HV suggest that the overarticulated stress contours derived by their theory can be ironed out by a Grid Simplification Convention (p. 266), which optionally deletes grid layers above the word layer. While this works for (15b), it seems unworkable in general, because it cannot achieve alternating rhythm in cases like (16) or (17). Summing up, the depth-of-embedding approaches considered here make correct predictions in commonplace examples, but appear to be inadequate when more complex cases are considered.

What DOES determine the relative prominence of non-sister stresses? Formally, I assume that this is due to a rule of Beat Addition, discussed in (32) and § 9.6. A Beat Addition rule can be appropriately formulated to add stresses in the correct positions, avoiding overgeneration.

The actual factors that condition Beat Addition are complex. The most salient of these is eurhythmy; Beat Addition derives (16a), and usually not (16b), because the former is more eurhythmic than the latter. In addition, it matters whether the relevant sequence precedes or follows the main stress; posttonic sequences in English are largely immune to Beat Addition, admitting at most an amplification of the phrase-final stress, whereas pretonic sequences freely allow Beat Addition (see § 9.6.2). Finally, there appears to be a **hierarchy of stressability** among grammatical categories (Ladd 1980), whereby verbs in particular resist Beat Addition.

The claims about Beat Addition made in this section are crucial to the arguments about the End Rule that follow. I return to Beat Addition in § 9.4.2 and § 9.6, proposing a formalization.

9.4. THE END RULE AT PHRASAL LEVELS

Consider now the question of how the End Rule applies at phrasal levels. In the bracketed-grid framework adopted here, an important problem arises: since the constituents being concatenated can differ in their internal complexity, they can differ in the height of their tallest grid column. Where these heights are the same, assigning phrasal stress by the End Rule is a simple matter: we simply form a new layer on top of what we already have, and label it with one version of the End Rule (right or left). This would suffice, for example, for the phrase *Jèsus wépt* in (2a) above. However, it is often the case that one of the concatenated phrases is internally more complex than the other, and thus has more grid marks in its highest column:

```
(18) a. (        ×  )              b.            (    ×  )
        (×   .)(×  )    (× )           (×  )     (×)  (× .)
        Mighty oaks  +  fell           John  +   saw Mary
```

Such cases pose two distinct problems. First, it is not clear how the Nuclear Stress Rule can promote the stress on *fell* in (18a) without violating the Continuous Column Constraint.

```
(19) * (               × )
       (        × )
       (×    .)(× ) (× )
       Mighty oaks fell
```

In the previous section we saw two possible answers to this problem that both fail: if we align the inputs at their top layers, or if we invoke Stress Equalization (12) to amplify whichever input is shorter, then we avoid Continuous Column Constraint violations, but we also generate incorrect secondary stress contours for broad classes of examples.

The second problem concerns what to do when the constituent that we wish to promote by a phrasal stress rule already has the taller column, as in (18b): Do we add an extra grid mark with the new domain, as in (20a); or do we simply add a new domain and let the existing grid columns serve to depict the stress contrast, as in (20b)?

```
(20) a. (              ×  )       b.
                (      ×  )          (    (      ×  ))
        (× ) (×)   (× .)             (× ) (×)   (× .)
        John saw Mary                John saw Mary
```

I first propose concrete answers to these questions, then try to justify them.

9.4.1 Making the Shorter Taller

Examples of the type "make the shorter taller" are illustrated in (18a). The first point to consider is that there are apparently no languages in which phrasal stress rules refer to how many grid marks appear in their inputs. A hypothetical case would be a dialect of English that stressed our examples as in (21):

```
(21) a.                          (      × )          (End Rule Right)
        (× .)     (× )           (× .) (× )
        Jesus  +  wept     →     Jesus wept

     b.                               (       ×      )    (End Rule Right,
          (       × )                 (       × )          only one landing
          (×    .)(× )    (× )        (×    .)(× ) (× )    site available)
          Mighty oaks  +  fell   →    Mighty oaks fell
```

In general, the assignment of prominence contours to sequences of phrases does not depend on their internal structure.[1] This is what we should expect, given the nature of rhythmic structure: a strong beat is not strong to some

1. This needs one qualification: function words, such as pronouns, typically do not take phrasal stress. We assume this is to be accounted for by phonologically cliticizing them onto neighboring full words, as in Hayes (1989a); thus they are not present as terminals for purposes of phrasal stress rules.

absolute degree, but only in relation to other beats in the same rhythmic structure (Liberman and Prince 1977, 262–63). Thus when two independent rhythmic structures are concatenated, no a priori prominence relation exists between them, and this relation must be determined instead by the stress rules. To make this possible, some kind of convention is needed for cases like (18a). The conventions of the preceding section (top-to-top alignment and Stress Equalization) were motivated by precisely this problem.

The second point to consider is just where the equalization conventions go wrong. For cases of the form (18a), where the shorter column of the input must be made the taller column of the output, they work perfectly well. It is in cases like (18b), where the column to be promoted is already taller, that they lead to wrong predictions; namely, they overapply Beat Addition in a wide range of contexts.

A reasonable conclusion to draw is that whatever equalization convention we propose should be limited to cases like (18a) and not to (18b). In other words, the convention applies only where it is required to avoid violating the Continuous Column Constraint:

(22) **Stress Equalization Convention (Revised)**
When two metrical constituents are concatenated, and their tallest grid columns are unequal, then grid marks are assigned to the shorter column if necessary to avoid violating the Continuous Column Constraint.

In essence, this revised convention claims that the End Rule works differently at the phrasal level (where constituents are freshly concatenated) than at the word level (where the whole string is usually present from the beginning). Rather than simply scanning the domain and promoting a peripheral element at the highest grid layer, the phrasal End Rule considers the separate constituents that form the input to the rule, and creates the minimal structure that promotes the rightmost (or leftmost) one. Note that the same kind of End Rule appears to be necessary for cyclic stress assignment, which also creates multiple stress levels. I will not address this issue here, however.

In this view, phrasal stress rules in bracketed grid theory are much like phrasal stress rules in pure tree theory, in that they make the right- or leftmost element strongest, irrespective of its internal structure. The only difference is that within bracketed grid theory, extra structure must sometimes be added to preserve well-formedness.

The notion "add grid layers" employed in (22) must be made more precise. The actual structural change I propose is as in (23):

(23) **Domain Generation**

```
a.            ( × )   =   b.                  ×
    ( × )  →  ( × )                 ×         ×
                                ( . . . )  →  (( . . . ))
```

That is, to a metrical domain whose head grid mark is the highest of its column, add a new grid mark with its own domain. In (23a) the convention is stated in the informal notation I have been using throughout, for convenience. In "official" notation, where metrical constituents actually bracket the terminal string (§ 3.5), Domain Generation would appear as in (23b).

Domain Generation can be thought of as a general-purpose structural change: it carries out the action of the Stress Equalization Convention (22), and as we shall see, it can also serve as the structural change of individual phonological rules such as Beat Addition.

In example (18a), *Mighty oaks fell,* Domain Generation applies to *fell,* so that the Nuclear Stress Rule (i.e. End Rule Right) may promote it without violating the Continuous Column Constraint:

```
(24) a. (        × )                inputs
        (×   .)(× )    (×)
        Mighty oaks  + fell

     b. (        × )(×)             Domain Generation
        (×   .)(× )(×)              (invoked by Stress Equalization (22))
        Mighty oaks fell

     c. (           ×)              End Rule Right
        (        × )(×)
        (×   .)(× )(×)
        Mighty oaks fell
```

9.4.2 Making the Taller Taller

The other type of example I am considering is where the phrase to be promoted is already taller, due to its internal structure. As we have seen, there are two choices for the output structure. In (25a), an entirely new grid layer is added, to depict the relative prominence of *John* versus *saw Mary.* In (25b), the two constituents are simply grouped together, with the existing grid columns serving to depict the difference in stress.

```
(25) a. (              ×  )      b.
               (       ×  )         (     (      ×  ))
        (× ) (×)   (× .)            (× ) (×)   (× .)
        John saw Mary               John saw Mary
```

What might decide between these two possibilities? One might imagine that (25b) should be preferred, since it represents the greater prominence of *Mary* less redundantly. But there is evidence elsewhere that we must in any event interpret grid column heights in relative, not absolute fashion. For example, the occurrences of the syllable *sing* in the following two utterances can be phonetically homophonous:

```
(26) a. (                          × )        b.          }
        (                    ×     )( × )                 } phrasal layers
        (          ×    ) (×       )( × )                 }
        (×     ) ( ×    ) (×       )( × )       ( × ) } word layer
        (× )(× ) ( ×  . ) (×    .  )( × )       ( × ) } foot layer
        Nineteen thousand linguists sing.       Sing.
```

It is true that (26b) might be assigned more layers, to represent various levels of prosodic structure. But since (26a) could be modified to be as complex as one likes, the basic point remains: main-stressed *sing* has the same stress level no matter how many grid marks we happen to need to represent it. That is, only the relative height of grid columns appears to be relevant to phrasal stress assignment. As noted in the preceding section, this relative interpretation is consistent with the role of grids as representing rhythmic structure.

Given this, then there can be no objection to letting (27a) and (27b) represent the same thing, because they depict the same relations of relative prominence.

```
(27) a.     ×       b.
            ×                ×
        × × ×            × × ×
```

Thus (25a) cannot be faulted on this score.

In the theory I have assumed here, the choice between (25a) and (25b) is dictated by the Faithfulness Condition (§ 3.7), repeated as (28) for convenience:

(28) **Faithfulness Condition** Grid marks must be in one-to-one correspondence with domains of which they are heads.

The form in (25a), with the extra grid mark added by the End Rule, obeys the Faithfulness Condition, but (25b) violates it, since the top grid mark over *Mary* must serve as the head of the two domains, *saw Mary* and *John saw Mary.*

Before continuing, it will be useful to characterize my version of the Faithfulness Condition more precisely, since with phrasal stress it is sometimes not immediately obvious whether the constraint is satisfied or not. The crucial notion in (28), namely "head," is given recursively in (29):

(29) **Head**

a. Where $\mathscr{C}_{max}$ is a metrical constituent not included within any other metrical constituent, designate the highest grid mark within $\mathscr{C}_{max}$, if there is one, as the head of $\mathscr{C}_{max}$.

b. Where $\mathscr{C}_n$ is a constituent that has a head, and $\mathscr{C}_{n-1}$ is its daughter, designate the highest grid mark within $\mathscr{C}_{n-1}$ that is not the head of some other constituent as the head of $\mathscr{C}_{n-1}$.

This definition is intended to assign the largest constituent the highest grid mark (29a), then move recursively down the bracketed structure to pair up the remaining constituents and grid marks.

Consider again examples (25a, b), repeated as (30). In (30b), ×'s are replaced by numbers and various ( ) with { } or [ ] for ease of reference.

```
(30) a. (            ×  )      b.
            (     ×  )           {    [       1  ]}
        (× ) (×)   (× .)         (2 ) (3)   (4 .)
        John saw Mary            John saw Mary
```

In (30a), the algorithm that defines heads will move smoothly down the layered structure, assigning grid marks to the appropriate domains. But in (30b), we cannot assign a head to the constituent [ ]: 1 is already taken, since it is the head of { }, and since 3 and 4 are tied, there is no single highest grid mark within [ ] to serve as a head. Since [ ] is headless, (30b) violates the Faithfulness Condition.

To sum up, I have suggested that the problem of "making the taller taller" is to be addressed as follows: the End Rule creates a new layer, whose head vacuously amplifies the already tallest grid column. This maintains a more restrictive theory, since all representations must respect the Faithfulness Condition. I will support the approach empirically below.

I can now provide a derivation for example (25). The initial stages would proceed as just discussed:

```
(31)                              (          ×  )
              (     ×  )               (     ×  )
     (× )     (×)  (× .)  →      (× ) (×)   (× .)
     John  +  saw Mary           John saw Mary
```

The rest of the derivation consists of amplifying the stress on *John* by Beat Addition, which I formulate as in (32) (for full discussion, see § 9.6):

(32) **Beat Addition** Apply Domain Generation (23) pretonically within a domain.

Like most other rules of rhythmic adjustment, Beat Addition is optional, and applies when it improves the degree of eurhythmy. In (31), the relevant domain is the entire sentence; the target constituent for Domain Generation is *John,* and eurhythmy is improved by providing alternating rhythm at the second grid layer.

```
(33) (            ×  )
     (× ) (       ×  )
     (× ) (×)   (× .)
     John saw Mary
```

Since Beat Addition is limited to pretonic position in English, we correctly derive a flat structure for right-branching compounds like (15). (Recall again that the absolute height of the first grid column is of no significance.)

```
(34)  ( ×                           )
      ( ×                  )
      ( ×          )
      ( ×     )
      ( × ) (×)(×  ) ( × )  (×  .)
      whale oil lamp stand dealer
```

The crucial point is that since Beat Addition is a rule, it can involve language-particular restrictions, such as the limitation to pretonic position (as we will see, Dutch lacks such a restriction). Thus it can derive the difference between the alternating rhythm in (17) and the flat rhythm of (15).

Summing up so far, I have proposed explicit answers to the technical problems involved in applying the End Rule at the phrasal level. In cases of the type "make the shorter taller" (§ 9.4.1), the shorter column in the input is amplified by Domain Generation (23), so that the End Rule can apply without violating the Continuous Column Constraint. In cases of the type "make the taller taller" (§ 9.4.2), a new layer is added, with a vacuous grid mark, to satisfy the Faithfulness Condition.

The descriptive result obtained is that Beat Addition can be formulated as a language-specific rule, thus avoiding the empirical problems that arise in theories that attempt to derive the effects of Beat Addition by automatic convention (§ 9.3). In particular, Beat Addition may be limited in English to pretonic position.

The sections that follow provide evidence for the approach just taken. § 9.5.1 shows that the use of Domain Generation in the "make the shorter taller" cases automatically derives the effects attributed by Hayes 1984 to a stipulated condition of Maximality. § 9.5.2 shows that the representations enforced by the Faithfulness Condition derive as an automatic consequence the phenomena attributed by Kager and Visch 1988 to a stipulated Strong Domain Principle.

9.5 MOVE X AND ITS RESTRICTIONS

The Rhythm Rule (§ 9.2.2) is expressed in bracketed grid theory with the rule schema Move X (4). In other frameworks (see § 3.5), it has been expressed as tree relabeling (Liberman and Prince 1977; Kiparsky 1979) and as various forms of tree restructuring (Rischel 1972; Hayes 1984; Giegerich 1985; Kager and Visch 1988). The Move X formalization is inherited from pure grid theory (Prince 1983a).

9.5.1 Maximality

Hayes 1984, 64, working in a tree-restructuring approach, collapsed the Rhythm Rule and Beat Addition into a single rule, paraphrased slightly in (35):

(35) **Rhythmic Adjustment**
In the configuration . . . X Y . . . Mainstress . . . , adjoin Y to X.

By "adjunction," it is meant that the sequence XY is made a constituent, with Y weak relative to X. Where the rule functions as Beat Addition, it carries out a rebracketing, such that the output stress contour can be simply read off the tree, using the tree-to-grid translation rules employed by Hayes. In (36), X is *Farrah*, the Y that adjoins as X's weak sister is *Fawcett*, and Mainstress is the stressed syllable of *Majors:*

(36)

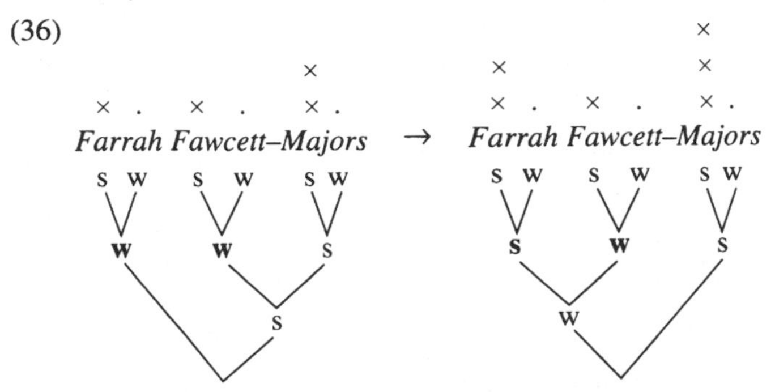

Rhythmic Adjustment acts as the Rhythm Rule when X and Y are sisters. Here, adjunction is vacuous for purposes of constituent structure but does render Y weak relative to X. In the following example, X is *Mary*, Y is *Ellen*, and Mainstress is the stressed syllable of *Mathers*.

(37)

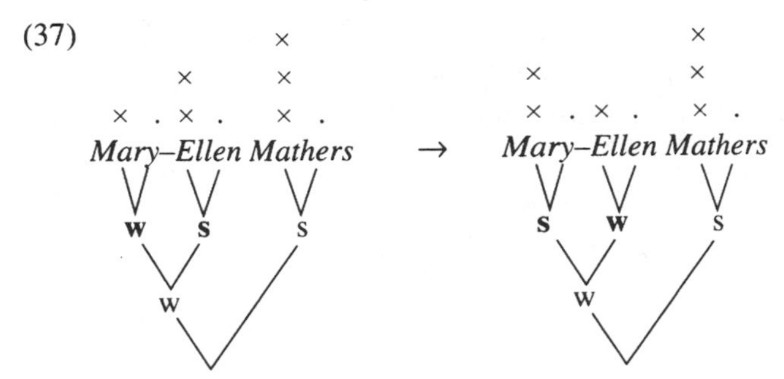

Hayes also observed a serious inadequacy in this analysis: when Rhythmic Adjustment is applied to certain left-branching structures, it overgenerates, as in the derivation in (38). Given the stress pattern of the input (cf. *overdóne, overdone stéak, overdone steak blúes*), the derivation begins with two applications of Rhythmic Adjustment in its Rhythm Rule incarnation:

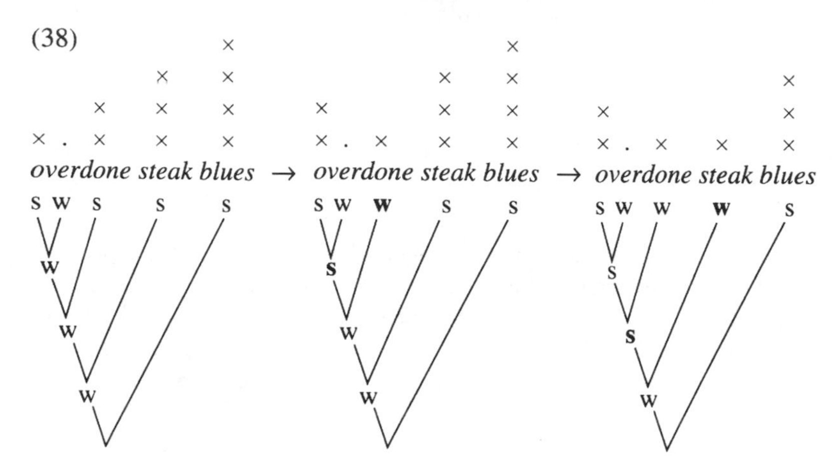

The crucial place where the rule fails is in (39): if we take *done* to be X and *steak* to be Y, we derive an ill-formed output.

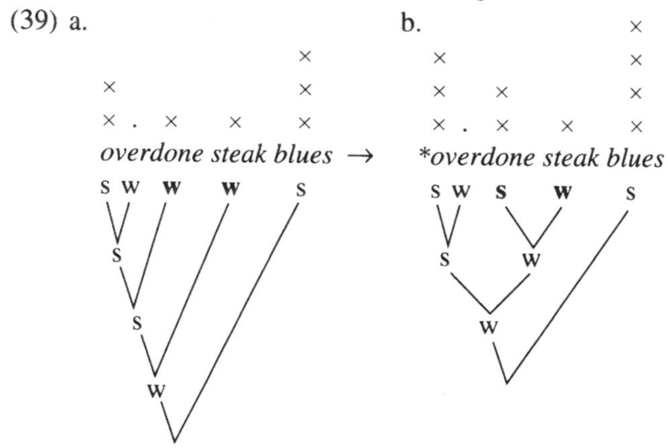

It is important to note that this output is ill-formed even though it is eurhythmic; essentially the same output stress contour can be achieved in other examples, such as *álmost hárd-boiled égg:*

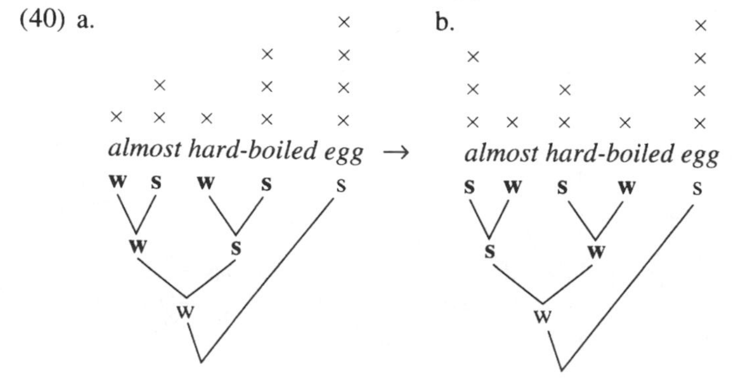

What rules out (39) is not its grid structure, but its bracketing. The problem is to determine just how bracketing blocks Rhythmic Adjustment.

In Hayes 1984, the crucial factor is claimed to be a principle of **Maximality,** stating that metrical rules must analyze their terms maximally. In (39), we cannot choose *done* as term X of Rhythmic Adjustment, since a larger choice (*overdone*) is available. Adjunction of *steak* to *overdone* is vacuous in its effects, so (39a) remains as the surface representation.

The problem with this account is that few if any other cases in metrical phonology have turned up that support Maximality as a general principle. For the one case I am aware of, Kager 1989, 156–58, Maximality is not the only possible account (cf. § 5.1.7). It seems likely, then, that Maximality cannot qualify as a general explanation of the facts of (39), and a better account would be welcome.

The phenomenon is in fact predicted by the principles assumed here, in particular by the Continuous Column Constraint together with the Faithfulness Condition. Consider first the construction of the basic bracketed grid for (39). This is a case of making the shorter taller, and as proposed above, we must invoke Domain Generation (23) to allow the End Rule to apply without violating the Continuous Column Constraint. Domain Generation applies once to *steak,* twice to *blues:*

```
(41) a.                              b.                          c. (            × )
       (     × )                        (      × ) (× )             (      × ) (× )
       (. ×) (× )     (× )  →           (× .) (× ) (× )  →          (× .) (× ) (× )
       overdone  +  steak               overdone steak              overdone steak

     d. (             × )                e. (             × ) (× )
        (      × ) (× )                     (      × ) (× ) (× )
     →  (× .) (× ) (× )      (× )  →        (× .) (× ) (× ) (× )  →
        overdone steak  +  blues            overdone steak blues

     f. (                  × )
        (             × ) (× )
        (      × ) (× ) (× )
        (× .) (× ) (× ) (× )
        overdone steak blues
```

Within the structure that results, the Rhythm Rule (Move X) can apply twice:

```
(42) a. (                  × )       b. (                  × )
        (             × ) (× )          (× ←——————— ) (× )
        (× ←— ) (× ) (× )    →          (×         ) (× ) (× )
        (× .) (× ) (× ) (× )            (× .) (× ) (× ) (× )
        overdone steak blues            overdone steak blues
```

This brings us to the crucial stage of the derivation. The eurhythmy of the output could be increased by moving the second grid mark on *steak* leftward onto *done.* But to do this would violate the Faithfulness Condition, since the domain *overdone* would have two marks and the second-layer domain on *steak* none:

```
(43) * (                    × )
       (×              ) ( × )
       (×      ×)(   ) ( × )
       (× .)(× ) ( × ) ( × )
        overdone steak blues
```

The form must therefore remain as in (42b), which is correct. Other examples would work the same way.

It can be seen, then, that the predictions of the Maximality principle fall out from the theory presented here. Note that both the crucial principles in the present theory—the Faithfulness Condition and the Continuous Column Constraint—can be independently motivated, as was argued in § 3.8.1 and § 3.4.2.

9.5.2 The Strong Domain Principle

Kager and Visch 1988 extended the approach of Hayes 1984 in exploring further properties of Rhythmic Adjustment in both English and Dutch. They find that many of the facts of English are replicated in Dutch, particularly the exclusion of forms comparable to **òverdòne steak blúes* (pp. 66–67). The Dutch data are in many respects more interesting than English, since in Dutch both the Rhythm Rule and Beat Addition are bidirectional rather than leftward-only. Kager and Visch arrive at a number of important results; here I focus only on their proposal that Rhythmic Adjustment is constrained by a **Strong Domain Principle,** which is stated as in (44):

(44) **Strong Domain Principle** No prosodic transformation may apply to the head of a strong domain.

The principle excludes Rhythmic Adjustment in two classes of examples. One type is exemplified by (45), due originally to Prince 1983a. Here, the relevant strong domain is *Tom Paine.* In principle, we could apply Rhythmic Adjustment (35) to adjoin *Paine* to *Tom,* with the main stress on *blues* serving as the trigger. But such an application turns out to be ill-formed:

(45)

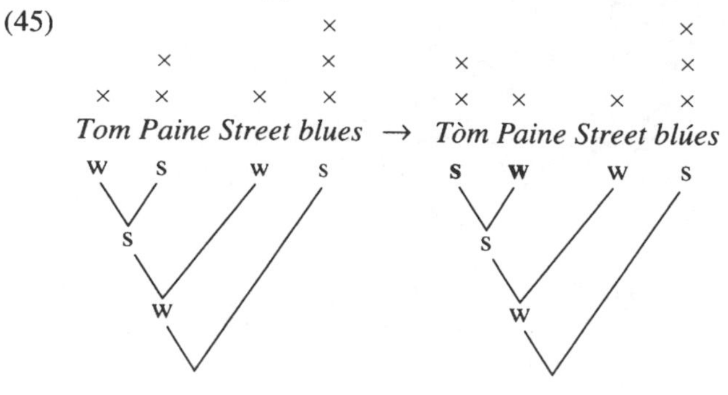

The Strong Domain Principle excludes this outcome straightforwardly, since to derive it Rhythmic Adjustment would have to adjoin the constituent *Paine,* which is the head of the strong domain *Tom Paine.*

The other class of Strong Domain examples in English arises when the relevant strong domain follows its weak sister. In (46), the relevant strong domain is *Fawcett-Majors;* thus *Majors* is the head of a strong domain, and Rhythmic Adjustment cannot apply to it:

(46)

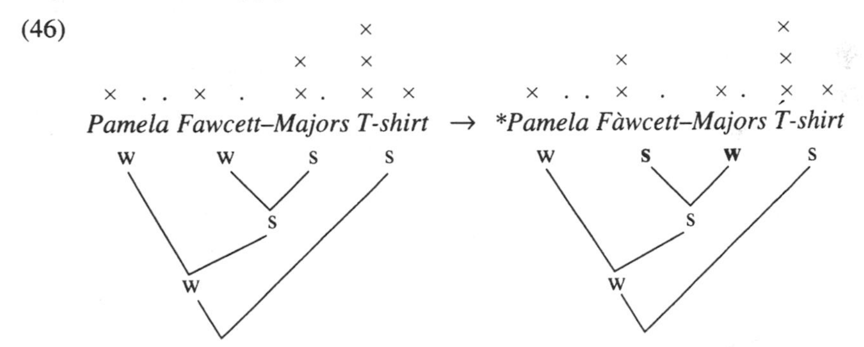

With a different derivation, however, it is possible to apply Rhythmic Adjustment to *Majors.* All that is necessary is to apply the rule first to the larger domain *Pamela Fawcett-Majors.* Once this is done, *Fawcett-Majors* is no longer a strong domain, and it may be internally relabeled:

(47) a. *Pamela Fawcett–Majors T-shirt* → b. *Pamela Fawcett–Majors T-shirt*

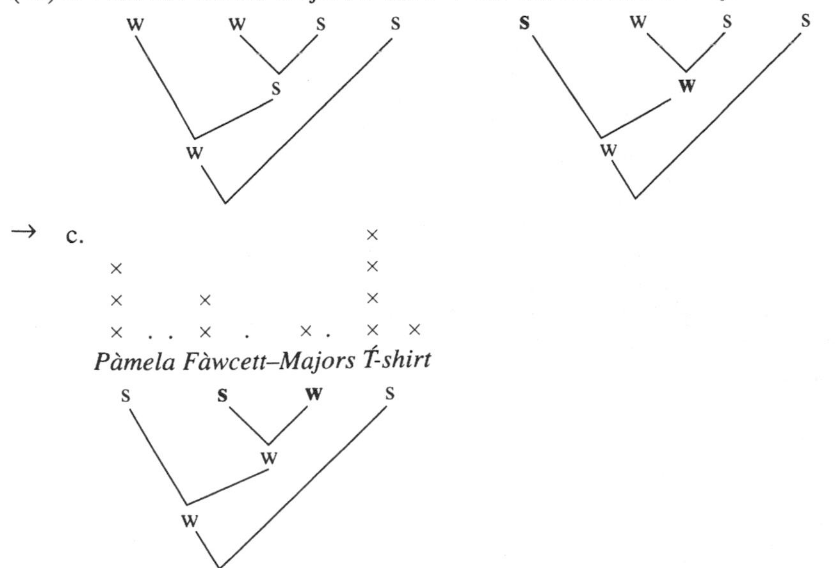

Thus the Strong Domain Principle can dictate the order of derivations (47), as well as block particular derivations (45).

As Kager and Visch show, the Strong Domain Principle is also supported by Dutch data. For instance, the English facts just reviewed can be replicated in Dutch. In addition, since Rhythmic Adjustment in Dutch is bidirectional (48a), one may also check for Strong Domain cases in the rightward direction. As (48b) shows, such cases are indeed ill-formed.

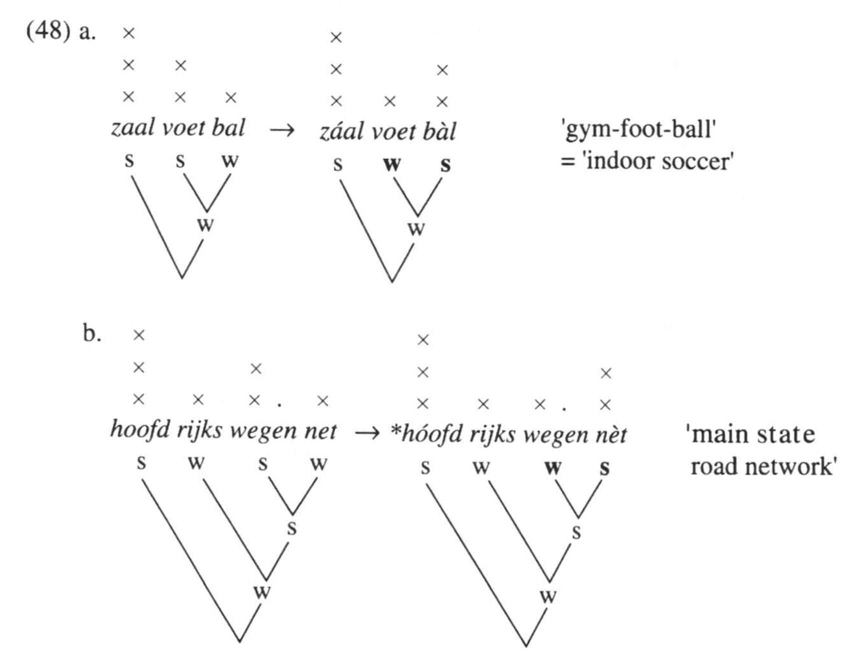

It is clear, then, that the Strong Domain Principle makes interesting predictions about both Dutch and English. I will now show that the phenomena it accounts for follow from more primitive formal principles of bracketed grid theory, namely the Faithfulness Condition and the Continuous Column Constraint.

Consider first example (45), *Tom Pàine Street blúes.* The proposed input grid in (49) can be justified as follows. Names in English receive rising stress, so there is a right-strong domain on *Tom Paine. Tom Paine,* in turn, is strong relative to *Street,* as with all *street* compounds. Because of the Faithfulness Condition, the domain *Tom Paine Street* must have its own (phonetically vacuous) head, which is the third grid mark over *Paine.* The grid column and domains over *blues* are the result of Domain Generation (23) and End Rule Right.

```
(49) (                        × )
     (       ×          ) ( × )
     (       × )          ( × )
     (×)   ( × ) ( × ) ( × )
     Tom Paine Street blues
```

Once the representation is set up, the blockage of the Rhythm Rule (in my terms, Move X) on *Tom Paine* follows straightforwardly from the Continuous Column Constraint: neither of the top two grid marks on *Paine* can move without creating a discontinuous grid column.

```
(50) a.  (                  × )         b.  (                    × )
         (      ×        ) ( × )            (× ←——           ) ( × )
         (× ←—— )          ( × )            (      × )         ( × )
         (×)  ( × ) ( × ) ( × )             (×)  ( × ) ( × ) ( × )
       * Tòm Paine Street blúes,          * Tòm Paine Street blúes
```

The example *Pamela Fawcett-Majors T-shirt* (46) has a similar account: at the crucial stage in the derivation, the boldface grid marks on *Majors* are blocked by the Continuous Column Constraint from moving leftward:

```
(51) (                          ×     )
     (                      ×  )(×    )
              (             ×  )(×    )
     (×  . .) (×   . )  (× . )(×)( × )
     Pamela Fawcett-Majors T-shirt
```

The actual derivation may continue as follows. Beat Addition (32) applies to amplify the stress on *Pamela* (52a), whereupon it becomes a possible landing site for Move X on the third grid layer (52b). Once the third grid mark on *Majors* has been moved, the Continuous Column Constraint will no longer prevent leftward movement of the second grid mark (52c):

```
(52) a. (                          ×     )     Beat Addition
        (                      ×  )(×    )
        (×      ) (            ×  )(×    )
        (×  . .) (×   . )  (× . )(×)( × )
        Pamela Fawcett-Majors T-shirt

     b. (                          ×     )     Move X (Layer 3)
        (× ←——————————————        )(×    )
        (×      ) (            ×  )(×    )
        (×  . .) (×   . )  (× . )(×)( × )
        Pamela Fawcett-Majors T-shirt

     c. (                          ×     )     Move X (Layer 2)
        (×                        )(×    )
        (×      ) (× ←——————      )(×    )
        (×  . .) (×   . )  (× . )(×)( × )
        Pamela Fawcett-Majors T-shirt
```

This completes the derivation.

The Dutch example (48b) is the mirror image of (49) and has the same account: here, the extra grid mark on *wegen* blocks rightward movement:

```
(53) (×                    )
     (×  ) (       ×       )
     (×  )        (×       )
     (×  ) (× )  (×  .) (×)
     hoofd rijks wegen net
     main  state road   network
```

In general, it appears that the effects of the Strong Domain Principle are duplicated in a theory positing two principles of an arguably more primitive formal character, namely the Faithfulness Condition and the Continuous Column Constraint.

Kager and Visch (1988) note that the Strong Domain Principle has effects that go beyond the cases discussed here; for example, it can properly block Destressing rules from applying to strong constituents. In fact, the Continuous Column Constraint and the Strong Domain Principle are quite similar in their content and predictions, in that both tend to prevent weakening of already strong elements. The Continuous Column Constraint is arguably the more general of the two, in that it also crucially regulates the application of the End Rule within words: the Continuous Column Constraint allows the End Rule to apply properly when it must favor foot heads over stray syllables in assigning word stress (§ 5.1.3). More generally, the Continuous Column Constraint predicts effects of the form "the strong get stronger," as well as effects of the form "only the weak get weaker"; whereas the Strong Domain Principle only accounts for the latter cases. This suggests that it is more desirable to adopt a theory in which the empirical effects of the Strong Domain Principle are attributed to the Continuous Column Constraint, rather than vice versa.

Other accounts of the Strong Domain facts have also appeared in the literature. The proposal of Prince 1983a, 40–41, relies on the assumption that *Tom Paine* is to be treated like a phrase, not a compound. This seems unlikely: for example, to account for the Strong Domain effect in *Ocean Àvenue Club mémbership list* one would have to assume that *Ocean Ávenue* is a phrase, despite its being completely parallel to the compound *Ócean Street* in its morphosyntactic structure. Hammond 1984a was the first to observe that a bracketed grid approach can structurally distinguish Strong Domain forms from parallel forms that do undergo the Rhythm Rule. But his proposal seems incomplete, in that it provides no formal account of two crucial notions: how the End Rule is to be applied, and how adjacency is defined. Moreover, Hammond's representations face a critical problem of interpretation, since in them a taller grid column can represent a weaker stress. Finally, the account of Gussenhoven 1991 is quite similar to my own; I believe that the present account is to be preferred because it does not depend on the presence of intonational pitch accents: the phenomena above persist even in positions (e.g. following a contrastive stress) where intonational pitch accents do not occur.

9.5.3 Summary

This section has shown how evidence from the Rhythm Rule justifies the approach to phrasal stress developed in § 9.4. In particular, the use of Domain Generation (23) to "make the shorter taller" gives us precisely the domains (e.g. the second layer on *steak* in (42b)) that allow us to derive Maximality from a primitive principle, namely the Faithfulness Condition. Further, my account of "making the taller taller," which conforms to the Faithfulness Condition by adding vacuous grid marks rather than retaining the shortest grid, makes it possible to derive the Strong Domain facts from the Continuous Column Constraint.

9.6 MORE ON BEAT ADDITION

9.6.1 Application of Beat Addition to Deeply Embedded Constituents

The version of Beat Addition I have assumed so far is: apply Domain Generation in pretonic position. In simple examples, this rule suffices, since the bracketed grid has a slot available to receive the new domain. Such an example would be (33), repeated as (54):

```
(54) (             ×  )        (            ×  )
             (     ×  )   →    (× ) (       ×  )
     (× ) (×)  (×  .)          (× ) (×)  (×  .)
     John saw Mary             John saw Mary
```

However, in more complex examples we run into difficulties. Consider (16), repeated as (55). The End Rule gives us the minimal structure of (55a), which should be amplified by Beat Addition to become the eurhythmic (55b), shown in schematic pure-grid representation. But as the full bracketed form (55c) shows, this violates the Faithfulness Condition. Specifically, in promoting the stress on *chunks*, the domain *chunks of banana* is deprived of a head; that is, the definition (29) fails to single out a head for this domain.

(55) a.
```
(                         ×  )
      (                   ×  )
            (             ×  )
(×   )( ×) (×  )(.     .  × .)
John's three chunks of banana
```

b.
```
                          ×
×                         ×
×             ×           ×
×       ×     ×     .   . × .
John's three chunks of banana
```

c.
```
(                         ×  )
(×   )(                   ×  )
(×   )      ((×   )       ×  )
(×   )( ×)( ×   )(.    .  × .)
John's three chunks of banana
```

In general, we will encounter this difficulty whenever we try to apply Beat Addition to the most deeply embedded weak constituent. Because of eurhythmy, this is not the most common environment for Beat Addition; but given suitable spacing of the syllables it is perfectly possible, as examples like (55) and (17) show.

This problem is not unique to the theory proposed here. It also arises in the adjunction theory of Hayes 1984: this theory cannot promote the stress on *chunks* because Beat Addition in English can only be leftward adjunction. But there is nothing to adjoin leftward to *chunks* that could promote its stress (*of* is a stressless clitic, and would not induce an additional grid mark if adjoined).

To solve this problem, I adapt a proposal made by Selkirk 1984 within a pure-grid framework, based on the convention in (56):

(56) **Textual Prominence Preservation Condition** (p. 56)
A text-to-grid alignment rule applying on a syntactic domain d_i is necessarily satisfied on that domain.

What this is intended to mean is that if the End Rule marks a particular syllable as the strongest of a domain, the output of all subsequent rules must be adjusted to keep that syllable strongest. Thus in the earlier example *Farrah Fawcett-Majors*, the application of Beat Addition to *Farrah* would in Selkirk's approach require further addition of a grid mark to *Majors* so that End Rule Right will continue to be satisfied on the domain as a whole:

(57)
```
                                                       ×
                     ×            ×                    ×
×  .   ×   .         × .          ×  .    ×   .        × .
Farrah Fawcett-Majors      →      Farrah Fawcett-Majors
```

As I will show, Selkirk's proposal can be adapted to a bracketed grid theory to provide a solution to the problem we are facing. To start, recall the claim that absolute heights of grid columns are phonetically irrelevant; thus (58a) is in principle homophonous with (58b):

(58) a.

```
(                          ×  )
    (                      ×  )
         (                 ×  )
(×    )( ×) ( ×   )(.    . × .)
John's three chunks of banana
```

b.

```
(                          ×  )
    (                      ×  )
         (                 ×  )
                  (        ×  )
(×    )( ×) ( ×   )(.    . × .)
John's three chunks of banana
```

Given this assumption, there is no harm in allowing representations to be augmented optionally with vacuous grid layers, where "vacuous" means that no change in relative column heights is induced. For example, when we add the boldface layer in (58b), we still retain essentially what we had in (58a), namely a grid in which *banana* is stronger than everything to its left. Let us adopt a formal convention to allow this operation:

(59) **Grid Expansion**
   a. Insert an empty grid layer.
   b. Resolve any Continuous Column Constraint violations that would result by applying Domain Generation (23).

Example (58b) can be derived from (58a) by precisely this convention. The layer that is inserted is the second, and Domain Generation is applied to the constituent *of banana* to avoid a Continuous Column Constraint violation.

My assumption is that Grid Expansion is always available as an option in metrical derivations. In most places it applies without concrete effects, merely creating an output grid that represents the same thing as the input. But since it expands the number of grid layers, Grid Expansion creates new locations where Beat Addition could apply. For example, to the new grid in (58b) we can apply Beat Addition as shown, thus satisfying eurhythmy and deriving the correct surface form:

(60)

```
(                          ×  )
    (                      ×  )
(×    )      (             ×  )
(×    )      ( ×  )(       ×  )
(×    )( ×) ( ×   )(.    . × .)
John's three chunks of banana
```

I propose, then, the mechanism for Beat Addition in (61):

(61) a. Grid Expansion (59), applied optionally anywhere in the derivation
   b. Beat Addition (32): Apply Domain Generation pretonically within a domain.

This proposal can now be compared with that of Selkirk 1984: I follow Selkirk in carrying out Beat Addition by amplifying both the target syllable and the tallest grid column; the latter creates vertical space to accommodate the former. Where my proposal differs is in the formal convention ((56) vs. (59)) that is used to expand the grid. Under Selkirk's account, the rules must be global in character: they "remember" which grid marks were assigned by text-to-grid alignment rules, and apply so as not to override the prominence relations that these grid marks encode. In the proposal made here, the rules need refer only to their input representations, not to earlier stages of the derivation.

To return to the original point, the proposal solves the problem for the Faithfulness Condition created by cases like (60), where a deeply embedded weak element undergoes Beat Addition.

One may wonder whether the formulation of Beat Addition in (61) does not overgenerate; for example, in a simple case like *John saw Mary* (62), might not *John* be promoted to become the strongest stress of the phrase? This turns out to be excluded. If *John* is promoted to the level of *Mary,* as in (62a), then definition (29) no longer yields a head for the domain *John saw Mary,* since it has no tallest grid column. If *John* is promoted even higher than *Mary,* then we get a mismatch: the smallest domain around *John* is paired with no grid mark, and the lowest grid mark on *Mary* has no domain. This is shown in (62b), which uses "official" representations (§ 3.5) for clarity.

(62)

a.
```
(                ×  )        ((×  )        ×  )
      (          ×  )   →     (×  ) (      ×  )
(×  ) (×)     (× .)           (×  ) (×)   (× .)
John  saw     Mary          * John  saw   Mary
```

b.
```
                                       ×1
                  ×1                   ×2                  ×3
                  ×2        →          ×4                  ×5
 ×3       ×4      ×5                   ×6          ×7      ×
```
$({}_1({}_3\text{John})_3({}_2({}_4\text{saw})_4({}_5\text{Mary})_5)_2)_1 \rightarrow ({}_1({}_2({}_4({}_6(\text{John}))_6)_4)_2({}_3({}_7\text{saw})_7({}_5\text{Mary})_5)_3)_1$

The result is that illegal applications of Beat Addition are excluded by the Faithfulness Condition.

The introduction of Grid Expansion into the system likewise does not lead to problems: since the addition of a grid layer must always be accompanied by the cloning of higher-layer grid marks, Grid Expansion cannot alter existing prominence relations, but only permit the addition of new ones, which is what we want.

9.6.2 Directional Effects in Beat Addition

I have stated that in English, Beat Addition is limited to pretonic position. For this reason, left-branching forms like (34), *whále oil lamp stand dealer,* resist

Beat Addition, whereas right-branching forms like (17a), *Jòhn bought five black cáts,* invite it.

A question worth pursuing further is: What is meant by "pretonic"? There are two reasonable interpretations. Pretonic might mean "preceding the main stress of the entire utterance." Such an interpretation might be called "maximal," since it is the largest available domain in which one determines whether a beat is pretonic or not. The other interpretation would be a "minimal" interpretation: one considers the SMALLEST domain in which the target beat is weak relative to another beat, and classifies the beat as pretonic or posttonic depending on its relative position within that domain.

Data already in the literature can help decide this question. Hayes 1984 adduces examples like (63) as further evidence for the Maximality principle:

(63)

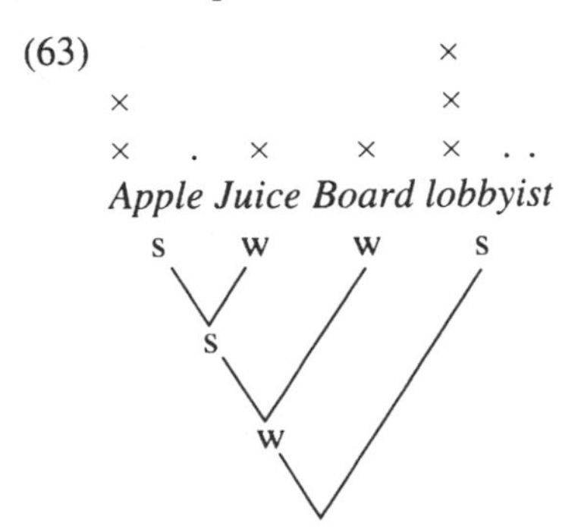

'lobbyist for the board that regulates apple juice'

In this example, it seems impossible to promote the stress on *Juice,* even though this would increase eurhythmy. This observation follows from Maximality (§ 9.5.1), since *Board* cannot be adjoined to *Juice,* the larger target *Apple Juice* being available.

Given that we have seen that other Maximality phenomena can be made to follow elsewhere from more primitive principles (§ 9.5.1), and that the adjunction approach to Beat Addition fails elsewhere (§ 9.6.1), there is good reason to find an alternative account of such examples. My proposal is to assume that "pretonic" and "posttonic" are defined minimally, as in (64):

(64) **Pretonic, Posttonic**

Let $\mathcal{G}_i$ be a grid column.

Let $\mathcal{C}$ be the smallest domain containing a grid column $\mathcal{G}_j$ that is taller than $\mathcal{G}_i$.

$\mathcal{G}_i$ is **pretonic** if it precedes $\mathcal{G}_j$; otherwise it is **posttonic.**

In simple examples where the minimal and maximal definitions coincide, it is easy to show that Beat Addition is limited in English to pretonic position. For example, as Kager and Visch 1988 point out, in the simple phrase *Apple Juice Board* it is impossible to promote the posttonic beat on *Juice: *Ápple*

Jùice Board. Now, in (65) *Apple Juice Board lobbyist,* the crucial grid column on *Juice* precedes the nuclear stress, but is posttonic by the minimal definition:

```
(65) (                    ×     )
     (×               ) (×      )
     (×         )       (×      )
     (×   .)( × ) ( × ) (×   . .)
     Apple Juice Board lobbyist
```

Specifically, column $\mathcal{G}_i$ is on *Juice,* the minimal constituent $\mathcal{C}$ containing a taller column is *Apple Juice,* the taller column $\mathcal{G}_j$ is on *Apple,* and $\mathcal{G}_i$ follows $\mathcal{G}_j$. Thus if we adopt the hypotheses that pretonic and posttonic positions are defined minimally, and that English Beat Addition is pretonic only, we correctly predict that Beat Addition cannot apply to *Juice* in (65).

If this strategy is correct, we should be able to get Beat Addition in contexts that are "maximally" posttonic, but minimally pretonic. In (66) is such an example, involving contrastive main stress on *Pamela:*

```
(66) a. (×                     )      b. (×                     )
        (×       )(          × )  →      (×       )(          × )
        (×       )     (     × )         (×       )(×) (      × )
        (×  . . )(×) ( × ) (× )          (×  . . )(×) ( × ) (× )
        Pamela's five black cats         Pamela's five black cats
        (e.g. not Amanda's)
```

(In both a. and b., *Pamela's* is printed in bold and *five black cats* in italics.)

Such examples are hard to verify objectively, since the intonational pitch accents that mark stress most saliently are excluded after the main stress. Nonetheless, the judgment that (66b) is well-formed seems plausible.

Finally, consider data from Dutch, where, as Kager and Visch (1988) observe, Beat Addition can apply posttonically as well as pretonically. If the approach taken here is correct, Dutch examples parallel to English (65) should be able to undergo Beat Addition. As Kager and Visch point out, this is indeed the case. In (67) is the derivation I propose for the relevant examples:

```
(67) a. (                    × )    '10 Mendelssohn Street'    KV 66
        (×                 )(× )
        (×          )       (× )
        (×   . )(× ) ( × )(× )
        Mendelssohn straat tien
        Mendelssohn street ten
```

```
             b. (                    × )             c. (                    × )
   →            (×                 )(× )      →         (×                 )(× )
  Grid          (×          )       (× )     Beat       (×          )       (× )
Expansion       (×     )            (× )   Addition     (×     )(× )        (× )
  (59)          (×   . )(× ) ( × )(× )                  (×   . )(× ) ( × )(× )
                Mendelssohn straat tien                 Mendelssohn straat tien
```

Since Dutch has posttonic Beat Addition, it can increase eurhythmy in such examples. The minimal domain in which Beat Addition is applied posttonically is the constituent *Mendelssohn.* The comparable cases in English (e.g. (65)) must remain with a rhythmic lapse.

Kager and Visch cleverly derive forms like (67) by first adjoining the syllable *-sohn* rightward to the syllable *straat.* This creates a constituent to which the leftward Rhythm Rule can then apply. My derivation is less ambitious; it simply relates (67) to the fact that Dutch allows posttonic Beat Addition.

One further observation seems relevant. The theory proposed here assigns different stress patterns to forms like (42b) *overdone steak blues* and (65) *Apple Juice Board lobbyist;* in the former the stresses are ranked 20431 while the latter ranks them 20331. In contrast, the theory of Hayes 1984 gives them identical 20331 contours. My judgment is that there is a slight difference between the two, going in the direction predicted by the newer theory; this agrees with the judgment of Prince 1983a, 44–45. Obviously, not too much weight can be attached to such subtle distinctions.

A final issue to be addressed is the fact that by adopting separate accounts for the Rhythm Rule and Beat Addition (i.e. (4) and (32)), we have lost a generalization made by the theory of Hayes 1984, which unifies the rules: Beat Addition and the Rhythm Rule go in the same direction(s) (either left only, or ambidextrous) within a single language. Thus, English with leftward-only rhythm differs from Danish, German, and Dutch, with bidirectional rules.

In defense of my approach, it can be noted that there are configurations in which it appears to be easier to apply Beat Addition than the Rhythm Rule; see Rischel 1972, 213–14, 221, for Danish; Visch 1989, 136–37, for Dutch. This means that the precise equation of the two made in Hayes 1984 is slightly suspect in any event.

Nevertheless, it still seems appropriate to try to explain the cross-linguistic correlations. I would venture a functional account. Consider first the following asymmetry: no language yet examined has only rightward rhythm and not leftward. Why should this be so? A reasonable fact to take into account is that in many languages, (e.g. English and Dutch) intonational pitch accents may not occur to the right of the nuclear stress. Since pitch accents are usually the strongest cue for stress (§ 2.1), this means that in the simpler (and surely far more common) cases, the phonetic cues for secondary stress before the main stress will be stronger than those for secondary stress after it. Assuming that it is the simple cases that lead language learners to set up a phonological rule, we would expect languages to exhibit a bias in favor of leftward rhythm.

More generally, I would claim that the likelihood of a language establishing a Rhythm Rule or Beat Addition in a given context is proportional to the strength of the phonetic cues for stress in that context. My casual impression is that Dutch has somewhat stronger cues for stress in posttonic position than English; this correlates with the fact that Dutch has developed rightward versions of Beat Addition and the Rhythm Rule.

Under this account, the Rhythm Rule and Beat Addition covary in their direction because they both arise under the same conditions, namely the presence of sufficiently salient cues for stress in a given context. Like other functional explanations, this one is not guaranteed to be exceptionless, and would admit the possibility of unusual cases where a language has Beat Addition but not the Rhythm Rule, or vice versa, in some particular direction.

Explanations based on data available to the language learner must also take into account principles of universal grammar. Consider that in examples with unusual structure, Beat Addition can be blocked before the nuclear stress (e.g. (65) **Àpple Juìce Board lóbbyist*), or can apply anyway after the main stress (e.g. (66) *Pámela's fìve black càts*). Such cases are counterexamples to the functional generalization made above, but they are the predicted outcome in a system where "pretonic" and "posttonic" are defined minimally, as above. I would suggest that a language learner determines whether her language has Beat Addition or the Rhythm Rule on the basis of simple examples where the categories "prenuclear/postnuclear" and "pretonic/posttonic" are equivalent; surely such examples are far more common. The formal grammar the learner constructs, however, is constrained by universal grammar to define "pretonic" and "posttonic" minimally, and thus yields the observed judgments when we present it with examples like (65)–(66) in which "pretonic/posttonic" and "prenuclear/postnuclear" do not coincide.

9.7 CONCLUSION

This chapter has attempted to develop an explicit theory of phrasal stress rules within bracketed grid theory, concentrating mostly on English. My proposals have covered the following areas.

First, I have maintained that phrasal stress rules respect the Faithfulness Condition and the Continuous Column Constraint. These assumptions lead to a particular view of how the End Rule works at the phrasal level: (a) Where it promotes an element whose grid column is shorter than that of its sister constituent, it adds sufficient grid columns by Domain Generation to produce an output that is compatible with the Continuous Column Constraint. (b) Where the End Rule promotes an element whose grid column is shorter than that of its sister, it adds a vacuous grid layer, to produce an output that is compatible with the Faithfulness Condition.

Arranging the End Rule to conform to these basic principles makes it possible to obtain more satisfactory accounts of two phenomena from the literature: the Maximality cases examined by Hayes 1984 and the Strong Domain Principle cases treated by Kager and Visch 1988.

I have also discussed Beat Addition and how to formulate it. First, I defended the need for such a rule against claims that its effects can be attributed to general conventions. Next, I showed that the adjunction account of Beat

Addition in Hayes 1984 cannot cover all cases. To handle this problem, I adapted the proposal of Selkirk 1984, making it nonglobal by expressing it with the formal convention of Grid Expansion (59). Finally, I discussed directional effects in Beat Addition, showing that effects attributed to Maximality by Hayes 1984 and Kager and Visch 1988 fall out naturally from the restriction of English Beat Addition to "pretonic" position. Here, the crucial assumption was that "pretonic" is defined on minimal constituents, not on the utterance as a whole.

Phrasal stress is clearly a less well understood area than word stress, and the proposals made in this section, based primarily on just two languages, must be counted as speculative. If they are correct, they constitute a case where rather mysterious surface data can be explained on the basis of simple, abstract principles.

• 10 •

Theoretical Synopsis and Conclusion

10.1 SYNOPSIS OF FORMAL PROPOSALS

For reference and as a spur to further research, the following material gives in outline the formal proposals of this book, cross-referenced to the sections of the book where they are discussed.

(1) **Well-Formedness Conditions on Bracketed Grids**

a. **Continuous Column Constraint**
A grid containing a column with a mark on layer $n + 1$ and no mark on layer n is ill-formed. Phonological rules are blocked when they would create such a configuration (§ 3.4.2).

b. **Faithfulness Condition**
Grid marks must be in one-to-one correspondence with the domains of which they are heads (§ 3.7, § 9.4.2).

(2) **Inventory of Basic Metrical Structures**

a. **Bounded**
  i. Moraic Trochee (§ 6.1)
  ii. Syllabic Trochee (§ 6.2, § 5.1.9)
  iii. Iamb (§ 6.3)

b. **Unbounded**
  i. End Rule (Right/Left) (§ 3.12, § 9.4)
  ii. Quantity-Sensitive, Unbounded Feet; (right/left) headed (§ 7.2)

(3) **Parameters of Parsing**

a. **Location** of layer being created (top down/bottom up) (§ 5.4.3)
b. **Direction** left-to-right vs. right-to-left (§ 3.10)
c. **Iterative vs. Non-Iterative** (§ 5.4)
d. **Persistent vs. Non-Persistent** (§ 5.4.1)
e. **Strong Local** (unmarked) **vs. Weak Local** (marked) (§ 8.1)
f. **(Strong/Weak) Prohibition on Degenerate Feet**
Degenerate feet (= /◡́/; § 5.1.9) are forbidden (entirely/in weak position) (§ 5.1.1). Under the weak ban, weak degenerate feet are removed at the end of the word phonology.

(4) **Principles of Parsing**

a. **Maximality**
Prosodic structure is created maximally (§ 5.1.9).

b. **Free Element Condition**
Rules of primary metrical analysis apply only to free elements (§ 5.4.1).

c. **Priority Clause**
If at any stage in foot parsing the portion of the string being scanned would yield a degenerate foot, the parse scans further along the string to construct a proper foot where possible (§ 5.1.6).

(5) **Syllable Theory**

a. **Stress-Bearing Unit** = the syllable, universally. Stress contrasts may not occur within heavy syllables, nor may syllables be split between feet (§ 3.9.1, § 5.6.2).

b. **Syllable Quantity** moraic theory (§ 3.9.2)
CV is always monomoraic, CVV bimoraic, CVC is variable across and within languages (§ 3.9.2, § 5.6.1).
Trimoraic syllables occur as a marked option (§ 6.3.2, § 6.3.8.1) and may optionally be treated as disyllabic /$\mu\mu + \mu$/ (§ 6.1.7, § 8.5.3).
Two-layered moraic theory, for languages where more than one weight criterion is invoked: § 7.3

c. **Syllable Prominence** (§ 7.1)
On a separate and temporary plane, project grid columns based on prominence properties (weight, H tone, vowel sonority, presence and voicing of onset). Prominence may be referred to by the End Rule, extrametricality, and destressing. Foot templates, being defined quantitatively, cannot access prominence (§ 7.1.1).

d. **Stray Consonants,** as in final CVC.C#, may have metrical effects due to the Peripherality Condition (§ 5.2.1).

(6) **Extrametricality Theory** (§ 3.11, § 5.2)

a. **Parameters**
i. **Constituent** (segment, syllable, foot, suffix, word, but not mora, due to ban on syllable-splitting (5a))
ii. **Edge** (right edges unmarked)
iii. **Clash Environments** are permitted (§ 5.2.2)

b. **Peripherality Condition**
Extrametrical elements must be peripheral (§ 3.11).
Peripherality is interpreted to allow extrametrical higher level constituents to dominate extrametrical lower level constituents; but not "chained" extrametricality (§ 5.2.1).

c. **Exhaustivity**
Extrametricality is blocked when it would exhaust its domain (§ 3.11).

d. **"Unstressable Word Syndrome"**
Extrametricality may be suspended, and unfooted syllables may be incorporated, when a strong ban on degenerate feet would violate culminativity (§ 5.3) or the Continuous Column Constraint (§ 5.4.4).

(7) **Metrical Transformations**

a. **Move X**
Move one grid mark at a time along its layer (§ 3.4.2, § 9.5). Where Move X resolves a stress clash, movement must take place along the row where the clash occurs.

b. **Destressing in Clash** a. $\times \rightarrow \emptyset$ / ___ $\times$ (§ 3.4.2)
b. $\times \rightarrow \emptyset$ / $\times$ ___

c. **Domain Generation** $(\times) \rightarrow \begin{matrix}(\times)\\(\times)\end{matrix}$ (§ 9.4.1, § 9.4.2)

d. **Grid Expansion** part of the structural change of Beat Addition (§ 9.6.1)

e. **"Pretonic"** and **"Posttonic"** are defined minimally (§ 9.6.2).

(8) **Cyclic Footing**
Foot Construction may be allowed to affect material from a former cycle if necessary to parse material added on a new cycle (§ 5.4.2).

This book is an attempt to analyze a large and representative set of data while staying strictly within the limits of the theory outlined above. Some cases where I fall short of this goal and have had to provide additional devices for particular analyses include the following: Stray Adjunction in Pirahã (§ 7.1.7), rule-governed loss of degenerate feet in Asheninca (§ 7.1.8.4), Reparsing in Estonian (§ 8.5), and limited persistence in Pacific Yupik (§ 8.8.4).

10.2 CONCLUSION

Further research on the world's stress systems will suggest ways in which the theory presented here should be revised. Almost certainly, its set of formal devices will need expansion. It would be nice if ways can be found to contract the formal arsenal while remaining compatible with the data. And it would be arrogant to deny the possibility that the whole approach might be inferior to some radically different alternative.

Irrespective of the fate of the theory, I hope to have encouraged other linguists in a methodological direction that may become increasingly fruitful as the field matures: that of establishing a close connection between abstract theoretical work and broad-based typological study. Work in linguistic typology has achieved remarkable coverage of languages, but much (not all) of it has focused on relatively superficial, directly observable properties of languages. Work at an abstract theoretical level, on the other hand, tends to focus on detailed properties of one language, or just a few. It is believed by some that such study will necessarily uncover fundamental properties of the language faculty. There is a reason to think this unlikely: hypothesized general principles in linguistics have almost always required almost immediate revision or amplification when extended to other languages. Phonological studies, which have access to perhaps the greatest breadth of language data, have shown that individual languages may develop an amazing variety of rich and idiosyncratic formal systems, which can hide their universally determined traits in subtle ways. Only broad typological work can establish which traits are properties of Language, and which are properties of individual languages.

A closer connection between theory and typology may also benefit typology. Linguistic theories, particularly parametric ones, naturally give rise to typological hypotheses, and make typological research more interesting and more fun.

This book has largely been an effort to explore the typological predictions of a theory. I hope that linguists who read it will feel encouraged to undertake similar investigations, both in stress and in other areas of linguistic structure.

REFERENCES

Abaev, Vasilii Ivanovich (1964) *A Grammatical Sketch of Ossetic,* International Journal of American Linguistics, v. 30, no. 4, pt. 2, Indiana University Center in Anthropology, Folklore and Linguistics, Bloomington.

Abbott, Miriam (1991) "Macushi," in Desmond C. Derbyshire and Geoffrey K. Pullum, eds., *Handbook of Amazonian Languages, Vol. 3,* Mouton de Gruyter, Berlin, pp. 23–160.

Abbott, Stan (1985) "A Tentative Multilevel Multiunit Phonological Analysis of the Murik Language," *Papers in New Guinea Linguistics* 22, 339–73. [= *Pacific Linguistics* A63, Australian National University, Canberra]

Abu-Salim, Issam M. (1980) "Epenthesis and Geminate Consonants in Palestinian Arabic," *Studies in the Linguistic Sciences* 10.2, Dept. of Linguistics, University of Illinois, Urbana, pp. 1–11.

Allen, George D. (1975) "Speech Rhythm: Its Relation to Performance Universals and Articulatory Timing," *Journal of Phonetics* 3, 75–86.

Allen, W. Sidney (1973) *Accent and Rhythm,* Cambridge Studies in Linguistics 12, Cambridge University Press, Cambridge.

Al-Mozainy, Hamza (1981) "Vowel Alternations in a Bedouin Hijazi Arabic Dialect: Abstractness and Stress," Doctoral dissertation, University of Texas, Austin.

Al-Mozainy, Hamza, Robert Bley-Vroman, and John McCarthy (1985) "Stress Shift and Metrical Structure," *Linguistic Inquiry* 16, 135–44.

Anderson, Stephen R. (1972) "Icelandic *u*-Umlaut and Breaking in a Generative Grammar," in Evelyn Scherabon Firchow, Kaaren Grimstad, Nils Hasselmo and Wayne A. O'Neil, eds., *Studies for Einar Haugen,* Mouton, The Hague, pp. 13–30.

——— (1981) "Why Phonology Isn't 'Natural'," *Linguistic Inquiry* 12, 493–539.

Archangeli, Diana (1984) "Extrametricality in Yawelmani," *The Linguistic Review* 4, 101–20.

——— (1986) "Extrametricality and the Percolation Convention," ms., Dept. of Linguistics, University of Arizona, Tucson.

Archangeli, Diana, and Douglas Pulleyblank (in press) *Grounded Phonology,* MIT Press, Cambridge, MA.

Arany, László (1898) *Hangsúly és ritmus,* Budapest.

Ariste, Paul (1968) *A Grammar of the Votic Language,* Uralic and Altaic series 68, Indiana University, Bloomington.

Árnason, Kristján (1980) *Quantity in Historical Phonology: Icelandic and Related*

Cases, Cambridge Studies in Linguistics 30, Cambridge University Press, Cambridge.

——— (1985) "Icelandic Word Stress and Metrical Phonology," *Studia Linguistica* 39, 93–129.

Austin, Peter (1981) *A Grammar of Diyari, South Australia,* Cambridge Studies in Linguistics 32, Cambridge University Press, Cambridge.

Austin, William M., John G. Hangin, and Peter M. Onon (1963) *Mongol Reader,* Uralic and Altaic Series 29, Indiana University, Bloomington.

Bach, Emmon (1975) "Long Vowels and Stress in Kwakiutl," *Texas Linguistic Forum* 2, 9–19.

Bach, Emmon, and Robert Harms (1972) "How Do Languages Get Crazy Rules?" in Robert P. Stockwell and Ronald K. S. Macaulay, eds., *Linguistic Change and Generative Theory,* Indiana University Press, Bloomington, pp. 1–21.

Bagemihl, Bruce (1991) "Syllable Structure in Bella Coola," *Linguistic Inquiry* 22, 589–646.

Balassa József (1890) "Hangsúly a magyar nyelvben," *Nyelvtudományi Közlemények* 21, 401–34.

Barker, Chris (1989) "Extrametricality, the Cycle, and Turkish Word Stress," in Junko Ito and Jeff Runner, eds., *Phonology at Santa Cruz,* Vol. 1, Syntax Research Center, University of California, Santa Cruz, pp. 1–34.

Barker, Muhammad Abd-al-Rahman (1963) *Klamath Dictionary,* University of California Publications in Linguistics 31, University of California Press, Berkeley.

——— (1964) *Klamath Grammar,* University of California Publications in Linguistics 32, University of California Press, Berkeley.

Bat-El, Outi (1992) "Parasitic Metrification in the Modern Hebrew Stress System," ms., Dept. of Linguistics, Tel-Aviv University.

Bates, Dawn (1992) "Simple Syllables in Spokane Salish," *Linguistic Inquiry* 23, 653–659.

Bečka, Jiri (1969) *A Study in Pashto Stress,* Dissertationes Orientales 12, Academia, Prague.

Beckman, Mary E. (1986) *Stress and Non-Stress Accent,* Foris, Dordrecht.

Beckman, Mary E., and Janet Pierrehumbert (1986) "Intonational Structure in Japanese and English," *Phonology Yearbook* 3, 255–309.

Beckman, Mary E., Maria G. Swora, Jane Rauschenberg and Kenneth DeJong (1990) "Stress Shift, Stress Clash, and Polysyllabic Shortening in a Prosodically Annotated Discourse," *Proceedings of the 1990 International Conference on Spoken Language Processing,* Vol. 1, 5–8.

Bell, Alan (1977) "Accent Placement and Perception of Prominence in Rhythmic Structures," in Hyman 1977a, pp. 1–13.

Benger, Janet (1984) "The Metrical Phonology of Cayuga," M.A. thesis, University of Toronto.

Berinstein, Ava E. (1979) *A Cross-Linguistic Study on the Perception and Production of Stress,* UCLA Working Papers in Phonetics 47, Dept. of Linguistics, UCLA.

Bickmore, Lee (1989) "Kinyambo Prosody," Ph.D. dissertation, University of California, Los Angeles.

Bickmore, Lee (1992) "Multiple Phonemic Stress Levels in Kinyambo," *Phonology* 9, 155–98.

Bing, Janet M. (1979) "Up the Noun Phrase: Another Stress Rule," in Engdahl and Stein 1979, pp. 14–31.

Birk, D. B. W. (1976) *The Malakmalak Language, Daly River (Western Arnhem Land),* Pacific Linguistics B45, Australian National University, Canberra.

Blake, Barry J. (1969) *The Kalkatungu Language: A Brief Description,* Australian Aboriginal Studies 20, Australian Institute of Aboriginal Studies, Canberra.

——— (1979a) *A Kalkatungu Grammar,* Pacific Linguistics B57, Australian National University, Canberra.

——— (1979b) "Pitta-Pitta," in R. M. W. Dixon and Barry J. Blake, eds., *Handbook of Australian Languages,* Vol. 1, John Benjamins, Amsterdam, pp. 182–242.

Blanc, Haim (1970) "The Arabic Dialect of the Negev Bedouins," *Proceedings of the Israeli Academy of Sciences and Humanities* 4.7, pp. 112–50.

Blevins, Juliette (1990) "Alternatives to Exhaustivity and Conflation in Metrical Theory," ms., Dept. of Linguistics, University of Texas, Austin.

——— (1991) "A Tonal-Metrical Analysis of Lithuanian Accentual Phonology," ms., Dept. of Linguistics, University of Texas, Austin.

Bloomfield, Leonard (1939) "Menomini Morphophonemics," in *Études phonologiques dédiées à la mémoire de N. S. Trubetzkoy* (*Travaux du Cercle Linguistique de Prague* 8), pp. 105–15. Reprinted in Charles F. Hockett, ed., *A Leonard Bloomfield Anthology,* Indiana University Press, Bloomington, pp. 351–62.

——— (1956) *Eastern Ojibwa: Grammatical Sketch, Texts and Word List,* University of Michigan Press, Ann Arbor.

——— (1962) *The Menomini Language,* Yale University Press, New Haven, CT.

——— (1975) *Menomini Lexicon,* Milwaukee Public Museum Publications in Anthropology and History 3, Milwaukee Public Museum, Milwaukee, WI.

Boas, Franz (1947) *Kwakiutl Grammar with a Glossary of the Suffixes,* ed. by Helene Boas Yampolsky with the collaboration of Zellig S. Harris, *Transactions of the American Philosophical Society,* N.S. 37, Part 3, pp. 201–377.

Bolinger, Dwight (1958) "A Theory of Pitch Accent in English," *Word* 14, 109–49.

Bolozky, Shmuel (1982) "Remarks on Rhythmic Stress in Modern Hebrew," *Journal of Linguistics* 18, 275–89.

Bolton, Thaddeus L. (1894) "Rhythm," *American Journal of Psychology* 6, 145–238.

Booij, Geert, and Jerzy Rubach (1984) "Morphological and Prosodic Domains in Lexical Phonology," *Phonology Yearbook* 1, 1–27.

——— (1985) "A Grid Theory of Stress in Polish," *Lingua* 66, 281–319.

Boxwell, Helen, and Maurice Boxwell (1966) "Weri Phonemes," in Stephen A. Wurm, ed., *Papers in New Guinea Linguistics* 5, 77–93. Australian National University, Canberra.

Brame, Michael (1973) "On Stress Assignment in Two Arabic Dialects," in Stephen R. Anderson and Paul Kiparsky, eds., *A Festschrift for Morris Halle,* Holt, Rinehart, and Winston, New York, pp. 14–25.

——— (1974) "The Cycle in Phonology: Stress in Palestinian, Maltese, and Spanish," *Linguistic Inquiry* 5, 39–60.

Breen, John Gavin (1973) *Bidyara and Gungabula: Grammar and Vocabulary,* Linguistic Communications 8, Monash University, Melbourne.

Breen, Gavan (1981) *The Mayi Languages of the Queensland Gulf Country,* AIAS new series 29, Australian Institute of Aboriginal Studies, Canberra.

Broadbent, Sylvia M. (1964) *The Southern Sierra Miwok Language,* University of California Publications in Linguistics 38, University of California Press, Berkeley.

Broselow, Ellen (1988) "Prosodic Phonology and the Acquisition of a Second Language," in Suzanne Flynn and Wayne O'Neil, eds., *Linguistic Theory in Second Language Acquisition,* Kluwer, Dordrecht, pp. 295–308.

Bruce, Gösta (1977) *Swedish Word Accents in Sentence Perspective,* Gleerup, Lund.

——— (1984) "Rhythmic Alternation in Swedish," in Claes-Christian Elert, Iréne Johansson and Eva Strangert, eds., *Nordic Prosody III,* Almqvist & Wiksell International, Stockholm, pp. 31–41.

Buckley, Eugene (1991) "Persistent and Cumulative Extrametricality in Kashaya," ms., Dept. of Linguistics, University of California, Berkeley.

Bunn, Gordon, and Ruth Bunn (1970) "Golin Phonology," *Pacific Linguistics* A23, Australian National University, Canberra, pp. 1–7.

Callaghan, Catherine A. (1987) *Northern Sierra Miwok Dictionary,* University of California Press, Berkeley.

Capell, Arthur (1962a) *Some Linguistic Types in Australia,* Oceania Linguistic Monographs, no. 7, University of Sydney.

——— (1962b) *The Polynesian Language of Mae (Emwae), New Hebrides,* Linguistic Society of New Zealand, Auckland.

Carlson, Lauri (1978) "Word Stress in Finnish," ms., Dept. of Linguistics, MIT, Cambridge, MA.

Chadwick, Neil (1975) *A Descriptive Study of the Djingili Language,* Research and Regional Studies 2, Australian Institute of Aboriginal Studies, Canberra.

Chafe, Wallace L. (1970) *A Semantically Based Sketch of Onondaga,* Indiana University Publications in Anthropology and Linguistics, Memoir 25, Indiana University, Bloomington.

——— (1977) "Accent and Related Phenomena in the Five Nations Iroquois Languages," in Hyman 1977a, pp. 169–81.

Chai, Nemia M. (1971) "A Grammar of Aklan," Doctoral dissertation, University of Pennsylvania, Philadelphia.

Chambers, J. K. (1978) "Dakota Accent," in Eung-Do Cook and Jonathan Kaye, eds., *Linguistic Studies of Native Canada,* University of British Columbia Press, Vancouver, pp. 3–18.

Chaski, Carole (1985) "Linear and Metrical Analyses of Manam," *Oceanic Linguistics* 25, 167–209.

Chlumský, Josef (1928) *Česká Kvantita, Melodie a Přízvuk,* Nákladem České Akademie Věd a Umění, Prague.

Chomsky, Noam (1965) *Aspects of the Theory of Syntax,* MIT Press, Cambridge, MA.

Chomsky, Noam, and Morris Halle (1968) *The Sound Pattern of English,* Harper and Row, New York.

Chomsky, Noam, Morris Halle, and Fred Lukoff (1956) "On Accent and Juncture in English," in Morris Halle, Horace G. Lunt, Hugh McLean, and Cornelis H. van Schooneveld, eds., *For Roman Jakobson,* Mouton, The Hague, pp. 65–80.

Chung, Sandra (1983) "Transderivational Relationships in Chamorro Phonology," *Language* 59, 35–66.

Churchward, C. Maxwell (1953) *Tongan Grammar,* Oxford University Press, London.

Churchyard, Henry (1990) "Tiberian Biblical Hebrew Vowel Reduction and Multiplanar Metrical Theories," ms., Dept. of Linguistics, University of Texas, Austin.

——— (1992) "The Tiberian Hebrew Rhythm Rule in the Typology of Rhythm Rules," ms., Dept. of Linguistics, University of Texas, Austin.

Clements, George N. (1976) "Vowel Harmony in Nonlinear Generative Phonology: An Autosegmental Model," paper distributed by the Indiana University Linguistics Club, Bloomington.

——— (1985) "The Geometry of Phonological Features," *Phonology Yearbook* 2, 223–52.

——— (1987) "Phonological Feature Representation and the Description of Epenthetic Stops," *Parasession on Autosegmental and Metrical Phonology,* Chicago Linguistic Society, pp. 29–50.

——— (1991) "The Structure of Vowel and Consonant Gestures," paper given at the University of Illinois conference on "The Organization of Phonology: Features and Domains," May 1991.

Clements, George N., and Kevin Ford (1979) "Kikuyu Tone Shift and its Synchronic Consequences," *Linguistic Inquiry* 10, 179–210.

Clements, George N., and John Goldsmith (1984) "Autosegmental Studies in Bantu Tone: Introduction," in Clements, George N. and John Goldsmith (eds.) *Autosegmental Studies in Bantu Tone,* Foris, Dordrecht, pp. 1–17.

Clements, George N., and Samuel Jay Keyser (1983) *CV Phonology: A Generative Theory of the Syllable,* MIT Press, Cambridge, MA.

Cohn, Abigail (1989) "Stress in Indonesian and Bracketing Paradoxes," *Natural Language and Linguistic Theory* 7, 167–216.

——— (1993) "The Initial Dactyl Effect in Indonesian," *Linguistic Inquiry* 24, 372–81.

Cole, Jennifer, and Charles Kisseberth, eds. (in press) *Frontiers in Phonology: Proceedings of the University of Illinois Conference on the Organization of Phonology: Features and Domains,* Center for the Study of Language and Information, Stanford, CA.

Collier, Ken, and Margaret Collier (1975) "A Tentative Phonemic Statement of the Apoze Dialect, Kela Language," in Richard Loving, ed., *Phonologies of Five Austronesian Languages,* Summer Institute of Linguistics, Ukarumpa, Papua New Guinea, pp. 129–61.

Collinder, Björn (1960) *Comparative Grammar of the Uralic Languages,* Almqvist and Wiksell, Stockholm.

Comrie, Bernard (1976) "Irregular Stress in Polish and Macedonian," *International Review of Slavic Linguistics* 1, 227–40.

——— (1981) *The Languages of the Soviet Union,* Cambridge University Press, Cambridge.

———, ed., (1987) *The World's Major Languages,* Oxford University Press, Oxford.

Cooper, Grosvenor, and Leonard Meyer (1960) *The Rhythmic Structure of Music,* University of Chicago Press, Chicago.

Couro, Ted, and Christina Hutcheson (1973) *Dictionary of Mesa Grande Diegueño,* Malki Museum Press, Banning, CA.

Cowan, Hendrik K. J. (1965) *Grammar of the Sentani Language,* Verhandelingen van het Koninklijk Instituut voor Taal- Land- en Volkenkunde 47, Martinus Nijhoff, The Hague.

Cowan, William (1982) "A Note on Phonological Change in Ojibwa," *Canadian Journal of Linguistics* 27, 41–46.

Crowhurst, Megan (1991a) "Demorification in Tübatulabal: Evidence from Initial Reduplication and Stress," *Northeastern Linguistic Society* 21, 49–63.

——— (1991b) "Minimality and Foot Structure in Metrical Phonology and Prosodic Morphology," Ph.D. dissertation, University of Arizona, Tucson.

Crowley, Terry (1981) "The Mpakwithi Dialect of Anguthimri," in Robert M. W. Dixon and Barry J. Blake, eds., *Handbook of Australian Languages,* Vol. 2, John Benjamins, Amsterdam, pp. 146–94.

——— (1982) *The Paamese Language of Vanuatu,* Pacific Linguistics B87, Australian National University, Canberra.

Darwin, Christopher J., and Donovan, Andrew (1980) "Perceptual Studies of Speech Rhythm: Isochrony and Intonation," in Jean Claude Simon, ed., *Spoken Language Generation and Understanding,* Reidel, Dordrecht, pp. 77–85.

Dauer, Rebecca M. (1983) "Stress-Timing and Syllable-Timing Reanalyzed," *Journal of Phonetics* 11, 51–62.

Davis, Alice Irene (1984) *Basic Colloquial Maithili,* Motilal Banarsidass, Delhi.

Davis, Stuart M. (1985) *Topics in Syllable Geometry,* Doctoral dissertation, University of Arizona, Tucson.

——— (1988) "Syllable Onsets as a Factor in Stress Rules," *Phonology* 5, 1–19.

——— (1989) "Stress, Syllable Weight Hierarchies, and Moraic Phonology," *Eastern States Conference on Linguistics* 6, 84–92.

Dayley, Jon P. (1989a) *Tümpisa (Panamint) Shoshone Grammar,* University of California Publications in Linguistics 115, University of California Press, Berkeley.

——— (1989b) *Tümpisa (Panamint) Shoshone Dictionary,* University of California Publications in Linguistics 116, University of California Press, Berkeley.

Dell, François (1984) "L'accentuation des phrases en français," in François Dell, Daniel Hirst, and Jean-Roger Vergnaud, eds., *Forme sonore du langage,* Hermann, Paris, pp. 65–122.

Denes, Peter B., and Elliot N. Pinson (1963) *The Speech Chain,* Bell Telephone Laboratories, Murray Hill, N.J.

den Os, Els, and René Kager (1986) "Extrametricality and Stress in Spanish and Italian," *Lingua* 69, 23–45.

Derbyshire, Desmond C. (1985) *Hixkaryana and Linguistic Typology,* SIL Publications in Linguistics 76, Summer Institute of Linguistics, Dallas.

Dixon, Robert M. W. (1977) *A Grammar of Yidiɲ,* Cambridge Studies in Linguistics 19, Cambridge University Press, Cambridge.

——— (1981) "Wargamay," in R. M. W. Dixon and Barry J. Blake, eds., *Handbook of Australian Languages,* Vol. 2, John Benjamins, Amsterdam, pp. 1–144.

——— (1983) "Nyawaygi," in R. M. W. Dixon and Barry J. Blake, eds., *Handbook of Australian Languages,* Vol. 3, John Benjamins, Amsterdam, pp. 430–531.

——— (1988) *A Grammar of Boumaa Fijian,* University of Chicago Press, Chicago.

Donegan, Patricia (1978) "The Natural Phonology of Vowels," Ph.D. dissertation, Ohio State University, Columbus. [Published 1985, Garland, New York.]

Donovan, Andrew, and Christopher J. Darwin. (1979) "The Perceived Rhythm of Speech," *Proceedings of the Ninth International Congress of Phonetic Sciences,* Vol. 2, Institute of Phonetics, Copenhagen, pp. 268–74.

Douglas, Wilfred H. (1958) *An Introduction to The Western Desert Language,* Oceania Linguistic Monographs 4, University of Sydney, Australia.

Dresher, B. Elan (1980) "Metrical Structure and Secondary Stress in Tiberian Hebrew," in C. Chapin, ed., *Brown Working Papers in Linguistics* 4, Providence, pp. 24—37.

——— (1989) Review of Halle and Vergnaud (1987b), *Phonology* 7, 171–88.

Dresher, B. Elan, and Jonathan Kaye (1990) "A Computational Learning Model for Metrical Phonology," *Cognition* 34, 137–95.

Dresher, B. Elan, and Aditi Lahiri (1991) "The Germanic Foot: Metrical Coherence in Old English," *Linguistic Inquiry* 22, 251–86.

Dryer, Matthew S. (1989) "Large Linguistic Areas and Language Sampling," *Studies in Language* 13, 257–92.

DuBois, Carl D. (1976) *Sarangani Manobo: An Introductory Guide, Philippine Journal of Linguistics,* Special Monograph Issue 6, Linguistic Society of the Philippines, Manila.

Dunn, Leone (1988) "Badimaya, a Western Australian Language," *Papers in Australian Linguistics,* No. 17, Pacific Linguistics A71, Australian National University, Canberra, pp. 19–149.

Echeverría, Max S., and Heles Contreras (1965) "Araucanian Phonemics," *International Journal of American Linguistics* 31, 132–35.

Eek, Arvo (1975) "Observations on the duration of some word structures: II," *Estonian Papers in Phonetics* 4, Academy of Sciences of the Estonian SSR, Institute of Language and Literatures, pp. 7–54.

——— (1982) "Stress and Associated Phenomena: A Survey with Examples from Estonian. I." *Estonian Papers in Phonetics 1980–1981,* Academy of Sciences of the Estonian SSR, Institute of Language and Literatures, pp. 20–58.

Elbert, Samuel, and Mary Kawena Pukui (1979) *Hawaiian Grammar,* University Press of Hawaii, Honolulu.

Elenbaas, Nine (1992) "Een vergelijking van twee ternaire analyses, getoetst aan het Sentani," Master's thesis, University of Utrecht.

Elenbaas, Nine, and René Kager (forthcoming) "Stress in Sentani," ms., University of Utrecht.

Engdahl, Elisabet, and Mark J. Stein, eds. (1979) *Papers Presented to Emmon Bach by His Students,* Graduate Linguistics Student Association, University of Massachusetts, Amherst.

England, Nora (1983) *A Grammar of Mam, a Mayan Language,* University of Texas Press, Austin.

Everett, Daniel L. (1988) "On Metrical Constituent Structure in Pirahã," *Natural Language and Linguistic Theory* 6, 207–46.

Everett, Daniel L., and Karen Everett (1984) "Syllable Onsets and Stress Placement in Pirahã," *West Coast Conference on Formal Linguistics* 3, 105–16.

Fairbanks, Constance (1981) "The Development of Hindi Oral Narrative Meter," Doctoral dissertation, Dept. of South Asian Language and Literature, University of Wisconsin, Madison.

——— (1987a) "Hindi Stress: A New Approach through Meter," ms., Dept. of Linguistics, University of Minnesota, Minneapolis.

——— (1987b) "More on Stress and Metrical Rhythm," ms., Dept. of Linguistics, University of Minnesota, Minneapolis.

Fant, Gunnar, Anita Kruckenberg, and Lennart Nord (1991) "Stress Patterns and Rhythm in the Reading of Prose and Poetry with Analogies to Music Performance," in Johan Sundberg, Lennart Nord, and Rolf Carlson, eds., *Music, Language, Speech, and Brain,* Macmillan, London, pp. 380–407.

Feldman, Harry (1978) "Some Notes on Tongan Phonology," *Oceanic Linguistics* 17, 133–39.

Fleming, Ilah, and Ronald K. Dennis (1977) "Tol (Jicaque) Phonology," *International Journal of American Linguistics* 43, 121–27.

Flexner, Stuart Berg, ed. (1987) *The Random House Dictionary of the English Language,* 2nd ed., Random House, New York.

Foley, William A. (1986) *The Papuan Languages of New Guinea,* Cambridge University Press, Cambridge.

Fong, Eugene A. (1979) "Vowel and Consonant Quantity in Abruzzese," *Orbis* 28, 277–89.

Foster, Michael (1982) "Alternating Weak and Strong Syllables in Cayuga Words," *International Journal of American Linguistics* 48, 59–72.

Fraisse, Paul (1974) *Psychologie du rythme,* Presses universitaires de France, Paris.

Franks, Steven L. (1985) "Extrametricality and Stress in Polish," *Linguistic Inquiry* 16, 144–51.

——— (1987) "Regular and Irregular Stress in Macedonian," *International Journal of Slavic Linguistics and Poetics* 35/36, 93–142.

——— (1989) "The Monosyllabic Head Effect," *Natural Language and Linguistic Theory* 7, 551–63.

——— (1991) "Diacritic Extrametricality vs. Diacritic Accent: A Reply to Hammond," *Phonology* 8, 145–61.

Freeland, Lucy S. (1951) *Language of the Sierra Miwok,* Memoir 6 of *International Journal of American Linguistics,* Indiana University, Bloomington.

Fromkin, Victoria (1977) "Putting the EmPHAsis on the Wrong SylLABle," in Hyman 1977a, pp. 15–26.

Fry, Dennis B. (1955) "Duration and Intensity as Physical Correlates of Linguistic Stress," *Journal of the Acoustical Society of America* 35, 765–69.

——— (1958) "Experiments in the Perception of Stress," *Language and Speech* 1, 120–52.

Furby, Christine (1974) "Garawa Phonology," *Pacific Linguistics,* Series A, Australian National University, Canberra.

Geraghty, Paul A. (1983) *The History of the Fijian Languages, Oceanic Linguistics* Special Publication 19, University of Hawaii Press, Honolulu.

Giegerich, Heinz (1981) "On the Nature and Scope of Metrical Structure," paper distributed by the Indiana University Linguistics Club, Bloomington.

——— (1983) "On English Sentence Stress and the Nature of Metrical Structure," *Journal of Linguistics* 19, 1–28.

——— (1985) *Metrical Phonology and Phonological Structure,* Cambridge Studies in Linguistics 43, Cambridge University Press, Cambridge.

Goddard, Ives (1979) *Delaware Verbal Morphology*, Garland Publishing, New York.
——— (1982) "The Historical Phonology of Munsee," *International Journal of American Linguistics* 48, 16–48.
Goddard, Ives, Charles F. Hockett, and Karl V. Teeter (1972) "Some Errata in Bloomfield's Menomini," *International Journal of American Linguistics* 38, 1–5.
Goldsmith, John (1976) "Autosegmental Phonology," Ph.D. dissertation, Massachusetts Institute of Technology, Cambridge. [Distributed by Indiana University Linguistics Club, Bloomington.]
——— (1990) *Autosegmental and Metrical Phonology*, Blackwell, Oxford.
Golston, Chris (1989) "Floating H (and L*) Tones in Ancient Greek," *Proceedings of the Arizona Phonology Conference* 2.
Grierson, George A. (1895) "On the Stress Accent in the Modern Indo-Aryan Vernaculars," *Journal of the Royal Asiatic Society*, 139–47.
Griffith, Teresa (1991) "Cambodian as an Iambic Language," ms., Department of Linguistics, University of California, Irvine.
Grimes, Barbara F. (1992) *Ethnologue: Languages of the World* (12th ed.), Summer Institute of Linguistics, Dallas, TX.
Gupta, Abha (1987) "Hindi Word Stress and the Obligatory Branching Parameter," *Parasession on Autosegmental and Metrical Phonology*, Chicago Linguistic Society, pp. 134–48.
Gussenhoven, Carlos (1984) *On the Grammar and Semantics of Sentence Accents*, Publications in Language Sciences 16, Foris, Dordrecht.
——— (1991) "The English Rhythm Rule as an Accent Deletion Rule," *Phonology* 8, 1–35.
——— (1993) "The Dutch Foot and the Chanted Call," *Journal of Linguistics* 29, 37–63.
Gussenhoven, Carlos, Dwight Bolinger, and Cornelia Keijsper (1987) *On Accent*, Indiana University Linguistics Club, Bloomington.
Gussmann, Edmund (1985) "The Morphology of a Phonological Rule: Icelandic Vowel Length," in Edmund Gussmann, ed., *Phono-Morphology: Studies in the Interaction of Phonology and Morphology*, Redakcja Wydawnictw Katolickiego Uniwersytetu Lubelskiego, Lublin, Poland, pp. 75–94.
Haas, Mary (1977) "Tonal Accent in Creek," in Hyman 1977a, pp. 195–208.
Hale, Kenneth (1985) "A Note on Winnebago Metrical Structure," *International Journal of American Linguistics* 51, 427–29.
Hale, Kenneth, and Josie White Eagle (1980) "A Preliminary Metrical Account of Winnebago Accent," *International Journal of American Linguistics* 46, 117–32.
Hall, Robert A., Jr. (1938) *An Analytical Grammar of the Hungarian Language*, *Language* Monograph no. 18, Linguistic Society of America, Baltimore.
Halle, Morris (1990) "Respecting Metrical Structure," *Natural Language and Linguistic Theory* 8, 149–76.
Halle, Morris, and George N. Clements (1983) *Problem Book in Phonology*, MIT Press, Cambridge, MA.
Halle, Morris, and Michael J. Kenstowicz (1989) "On Cyclic and Noncyclic Stress," ms., Dept. of Linguistics, Massachusetts Institute of Technology, Cambridge, MA. [Draft version of Halle and Kenstowicz 1991.]

——— (1991) "The Free Element Condition and Cyclic vs. Noncyclic Stress," *Linguistic Inquiry* 22, 457–501.

Halle, Morris, and Paul Kiparsky (1977) "Towards a Reconstruction of the Indo-European Accent," in Hyman 1977a, pp. 209–38.

——— (1981) "Review Article: *Histoire de l'accentuation slave* by Paul Garde," *Language* 57, 150–81.

Halle, Morris, and K. P. Mohanan (1985) "Segmental Phonology of Modern English," *Linguistic Inquiry* 16, 57–116.

Halle, Morris, and Jean-Roger Vergnaud (1978) "Metrical Structures in Phonology," ms., Dept. of Linguistics, Massachusetts Institute of Technology, Cambridge, MA.

——— (1987a) "Stress and the Cycle," *Linguistic Inquiry* 18, 45–84.

——— (1987b) *An Essay on Stress,* MIT Press, Cambridge.

Hammond, Michael (1984a) "Constraining Metrical Theory: A Modular Theory of Rhythm and Destressing," Doctoral dissertation, University of California, Los Angeles. [Distributed by Indiana University Linguistics Club, Bloomington.]

——— (1984b) "Metrical Structure in Lenakel and the Directionality-Dominance Hypothesis," in *Papers from the Minnesota Regional Conference on Language and Linguistics.*

——— (1986) "The Obligatory Branching Parameter in Metrical Theory," *Natural Language and Linguistic Theory* 4, 185–228.

——— (1987) "Hungarian Cola," *Phonology Yearbook* 4, 267–69.

——— (1989) "Lexical Stress in Macedonian and Polish," *Phonology* 6, 19–38.

——— (1990a) "Deriving Ternarity," ms., Dept. of Linguistics, University of Arizona, Tucson.

——— (1990b) *Metrical Theory and Learnability,* ms., Dept. of Linguistics, University of Arizona, Tucson.

Hansen, Kenneth C., and L. E. Hansen (1969) "Pintupi Phonology," *Oceanic Linguistics* 8, 153–70.

——— (1978) *The Core of Pintupi Grammar,* Institute for Aboriginal Development, Alice Springs, Northern Territory, Australia.

Harms, Robert T. (1964) *Finnish Structural Sketch,* Uralic and Altaic series 42, Indiana University, Bloomington.

——— (1966) "Stress, Voice and Length in Southern Paiute," *International Journal of American Linguistics* 32, 228–35.

——— (1981) "A Backwards Metrical Approach to Cairo Arabic Stress," *Linguistic Analysis* 7, 429–50.

——— (1985) "The Locus of Hampered Voice in Southern Paiute," *International Journal of American Linguistics* 51, 438–41.

Harrell, Richard S. (1957) *The Phonology of Colloquial Egyptian Arabic,* American Council of Learned Societies, New York.

——— (1960) "A Linguistic Analysis of Egyptian Radio Arabic," in Charles A. Ferguson, ed., *Contributions to Arabic Linguistics,* Harvard University Press, Cambridge, pp. 3–77.

Harris, James (1983) *Syllable Structure and Stress in Spanish: A Nonlinear Analysis, Linguistic Inquiry* Monograph 8, MIT Press, Cambridge, MA.

——— (1989) "A Podiatric Note on Secondary Stress in Spanish," ms., Dept. of Linguistics, Massachusetts Institute of Technology, Cambridge.

——— (1992) "Spanish Stress: The Extrametricality Issue," paper distributed by Indiana University Linguistics Club, Bloomington.

Hata, Kazue, and Yoko Hasegawa (1989) "The perception of the low-high (LH) tonal sequence," *Journal of the Acoustical Society of America* 86, S35.

Hawkins, W. Neil (1950) "Patterns of Vowel Loss in Macushi (Carib)," *International Journal of American Linguistics* 16, 87–90.

Hayes, Bruce (1979) "Extrametricality," *MIT Working Papers in Linguistics* 1, 77–86.

——— (1981) "A Metrical Theory of Stress Rules," Doctoral dissertation (1980), Massachusetts Institute of Technology, Cambridge. [Revised version distributed by Indiana University Linguistics Club, Bloomington, and published by Garland Press, New York 1985.]

——— (1982a) "Metrical Structure as the Organizing Principle of Yidin^y Phonology," in Harry van der Hulst and Norval Smith, eds., *The Structure of Phonological Representations, Part I,* Foris, Dordrecht, pp. 97–110.

——— (1982b) "Extrametricality and English Stress," *Linguistic Inquiry* 13, 227–76.

——— (1984) "The Phonology of Rhythm in English," *Linguistic Inquiry* 15, 33–74.

——— (1985) "Iambic and Trochaic Rhythm in Stress Rules," in M. Niepokuj et al., eds., *Berkeley Linguistics Society* 13, pp. 429–46.

——— (1986a) "Inalterability in CV Phonology," *Language* 62, 321–51.

——— (1986b) "Assimilation as Spreading in Toba Batak," *Linguistic Inquiry* 17, 467–99.

——— (1986c) Review of Giegerich (1985), *Journal of Linguistics* 22, 229–35.

——— (1987) "A Revised Parametric Metrical Theory," *Northeastern Linguistic Society* 17, 274–89.

——— (1988) "Metrics and Phonological Theory," in Frederick Newmeyer, ed., *Linguistics: The Cambridge Survey, Vol. 2, Linguistic Theory: Extensions and Implications,* Cambridge University Press, Cambridge, pp. 220–49.

——— (1989a) "The Prosodic Hierarchy in Meter," in Paul Kiparsky and Gilbert Youmans, eds., *Rhythm and Meter,* Academic Press, Orlando, FL, pp. 201–60.

——— (1989b) "Compensatory Lengthening in Moraic Phonology," *Linguistic Inquiry* 20, 253–306.

——— (1990) "Diphthongization and Coindexing," *Phonology* 7, 31–71.

——— (1992) "Metrical Phonology," in William Bright, ed., *International Encyclopedia of Linguistics,* Oxford University Press, London, Vol. 2, pp. 424–27.

——— (forthcoming) "Weight of CVC May Vary by Context," to appear in Cole and Kisseberth (in press).

Hayes, Bruce, and May Abad (1989) "Reduplication and Syllabification in Ilokano," *Lingua* 77, 331–74.

Hayes, Bruce, and Aditi Lahiri (1991a) "Bengali Intonational Phonology," *Natural Language and Linguistic Theory* 9, 47–96.

——— (1991b) "Durationally-Specified Intonation in English and Bengali," in Johan Sundberg, Lennart Nord, and Rolf Carlson, eds., *Music, Language, Speech, and Brain,* Macmillan, London, pp. 78–91.

Hayes, Bruce, and Stanisław Puppel (1985) "On the Rhythm Rule in Polish," in Harry van der Hulst and Norval Smith, eds., *Advances in Nonlinear Phonology,* Foris, Dordrecht, pp. 59–81.

Hercus, Louise A. (1986) *Victorian Languages: A Late Survey,* Pacific Linguistics B77,

Australian National University, Canberra. [Revised and enlarged version of *The Languages of Victoria: A Late Survey* (in two vols.), Australian Institute of Aboriginal Studies, Canberra, 1969.]

Herrfurth, Hans (1964) *Lehrbuch des modernen Djawanisch,* Veb Verlag Enzyklopädie, Leipzig.

Hess, Thomas (1976) *Dictionary of Puget Salish,* University of Washington Press, Seattle.

Hewitt, Mark S. (1991) "Binarity and Ternary in Alutiiq," *Proceedings of the Arizona Phonology Conference* 4, Dept. of Linguistics, University of Arizona, Tucson.

Hill, Kenneth, and Jane Hill (1968) "Stress in the Cupan (Uto-Aztecan) Languages," *International Journal of American Linguistics* 34, 233–41.

Hint, Mati (1973) *Eesti Keele Sõnafonoloogia I,* Eesti NSV Teaduste Akadeemia, Tallinn, Estonia.

Hockett, Charles F. (1942) "The Position of Potowatomi in Central Algonquian," *Papers of the Michigan Academy of Science, Arts, and Letters* 28, 537–42.

——— (1948) "Potowatomi I: Phonemics, Morphophonemics, and Morphological Survey," *International Journal of American Linguistics* 14, 1–10.

——— (1981) "The Phonological History of Menominee," *Anthropological Linguistics* 23, 51–87.

Hoff, Berend (1968) *The Carib Language,* Verhandelingen van het Koninklijk Instituut voor Taal-, Land-, en Volkenkunde 55, M. Nijhoff, The Hague.

Hudson, Joyce (1978) *The Core of Walmatjari Grammar,* Australian Institute of Aboriginal Studies, Canberra.

Huffman, Franklin E. (1970) *Cambodian System of Writing and Beginning Reader,* Yale University Press, New Haven.

Huss, Volker (1975) "Neutralisierung englischer Akzentunterschiede in der Nachkontur," *Phonetica* 32, 278–91.

Hutchinson, Sandra Pinkerton (1974) "Spanish Vowel Sandhi," *Parasession on Natural Phonology,* Chicago Linguistic Society, pp. 184–192.

Hyman, Larry, ed. (1977a) *Studies in Stress and Accent,* Southern California Occasional Papers in Linguistics 4, Dept. of Linguistics, University of Southern California, Los Angeles.

——— (1977b) "On the Nature of Linguistic Stress," in Hyman 1977a, pp. 37–82.

——— (1985) *A Theory of Phonological Weight,* Publications in Language Sciences 19, Foris, Dordrecht.

Inkelas, Sharon (1989) "Prosodic Constituency in the Lexicon," Doctoral dissertation, Stanford University. [Published 1990 by Garland Press, New York.]

Inkelas, Sharon, and Draga Zec (1988) "Serbo-Croatian Pitch Accent: The Interaction of Tone, Stress, and Intonation," *Language* 64, 227–48.

Irshied, Omar, and Michael J. Kenstowicz (1984) "Some Phonological Rules of Bani-Hassan Arabic: A Bedouin Dialect," *Studies in the Linguistic Sciences* 14.1, Dept. of Linguistics, University of Illinois, Urbana, pp. 109–47.

Itkonen, Erkki (1955) "Ueber die Betonungsverhältnisse in den finnisch-ugrischen Sprachen," *Acta Linguistica Academiae Scientiarum Hungaricae* 5, 21–23.

Ito, Junko (1986) "Syllable Theory in Prosodic Phonology," Doctoral dissertation, University of Massachusetts, Amherst. [Distributed by Graduate Linguistic Student Association, Dept. of Linguistics, University of Massachusetts.]

——— (1989) "A Prosodic Theory of Epenthesis," *Natural Language and Linguistic Theory* 7, 217–59.

Ito, Lucille I. (1989) "Manam Stress: the Cycle and Extrametricality," in Junko Ito and Jeff Runner, eds., *Phonology at Santa Cruz,* Vol. 1, Syntax Research Center, University of California, Santa Cruz, pp. 35–59.

Jackson, Michel (1987) "A Metrical Analysis of the Pitch Accent System of the Seminole Verb," in Pamela Munro, ed., *Muskogean Linguistics,* UCLA Occasional Papers in Linguistics 6, Dept. of Linguistics, University of California, Los Angeles, pp. 81–95.

Jacobs, Haike (1989) "Historical Studies in the Nonlinear Phonology of French," Ph.D. dissertation, University of Nijmegen.

——— (1990) "On Markedness and Bounded Stress Systems," *The Linguistic Review* 7, 81–119.

Jacobson, Steven A. (1984) "The Stress Conspiracy and Stress-Repelling Bases in the Central Yup'ik and Siberian Yupik Eskimo Languages," *International Journal of American Linguistics* 50, 312–24.

——— (1985) "Siberian Yupik and Central Yupik Prosody," in Krauss 1985a, pp. 25–45.

Jakobson, Roman (1931) "Die Betonung und ihre Rolle in der Wort- und Syntagmaphonologie," *Travaux du cercle linguistique de Prague* 4. Reprinted in R. Jakobson (1962) *Selected Writings I: Phonological Studies,* Mouton, The Hague, pp. 117–36.

——— (1962) "Contributions to the Study of Czech Accent," in *Selected Writings I: Phonological Studies,* Mouton, The Hague, pp. 614–25.

Jeanne, Laverne Masavesya (1982) "Some Phonological Rules of Hopi," *International Journal of American Linguistics* 48, 245–70.

Jha, Subhadra (1940–44) "Maithili Phonetics," *Indian Linguistics* 8, 435–59.

——— (1958) *The Formation of the Maithili Language,* Luzac, London.

Jóhannsson, Johannes L. L. (1924) *Nokkrar sögulegar athuganir um helztu hljóðbreytingar o. fl. í íslenzku,* Felagsprentsmidjan, Reykjavík.

Jones, Daniel (1956) *Everyman's English Pronouncing Dictionary,* J. M. Dent and Sons, London.

Jones, Daniel, and Dennis Ward (1969) *The Phonetics of Russian,* Cambridge University Press, Cambridge.

Jones, W. E. (1971) "Syllables and Word Stress in Hindi," *Journal of the International Phonetic Association* 1, 74–78.

Kager, René (1989) *A Metrical Theory of Stress and Destressing in English and Dutch,* Linguistic Models 14, Foris, Dordrecht.

——— (1992a) "Are There Any Truly Quantity-Insensitive Systems?" *Berkeley Linguistics Society* 18.

——— (1992b) "Shapes of the Generalized Trochee," *West Coast Conference on Formal Linguistics* 11.

——— (1993) "Alternatives to the Iambic–Trochaic Law," *Natural Language and Linguistic Theory,* 11, 381–432.

Kager, René, and Ellis Visch (1988) "Metrical Constituency and Rhythmic Adjustment," *Phonology* 5, 21–71.

Kahn, Daniel (1976) "Syllable-based Generalizations in English Phonology," Ph.D. dissertation, Massachusetts Institute of Technology, Cambridge. [Distributed by the Indiana University Linguistics Club, Bloomington.]

Kaisse, Ellen (1982) "On the Preservation of Stress in Modern Greek," *Linguistics* 20, 59–82.

——— (1985) "Some Theoretical Consequences of Stress Rules in Turkish," *Chicago Linguistic Society* 21, 199–209.

Kalectaca, Milo (1978) *Lessons in Hopi,* University of Arizona Press, Tucson.

Kálmán, Béla (1965) *Vogul Chrestomathy,* Uralic and Altaic Series 46, Indiana University, Bloomington.

Kamprath, Christine (1987) "Suprasegmental Structures in a Raeto-Romansch Dialect: A Case Study in Metrical and Lexical Phonology," Ph.D. dissertation, University of Texas, Austin.

Kaplan, Lawrence D. (1985) "Seward Peninsula Inupiaq Consonant Gradation and its Relationship to Prosody," in Krauss 1985a, pp. 191–210.

Katada, Fusa (1990) "On the Representation of Moras: Evidence from a Language Game," *Linguistic Inquiry* 21, 641–46.

Kaye, Jonathan (1973) "Odawa Stress and Related Phenomena," in Glyne L. Piggott and Jonathan Kaye, eds., *Odawa Language Project: Second Report,* Centre for Linguistic Studies, University of Toronto, pp. 42–50.

Kelkar, Ashok R. (1968) *Studies in Hindi-Urdu I: Introduction and Word Phonology,* Deccan College, Poona.

Kenstowicz, Michael J. (1970) "On the Notation of Vowel Length in Lithuanian," *Papers in Linguistics* 3, 73–113.

——— (1979) "Vowel Harmony and Metathesis in Palestinian Arabic," ms., Dept. of Linguistics, University of Illinois, Urbana.

——— (1980) "Notes on Cairene Arabic Syncope," *Studies in the Linguistic Sciences* 10.2, Dept. of Linguistics, University of Illinois, Urbana, pp. 39–53.

——— (1981) "The Metrical Structure of Arabic Accent," paper delivered at the UCLA–USC Conference on Nonlinear Phonology, Lake Arrowhead, Calif.

——— (1983) "Parametric Variation and Accent in the Arabic Dialects," *Chicago Linguistic Society* 19, 205–13.

——— (1986) "Notes on Syllable Structure in Three Arabic Dialects," *Revue québécoise de linguistique* 16, 101–28.

——— (1991) "On Metrical Constituents: Unbalanced Trochees and Degenerate Feet," to appear in Cole and Kisseberth, in press.

——— (1994) *Phonology in Generative Grammar,* Blackwell, Oxford.

Kenstowicz, Michael J., and Kamal Abdul-Karim (1980) "Cyclic Stress in Levantine Arabic," *Studies in the Linguistic Sciences* 10.2, Dept. of Linguistics, University of Illinois, Urbana.

Kenstowicz, Michael J., and Charles Kisseberth (1979) *Generative Phonology: Description and Theory,* Academic Press, New York.

Kenstowicz, Michael J., and Charles Pyle (1973) "On the Phonological Integrity of Geminate Clusters," in Michael J. Kenstowicz and Charles Kisseberth, eds., *Issues in Phonological Theory,* Mouton, The Hague, pp. 27–43.

Kenyon, John S., and Thomas A. Knott (1944) *A Pronouncing Dictionary of American English,* G. and C. Merriam, Springfield, Mass.

Kerek, Andrew (1971) *Hungarian Metrics: Some Linguistic Aspects of Iambic Verse,* Indiana University Publications, Uralic and Altaic Series 117, Mouton, The Hague.

Kettunen, Lauri (1938) *Livisches Wörterbuch mit grammatischer Einleitung,* Suomalais-Ugrilainen Seura, Helsinki.

Key, Harold H. (1961) "Phonotactics of Cayuvava," *International Journal of American Linguistics* 27, 143–50.

——— (1967) *Morphology of Cayuvava,* Janua Linguarum, Series practica 53, Mouton, The Hague.

Key, Mary Ritchie (1968) *Comparative Tacanan Phonology,* Janua Linguarum, Series practica 50, Mouton, The Hague.

Keyser, Samuel Jay, and Wayne O'Neil (1985) *Rule Generalization and Optionality in Language Change,* Foris, Dordrecht.

Khubchandani, Lachman M. (1969) "Stress in Sindhi," *Indian Linguistics* 30, 115–18.

Kiparsky, Paul (1966) "Ueber den Deutschen Akzent," *Studia Grammatica* 7, 69–98.

——— (1979) "Metrical Structure Assignment is Cyclic," *Linguistic Inquiry* 10, 421–41.

——— (1982a) "Lexical Phonology and Morphology," in I. Yang, ed., *Linguistics in the Morning Calm,* Hanshin, Seoul, pp. 3–91.

——— (1982b) "The Lexical Phonology of Vedic Accent," ms., Dept. of Linguistics, Stanford University, Stanford, CA.

——— (1982c) *Explanation in Phonology,* Foris, Dordrecht.

——— (1984) "On the Lexical Phonology of Icelandic," in Claes-Christian Elert, Iréne Johansson and Eva Strangert, eds., *Nordic Prosody III: Papers from a Symposium,* Almqvist & Wiksell, Stockholm, pp. 135–64.

——— (1993) "Blocking in Non-derived Environments," in Ellen Kaisse and Sharon Hargus, eds., *Studies in Lexical Phonology,* Academic Press, San Diego, pp. 277–313.

Kirchner, Robert (1990) "Phonological Processes without Phonological Rules: Yidiny Apocope and Penultimate Lengthening," *Northeastern Linguistic Society* 21, 203–16.

——— (1992) "Yidiɲ Prosody in Harmony-Theoretic Phonology," ms., Dept. of Linguistics, University of California, Los Angeles.

Kisseberth, Charles (1970) "On the Functional Unity of Phonological Rules," *Linguistic Inquiry* 1, 291–306.

Klatt, Dennis (1975) "Vowel Lengthening is Syntactically Determined in a Connected Discourse," *Journal of Phonetics* 3, 129–40.

Knudson, Lyle M. (1975) "A Natural Phonology and Morphophonemics of Chimalapa Zoque," *Papers in Linguistics* 8, 283–346.

Krauss, Michael (1975) "St. Lawrence Island Eskimo Phonology and Orthography," *Linguistics* 152, 39–72.

Krauss, Michael, ed. (1985a) *Yupik Eskimo Prosodic Systems: Descriptive and Comparative Studies,* Alaska Native Language Center, Fairbanks.

——— (1985b) "Introduction," in Krauss 1985a, pp. 1–6.

——— (1985c) "A History of the Study of Yupik Prosody," in Krauss 1985a, pp. 7–23.

——— (1985d) "Supplementary Notes on Central Siberian Yupik Prosody," in Krauss (1985a), pp. 47–50.

Krueger, John R. (1961) *Chuvash Manual,* Uralic and Altaic series 7, Indiana University, Bloomington.

Ladd, D. Robert (1978) "Stylized Intonation," *Language* 54, 517–40.

——— (1980) *The Structure of Intonational Meaning: Evidence from English,* Indiana University Press, Bloomington.

Ladefoged, Peter (1967) *Three Areas of Experimental Phonetics,* Oxford University Press, Oxford.

——— (1990) "The Revised International Phonetic Alphabet," *Language* 66, 550–52.

Lahiri, Aditi, and Jacques Koreman (1988) "Syllable Weight and Quantity in Dutch," *West Coast Conference on Formal Linguistics* 7, 217–28.

Lahiri, Aditi, and Harry van der Hulst (1988) "On Foot Typology," *North Eastern Linguistic Society* 18, vol. 2, pp. 286–99.

Lakó, Gy. (1957) "Nordmansische Sprachstudien," *Acta Linguistica Academiae Scientiarum Hungaricae* 6, 347–423.

Langdon, Margaret (1970) *A Grammar of Diegueño (The Mesa Grande Dialect),* University of California Publications in Linguistics 66, University of California Press, Berkeley and Los Angeles.

Langendoen, D. Terence (1968) *The London School of Linguistics: A Study of the Linguistic Theories of B. Malinowski and J. R. Firth,* MIT Press, Cambridge, MA.

Larsen, Raymond S., and Eunice V. Pike (1949) "Huasteco Intonations and Phonemes," *Language* 25, 268–77.

Lea, Wayne (1977) "Acoustic Correlates of Stress and Juncture," in Hyman 1977a, pp. 83–119.

Lee, Gregory (1969) "English Word-Stress," *Chicago Linguistic Society* 5, 389–406.

Leer, Jeff (1985a) "Prosody in Alutiiq (The Koniag and Chugach dialects of Alaskan Yupik)", in Krauss 1985a, pp. 77–133.

——— (1985b), "Evolution of Prosody in the Yupik Languages," in Krauss 1985a, pp. 135–57.

——— (1985c) "Toward a Metrical Interpretation of Yupik Prosody," in Krauss 1985a, pp. 159–72.

——— (1989) "Prosody in Chugach Alutiiq," ms., Alaska Native Language Center, Fairbanks.

Lehiste, Ilse (1965) "The Function of Quantity in Finnish and Estonian," *Language* 41, 447–56.

——— (1970) *Suprasegmentals,* MIT Press, Cambridge.

——— (1977) "Isochrony Reconsidered," *Journal of Phonetics* 5, 253–63.

Lerdahl, Fred, and Ray Jackendoff (1983) *A Generative Theory of Tonal Music,* MIT Press, Cambridge.

Leskinen, Heikki (1984) "Ueber die Phonemsystem der Karelischen Sprache," in Péter Hajdú and László Honti, eds., *Studien zur Phonologischen Beschreibung uralischer Sprachen,* Akadémiai Kiadó, Budapest, pp. 247–57.

Levin, Juliette (1985a) "A Metrical Theory of Syllabicity," Doctoral dissertation, Massachusetts Institute of Technology, Cambridge.

——— (1985b) "Evidence for Ternary Feet and Implications for a Metrical Theory of Stress Rules," ms., Dept. of Linguistics, University of Texas, Austin.

——— (1988a) "Generating Ternary Feet," *Texas Linguistic Forum* 29, 97–113.

——— (1988b) "Bidirectional Foot Construction as a Window on Level Ordering," in Michael Hammond and Michael Noonan, eds., *Theoretical Morphology,* Academic Press, Orlando, FL, pp. 339–52.

Levinsohn, Stephen H. (1976) *The Inga Language,* Janua linguarum, Series practica 188, Mouton, The Hague.

Lewis, M. B. (1947) *Teach Yourself Malay,* English Universities Press, London.

Liberman, Mark (1975) *The Intonational System of English,* Doctoral dissertation, Massachusetts Institute of Technology, Cambridge. [Distributed by Indiana University Linguistics Club, Bloomington.]

Liberman, Mark, and Alan Prince (1977) "On Stress and Linguistic Rhythm," *Linguistic Inquiry* 8, 249–336.

Lichtenberk, Frantisek (1983) *A Grammar of Manam,* Oceanic Linguistics Special Publication No. 18, University of Hawaii Press, Honolulu.

Lieberman, Philip (1968) "Direct Comparison of Subglottal and Esophageal Pressure during Speech," *Journal of the Acoustical Society of America* 43, 1157–64.

Lieberman, Philip, John D. Griffiths, Jere Mead, and Ronald Knudson (1967) "Absence of Syllabic 'Chest Pulses'," *Journal of the Acoustical Society of America* 41, 1614.

Lindblom, Björn, and Karin Rapp (1973) *Some Temporal Regularities of Spoken Swedish,* Publication 21, Institute of Linguistics, University of Stockholm.

Lombardi, Linda, and John McCarthy (1991) "Prosodic Circumscription in Choctaw Morphology," *Phonology* 8, 37–72.

Lotz, John (1939) *Das ungarische Sprachsystem,* Stockholm.

Lunt, Horace (1952) *A Grammar of the Macedonian Literary Language,* Skopje.

Lynch, John D. (1974) "Lenakel Phonology," Doctoral dissertation, University of Hawaii.

——— (1977) *Lenakel Dictionary,* Pacific Linguistics C55, Australian National University, Canberra.

——— (1978) *A Grammar of Lenakel,* Pacific Linguistics B55, Australian National University, Canberra.

——— (1982) "Southwest Tanna Grammar and Vocabulary," in J. Lynch, ed., *Papers in the Linguistics of Melanesia* 4, Australian National University, Canberra, pp. 1–91.

Lytkin, V. I. (1961) *Komi-iaz'vinskii dialekt,* Izdatel'stvo Akademii Nauk SSSR, Moscow.

Maddieson, Ian (1985) "Phonetic Cues to Syllabification," in Victoria A. Fromkin, ed., *Phonetic Linguistics: Essays in Honor of Peter Ladefoged,* Academic Press, Orlando, FL, pp. 203–21.

Maddieson, Ian, and Peter Ladefoged (1993) "Phonetics of Partially Nasalized Consonants," in Marie Huffman and Rena Krakow, eds., *Nasals, Nasalization, and the Velum,* Academic Press, San Diego, pp. 251–301.

Maldonado Andres, Juan, Juan Ordonez Domingo, and Juan Ortiz Domingo (1986) *Diccionario mam,* Talleres graficos del Centro de Reproducciones de la Universidad Rafael Landivar, Guatemala.

Malikouti-Drachmann, Angelika, and Gaberell Drachmann (1981) "Slogan Chanting and Speech Rhythm in Greek," in Wolfgang Dressler, Oskar E. Pfeiffer and John R. Rennison, eds., *Phonologica 1980,* Institut für Sprachwissenschaft der Universität Innsbruck, Austria, pp. 283–92.

Marantz, Alec (1982) "Re: Reduplication," *Linguistic Inquiry* 13, 435–82.

Marsh, James (1969) "Mantjiljara Phonology," *Oceanic Linguistics* 8, 131–52.

Martens, Mary, and Salme Tuominen (1977) "A Tentative Phonemic Statement in Yil in West Sepik District," in Richard Loving, ed., *Phonologies of Five Papua New Guinea Languages,* Workpapers in Papua New Guinea Languages 19, Summer Institute of Linguistics, Ukarumpa, Papua New Guinea, pp. 29–48.

Martin, Jack (1992) "In Support of the Domino Condition," ms., Program in Linguistics, University of Michigan, Ann Arbor.

Mascaró, Joan (1975) *Catalan Phonology and the Phonological Cycle,* Doctoral dissertation, Massachusetts Institute of Technology. [Distributed 1978 by the Indiana University Linguistics Club, Bloomington, IN.]

Matteson, Esther (1965) *The Piro (Arawakan) Language,* University of California Publications in Linguistics 22, University of California Press, Berkeley and Los Angeles.

McArthur, Harry, and Lucille McArthur (1956) "Aguacatec Mayan Phonemes in the Stress Group," *International Journal of American Linguistics* 22, 72–76.

McCarthy, John (1978) "On Stress and Syllabification," ms., Dept. of Linguistics, Massachusetts Institute of Technology, Cambridge, MA. [Published as McCarthy 1979a.]

——— (1979a) "On Stress and Syllabification," *Linguistic Inquiry* 10, 443–65.

——— (1979b) "Formal Problems in Semitic Phonology and Morphology," Ph.D. dissertation, Massachusetts Institute of Technology, Cambridge. [Distributed by Indiana University Linguistics Club, Bloomington.]

——— (1982) "Prosodic Structure and Expletive Infixation," *Language* 58, 574–90.

——— (1986) "OCP Effects: Gemination and Antigemination," *Linguistic Inquiry* 17, 207–63.

McCarthy, John, and Alan Prince (1986) "Prosodic Morphology," ms., Dept. of Linguistics, University of Massachusetts, Amherst, and Program in Linguistics, Brandeis University, Waltham, Mass.

McCarthy, John, and Alan Prince (1990) "Foot and Word in Prosodic Morphology: The Arabic Broken Plural," *Natural Language and Linguistic Theory* 8, 209–83.

McCawley, James (1968) *The Phonological Component of a Grammar of Japanese,* Monographs on Linguistic Analysis 2, Mouton, The Hague.

——— (1974) Review of Chomsky and Halle, *The Sound Pattern of English, International Journal of American Linguistics* 40, 50–88.

McDonald, M., and Stephen A. Wurm (1979) *Basic Materials in Wangkumara (Galali): Grammar, Sentences and Vocabulary,* Pacific Linguistics B65, Australian National University, Canberra.

McElhanon, K. A. (1970) *Selepet Phonology,* Pacific Linguistics B14, Australian National University, Canberra.

McHugh, Brian (1990) "Cyclicity in the Phrasal Phonology of Kivunjo Chaga," Doctoral dissertation, University of California, Los Angeles.

Meerendonk, M. (1949) *Basic Gurkhali Grammar* (4th ed.), Sen Wah Press and Co., Singapore.

Meiklejohn, Percy, and Kathleen Meiklejohn (1958) "Accentuation in Sarangani Manobo," *Studies in Philippine Linguistics,* Oceania Linguistic Monographs, No. 3, University of Sydney, Australia, pp. 1–3.

Mester, R. Armin (1992) "The Quantitative Trochee in Latin," ms., Board of Studies in Linguistics, University of California, Santa Cruz, CA. [To appear in *Natural Language and Linguistic Theory.*]

Michelson, Karin (1983) "A Comparative Study of Accent in the Five Nations Iroquoian Languages," Doctoral dissertation, Harvard University, Cambridge, Mass.

——— (1984) "The Representation of Vowel Length in Seneca," paper given at the 1984 meeting of the Linguistic Society of America.

——— (1988) *A Comparative Study of Lake-Iroquoian Accent,* Kluwer Academic Publishers, Dordrecht.

——— (1989) "Theoretical Consequences of Composite Vowel Length in Seneca," ms., Dept. of Linguistics, State University of New York, Buffalo.

Miner, Kenneth L. (1979) "Dorsey's Law in Winnebago-Chiwere and Winnebago Accent," *International Journal of American Linguistics* 45, 25–33.

——— (1981) "Metrics, or Winnebago Made Harder," *International Journal of American Linguistics* 47, 340–42.

——— (1989) "Winnebago Accent: The Rest of the Data," *Anthropological Linguistics* 31, 148–72.

Mitchell, T. F. (1960) "Prominence and Syllabification in Arabic," *Bulletin of the School of Oriental and African Studies* 23, 369–89. [Reprinted in Mitchell 1975a, pp. 75–98.]

——— (1975a) *Principles of Firthian Linguistics,* Longmans, London.

——— (1975b) "Not of the Letter, but of the Spirit; for the Letter Killeth, but the Spirit Giveth Life," in Mitchell 1975a, pp. 33–74.

Mithun, Marianne, and Reginald Henry (1982) *Watę̨wayę̨stanih, A Cayuga Teaching Grammar,* Woodland Indian Cultural Educational Centre, Brantford, Ontario.

Miyaoka, Osahito (1971) "On Syllable Modification and Quantity in Yuk Phonology," *International Journal of American Linguistics* 37, 219–26.

——— (1985) "Accentuation in Central Alaskan Yupik," in Krauss 1985a, pp. 51–75.

Mohanan, K. P. (1979) "Word Stress in Hindi, Malayalam, and Sindhi," oral presentation, Dept. of Linguistics and Philosophy, Massachusetts Institute of Technology, Cambridge. [Partial summary in Hayes 1981, pp. 79–80.]

——— (1982) "Lexical Phonology," Doctoral dissertation, Massachusetts Institute of Technology, Cambridge.

——— (1986) *The Theory of Lexical Phonology,* Reidel, Dordrecht.

Mohanan, Tara (1989) "Syllable Structure in Malayalam," *Linguistic Inquiry* 20, 589–625.

Morton, J., and W. Jassem (1965) "Acoustical Correlates of Stress," *Language and Speech* 8, 159–81.

Munro, Pamela (1977) "Towards a Reconstruction of Uto-Aztecan Stress," in Hyman 1977a, pp. 303–26.

Munro, Pamela, and Charles Ulrich (1984) "Structure-Preservation and Western Muskogean Rhythmic Lengthening," *West Coast Conference on Formal Linguistics* 3, 191–202.

Mürk, Harry (1981) "The Grade Alternation System in Estonian Morphophonology," ms., Dept. of Linguistics, University of Toronto.

Myers, Scott (1991) "Persistent Rules," *Linguistic Inquiry* 22, 315–44.

Nacaskul, Karnchana (1978) "The Syllabic and Morphological Structure of Cambodian Words," *Mon-Khmer Studies* 7, 183–200.

Nakatani, Lloyd, and Carletta H. Aston (1978) "Acoustic and Linguistic Factors in Stress Perception," ms., ATT Bell Laboratories, Murray Hill, N.J. [Summary in Beckman 1986, 60–62.]

Nespor, Marina (1988) "Aspects of the Interaction between Prosodic Phonology and the Phonology of Rhythm," in Pier Marco Bertinetto and Michele Loporcaro, eds., *Certamen Phonologicum,* Rosenberg and Sellier, Turin, pp. 189–230.
Nespor, Marina, and Irene Vogel (1986) *Prosodic Phonology,* Foris, Dordrecht.
——— (1989) "On Clashes and Lapses," *Phonology* 6, 69–116.
Newman, Paul (1987) "Hausa and the Chadic Languages," in Comrie (1987), pp. 705–23.
Newman, Stanley S. (1944) *The Yokuts Languages of California,* Viking Fund Publications in Anthropology, New York.
——— (1946) "On the Stress System of English," *Word* 2, 171–87.
Newton, Robert P. (1975) "Trochaic and Iambic," *Language and Style* 8, 127–56.
Nicholson, Ray, and Ruth Nicholson (1962) "Fore Phonemes and Their Interpretation," in Arthur Capell and Stephen A. Wurm, eds., *Studies in New Guinea Linguistics,* Oceania Linguistic Monographs 6, University of Sydney, Australia, pp. 128–48.
Nicklas, Thurston Dale (1972) "The Elements of Choctaw," Ph.D. dissertation, University of Michigan, Ann Arbor.
——— (1975) "Choctaw Morphophonemics," in James M. Crawford, ed., *Studies in Southeastern Indian Languages,* University of Georgia Press, Athens, pp. 237–50.
Nooteboom, Sibout G. (1972) "Production and Perception of Vowel Duration," Doctoral dissertation, University of Utrecht.
Oates, William J., and Lynette Frances Oates (1964) "Gugu-Yalanji Linguistic and Anthropological Data," in *Gugu-Yalanji and Wik-Munkan Language Studies,* Australian Institute of Aboriginal Studies, Canberra, pp. 1–17.
Odden, David (1979) "Principles of Stress Assignment: A Crosslinguistic View," *Studies in the Linguistic Sciences* 9.1, 157–75, Dept. of Linguistics, University of Illinois.
——— (1987) "Ordering Paradoxes in Lexical Phonology," in Brian D. Joseph and Arnold M. Zwicky, eds., *A Festschrift for Ilse Lehiste,* Ohio State University Working Papers in Linguistics 35, pp. 21–28.
Ohala, John J. (1977) "The Physiology of Stress," in Hyman 1977a, pp. 145–68.
Ohala, Manjari J. (1977) "Stress in Hindi," in Hyman 1977a, pp. 327–38.
Orešnik, Janez (1971) "On the Phonological Boundary between Constituents of Modern Icelandic Compound Words," *Linguistica* (Ljubljana) 11, 51–59. [Reprinted 1985 in *Studies in the Phonology and Morphology of Modern Icelandic: A Selection of Essays,* H. Buske, Hamburg, pp. 49–57.]
Osborn, Henry (1966) "Warao I: Phonology and Morphophonemics," *International Journal of American Linguistics* 32, 108–23.
Oswalt, Robert L. (1961) "A Kashaya Grammar (Southwestern Pomo)," Doctoral dissertation, University of California, Berkeley.
——— (1988) "The Floating Accent of Kashaya," in William Shipley, ed., *In Honor of Mary Haas,* Mouton de Gruyter, Berlin, pp. 611–22.
Owens, Jonathan (1980) "The Syllable as Prosody: A Reanalysis of Syllabification in Eastern Libyan Arabic," *Bulletin of the School of Oriental and African Studies* 43, 277–87.
——— (1984) *A Short Reference Grammar of Eastern Libyan Arabic,* Otto Harrassowitz, Wiesbaden.
Pandey, Pramod Kumar (1989) "Word Accentuation in Hindi," *Lingua* 77, 37–73.

Paradis, Carole (1988) "On Constraints and Repair Strategies," *Linguistic Review* 6, 71–97.

Payne, David L. (1981) *The Phonology and Morphology of Axininca Campa,* Summer Institute of Linguistics and University of Texas, Arlington, TX.

——— (1983) "Notas fonológicas y morfofonémicas sobre el ashéninca del Pichis," in David L. Payne and Marlene Ballena Dávila, eds., *Estudios lingüísticos de textos ashéninca,* Serie Lingüistica Peruana 21, Instituto Lingüístico de Verano, Yarinacocha, Pucallpa, Peru, pp. 101–12.

Payne, David L., Judith K. Payne, and Jorge Sanchez Santos (1982) *Morphologia, fonologia y fonetica del asheninca del Apurucayali,* Serie Lingüistica Peruana 18, Instituto Lingüístico de Verano, Yarinacocha, Pucallpa, Peru.

Payne, Judith (1990) "Asheninca Stress Patterns," in Doris L. Payne, ed., *Amazonian Linguistics,* University of Texas Press, Austin, pp. 185–209.

Pesetsky, David (1979) "Menomini Quantity," *MIT Working Papers in Linguistics,* vol. 1, pp. 115–39.

Peterson, G. E. (1958) "Some Observations on Speech," *Quarterly Journal of Speech* 44, 402–12.

Phinnemore, Thomas R. (1985) "Ono Phonology and Morphophonemics," *Papers in New Guinea Linguistics* 22, 173–214. [= *Pacific Linguistics* A63, Australian National University, Canberra]

Piera, Carlos (1980) "Spanish Verse and the Theory of Meter," Doctoral dissertation, Program in Romance Linguistics and Literature, University of California, Los Angeles.

Pierrehumbert, Janet (1980) "The Phonology and Phonetics of English Intonation," Ph.D. dissertation, Massachusetts Institute of Technology, Cambridge. [Distributed by Indiana University Linguistics Club, Bloomington.]

Pierrehumbert, Janet, and Mary Beckman (1988) *Japanese Tone Structure,* MIT Press, Cambridge, Mass.

Pierrehumbert, Janet, and Julia Hirschberg (1990) "The Meaning of Intonation Contours in the Interpretation of Discourse," in Philip Cohen, Jerry Morgan, and Martha Pollack, *Intentions in Communication,* MIT Press, Cambridge, Mass., pp. 271–311.

Piggott, Glyne L. (1978) "Algonquin and Other Ojibwa Dialects: A Preliminary Report," in William Cowan, ed., *Papers of the Ninth Algonquian Conference,* Carleton University, Ottawa, pp. 160–87.

——— (1980) *Aspects of Odawa Morphophonemics,* Garland Publishing, New York.

——— (1983) "Extrametricality and Ojibwa Stress," *McGill Working Papers in Linguistics,* 1, 80–117.

Pike, Kenneth (1964) "Stress Trains in Auca," in D. Abercrombie, Dennis B. Fry, P. A. D. MacCarthy, N. C. Scott, J. L. M. Trim, eds., *In Honour of Daniel Jones,* Longmans, London, pp. 425–31.

Pike, Kenneth, and Eunice Pike (1947) "Immediate Constituents of Mazateco Syllables," *International Journal of American Linguistics* 13, 78–91.

Pike, Kenneth, and Graham Scott (1963) "Pitch Accent and Non-Accented Phrases in Fore (New Guinea)," *Zeitschrift für Phonetik* 16, 179–89.

Poser, William J. (1984) "The Phonetics and Phonology of Tone and Intonation in Japanese," Doctoral dissertation, Massachusetts Institute of Technology, Cambridge.

——— (1986) "Invisibility," *GLOW Newsletter* 16, 63–64.

——— (1989) "The Metrical Foot in Diyari," *Phonology* 6, 117–48.

——— (1990) "Evidence for Foot Structure in Japanese," *Language* 66, 78–105.

Prince, Alan (1975) "The Phonology and Morphology of Tiberian Hebrew," Doctoral dissertation, Massachusetts Institute of Technology, Cambridge.

——— (1976a) "Applying Stress," ms., Dept. of Linguistics, University of Massachusetts, Amherst.

——— (1976b) "Stress," ms., Dept. of Linguistics, University of Massachusetts, Amherst.

——— (1980) "A Metrical Theory for Estonian Quantity," *Linguistic Inquiry* 11, 511–62.

——— (1983a) "Relating to the Grid," *Linguistic Inquiry* 14, 19–100.

——— (1983b) "Hierarchy without Constituency in Stress Theory," paper given at the A. P. Sloan Foundation Workshop in Phonology, University of Massachusetts, Amherst.

——— (1984) "Phonology with Tiers," in Mark Aronoff and Richard T. Oehrle, eds., *Language Sound Structure: Studies in Phonology Presented to Morris Halle by his Teacher and Students,* MIT Press, Cambridge, Mass., pp. 234–44.

——— (1985) "Improving Tree Theory," *Berkeley Linguistics Society* 11, 471–90.

——— (1990) "Quantitative Consequences of Rhythmic Organization," *Parasession on the Syllable in Phonetics and Phonology,* Chicago Linguistic Society, pp. 355–98.

Pulleyblank, Douglas (1986a) "Rule Application on a Noncyclic Stratum," *Linguistic Inquiry* 17, 573–80.

——— (1986b) *Tone in Lexical Phonology,* Reidel, Dordrecht.

Rappaport, Malka (1984) "Issues in the Phonology of Tiberian Hebrew," Doctoral dissertation, Massachusetts Institute of Technology, Cambridge.

Repetti, Lori (1989) *The Bimoraic Norm of Tonic Syllables in Italo-Romance,* Doctoral dissertation, Program in Romance Linguistics, University of California, Los Angeles.

Rhodes, Richard (1985) "Lexicography and Ojibwa Vowel Deletion," *Canadian Journal of Linguistics* 30, 453–71.

Rice, Curtis (1988) "Stress Assignment in the Chugach Dialect of Alutiiq," *Chicago Linguistics Society* 24, 304–15.

——— (1989) "An Autosegmental Analysis of Secondary Stress in Chugach Alutiiq," paper presented at the Annual Meeting of the Linguistic Society of America, New Orleans.

——— (1990) "Pacific Yup'ik: Implications for Metrical Theory," *Coyote Papers,* Dept. of Linguistics, University of Arizona, Tucson.

——— (1992) "Binarity and Ternarity in Metrical Theory: Parametric Extensions," Doctoral dissertation, University of Texas, Austin.

Rischel, Jørgen (1972) "Compound Stress in Danish without a Cycle," *Annual Report of the Institute of Phonetics* 6, University of Copenhagen, pp. 211–28.

Robbins, Scarlett (1991) "Lexicalized Metrical Foot Structure in Maidu," in Armin Mester and Scarlett Robbins, eds., *Phonology at Santa Cruz,* vol. 2, Syntax Research Center, University of California, Santa Cruz, pp. 95–116.

Roberts, John R. (1987) *Amele,* Biddles, Guildford, England.

Roca, Iggy (1986) "Secondary Stress and Metrical Rhythm," *Phonology Yearbook* 3, 341–70.
——— (1992) "Constraining Extrametricality," *Phonologica 1988,* Cambridge University Press, Cambridge.
Rohlfs, Gerhard (1949) *Historische Grammatik der italienischen Sprache und ihrer Mundarten,* Francke Verlag, Bern.
Rombandeeva, Evdokiia Ivanovna (1973) *Mansijskij (Vogul'skij) Jazyk,* Isdatel'stvo "Nauka", Moscow.
Ross, John R. (1972) "A Reanalysis of English Word Stress (Part I)," in Michael Brame, ed., *Contributions to Generative Phonology,* University of Texas Press, Austin, pp. 229–323.
Russom, Geoffrey (1987) *Old English Meter and Linguistic Theory,* Cambridge University Press, Cambridge.
Sag, Ivan, and Mark Liberman (1975) "The Intonational Disambiguation of Indirect Speech Acts," *Chicago Linguistic Society* 11, 487–97.
Saint, Rachel, and Kenneth Pike (1962) "Auca Phonemics," in Catherine Peeke and Benjamin Elson, eds., *Studies in Ecuadorian Indian Languages: I,* Summer Institute of Linguistics, Norman, Okla., pp. 2–30.
Saksena, Baburam (1971) *Evolution of Awadhi,* Motilal Banarsidass, Delhi.
Sapir, Edward (1930) *Southern Paiute, A Shoshonean Language,* Proceedings of the American Academy of Arts and Sciences 65, 1–296.
——— (1949) "The Psychological Reality of Phonemes," in David G. Mandelbaum, ed., *Selected Writings of Edward Sapir in Language, Culture, and Personality,* pp. 46–60. [Published originally as (1930) "La réalité psychologique des phonèmes," *Journal de psychologie normale et pathologique* 30, 247–65.]
Sapir, Edward, and Morris Swadesh (1960) *Yana Dictionary,* University of California Publications in Linguistics 22, University of California Press, Berkeley.
Sauzet, Patrick (1989) "L'accent du grec ancien et les relations entre structure métrique et réprésentation auto-segmentale," *Langages* 95, 81–113.
Schein, Barry, and Donca Steriade (1986) "On Geminates," *Linguistic Inquiry* 17, 691–744.
Schmerling, Susan (1976) *Aspects of English Sentence Stress,* University of Texas Press, Austin.
Schürmann, Clamor Wilhelm (1844) *A Vocabulary of the Parnkalla Language,* George Dehane, Adelaide.
Schütz, Albert J. (1978) "English Loanwords in Fijian," in Schütz, Albert J., ed., *Fijian Language Studies: Borrowing and Pidginization,* Bulletin of the Fiji Museum 4, Fiji Museum, Suva.
Schütz, Albert J. (1985) *The Fijian Language,* University of Hawaii Press, Honolulu.
Scorza, David (1985) "A Sketch of Au Morphology and Syntax," *Papers in New Guinea Linguistics* 22, 215–73. [= *Pacific Linguistics* A63, Australian National University, Canberra]
Scott, Graham (1978) *The Fore Language of Papua New Guinea,* Pacific Linguistics B47, Australian National University, Canberra.
Scott, Norman C. (1948) "A Study in the Phonetics of Fijian," *Bulletin of the School of Oriental and African Studies* 12, 737–52.

Seaman, P. David (1985) *Hopi Dictionary,* Dept. of Anthropology, Northern Arizona University, Flagstaff.

Sebeok, Thomas, and Francis Ingemann (1961) *An Eastern Cheremis Manual,* Uralic and Altaic series 5, Indiana University, Bloomington.

Seeger, Pete (1954) *American Folk Songs for Children,* Folkways Record and Service Corp., New York. [sound recording]

Seiler, Hansjakob (1957) "Die phonetischen Grundlagen der Vokalphoneme des Cahuilla," *Zeitschrift für Phonetik und allgemeine Sprachwissenschaft* 10, 204–23.

——— (1965) "Accent and Morphophonemics in Cahuilla and Uto-Aztecan," *International Journal of American Linguistics* 31, 50–59.

——— (1967) "Structure and Reconstruction in some Uto-Aztecan Languages," *International Journal of American Linguistics* 33, 135–47.

——— (1977) *Cahuilla Grammar,* Malki Museum Press, Banning, Calif.

Seiler, Hansjakob, and Kojiro Hioki (1979) *Cahuilla Dictionary,* Malki Museum Press, Banning, Calif.

Selkirk, Elisabeth O. (1980a) "Prosodic Domains in Phonology: Sanskrit Revisited," in Mark Aronoff and Mary-Louise Kean, eds., *Juncture,* Anma Libri, Saratoga, Calif., pp. 107–29.

——— (1980b) "The Role of Prosodic Categories in English Word Stress," *Linguistic Inquiry* 11, 563–605.

——— (1981) "Epenthesis and Degenerate Syllables in Cairene Arabic," *MIT Working Papers in Linguistics* 3, 209–32.

——— (1982) "The Syllable," in Harry van der Hulst and Norval Smith, eds., *The Structure of Phonological Representations (Part II),* Foris, Dordrecht, pp. 337–83.

——— (1984) *Phonology and Syntax: The Relation between Sound and Structure,* MIT Press, Cambridge, Mass.

——— (1988) "A Two-Root Theory of Length," paper presented at the Nineteenth meeting of the Northeastern Linguistic Society, Cornell University, Ithaca, N.Y. [Draft of Selkirk 1990.]

——— (1990) "A Two-Root Theory of Length," *University of Massachusetts Occasional Papers in Linguistics* 14.

Sharma, Aryendra (1969) "Hindi Word-Accent," *Indian Linguistics* 30, 115–18.

Shaw, Patricia A. (1985a) "Modularisation and Substantive Constraints in Dakota Lexical Phonology," *Phonology Yearbook* 2, 173–202.

——— (1985b) "Coexistent and Competing Stress Rules in Stoney (Dakota)," *International Journal of American Linguistics* 51, 1–18.

Shipley, William F. (1964) *Maidu Grammar,* University of California Publications in Linguistics 41, University of California Press, Berkeley and Los Angeles.

Shryock, Aaron (1993) "A Metrical Analysis of Stress in Cebuano," *Lingua* 91, 103–48.

Shukla, Shaligram (1981) *Bhojpuri Grammar,* Georgetown University Press, Washington, D.C.

Sjögren, Joh. Andreas (1861) *Livische Grammatik nebst Sprachproben,* Kaiserlichen Akademie der Wissenschaften, St. Petersburg.

Smith, Norval, Roberto Bolognesi, Frank van der Leeuw, Jean Rutten, and Heleen de Wit (1989) "Apropos of the Dutch Vowel System 21 Years On," in Hans Bennis and

Ans van Kemenade, eds., *Linguistics in the Netherlands 1989,* Foris, Dordrecht, pp. 133–42.

Solan, Lawrence (1979) "A Metrical Analysis of Spanish Stress," in William W. Cressey and Donna Jo Napoli, eds., *Linguistic Symposium on Romance Languages 9,* Georgetown University Press, Washington, D.C., pp. 90–104.

Sovijärvi, Antti (1956) *Ueber die Phonetischen Hauptzüge der finnischen und der ungarischen Hochsprache,* Otto Harrassowitz, Wiesbaden.

Spring, Cari (1989a) "Cayuvava Dactyls," ms., Dept. of Linguistics, University of Arizona, Tucson.

——— (1989b) "The Ternary Foot in Axininca Stress and Morphology," ms., Dept. of Linguistics, University of Arizona, Tucson.

——— (1990) "Implications of Axininca Campa for Prosodic Morphology and Reduplication," Doctoral dissertation, Dept. of Linguistics, University of Arizona, Tucson.

Steriade, Donca (1982) "Greek Prosodies and the Nature of Syllabification," Doctoral dissertation, Massachusetts Institute of Technology, Cambridge.

——— (1984) "Glides and Vowels in Romanian," *Berkeley Linguistics Society 10,* pp. 47–64.

——— (1987) "Locality Conditions and Feature Geometry," *North Eastern Linguistic Society* 17, 595–617.

——— (1988a) "Review Article: Clements and Keyser, *CV Phonology,*" *Language* 64, 118–29.

——— (1988b) "Greek Accent: A Case for Preserving Structure," *Linguistic Inquiry* 19, 271–314.

——— (1991) "Moras and Other Slots," in *Proceedings of the Formal Linguistics Society of Midamerica* 1, 254–80.

Stetson, Raymond H. (1928) *Motor Phonetics,* North Holland, Amsterdam. [2nd edition 1951; republished 1988 with commentary by J. A. Scott Kelso and Kevin G. Munhall, Little, Brown and Company, Boston.]

Stockwell, Robert P. (1972) "The Role of Intonation: Reconsiderations and Other Considerations," in Dwight Bolinger, ed., *Intonation,* Penguin Books, Harmondsworth, England, pp. 87–109.

Stowell, Timothy (1979) "Stress Systems of the World, Unite!" *MIT Working Papers in Linguistics,* vol. 1, pp. 51–76.

Street, John C. (1963) *Khalkha Structure,* Uralic and Altaic series 24, Indiana University, Bloomington.

Strehlow, Theodor G. H. (1945) *Aranda Phonetics and Grammar,* Australian National Research Council, Sydney.

Susman, Amelia (1943) "The Accentual System of Winnebago," Doctoral dissertation, Columbia University, New York.

Szinnyei, Josef (1912) *Ungarische Sprachlehre,* Göschen, Berlin.

Tanaka, Shin-ichi (1990) "Old English as a Mora-Counting Language: Stress and its Relation to High Vowel Deletion," *Tsukuba English Studies* 9, 39–60.

Tauli, Valter (1954) "The Origin of the Quantitative System in Estonian," *Journal de la société finno-ougrienne* 57, 1–19.

Teeter, Karl V. (1971) "The Main Features of Malecite-Passamaquoddy Grammar," in

Jesse Sawyer, ed., *Studies in American Indian Languages,* University of California Publications in Linguistics 65, University of California Press, Berkeley, pp. 191–249.

Teeter, Karl V., and Philip LeSourd (1983) "Vowel Length in Malecite," in William Cowan, ed., *Actes du quatorzième congrès des algonquinistes,* Carleton University, Ottawa, pp. 245–48.

Thompson, David A. (1976) "A Phonology of Kuuku-Yaʔu," in Peter Sutton, ed., *Languages of Cape York,* Australian Institute of Aboriginal Studies, Canberra, pp. 213–35.

Thompson, Henry (1980) "Stress and Salience in English," Doctoral dissertation, University of California, Berkeley.

Tiwari, Udai Narain (1960) *The Origin and Development of Bhojpuri,* Asiatic Society Monograph 10, Asiatic Society, Calcutta.

Trager, George L., and Henry Lee Smith, Jr. (1951) *An Outline of English Structure,* Battenburg Press, Norman, Oklahoma.

Tranel, Bernard (1992) "CVC Light Syllables, Geminates, and Moraic Theory," *Phonology* 8, 291–302.

Trubetzkoy, Nikolay Sergeevich (1969) *Principles of Phonology,* translated by Christiane A. M. Baltaxe, University of California Press, Berkeley.

Tryon, Darrell T. (1967a) *Nengone Grammar,* Pacific Linguistics B6, Australian National University, Canberra.

——— (1967b) *Dehu Grammar,* Pacific Linguistics B7, Australian National University, Canberra.

——— (1970) *An Introduction to Maranungku,* Pacific Linguistics B15, Australian National University, Canberra.

Tyhurst, James J. (1987) "Accent Shift in Seminole Nouns," in Pamela Munro, ed., *Muskogean Linguistics,* UCLA Occasional Papers in Linguistics 6, pp. 161–70.

Ulrich, Charles (1986) "Choctaw Morphophonology," Ph.D. dissertation, University of California, Los Angeles.

Vance, Timothy (1987) *Introduction to Japanese Phonology,* State University of New York Press, Albany.

van der Hulst, Harry (1984) *Syllable Structure and Stress in Dutch,* Foris, Dordrecht.

van der Hulst, Harry, and John van Lit (1988) "Dutch Stress," ms., University of Leiden.

van Katwijk, Albert (1974) *Accentuation in Dutch: An Experimental Linguistic Study,* van Gorcum, Amsterdam.

Vanderslice, Ralph, and Peter Ladefoged (1972) "Binary Suprasegmental Features and Transformational Word-Accentuation Rules," *Language* 48, 819–38.

Vincent, Nigel (1987) "Italian," in Comrie 1987, pp. 279–302.

Visch, Ellis (1989) "A Metrical Theory of Rhythmic Stress Phenomena," Doctoral dissertation, University of Utrecht.

Voegelin, Charles (1935) *Tübatulabal Grammar,* University of California Publications in American Archaeology and Ethnology, Vol. 34, no. 2, University of California Press, Berkeley, pp. 55–189.

Voegelin, Charles, and Florence Voegelin (1977) "Is Tübatulabal De-Acquisition Relevant to Theories of Language Acquisition?" *International Journal of American Linguistics* 43, 333–38.

Wald, Benji (1987) "Swahili and the Bantu Languages," in Comrie 1987, pp. 991–1014.
Ward, Gregory, and Julia Hirschberg (1985) "Implicating Uncertainty: The Pragmatics of Fall-Rise Intonation," *Language* 61, 747–76.
Weeda, Don (1989) "Trimoraicity in Central Alaskan Yupik and Elsewhere," ms., Dept. of Linguistics, University of Texas, Austin.
——— (1990) "Foot Extrametricality in Central Alaskan Yupik," ms., Dept. of Linguistics, University of Texas, Austin.
Welden, Ann (1980) "Stress in Cairo Arabic," *Studies in the Linguistic Sciences* 10.2, Dept. of Linguistics, University of Illinois, Urbana, pp. 99–120.
Wheeler, Deirdre (1979a) "A Metrical Analysis of Stress and Related Processes in Southern Paiute and Tübatulabal," *University of Massachusetts Occasional Papers in Linguistics* 5, 145–75.
——— (1979b) "A Historical Explanation for Final Stress in Tübatulabal," in Engdahl and Stein 1979, pp. 222–32.
White Eagle, Josie (1982) "Teaching Scientific Inquiry and the Winnebago Language," *International Journal of American Linguistics* 48, 306–19.
Whorf, Benjamin Lee (1946) "The Hopi Language, Toreva Dialect," in Cornelius Osgood, ed., *Linguistic Structures of Native America,* Viking Fund Publications in Anthropology 6, Viking Fund, New York, pp. 158–83.
Wightman, Colin W., Stefanie Shattuck-Hufnagel, Mari Ostendorf, and Patti J. Price (1992) "Segmental Durations in the Vicinity of Prosodic Phrase Boundaries," *Journal of the Acoustical Society of America* 91, 1707–17.
Winstedt, Richard O. (1927) *Malay Grammar,* Oxford University Press, Oxford.
Wolfram, Walt, and Robert Johnson (1982) *Phonological Analysis: Focus on American English,* Center for Applied Linguistics, Georgetown University, Washington, D.C.
Woodbury, Anthony (1981) "Study of the Chevak Dialect of Central Yupik Eskimo," Ph.D. dissertation, University of California, Berkeley.
——— (1985a) "Graded Syllable Weight in Central Alaskan Yupik Eskimo (Hooper Bay-Chevak)" *International Journal of American Linguistics* 51, 620–23.
——— (1985b) "Meaningful Phonological Processes: A Consideration of Central Alaskan Yupik Eskimo Prosody," ms., University of Texas, Austin. [Draft of Woodbury (1987).]
——— (1987) "Meaningful Phonological Processes: A Consideration of Central Alaskan Yupik Eskimo Prosody," *Language* 63, 685–740.
——— (1989) Phrasing and Intonational Tonology in Central Alaskan Yupik Eskimo: Some Implications for Linguistics in the Field," in John Dunn, ed., *1988 Mid-America Linguistics Conference Papers,* University of Oklahoma, Norman, pp. 3–40.
——— (1990) "Tonal Association in the Central Alaskan Yupik Eskimo Intonational Word," ms., Dept. of Linguistics, University of Texas, Austin. [Paper delivered at the 1989 meeting of the Linguistic Society of America, Washington, D.C.]
Woodrow, Herbert (1909) "A Quantitative Study of Rhythm," *Archives of Psychology* (New York) 14, 1–66.
——— (1951) "Time Perception," in S. S. Stevens, ed., *Handbook of Experimental Psychology,* Wiley, New York, pp. 1234–36.
Yadav, Ramawatar (1984) *Maithili Phonetics and Phonology,* Selden & Tamm, Mainz.

Yip, Moira J. W. (1980) *The Tonal Phonology of Chinese,* Doctoral dissertation, MIT, Cambridge, MA. [Distributed 1980 by the Indiana University Linguistics Club, Bloomington, IN.]

——— (1992) "On Quantity-Insensitive Languages," ms., Program in Linguistics, University of California, Irvine.

Zebek, Schalonow, and Johannes Schubert (1961) *Mongolisch-Deutsches Wörterbuch,* Veb Verlag Enzyklopädie, Leipzig.

Zec, Draga (1988) *Sonority Constraints on Prosodic Structure,* Doctoral dissertation, Stanford University, Stanford, CA.

Zigmond, Maurice L., Curtis G. Booth, and Pamela Munro (1990) *Kawaiisu: A Grammar and Dictionary,* University of California Publications in Linguistics 119, University of California Press, Berkeley and Los Angeles.

INDEX OF NAMES

Abad, May 123
Abaev, Vasilii Ivanovich 261
Abbott, Miriam 208
Abbott, Stan 297
Abdul-Karim, Kamal 125, 129, 181, 228
Abu-Salim, Issam M. 125
Al-Mozainy, Hamza 42, 181, 239
Allen, George D. 79
Allen, W. Sidney 40, 67, 69, 91, 92, 112, 120, 121, 180
Anderson, Stephen R. 81, 197
Arany, László 330
Archangeli, Diana 57, 58, 59, 108, 204, 306
Ariste, Paul 201
Aston, Carletta 7, 11
Austin, Peter 122, 199
Austin, William M. 297
Árnason, Kristján 188–98

Bach, Emmon 220, 297
Bagemihl, Bruce 110
Balassa, József 330
Barker, Chris 57, 60, 113, 262
Barker, Muhammad Abd-al-Rahman 280n, 279–81
Bat-El, Outi 32
Bates, Dawn 110
Beckman, Mary 5, 7, 372
Bečka, Jiri 32
Bell, Alan 79, 81
Benger, Janet 222, 225
Berinstein, Ava 5, 7
Bickmore, Lee 25, 69
Bing, Janet M. 375
Birk, D. B. W. 203
Blake, Barry J. 122, 201
Blanc, Haim 226–27, 283
Blevins, Juliette 49, 64, 66, 113, 119, 167, 168, 201, 205, 278, 279. *See also* Levin, Juliette
Bley-Vroman, Robert 42, 42n, 181, 239
Bloomfield, Leonard 216–19, 221, 269
Boas, Franz 297
Bolinger, Dwight 6, 369
Bolognesi, Roberto 306
Bolozky, Shmuel 98, 198, 367
Bolton, Thaddeus L. 79–80
Booij, Geert 170, 204
Booth, Curtis 181
Boxwell, Helen 265
Boxwell, Maurice 265
Brame, Michael 125, 129, 185
Breen, Gavan 199–200
Broadbent, Sylvia M. 113, 250, 261
Broselow, Ellen 70
Bruce, Gösta 84, 97, 117, 184, 367
Buckley, Eugene 57, 74, 108n, 260
Bunn, Gordon 278
Bunn, Ruth 278

Callaghan, Catherine A. 261
Capell, Arthur 199, 205
Carlson, Lauri 7, 329
Chadwick, Neil 202
Chafe, Wallace L. 59, 222, 225–26, 266
Chai, Nemia M. 265, 305
Chambers, J. K. 267
Chaski, Carole 182
Chlumský, Josef 203
Chomsky, Noam xiii, 9, 55, 108n, 181, 350, 367, 373
Chung, Sandra 7, 59, 148, 205
Churchward, C. Maxwell 182
Churchyard, Henry 42, 117, 263, 368
Clements, George N. 12, 34, 43, 106, 119, 229, 357, 359, 360n

Cohn, Abigail 49n, 97, 98n, 170
Collier, Ken 205
Collier, Margaret 205
Collinder, Björn 329
Comrie, Bernard 3, 204, 205
Contreras, Heles 266
Cooper, Grosvenor 37, 80
Couro, Ted 181
Cowan, Hendrik K. J. 330–31, 333
Cowan, William 221
Crowhurst, Megan 53, 263, 265, 299, 301, 304
Crowley, Terry 103, 178–79, 198

Darwin, Christopher J. 31
Dauer, Rebecca M. 30
Davis, Alice Irene 149, 161
Davis, Stuart M. 272–73, 273, 277, 285, 306
Dayley, Jon P. 180
de Wit, Heleen 306
DeJong, Kenneth 372
Dell, François 24, 313, 371–72
den Os, Els 59
Denes, Peter B. 276
Dennis, Ronald K. 182
Derbyshire, Desmond C. 205–8
Dixon, Robert M. W. 25, 40, 140–42, 145, 180, 260
Donegan, Patricia 349n
Donovan, Andrew 31
Douglas, Wilfred H. 62
Drachmann, Gaberell 99, 204, 367
Dresher, B. Elan 55, 82, 118, 119, 188, 224
Dryer, Matthew S. 269
DuBois, Carl 179
Dunn, Leone 198
Dyen, Isidore 305

Echeverría, Max S. 266
Eek, Arvo 319, 321
Elbert, Samuel 148, 181
Elenbaas, Nine 314–15, 330, 333
England, Nora 281–83
Everett, Daniel L. 48, 272–73, 285–87
Everett, Karen 48, 272–73, 285

Fairbanks, Constance 163–67
Fant, Gunnar 80
Feldman, Harry 182
Fleming, Ilak 182
Flexner, Stuart Berg 111
Foley, William 3, 330
Fong, Eugene A. 148
Ford, Kevin 43, 357
Foster, Michael 122, 222–24, 301
Fraisse, Paul 79
Franks, Steven L. 57, 59, 204, 205
Freeland, Lucy S. 112, 113, 261
Fromkin, Victoria 42
Fry, Dennis 6–7
Furby, Christine 202

Geraghty, Paul A. 145
Giegerich, Heinz 181, 188, 200, 375, 382
Goddard, Ives 211–13, 219, 222
Goldsmith, John 2, 42, 179, 203, 360, 364
Golston, Chris 181, 365
Grierson, George A. 163–64
Griffith, Teresa 261
Griffiths, John D. 6
Grimes, Barbara 3
Gupta, Abha 277
Gussenhoven, Carlos 18, 22, 367, 369, 391
Gussman, Edmund 188, 189, 192, 193

Haas, Mary 64, 66, 67
Hale, Kenneth 25, 122, 346–53, 364
Hall, Robert A., Jr. 330
Halle, Morris xiii, 3, 9, 27, 32, 38–42, 42n, 49, 55, 57, 60, 64, 65, 67, 71, 75–77, 77n, 87, 108n, 109, 113–15, 119, 122, 130, 167, 170, 176n, 178, 181, 182, 185, 187–88, 204, 216, 225, 243, 260, 262, 263, 266, 273, 275, 278, 279, 281, 285, 297, 305–7, 309, 311, 313, 314, 328–29, 333, 335, 345, 346, 350, 351–55, 355n, 358–60, 367, 372, 373–74, 376
Hammond, Michael 3, 36, 38, 41, 62, 76, 113, 117, 167, 174, 176n, 176–78, 197, 199, 204, 249, 259, 262, 279, 280n, 281, 285, 293, 308, 313, 330, 333, 390
Hangin, John G. 297
Hansen, Kenneth C. 62–63
Hansen, L. E. 62–63
Harms, Robert T. 69, 121, 122n, 220, 268, 322, 329
Harrell, Richard S. 67, 69, 71, 130–31
Harris, James 57, 58, 59, 94, 96, 97, 109, 181
Hasegawa, Yoko 357
Hata, Kazue 357
Hawkins, W. Neil 208
Hayes, Bruce 12, 18, 24, 28, 29, 30, 38, 43, 52–57, 59–62, 64, 67, 75–77, 82, 91, 93,

96, 97, 109, 109n, 111, 112, 114n, 122, 123, 130, 139, 149, 163, 175, 181, 194, 200, 204, 230, 260, 261, 265, 266, 279n, 285, 299, 305, 307, 309, 313, 324, 329, 341, 367–69, 371, 372, 377n, 382–86, 392, 395, 397–99
Henry, Reginald 223–24
Hercus, Louise A. 306
Herrfurth, Hans 262
Hess, Thomas 297
Hewitt, Mark S. 333, 345
Hill, Jane 32
Hill, Kenneth 32
Hint, Mati 56, 175, 316, 317, 319n, 321, 322
Hioki, Kojiro 132
Hirschberg, Julia 10
Hockett, Charles F. 218, 219, 221
Hoff, Berend 83, 208
Hudson, Joyce 100
Huffman, Franklin E. 261
Huss, Volker 8
Hutcheson, Christina 181
Hutchinson, Sandra Pinkerton 43
Hyman, Larry 25, 52, 53, 61, 205, 299

Ingemann, Francis 296
Inkelas, Sharon 57, 69, 170, 204, 278
Irshied, Omar 42, 230, 239, 366
Itkonen, Erkki 199, 297
Ito, Junko 29, 52, 53, 57, 59, 103, 109, 114, 149, 214, 237, 237n
Ito, Lucille I. 182

Jackendoff, Ray 26–30, 37, 80
Jackson, Michel 64, 185
Jacobs, Haike 50, 73, 266
Jacobson, Steven A. 239–56, 241n, 304
Jakobson, Roman 49, 203
Jeanne, Laverne Masavesya 111, 261
Jha, Subhadra 149–62
Jóhannsson, Johannes L. L. 192
Johnson, Robert 2
Jones, Daniel 5, 23
Jones, W. E. 278

Kager, René 3, 59, 86, 87, 102n, 120, 188, 224, 260, 264, 266, 305, 306, 330, 333, 342, 343, 345, 367, 368, 370, 373, 382, 385, 386–99
Kahn, Daniel 12, 48, 109
Kaisse, Ellen 43, 262
Kalectaca, Milo 261
Kálmán, Béla 200
Kamprath, Christine 181
Kaplan, Lawrence D. 83, 302
Katada, Fusa 53
Kaye, Jonathan 55, 216, 217, 230
Keijsper, Cornelia 369
Kelkar, Ashok R. 276, 278
Kenstowicz, Michael 2, 42, 67, 71, 87, 106, 109, 115, 115n, 119, 125–30, 181, 182, 187–88, 226–30, 237, 239, 305, 366
Kenyon, John S. 5, 111
Kerek, Andrew 330
Kettunen, Lauri 200
Key, Harold H. 309
Key, Mary Ritchie 202
Keyser, S. Jay 43, 106, 118, 359
Khubchandani, Lachman M. 278
Kiparsky, Paul 12, 13, 18, 32, 49, 54, 82, 103, 109, 115, 185, 186, 188, 191, 256, 278, 297, 370, 382
Kirchner, Robert 149, 260
Kisseberth, Charles 2, 114
Klatt, Dennis 7, 100
Knott, Thomas A. 5, 111
Knudson, Lyle M. 104
Knudson, Ronald 6
Koreman, Jacques 305
Krauss, Michael 239–58, 241n
Kruckenberg, Anita 80
Krueger, John R. 296

Ladd, D. Robert 5, 7, 11, 18, 375, 376
Ladefoged, Peter 3, 6, 9, 22, 338
Lahiri, Aditi 18, 61, 82, 118, 224, 305, 306
Lakó, Gy. 200
Langdon, Margaret 181
Langendoen, D. Terence 228–38
Larsen, Raymond S. 296
Lea, Wayne 5, 23
Leer, Jeff 48, 239, 241, 243, 302, 304, 333–45
Lehiste, Ilse 5, 9, 31, 40, 276
Lerdahl, Fred 26, 27, 28, 30, 37, 80
Leskinen, Heikki 330
LeSourd, Philip 215–16
Levin, Juliette 51, 132, 133, 138, 185, 225, 279, 300, 307, 309, 311, 312–13. *See also* Blevins, Juliette
Levisohn, Stephen H. 181
Lewis, M. B. 263
Liberman, Mark 3, 8, 10, 16, 18, 24–30, 33, 38, 57, 108, 108n, 112, 114n, 179, 367, 369–72, 374, 378, 382

Lichtenberk, Frantisek 182
Lieberman, Philip 6
Lindblom, Björn 101
Lombardi, Linda 209, 286
Lotz, John 330
Lukoff, Fred 367
Lunt, Horace 205, 367
Lynch, John D. 167–78, 180, 355, 367
Lytvin, V. I. 297

Maddieson, Ian 220, 338
Maldonado Andres, Juan 281–82
Malikouti-Drachmann, Angelika 99, 204, 367
Marantz, Alec 123
Marsh, James 365
Martens, Mary 93
Martin, Jack 64–67, 119, 359n
Mascaró, Joan 23
Matteson, Esther 201, 367
McArthur, Harry 297
McArthur, Lucille 297
McCarthy, John 3, 40, 42, 42n, 47, 51–53, 55, 57, 62, 67–71, 77, 78, 82, 86, 87, 88, 101, 109, 114, 123, 131, 173, 180, 181, 209, 228, 239, 245, 255, 260, 263, 285, 286, 296, 307
McCawley, James 48
McDonald, M. 122, 202
McElhanon, K. A. 201
McHugh, Brian 357, 365
Mead, Jere 6
Meerendonk, M. 93
Meiklejohn, Kathleen 262
Meiklejohn, Percy 262
Mester, R. Armin 50, 79, 92, 114, 115, 149
Meyer, Leonard 37, 80
Michelson, Karin 47, 83, 88, 111, 222–26, 266, 301, 355
Miner, Kenneth L. 268, 346–64, 349n
Mitchell, T. F. 67, 69, 130, 131, 228–31
Mithun, Marianne 223–24
Miyaoka, Osahito 25, 40, 239, 240, 255, 258, 259
Mohanan, K. P. 57, 92–93, 185
Mohanan, Tara 93, 299n, 305
Morton, J. 6, 7
Munro, Pamela 64, 139n, 181, 209–11, 265
Myers, Scott 114
Mürk, Harry 318

Nacaskul, Karnchana 261
Nakatani, Lloyd 7, 11
Nespor, Marina 10, 24, 29, 37, 40, 170, 367, 368, 369, 371, 372
Newman, Paul 303
Newman, Stanley S. 367
Newton, Robert P. 81
Nicholson, Ray 279n
Nicholson, Ruth 279n
Nicklas, Thurston Dale 209, 210
Nooteboom, Sibout G. 306
Nord, Lennart 80

Oates, Lynette Frances 204
Oates, William J. 204
Odden, David 185, 297
Oehrle, Richard T. 112
Ohala, John 6
Ohala, Manjari J. 162
O'Neil, Wayne 118
Onon, Peter M. 297
Ordonez Domingo, Juan 281, 282
Orešnik, Janez 191
Ortiz Domingo, Juan 281, 282
Osborn, Henry 203
Ostendorf, Mari 7, 100
Oswalt, Robert L. 260
Owens, Jonathan 228, 229, 238

Pandey, Pramod Kumar 162–63
Paradis, Carole 114
Payne, David L. 274, 288, 289, 293, 295
Payne, Judith K. 274, 288–96
Pesetsky, David 218
Peterson, Gordon 6
Phinnemore, Thomas 200–201
Piera, Carlos 28
Pierrehumbert, Janet 7, 10–16, 184, 217, 345, 370
Piggott, Glyne L. 216–18
Pike, Eunice V. 51, 296
Pike, Kenneth 51, 182–88, 184n, 279n, 365
Pinson, Elliot N. 276
Poser, William J. 57, 87, 115, 182, 188, 199
Price, Patti J. 7, 100
Prince, Alan 3, 7, 8, 24–30, 33–36, 38, 40, 42, 46, 47, 50–53, 55, 57, 59–62, 64, 67, 69–71, 77–79, 82, 86–88, 93, 96, 101–3, 105, 108, 108n, 112, 114, 114n, 115, 123, 146–49, 175, 178, 179, 222–25, 228, 260, 263, 285, 296- 300, 311, 316–29, 319n, 365, 367, 369–72, 374, 378, 382, 386, 390, 397

Pukui, Mary Kawena 148, 181
Pulleyblank, Douglas 57, 58, 185
Puppel, Stanisław 97, 204
Pyle, Charles 230

Rapp, Karin 101
Rappaport, Malka 263
Rauschenberg, Jane 372
Repetti, Lori 245
Rhodes, Richard 216, 217
Rice, Curtis 48, 79, 333, 335, 345
Rischel, Jørgen 370–71, 382, 397
Robbins, Scarlett 261
Roberts, John R. 297
Roca, Iggy 57, 58, 96
Rohlfs, Gerhard 148
Rombandeeva, Evdokiia Ivanovna 200
Ross, John R. 112, 121
Rubach, Jerzy 170, 204
Russom, Geoffrey 118
Rutten, Jean 306

Sag, Ivan 16
Saint, Rachel 182, 184n
Saksena, Baburam 179
Sanchez Santos, Jorge 288, 289, 295
Sapir, Edward 121, 266, 297
Sauzet, Patrick 57, 181, 365
Schein, Barry 122
Schmerling, Susan 375
Schubert, Johannes 297
Schürmann, Clamor Wilhelm 205
Schütz, Albert J. 142–47, 367
Scorza, David 297
Scott, Graham 279n
Scott, Norman C. 40, 143, 144
Seaman, P. David 261
Sebeok, Thomas 296
Seeger, Pete 27
Seiler, Hansjakob 132–40, 135n
Selkirk, Elisabeth O. 10, 12, 25, 29, 38, 40, 67, 86n, 109, 170, 237, 299, 305, 367–69, 371, 372, 374, 392–94, 399
Sharma, Aryendra 278
Shattuck-Hufnagel, Stefanie 7, 100
Shaw, Patricia A. 267, 267n
Shipley, William F. 261
Shryock, Aaron 265
Shukla, Shaligram 278
Sigurjonsdottir, Sigriður 189, 191, 192, 193n
Sjögren, Joh. Andreas 200

Smith, Henry Lee, Jr. 367
Smith, Norval 306
Solan, Lawrence 371
Sovijärvi, Antti 329, 330
Spring, Cari 288, 293, 309
Steriade, Donca 34, 53, 54, 57, 59, 89, 94, 106, 108n, 109, 115, 122, 181, 188, 252, 299, 300, 301, 303–5, 343
Stetson, Raymond H. 5–6
Stockwell, Robert P. 9, 22
Stowell, Timothy 215, 216, 225
Street, John C. 297
Strehlow, Theodor G. H. 306
Susman, Amelia 346–64
Swadesh, Morris 297
Swora, Maria G. 372
Szinnyei, Josef 330

Tanaka, Shin-ichi 118
Tauli, Valter 318
Teeter, Karl V. 215, 216, 219
Thompson, David A. 296
Thompson, Henry 23
Tiwari, Udai Narain 278
Trager, George L. 367
Tranel, Bernard 299
Trubetzkoy, Nikolay Sergeevich 302, 306
Tryon, Darrell T. 199, 200, 203
Tuominen, Salme 93
Tyhurst, James J. 42, 64

Ulrich, Charles 209–11

van der Hulst, Harry 52, 111, 116, 118, 306
van der Leeuw, Frank 306
van Katwijk, Albert 6
van Lit, John 111, 306
Vance, Timothy 303
Vanderslice, Ralph 9, 22
Vergnaud, Jean-Roger xiii, 3, 27, 32, 38–42, 42n, 49, 55, 57, 60, 64, 65, 67, 71, 75, 76–77, 77n, 109, 113, 114, 119, 122, 167, 170, 176n, 178, 188, 204, 216, 225, 260, 262, 263, 266, 273, 275, 279, 281, 285, 297, 299, 305–7, 309, 311, 313, 314, 329, 335, 346, 351–55, 355n, 359, 360, 367, 372–74, 376
Vincent, Nigel 23
Visch, Ellis 367–69, 371, 382, 386–90, 396–99
Voegelin, Charles 25, 263, 265
Voegelin, Florence 265

Vogel, Irene 10, 24, 29, 37, 40, 170, 367–69, 371, 372

Wald, Benji 205
Ward, Dennis 23
Ward, Gregory 10
Weeda, Don 243, 253, 259
Welden, Ann 71
Wheeler, Deirdre 263, 265, 299
White Eagle, Josie 122, 346–53, 364
Whorf, Benjamin Lee 261
Wightman, Colin 7, 100
Winstedt, Richard O. 263
Wolfram, Walt 2
Woodbury, Anthony 25, 239–59, 304
Woodrow, Herbert 79–81
Wurm, Stephen A. 122, 202

Yadav, Ramawatar 149
Yip, Moira J. W. 23, 102n

Zebek, Schalonow 297
Zec, Draga 52, 53, 69, 278, 297
Zigmond, Maurice 181

INDEX OF LANGUAGES AND LANGUAGE FAMILIES

Aguacatec 297
Aklan **265**, 305
Algonquian family 83, 84, **211–22**, 302
  Central 216, 218, 221
  Eastern 211, 215, 221
Altaic family 297
Alutiiq. *See* Yupik: Pacific
Amele 297
Anguthimri 100, 102, 103, 111, **198**
Arabic
  Bani-Hassan Bedouin 42, 74, 95, 98–99, 106, 107, 108, 114, 182, 239, 315, **366**
  Bedouin Hijazi 42, 181, 239
  Cairene 36, 60, **67–71**, 72, 87, 88, 89, 90, 91, 93, 107, 117, 119, 128, 130, 131, 132, 188, 231, 272
  Cairene Classical 67–69, 127, 130
  Classical 131, 132, **296**
  Cyrenaican Bedouin 42, 43, 88, 107, 114, 215n, **228–39**, 366
  Egyptian Radio 37, 71, 78, 106, **130–32**, 180, 228
  Lebanese 88, 180, **181**
  Negev Bedouin 107, 114, **226–28**, 232, **283–85**
  Palestinian 78, 88, 106, 107, 121, **125–30**, 130, 131, 132, 164, 180, 181, 185, 228, 239, 366
  Tripoli 228
Arabic family 42, 57, 70, 78, 107, 108n, 123, 125, 218, 227–29, 233, 237, 267n, 271, 303
Aranda, Western 306
Araucanian 84, 88, **266**, 268
Arawakan family 201, 288
Asheninca 42, 60, 88, 98, 111, 114, 119, 262, 271, 273, 274, **288–96**, 403
  Apurucayali dialect 288, 289, 293, 295
  Pichis dialect 288, 303
Au **297**, 299
Auca 46, 87, 89, 90, 91, 93, 114, **182–88**, 184n, 315, **365**
Austronesian family 142, 148, 167, 178, 179, 180, 181, 182, 199, 203, 205, 262, 263
Awadhi 78, 88, 108, 167, **179–80**

Badimaya 95, 100, 103, 111, **198**
Balto-Finnic family 200, 201, 329
Bantu family 357, 360, 365
Bengali 61, 162
Bergüner-Romansh 181
Bhojpuri 167, **278**
Bidyara 48, 88, 100, **199**

Cahuilla 37, 89, 90, 91, 93, 99, 101, 114, 117, 118, 121, **132–40**, 180, 185, 303
  Mountain dialect 139n
Cambodian 84, 88, **261–62**
Carib, Surinam 83, 208
Cariban family 83, 205, 208
Catalan 23
Cavineña 88, 98, **202**
Cayuga 25, 83, 84, 88, 122, **222–25**, 226, 227, 232, 259, **301–2**, 303, 305
Cayuvava 59, 88, 105, 114, 117, 205, **309–14**, 315, 322, 331, 333, 359, 367
Cebuano 265
Chaga 357, 365
Chamorro 7, 59, 148, 204
Cheremis, Eastern 296
Cheremis, Western **297**, 303
Chickasaw 83, 84, 88, **209–11**, 241, 269, 303
Chinese, Mandarin 23
Chiwere 355–56

Choctaw 83, 84, 88, **209–11**, 241, 269, 303
Chugach, 333, 345
Chuvash **296**, 302, 303
Creek. *See* Seminole/Creek
Cupeño 32
Czech 89, 99, 102, 117, **203**

Dakota 88, 107, **267**, 268
  Stoney dialect 267n
Dalabon 89, 100, **199**
Daly family 200, 203
Danish **370–71**, 397
Dehu 89, 100–1, **199**
Delaware 211. *See also* Munsee; Unami
Diegueño 88, **181**
Diyari 87, 88, 100, 122, **199**
Djingili 59, 98, 103, **202**, 204
Dutch 18, 111, 182, **305–6**, 368, 369, 370, 382, **386–97**

English 5, 6, 7, **9–22**, 23, 24, 25, 29, 31, 32, 33, 35, 37, 38, 43–45, 49, 57, 59, 60, 61, 81, 88, 97, 108n, 109, 111, 112, 114, 114n, 115, 117, 121, 148, 158, 175, 179, **181**, 184, 192n, 306, 307, 313, 367, 368, 369, 370, 371, **373–99**. *See also* Middle English; Old English
  American 10
Eskimo, Seward Peninsula Inupiaq 83, 302
Eskimo family 83, 239. *See also* Yupik
Estonian 56, 57, 101, 102, 103, 104–5, 114, 164, 164n, 175, 271, 303, 315, **316–29**, 319n, 403

Fijian 47, 88, 98, 111, 114, **142–47**, 148, 181, 367
  Boumaa dialect 142, 144
  other dialects 142, 144, 145
  Standard dialect 145
Finnish 7, 102, 114, 315, 322, **329–30**, 371
Finno-Ugric family 199, 296, 297, 315, 316, 329
Fore 279n
French 24, 313, 371

Garawa 103, 119, **202–3**
German 88, **181**, **200**, 370, 397
Goilalan family 265
Golin 271, 278, **279**, 297, 299
Greek, Ancient 49, 57, **181**, 299, 303, 305, 365
Greek, Modern 43, 46, 98, 99, **204**, 367
Gugu-Yalanji 204
Gum family 297
Gungabula 48, 88, 100, **199**
Gurkhali 93
Guwinyguan family 199

Hausa 303
Hawaiian 88, 148, **181**
Hebrew, Biblical. *See* Hebrew, Tiberian
Hebrew, Modern 32, 98, 198, 367
Hebrew, Tiberian 42, 117, 119, 173, 182, 245, **263**, 303
Hindi 57, 59, 78, 95, **162–67**, 164n, 178, 271, **276–78**, 303, 318
  Eastern dialect 162
Hixkaryana 83, 88, 95, 110, **205–8**, 269, 303
Hokan family 182, 297
Hopi 49, 88, 111, **261**, 303
Huasteco **296**, 303
Hungarian 54, 89, 99, 102, 114, 119, 200, 315, 329, **330**
Huon, Western, family 200, 201

Icelandic 37, 83, 84, 89, 99, 100, 101, 103, 114, 144, 175, **188–98**, 189n, 192n, 193n, 313
Ilokano 123
Indo-Aryan family 93, 149, 162, 163, 179, 278
Indo-European family 32, 297, 299
Indonesian 97, 49n
Inga 89, 95, 98, **181**
Inupiaq Eskimo, Seward Peninsula 83, 302
Iranian family 261
Iroquoian family 83
  Lake subfamily **222–26**, 269
Italian 23, 24, 37, 148, 245, 306, 367, 371
  Abruzzese 148

Japanese 182, 303, 357
Javanese 108, **262–63**, 303

Kalkatungu 122
Karawic family 202
Karelian 100, 114, 315, 329, **330**
Kashaya 74, 83, **260**, 269, 303
Kawaiisu 88, **181**
Kela 89, **205**
Kikuyu 185, 357
Kimatuumbi 185
Kinyambo 25, 69
Klamath 76, 89, 95, 178, 181, **279–81**, 303

Komi **297**, 298, 303
Korean 299n
Kuuku-Yaʔu **296**, 303
Kwakw'ala **297**, 303

Lake Iroquoian family **222–26**, 269
Lappish, Central Norwegian 100–101, 199
Latin 50, 51, 52, 59, 88, **91–92**, 94, 108n, 111, 112, 114, 115n, 118, 120–21, 149, 180–81, 239, 279–80, 281, 303, 330
  Early 78, **180**
Lenakel 37, 46, 59, 76, 89, 95, 97, 99, 114, 119, **167–78**, 176n, 180, 204, 303, 355, 367, 371
Lithuanian 42n, 49, 271, 278, 297
Livonian 99, **200**
Lower Sepik family 297
Lushootseed **297**, 299

Macedonian 58, 59, 89, **205**, 367
Macushi 83, 84, **208**, 269
Madimadi 306
Mae 89, **205**
Maidu 83, 95, **261**, 269
Maithili 78, 88, 98, 99, 108, 114, 119, **149–62**, 167, 245, 278, 303
Malakmalak 203
Malay 108, **263**
Malayalam 76, **92–93**, 94, 95, 118, 180, 298, 299n, 303
Malecite-Passamaquoddy 36, 119, **215–16**, 221
Mam 88, 119, 181, **281–83**
Manam 78, **182**
Mandarin Chinese 23
Manobo, Sarangani 88, 108, **179–80**, **262–63**, 303
Mansi 100, 102, 103, **200**
Mantjiltjara 100, 103, 114, 315, **365–66**
Maranungku 89, 99, **200**
Mayan family 281, 296, 297
Mayi 100, **200**
Menomini 57, 83, 88, **218–21**, 269, 303
Middle English 148
Miwok, Sierra. *See* Sierra Miwok
Miwok-Costanoan family 261
Mohawk 47, 83, 84, 88, 111, 355
Mohican 222
Mongolian, Khalkha 76, 178, **297**
Munsee 36, 83, 88, **211–15**, 216, 222, 227, 232, 216
Mura family 285
Murik 297
Muskogean family 64, 209

Nengone 40, 89, 98, **203**
Nyawaygi 104–5, **180**, 303

Ojibwa, Eastern 36, 98, 99, 119, 121, **216–18**, 221, 269, 299n, 303
Ojibwa, Maniwaki 149, **218**
Ojibwa, Southern 222
Old English 88, **117–18**
Ono 89, 99, **200–201**
Onondaga 55, 59, 83, 222, **226–66**, 267, 268
Ossetic 84, 88, **261**, 302, 303

Paamese 59, 88, 90, 91, 119, **178–79**
Pacific Yupik. *See* Yupik: Pacific
Paiute, Southern 83, 88, 111, 121, 122, 123, **266**, 268
Pama-Nyungan family 62, 140, 180, 198, 199, 201, 204, 205, 296, 365
Paman family 198
Parnkalla 89, **205**
Pashto 32
Passamaquoddy 36, 119, **215–16**, 221
Penutian family 261
Persian 158
Pintupi 40, **62–64**, 100, 102, 103, 116, 117
Pirahã 48, 59, 109, 271, 272, **285–87**, 306, 403
Piro 89, 99, 102, **201**, 367
Pitta-Pitta 48, 88, 100, 122, **201**
Polish 31, 46, 59, 89, 97, 98–99, **204**
Pomoan family 260
Potawatomi 83, **221**, 269, 303

Quechuan family 181

Romanian 57, 59, 89, 94, **181**
Romansh, Bergüner 181
Russian 23, 42n, 297, 344

Salishan family 297
Sanskrit 42n, 297
Sarangani Manobo 88, 108, **179–80**, **262–63**, 303
Selepet 89, 100, **201**
Seminole/Creek 36, 42, 60, **64–67**, 70, 72, 88, 90, 91, 93, 111, 117, 119, 185, 188, 209, 262, 303, 355n
Seneca 25, 83, 222, **225–26**, 266, 268, 269, 365

Sentani 89, 95, 114, 182, 315, **330–33**
Serbo-Croatian 49, 69, 271, **278–79**
Shoshone, Tümpisa 88, 117, 118, **180**, 303
Sierra Miwok 25, 83, 88, 113, **261**, 269, 303
  Central 112, 113, 261
  Northern 261
  Southern 113, **250–51**, 261
Sindhi **278**
Siouan family 267, 268, 346, 355–56
Slavic family 205
Southern Paiute 83, 88, 111, 121, 122, 123, **266**, 268
Southwest Tanna 46, 89, **180**
Spanish 31, 37, 57, 59, 89, 94, 94n, 96, 97, 102, 114, **181**, 198, 371
Sukuma 338
Swahili 205
Swedish 80, 84, 97, 117, 367

Tacanan family 202
Tiberian Hebrew. *See* Hebrew, Tiberian
Tibetan, Lhasa 297
Tol 89, 95, **182**
Tongan 84, 148, **182**
Torricelli family 93, 297
Tübatulabal 25, **263–65**, 299
Turkic family 296
Turkish 60, 76, 95, 113, 178, **262**, 263, 303

Unami 36, 42, 83, 88, **211–15**, 216, 222, 227, 232
Uralic family 200
Uto-Aztecan family 132, 180, 181, 261, 263, 266

Vogul. *See* Mansi
Votic 99, 102, 103, **201–2**, 303

Wakashan family 297
Walmatjari 100
Wangkumara 48, 88, 100, 122, **202**
Warao 89, 98, **203**, 204
Wargamay 36, 37, 84, 88, 90, 91, 93, 114, **140–42**, 141n, 180, 188, 303
Weri 265
West Barkly family 202
Western Aranda 306
Western Huon family 200, 201
Winnebago 69, 88, 114, 122–23, 179, 262, 311, 312, 314, 315, 345, **346–65**

Yana **297**, 303
Yawelmani 59, 204, 367
Yidiɲ 25, 83, 88, 149, **260**, 269, 303
Yil 93
Yuman family 181
Yupik
  General Central **248–51**, 252, 253, 254, 255, 258
  Central Alaskan 25, 42, 83, 114, 239, 240, 241n, **242–59**, 269, 271, 302, 303, 337, 338, 343, 344
  Central Siberian 240
  Chaplinski 240
  Chevak 240, 247, **251–52**, 253, 255, 257, 259, 260
  Coastal **255–56**, 260
  Hooper Bay 240, 255
  Hooper Bay/Chevak 240
  Inland 255–56
  Kotlik 240, 255
  Naukanski 240
  Norton Sound 240, **242–47**, 248, 249, 251, 252, 253, 254, 255, 260, 303, 334
  Nunivak Island 240, 255
  Pacific 42, 48, 83, 114, 240, 262, 302, 303, 304, 311, 312, 314, 315, **333–46**, 403
    Chugach dialect 333, 345
    Koniag dialect 333, 344, 345
  St. Lawrence Island 51, 52, 83, 179, 211, **240–42**, 241n, 243, 247, 254, 258, 269, 303
  Unaliq 240, 254, 255
Yupik family 25, 88, 121, 185, 239, 240, 241, 244, 247, 259, 302, 305, 315, 333, 346

Zoque, Chimalapa 83, 95, 102, **104**

INDEX OF SUBJECTS

absolute stress levels, absence 14, 377–80, 382, 393
  in grid notation 29
  redefined as "strongest of domain" 29
amphibrachs 313
  in Cayuvava 313
  relating binary and ternary systems 315
  suitability for Estonian 328
anapests, final, 98
"Arab" rule 121
assimilation, absence in stress systems 26, 30
autosegmental phonology 2. *See also* pitch accent; tone
  and geminate integrity 230
  applied to grid marks 42
  English /t/ Insertion 12
  floating backness autosegment 229, **237**
avoidance clauses 59
awareness of stress 9, 29

Beat Addition 371–77, 391–98
  application to deeply embedded constituents 391–94
  conditioning factors 376
  directional effects 394–98
  in Dutch 386; Garawa 203; Maithili 153; Ono 201; Piro 201; Sierra Miwok 113
  with Grid Expansion 393
binary feet 55, 58
bottom-up stressing 36, **116–17**
  in Malayalam 118; Tümpisa Shoshone 118, **180**
  list of languages 36
boundaries. *See* cohering vs. non-cohering suffixes; juncture
boundary tones
  in Auca 184; Central Alaskan Yupik 259; English 10

bounded stress systems 32
bracketed grids 38–39. *See also* grids
  arguments for 41–48
  compared with other representations 38
  effects on Rhythm Rule 43–45, 385
  governed by Faithfulness Condition 41
  well-formedness conditions 400
bracketing in metrical structure. *See* bracketed grids
branching 60
breath pulse 5–6

canonical foot 86–87
clash 36
  and Move X 35
  as motivation for Incorporation 120
  condition on extrametricality rules 108
  double, in General Central Yupik 249; Cayuga 224; Estonian 326; Hungarian 330
  in Liberman and Prince 372
  irrelevant in moraic grids 300
  irrelevant in prominence grids 274
clitic stress
  and Free Element Condition 87, 115n
  clitics as part of domain of foot construction 367
  clitics exempt from minimal word requirements 66, 88
  culminativity not imposed on clitics 24
  in Central Alaskan Yupik 257–58; Latin 115n; Lenakel 170; Pacific Yupik 344
  invisibility of clitics to phrasal stress 377n, 392
close contact, correlation of 306
coherence, metrical 114
  in Cayuga 224
  in Trochaic Shortening 147

cohering vs. non-cohering affixes
in English 32; Icelandic 190–91; Lenakel 170; Seminole/Creek 64; Winnebago 350
cola 119
compensatory processes
and two-layer theory of quantity 300
in Maithili 157; Seneca 225
compound stress
cross-linguistically 368
in Auca 185; Dutch 368; English 60, 369, 388; Icelandic **190–93**, 192n; Lenakel 171; Winnebago 350
in English long compounds 374–75
conflation 119
conscious awareness of stress 9, 29
Continuous Column Constraint **34–37**, 400
and destressing 36–37
and End Rules 36, 61
and Grid Expansion 393
and moraic grids 300
and Move X 35–36, 385–86, 388–90
and prominence grids 274
in phrasal stress 376–79
in Cairene Arabic 70; Seminole/Creek 66; Wargamay 141–42
and Strong Domain Principle, compared 386–90
and top-down stressing 117
rhythm, basis in 34
contrastive stress. *See* focus stress
correlation of close contact 306
counting in phonology
foot inventories 307, 322
four syllables = two disyllabic feet 127, 166
in ternary stress 307
moras, not segments counted 50, 53, 304–5
culminativity 24–25. *See also* Unstressable Word Syndrome
claimed exceptions 24–25
in Cayuga 225; Seneca 225
expected under metrical account 29
grammatical words exempt 24
phrasal, not word level, in French and Italian 24
CVC syllables, quantity varies
across contexts 120–21
across languages 51
CVVC syllables and quantity, of coda consonants 303
cycle, phonological
and Free Element Condition 115
in Auca 185; Cahuilla 133–34; Central Alaskan Yupik 257–58; Choctaw and Chickasaw 210; Diyari 199; English 26, 37; Lenakel 176; Pacific Yupik 344; Palestinian Arabic 129
metrical rebracketing 186, 187–88
planes 188, 275
SPE phrasal stress 373
type of End Rule required 378

degenerate feet 86–105
and Priority Clause 93–95
and unbounded stress systems 92–93, **298**
ban (*see* degenerate feet: prohibition)
defined 86, **102**
defined for syllabic trochee systems 101–5
in Cahuilla 137; Choctaw 210; Chickasaw 210; Hopi 111; Icelandic 188–96; Maithili 161–62; Malakmalak 203; Malayalam 92–93; Munsee 213; Negev Bedouin Arabic 228; Ojibwa dialects 217–18; Pacific Yupik 344; Spanish 94n; Unami 213
Incorporation, trigger for 111
loss by rule 294
loss in weak position 87
in Cahuilla 134; Chevak Yupik 251; Diyari 199; Klamath 281; Maithili 158; Pacific Yupik 341
prohibition 87, 400
affects word layer labeling 89–91
stage of derivation 87
stem structure constraint, in Yupik 241n
strong vs. weak 87
word minima 87–89
proposed
for Bani-Hassan Arabic 366; Winnebago 353
in weak position 98–101
removed under persistent footing 115
repair 95–98
word layer labeling 188
degrees of stress 14. *See also* absolute stress levels
alternative of reliance on vowel quality 14
multiple 25
derived in complex structures 21, 369
in English 14, 15, 17, 20
in principle unbounded 22
upwardly-defined natural classes 28
destressing
and Continuous Column Constraint 36–37

and Faithfulness Condition 37n
in Asheninca 293, 294
and Strong Domain Principle 390
final position, proposed
in Central Alaskan Yupik 258–59; Old English 118; Pacific Yupik 343
in clash **37**, 403
in Asheninca 293; Auca (HV) 187; Cahuilla 136; Chevak Yupik 251; Egyptian Radio Arabic 132; English 114n; General Central Yupik 248–53; Icelandic 193–94; Lenakel 173; Pacific Yupik 341; Sentani 332, 333; Southern Sierra Miwok 250; Spanish 97; Wargamay 141; Winnebago 352
initial dactyl effect 97
iterative
in General Central Yupik 250; Southern Sierra Miwok 250
non-maximal feet, proposed for 313
phrasal 371
in French and Italian 24
prominence-based 272, 276, 290
resolves clash **36–37**, 249, 259, 293, 313
Trigger Prominence, guided by 197n
diagnostics for stress 10–23. *See also* reduction
attraction of intonational tones 10–11, 16–18
in Auca 184; Cahuilla 138; Klamath 280
in Pirahã 285
languages other than English 22–23
mutual consistency 12, 15, 17, 19–21
Rhythm Rule 18–21
segmental rules 12–16
in Maithili 150; Mam 282; Pirahã 285
verse 163
vowel quality
in Cahuilla 137; English 12–16; Pintupi 63; Mam 282
diphthongs
breaking, in Fijian dialects 145
heavy in Dutch 305
in Warao 203
patterning as disyllabic
in Cayuga 224; Seneca 225
rising, in Winnebago 362, 349n
short
in Auca 185; Fijian 145; Pacific Yupik 304
tolerate clash in Estonian 322
direction of parsing **40**, 113. *See also* foot: construction: bidirectional; iambs: right-to-left
asymmetries 265
change, in Latin and Lebanese Arabic 180
in Fijian 144
directional effects, in Beat Addition 376
Domain Generation **378**, 403
Beat Addition, structural change 381
End Rule, part of structural change 379
evidence for domains created by 385–86
in Grid Expansion 393
overgeneration blocked 394
dominance, in Hayes (1981) 75
Domino Condition 353–55
Dorsey's Law (Winnebago) 351–65
formulated 351
output monosyllabic 362n
double upbeats 311
absent in Pacific Yupik 335; Winnebago 359
in Cayuvava 311; Sentani 331
downstep
in English intonation 10
in Winnebago 349
duration. *See also* Iambic/Trochaic Law: and segmental phonology; phonetic final lengthening
cue for stress 6
integrated with loudness 7
where pitch is flat 8, 11
where vowel length is phonemic 7
cues multiple phonological entities 7
greater in unstressed syllables 7
rhythmic experiments 79
threshold ratio for iambic rhythm 81, 84
verse recitation 80
vowels, in closed and open syllables 220
word length effects 101

emphatic stress
and breath pulses 6
in Estonian 319n, 322; Finnish 7
empty phonological positions, in Central Alaskan Yupik 253, 259n
End Rules 61
and Continuous Column Constraint **36**, 70
creating cola 119
in cyclic stress assignment 378
initial or final word stress 61
phrasal 368–70, 376–82
in Icelandic compounds 194
prominence-based 273–76. *See also* prominence; prominence grids
top-down construction 116

epenthesis. *See also* Dorsey's Law
and Domino Condition 353–55
blocked in geminates 230
creates degenerate feet, in Negev Bedouin Arabic 228
"economy" 237
in Cyrenaican Bedouin Arabic 215n, 230; Early Cayuga 222; Eastern Ojibwa 216; Lenakel 167, 355; Mohawk 355; Negev Bedouin Arabic 228; Winnebago 365
to fulfill minimal word requirement 111
eurhythmy **372–73**, 395
governing Beat Addition 371
in Icelandic 194
varying across languages 372
even iambs 76, 266–68
excluded under theory 73
generate unattested systems 72–73, **74**
in Halle (1990) 77n
proposed by Jacobs as marked option 73
proposed for Central Alaskan Yupik 243–44, 247; Pacific Yupik 345; Araucanian 266; Dakota 267; Onondaga 266; Seneca 225–26; Southern Paiute 266; Yidiɲ 260
treated as standard iambs 267
exhausting domains with extrametricality, prohibition 58, 277
exhaustive parsing
and stressless feet 109n
basis of Trisyllabic Shortening 148
basis of Trochaic Shortening 146, 147
eschewed as rigid requirement for feet 2, 64, **109–10**
extrametricality **56–60**, **105–8**, 401. *See also* Peripherality Condition; unsyllabified segments
and labeling rules 59
and prominence theory 272, 276
and Stray Erasure 59
and weak local parsing 314
annulment, in Pirahã 285
avoidance clauses 59
"chained" excluded 107
proposed for English 108n; Kashaya 260; Latin 108n
constituents only 57
in clash 108
proposed for Wargamay 142
left edge, proposed for Kashaya 260; Western Aranda 306; Winnebago 352
lexical classes, characterizing 59
loss under affixation 59
notation 57, **58**
of clitics, in Latin 115n
of consonants 57
in Estonian 56–57
of feet 77–78
and stray consonants 106–7
in Bani-Hassan Arabic 366; Cayuga 222; Central Alaskan Yupik 259; Cyrenaican Bedouin Arabic 232; Early Latin 180; Eastern Ojibwa 216; Egyptian Radio Arabic 130, 131; Hindi 165; Javanese 263; Maithili 153; Malay 263; Manam 182; Munsee 211; Negev Bedouin Arabic 227; Paamese 178; Palestinian Arabic 127; Sarangani Manobo 263; Turkish 262; Unami 211
of initial vowels, in Western Aranda 306
of moras
excluded 58
proposed for Cairene Arabic 69; Winnebago 352
of suffixes, in Onondaga 266
of syllables 58
creating double troughs 73
in Asheninca 288; Auca 187; Cayuvava 310; Dakota 267n; Djingili 202; English **60–61**, 112, 148; Hindi 277; Hixkaryana 206; Hopi **111**, 261; Kela 205; Klamath 280; Latin **92**, 111; Lenakel 168; Macedonian 58, 205; Mae 205; Modern Greek 204; Paamese 178; Parnkalla 205; Pirahã 285; Sierra Miwok 112; Southern Paiute 266; Turkish 113, 262; Warao 203; Western Cheremis 297
in prominence grids 273
of words 369
on multiple levels 105–8
penultimate lengthening, mechanism of 148
prohibition on exhausting domain
in prominence grids 273
in regular grids **58**, 277
revocation 112 (*see also* Incorporation)
in Sierra Miwok 113
right edges, preference for **57–58**, 74
unattested
for syllabic trochees 78
for unbounded feet 299
extrasystemic explanation 81

Faithfulness Condition 41, **380**, 400. *See also* migration of stress
  and destressing 37n
  and End Rules 380–81
  and Move X 43–45, 385–86
  and Strong Domain Principle 386–90
  and syllable merger 43
  blocks overgeneration 394
  definition of head 380
  in Unami 214; Winnebago 365
final anapests 98
final lengthening. *See* phonetic final lengthening
fixed stress 31
focus stress
  in English 11, 369; Maithili 158
foot 40, 45–47. *See also* degenerate feet; foot templates
  and minimal word 47–48
  and Prosodic Morphology 47, 78
  and segmental rules
    in English 12; Pacific Yupik 342
  and tonal rules, in Pacific Yupik 344–45
  construction
    at phrasal level 367 (*see also* clitic stress)
    bidirectional (*see also* direction of parsing)
      in Auca 188; Lenakel 168; Modern Greek 204; Polish 204
    exhaustive and non exhaustive 2, 108–10, 308–9
  defines overlong syllables, in Estonian 320
  degenerate (*see* degenerate feet)
  extrametricality (*see* extrametricality: of feet)
  inventors 40
  stressless 109n
foot templates (*see also* amphibrachs; even iambs; iambs; moraic trochees; syllabic trochees; unbounded feet; uneven trochees)
  and Iambic/Trochaic Law 81
  common basis of variable stress patterns 328
  defined by quantity, not prominence 272
  in Hayes (1981) 75
  in Prosodic Morphology 78
  limited to binary and unbounded 55, 58
  proposed /´– –/ 82
  ternary 316
  uniform choice within a language 55, 315
  unifying principle of stress system 59
fortis consonants
  in Pacific Yupik 341
  reinterpretation of geminates 299n
Free Element Condition **115**, 401
  and cyclic stress 115
  and persistent footing 115
  and Stray Adjunction 109
  and uneven trochees 115n
free stress 31
function words. *See also* clitic stress
  exempt from minimal word requirements 66, 88
functions of stress 31

geminates, not making quantity 299, 302
  treated as fortis consonants 299n
grammatical words. *See also* clitic stress
  exempt from minimal word requirements 66, 88
Grid Expansion 393
  and Textual Prominence Preservation Condition 394
  no overgeneration 394
Grid Simplification Convention (HV) 376
grid theory. *See also* bracketed grids
  pure 38
grids 26–30
  bracketed (*see* bracketed grids)
  define natural classes of syllables 28
  exemplified 27, 28
  relevance of rows 27, 28
    for prominence grids 273
grouping. *See* bracketed grids

head, of metrical constituent 41, **380**
heavy syllables. *See also* quantity (of syllables)
  symbol xiii
  trochaic profile 69
hierarchy in metrical structure 28
  involves grouping 37, 38
  yielding multiple stress levels 30
historical change
  in Algonquian languages 221; Arabic 128, 227–28; Cayuga 224; Central Alaskan Yupik dialects 256; Latin 180; Winnebago 355–56
  in weak vs. strong local parsing 315
  source of superheavy syllables, in Estonian 318; Hindi 163

Iambic Gemination 83
Iambic Lengthening 82–83
  blocked in closed syllables 303
  blocked in final position 269
  enforces Iambic/Trochaic law 82–83
  list of cases 83
  often non-neutralizing 269
Iambic Shortening **112**, 120–21
  of CVC syllables 120
Iambic/Trochaic Law 79–85
  and metrical coherence 114
  and music 80
  and Revised Obligatory Branching feet 178
  and segmental phonology 77, 82–85 (*see also* Iambic Gemination; Iambic Lengthening; reduction; Trisyllabic Shortening; Trochaic Shortening)
  and Stray Adjunction 109
  and verse 80–81
  exclusion of /⏑ ⏑́/ in Cambodian 262
  experimental evidence 79–81
  foot inventories governed by 81–82
  formulated 80
  in Cayuga 301; Yidiɲ 260
  in languages without quantity 268
  phonetic vs. phonological effects 211n
iambs 65, 205–70
  Americas, common in 269
  canonical 86, 205
  even (*see* even iambs)
  frequent lack of higher grid layers 269
  Hayes (1981) 75
  in Araucanian 266; Asheninca 288–96; Cambodian 261–62; Carib of Surinam 208; Cayuga, **222–25**, 301–2; Central Alaskan Yupik 242–59; Chevak Yupik 251–52; Chickasaw 209–11; Choctaw 209–11; Common Mississippi Valley Siouan 268, 356; Cyrenaican Bedouin Arabic 228–39; Dakota 267; Eastern Ojibwa 216–18; General Central Yupik 248–53; Hixkaryana 110, 205–8; Hopi 111, **261**; Kashaya 260; Macushi 208; Madimadi 306; Maidu 261; Malecite-Passamaquoddy 215–16; Maniwaki Ojibwa 218; Menomini 218–21; Munsee 211–15; Negev Bedouin Arabic 226–28, 283–85; Norton Sound Yupik 242–47; Onondaga 226, 266–67; Ossetic 261; Pacific Yupik 302, 303, 304, **333–46**; Potawatomi 221; Seminole/Creek 64–67; Seneca 225–26; Seward Peninsula Inupiaq Eskimo 302; Sierra Miwok 112, **261**; Southern Paiute 121, **266**; St. Lawrence Island Yupik **240–42**, 303; Unami 211–15; Winnebago 346–65; Yidiɲ 260
  in Prosodic Morphology 78, 123
  languages with/without degenerate words 88
  obey Iambic/Trochaic Law 82
  right-to-left, proposed for (Tiberian) Biblical Hebrew 263; Aklan 265; Cebuano 265; Javanese 262–63; Malay 263; Sarangani Manobo 262–63; Turkish 262; Tübatulabal 263–65; Weri 265
Incorporation 111–12. *See also* Unstressable Word Syndrome
  and top-down stressing 117–18
  in Asheninca 291, 295; Hopi 261; Old English 117–18; Southern Paiute 266
  indistinguishable from extrametricality revocation 112
initial dactyl effect 96–98. *See also* final anapests
International Phonetic Alphabet **3**, 189
intonation 10–11, 16–18. *See also* boundary tones; intonational tunes; pitch accent
  and secondary stress in Sierra Miwok 113
  basis of stress assignment, in Hixkaryana 207
  evidence for nuclear stress in English 11
  in Klamath 281; Malayalam 93
  proposed to influence phrasal stress 369
intonational phrase
  as culminative domain, in English 24
  in Central Alaskan Yupik 258, 259; English 10
intonational tunes 10
  alignment governed by stress 11
  freedom of combination 10
  in Auca 184; Chamorro 7; Dutch 18; English 7, 10, 16, 18; Hixkaryana 207
IPA **3**, 189
isochrony 30
iterative foot construction 113

judgments of stress
  blurred in complex utterances 22
  disagreements for English 5

in Asheninca 296
intersubjective reliability 23
SPE controversy 9
well-formedness judgments mismatch speech production 372
juncture. *See also* cohering vs. non-cohering affixes
in Lenakel 170–74
unjustified by morphology (*see* pseudo-compounds)

labeling rules 60, 61
based on branching, proposed 60
Hayes (1981) 76
in Cairene Arabic 71; English 60; Wargamay 142
layer, grid 27
lengthening. *See* Iambic Lengthening; phonetic final lengthening; Unstressable Word Syndrome
in trochaic languages 83–84, 148–49
levels (Lexical Phonology) 32
in Auca **185**, 90; Cahuilla 133; English 32, 148; Seminole/Creek 64; Turkish 113
Lexical Phonology. *See* levels
lexically listed stress
and Free Element Condition 115
as extrametricality 59
diacritic prominence 297
in Cahuilla 133; Djingili 202; English 112; Fijian 144; Maidu 261; Modern Greek 204; Pacific Yupik 344; Spanish 94n; Warao 203
proposed for Lenakel 177
treated with prelisted metrical structure 31
light syllable. *See also* quantity (of syllables)
symbol xiii
locality **33–34**, 307. *See also* counting in phonology
and foot inventories 307
explicated in grid notation 33
in Hayes (1981) 307
in Lenakel destressing 174
in segmental phenomena 34
in weak local parsing 308, 315–16
preantepenultimate stress, in Hindi 166; Palestinian Arabic 127
Steriade (1987) 252
loudness
in rhythmic experiments 79
in stress perception 6
integrated across duration, effect on stress perception 7
perception affected by duration 81

markedness
of alternating stress patterns 47
of English stress patterns 112
of even iambs 73
of left edge extrametricality 57
of proposed labeling conventions 76
of ternary alternation 307
of trimoraic syllables 401
of weak local parsing 308
maximal foot
as canonical 77
construction 102–3
in Estonian 320
Maximality (proposed condition on tree-manipulating rules) 383–85
effects derived by Faithfulness Condition 386
effects derived by minimal definition of pretonic 395
metrical grids. *See* bracketed grids; grids
metrical structure. *See also* bracketed grids; grids
as organizing principle of phonologies 46, 82
migration of stress
under foot loss, in Asheninca 295
under vowel deletion 41–43
list of cases 42
under vowel devoicing, in Pacific Yupik 344
minimal, definitions for pretonic and posttonic 395–97
minimal word 47–48, 87–89
correlates with treatment of leftover material in foot parsing 90
disyllabic, in syllabic trochee languages 104–5
domain of End Rule in Pirahã 285
in Cayuvava 310; Eastern Ojibwa 217; Estonian 325; Fijian 144; Malayalam 93; Seminole/Creek 66; Winnebago 361–62
in syllabic trochee languages 103
in two-layer moraic theory 300
list of languages with minimal word constraint 88
list of languages without a minimal word constraint 88

moraic conservation 53
in two-layer moraic theory 301
moraic theory. *See* moras
moraic trochees 69, 125–82
as substitute for syllabic trochees 101
canonical 86
in Aklan 265; Ancient Greek 181; Awadhi 179–80; Bani-Hassan Arabic 366; Bedouin Hijazi Arabic 181; Bergüner-Romansh 181; Bhojpuri 278; Cahuilla 118, **132–40**; Cairene Arabic, **67–71**, 89; Cebuano, 265; Diegueño 181; Early Latin, 180; Egyptian Radio Arabic, 130–32; English 112, **181**; Fijian 142–49; German 181; Hawaiian 181; Hindi 162–67; Inga 181; Kawaiisu 181; Klamath 181, **279–81**; Latin, **91–92**, 111, 115n; Lebanese Arabic 181; Lenakel 167–78; Maithili 149–62; Mam 281–83; Manam 182; Nyawaygi 180; Old English 117–18; Paamese 90, **178–79**; Palestinian Arabic 125–30; Romanian 94, **181**; Sarangani Manobo 179–80; Sentani 330–33; Southwest Tanna 180; Spanish 94, **181**; (Tiberian) Biblical Hebrew 263; Tol 182; Tongan 182; Turkish 262; Tübatulabal 263–65; Tümpisa Shoshone 118, **180**; Wargamay 90, **140–42**, 141n; Weri 265
in HV 77
in Prosodic Morphology 78, 123
languages with degenerate words 89
languages without degenerate words 88
obey Iambic/Trochaic Law 81–82
substitute for quantity-sensitive, left-dominant feet 75
moras **52–54**, 120–21. *See also* quantity; Weight by Position
shared between segments in Pacific Yupik 337; Sukuma 338
symbol xiii
morphological stress systems 32
morphology
basis for bracketing in grids 40
influence on stress 31–32 (*see also* levels)
in Auca 185; Bani-Hassan Arabic 366; Cahuilla 133; Cayuvava 312; Choctaw and Chickasaw 210; English 60; Estonian 316; Fijian 143; Icelandic 189; Koniag Pacific Yupik 333, 344; Paamese 179; Turkish 113; Votic 202; Winnebago 350
Move X 35–36, 370–71, **382–91**. *See also* Rhythm Rule
and Continuous Column Constraint 35–36, 388–90
and Faithfulness Condition 43–45, **385–90**
and Stray Adjunction 109
direction 370, 397
formulated 35, 370, 403
in Lenakel 173; Negev Bedouin Arabic 228; Pirahã 286
inapplicability to strong beats 35–36, 388–90
language-particular restrictions 370–71
music, *See* Iambic/Trochaic Law: and music; rhythm: music

natural classes of stress rules 45–47
nonpersistent footing. *See* persistent footing
Nuclear Stress Rule, in English 368–69
nucleús (syllable constituent) **51**, 299

Obligatory Branching Parameter 54, **76**
HV 87
list of proposed cases 76
Revised 76
onsets
irrelevant to quantity 51
in moraic theory 52
location in moraic theory 53–54
relevance to stress, Davis (1988) 306
ordering. *See* rule ordering
overlong syllables. *See* superheavy syllables

paradoxes, ordering. *See* rule ordering: paradoxes
parameters. *See also* foot templates; parametric metrical theory
alternating stress in pure grid theory 71
direction of parsing 113
Foot Parsing Locality Parameter 308
illustrated 54
iterativity 113
number of grid layers 119
parsing, listed 400
persistent vs. nonpersistent footing 114
properties of rules vs. grammars 55
restructuring in cyclic stress 115
strong vs. weak prohibition on degenerate feet 87

top-down vs. bottom-up stressing 116
weak vs. strong local parsing 308
parametric metrical theory 54–56. *See also* parameters
alternative proposals 75–77
as theory of foot shapes 62
asymmetries, accounting for 74
Hayes (1981) 75–76
HV 76–77
illustrated 54–55
typology, as account of 403–4
parsing
modes of 113–18
of feet (*see* foot: construction)
parameters 400
principles 401
stress assignment as 40
parsing feet, distinguished from surface feet (Kager) 345
peak first 46, 72
perception, 6–7, 30
Peripherality Condition **58**, 401. *See also* extrametricality: "chained" excluded
and Prosodic Circumscription theory 286
and Stray Adjunction 109
and stray consonants 106–7
extrametricality on multiple levels 106, 107
in Cayuga 223; Cyrenaican Bedouin Arabic 233; Eastern Ojibwa 217; Hindi 165; Javanese 263; Maithili 156; Negev Bedouin Arabic 227; Paamese 178–79; Palestinian Arabic 128; Sarangani Manobo 263
loss of extrametricality under affixation 59
proposed exceptions
in Paamese 179; Kashaya 260
persistent footing 114–15
and conspiracies 198
and destressing 249
initial dactyl effect 97
list of cases 114
list of nonpersistent cases 114
phonetic final lengthening
and degenerate feet 100
in Icelandic 188–96
phonetics
of stress 5–23
absence of single consistent cue 5, 8
organized by metrical structure 8
parasitic character 7
perception 6–7
of weighted consonants, in Yupik languages 247
proposed basis for number of stress levels 22
rules of
in Central Alaskan Yupik 258
pitch interpolation in English 10
phrasal stress 367–99. *See also* End Rules; Move X; Beat Addition
pitch. *See also* intonation; tone
serves to cue multiple phonological entities 7
use as cue for stress dependent on intonation system 7, 11
pitch accent (unit of intonational analysis) 10
absent after main stress 396, 397
association based on stress 28, 370
defined 259
effects on stress perception 6
in Auca 184; Malayalam 93; Yupik 259
proposed as basis of phrasal stress assignment 369, 391
pitch accent languages 49
Ancient Greek 181; Cayuga 225; Chickasaw 209; Choctaw 209; Kashaya 260; Kinyambo 25; Lithuanian 42, 49; Menomini 221; Seminole/Creek 67; Seneca 225, 365; Winnebago 349
diagnosis 49–50
need for both metrical and tonal analysis 365
planes, metrical
in prominence grids 274–76
proposed for cyclic stress 188, 275
proposed for Tiberian Hebrew 263
posttonic, defined minimally 395
pretonic, defined minimally 395
Priority Clause **93–95**, 401
creates unfooted syllables 309
in right-to-left iambic systems 265
in Turkish 262; Tübatulabal 263
languages requiring, listed 95
limits applicability of Incorporation 112
prohibition on degenerate feet. *See* degenerate feet: prohibition on
prominence 270–96. *See also* prominence grids
and destressing 272, 276
in Asheninca 294; General Central Yupik 249; Sentani 333
and extrametricality 276

prominence (*continued*)
and unbounded stress systems 296–99
defined 271
diacritic 297
distinguished from quantity 270–73
in Asheninca 288, 296; Pirahã 287
End Rule governed by 275–76
ignored by foot construction **272–73**, 298
in HV 273
phonetic basis 276
vowel sonority-based 291, 292
in Asheninca 292
prominence grids **273–76**, 401. *See also* prominence
and extrametricality 277
and metrical grids 274
in Asheninca 294; Bhojpuri 278; Hindi 276; Klamath 280–81; Mam 283; Negev Bedouin Arabic 284; Pirahã 286; Sindhi 278
in unbounded stress systems 297–98
temporary character 274–75
tone-based 279, 297
vowel sonority-based 290;
proper foot 86–87
Prosodic Hierarchy
basis for bracketing in grids 40, 369
fewer layers than metrical structure 380
in Lenakel 170–71
Prosodic Morphology 47, 78
in Arabic 78; Sierra Miwok 261
splits syllables among feet 123
Prosody Shift, in Chaga 364–65
pseudo-compounds
in Estonian 316; Icelandic 197; Lenakel 175–76

quantity (of syllables) 50–54. *See also* CVC syllables; CVVC syllables and quantity; onsets; prominence; Weight by Position
alone relevant to foot construction 272
and fortis consonants 299n
and persistent stressing 249
and segment extrametricality 57
and typology of alternating stress rules 72–73
associated with phonemic vowel length 205
associated with quantity-sensitive feet 102
based on duration 271
basis of prominence 276–77
diacritic, proposed for Aklan 305
direct vs. indirect reference in footing 329
distinguished from prominence (*see* prominence)
dual criteria 299–305 (*see also* two-layer moraic theory)
in Ancient Greek 299
in Cahuilla 132; Cairene Arabic 68; Dakota 267n; Dutch 306; Estonian 56; Lenakel 167; Madimadi 306; Maithili 155–58; Polish 99; Winnebago 349, 363
in HV 76, 273
in syllabic trochee languages 101–2
moraic theory **52–54**, 120–21
syllabic constituency theory **51**, 76
three degrees
in Choctaw and Chickasaw 211; Estonian 318–19
proposed for Central Alaskan Yupik 249
quantity sensitivity
Hayes (1981) 75
HV 76

Recoverability Condition 313
counterexamples 335, 359
in Cayuvava 313–14
unbounded constituents, governs 314n
reduction, of vowels
and Iambic/Trochaic Law 84–85
correlated with distinction of quantity 85
in iambic languages, list of cases 84
in Russian 23
in trochaic languages 85
Bhojpuri 278; Catalan 23; English 12; Icelandic 193; Italian 23; Maithili 150, 151, **161**; Mam 282; Tiberian Hebrew 263
relative prominence. *See also* absolute stress levels
preservation under embedding 369
Reparsing, in Estonian 326
revocation of extrametricality. *See* extrametricality: revocation
rhyme (syllable constituent) 49, 51, 299
defines stress-bearing elements in HV 77
rhythm
experimental work 79
grouping in 37
limits in language 30–31
music 27–28, 80
grouping 37
two against three structures 27n
natural physical correlates 9

no invariant physical realization 8
non-hierarchical forms 28
relative rather than absolute prominence 380
structure 26–28
stress as linguistic manifestation 1, 8, 26–31
verse 28, 80–81, 163
Rhythm Rule. *See also* Move X
Beat Addition, proposed as unified with 397
cannot retract stress onto stressless syllables 19
direction 370, 397
in Dutch 386
Rhythmic Adjustment, proposed tree-manipulating rule 383
in Dutch 386
rhythmic distribution, of stress 25, 30
Rhythmic Lengthening. *See also* Iambic Lengthening
in Choctaw and Chickasaw 209
rhythmic stress systems 31–32
source of bracketing in 40
rhythmic structure 26–28. *See also* rhythm
right-to-left iambs. *See* iambs: right-to-left
rule ordering
historical change in 256
in Asheninca 290, 293, 295; Auca 185; Cayuga 222; Central Alaskan Yupik dialects 253; Cyrenaican Bedouin Arabic 231; English phrasal stress 387; Estonian 327; Icelandic 194; Sentani 333; Winnebago 363
paradoxes
in Central Alaskan Yupik 243; Coastal Central Alaskan Yupik 256; Cyrenaican Bedouin Arabic 231; Winnebago 352
proposed for phonetic rules, Central Alaskan Yupik 259
top-down vs. bottom-up stressing 116

segmental phonology. *See* Iambic/Trochaic Law: and segmental phonology
Silbenschnittkorrelation 306
simultaneous rule application, in Cyrenaican Bedouin Arabic 232
splitting syllables across feet 121–23
in Prosodic Morphology 123
outcome for CVC 123
prohibition, stated 50
proposed for Cahuilla 138; Southern Paiute 121; Winnebago 349, 353, 358
Stray Adjunction 108–10
in Pirahã 285
stray consonants. *See* unsyllabified segments
Stray Erasure 59
Stray Mora Association
in Cahuilla 139; Central Alaskan Yupik 245–46
Stray Syllable Adjunction. *See* Stray Adjunction
stress
as organizing principle 8, 9, 21
functions 31
parallels with rhythmic structure 25, 26–30, 79–85
stress clash. *See* clash
Stress Equalization Convention, in HV 373–76
revised 378
Stress Subordination Convention (*SPE*) 373
stress-bearing unit
language-particular in HV 49, 77
schwa not stress-bearing in Indonesian 49n
syllable as 49–50
stressless feet 109n
Strong Domain Principle 386–90
strong local parsing. *See* weak local parsing
strong vs. weak prohibitions on degenerate feet 87. *See also* degenerate feet
superheavy syllables. *See also* vowel length: three-way distinctions
equivalent to disyllabic sequences
in Estonian 325; Hindi 163
in Cairene Arabic 67; Cyrenaican Bedouin Arabic 233; Egyptian Radio Arabic 130; Estonian 320–21, 324; Negev Bedouin Arabic 226; Palestinian Arabic 125; St. Lawrence Island Yupik 241
symbol xiii
treated as feet, in Estonian 318
surface well-formedness conditions 114. *See also* persistent footing
syllabic trochees 63, 182–205
absence of foot extrametricality 78
degenerate feet in 101–5
distinct from moraic trochees 102
generalized to other structures 105
in Estonian 320, 326
Hayes (1981) 75
in Anguthimri 198; Auca 90, **182–88**, 365; Badimaya 198; Bidyara/Gungabula 199; Cavineña 202; Cayuvava 309–14;

syllabic trochees (*contiued*)
Central Norwegian Lappish 199; Czech 203; Dalabon 199; Dehu 199; Diyari 199; Djingili 202; Early Cayuga 224; Estonian 316–29; Finnish 329; Garawa 202–3; German 200; Gugu-Yalanji 204; Hungarian 330; Icelandic 188–98; Inga 181; Karelian 330; Kela 205; Livonian 200; Macedonian 205; Mae 205; Malakmalak 203; Mansi 200; Mantjiltjara 365; Maranungku 200; Mayi 200; Modern Greek 204; Mohawk 111; Nengone 203; Ono 200–1; Onondaga 266; Parnkalla 205; Pintupi **62–64**, 116; Pirahã 285–87; Piro 201; Pitta-Pitta 201; Polish 204; Selepet 201; Swedish 367; Votic 201–2; Wangkumara 202; Warao 203; Weri 265; Yidiɲ 260
in languages with quantity oppositions 102
in Prosodic Morphology 78
lack of vowel reduction 85
languages with monosyllables 89
languages without degenerate words 88
obey Iambic/Trochaic Law 81
proposed for Modern Hebrew 198; Paamese 179; Spanish 198; Wargamay 141n
syllable. *See also* splitting syllables across feet
as stress-bearing unit 49–50
in Cahuilla 138
views of HV 49
relevance to stress rules 48
symbol xiii
syllable merger, with retention of stress 43, 122
in Cayuga 224; Hindi 164; Maithili 150; Mam 283; Seneca 225; Wangkumara 202; Winnebago 364
in tonal systems 43
syllable structure 51–54, 120–21. *See also* moras; onsets; rhyme; unsyllabified segments
asymmetries 79
created maximally 103
governing English segmental phonology 12
in bracketed grids 39
in Estonian 326
syncopation 27
syncope. *See* migration of stress: under vowel deletion
syntactic effects on stress 21
templates. *See* foot templates
ternary stress. *See* weak local parsing
Textual Prominence Preservation Condition 392
reformulated with Grid Expansion 394
tonal rules. *See also* intonational tunes; pitch accent; pitch accent languages; Prosody Shift
alternative account of stress migration 42n
in Pirahã 285; Winnebago 363, 364
Mandarin tone reduction 23
Pacific Yupik tone docking 344–45
relinking of stranded tones 43
tone shift, in Chaga 357; Japanese 357; Kikuyu 357
tone. *See also* boundary tones; intonation; pitch accent; pitch accent languages; tonal rules; tone-bearing units
accounting for exceptions to Culminativity 25
and deforestation 188
basis for prominence 276, 278, 297
proposed for Fore 279n
docking in heavy syllables 69
tonal stability 42
tone-bearing units 49
top-down stressing 116–17
creates unstressable words 117–18
list of cases 117
obscures diagnosis of foot type 198
top-layer alignment, theory of phrasal stress 373–74
trees, metrical 38
as model of phrasal stress 378
require Stray Adjunction convention 109
restructuring rules 382
Trigger Prominence, principle of destressing 197n
Trisyllabic Shortening 148. *See also* Trochaic Shortening
in Cayuga 224
Trochaic Shortening 84, **145–49**. *See also* Trisyllabic Shortening
formulated 146
list of examples 148
trochees. *See* moraic trochees; syllabic trochees
trough first 46, 72
tunes, in intonation. *See* intonational tunes
two-layer moraic theory 299–305
in Algonquian languages 302; Cayuga 301–

2; Chuvash 302; Menomini 219; Ossetic 302; Southern Paiute 268; Yupik 302–4
two-root theory of length 305
typology of stress 24–31, 26–33, 403
  asymmetries 45–47, **71–74**
    parsing direction 266
    Rhythm Rule direction 397
  bounded vs. unbounded 32–33
  free vs. fixed 31
  from viewpoint of pure grid theory 71–74
  matched against theory 73
  rhythmic vs. morphological 31–32

unbounded feet 296–99. *See also* unbounded stress systems
  and Recoverability Condition 314n
  in Malayalam 93; Southern Sierra Miwok 250–51
unbounded stress systems 32, 296–99. *See also* unbounded feet
  Classical Arabic 131
  list of cases 296
  multiple analyses 33, 42n
uneven trochees **75–76**, 91
  and Trochaic Shortening 149
  counterarguments
    from foot extrametricality 78
    generate unattested systems 74
    Mester (1992) 79
  proposed for Bani-Hassan Arabic 366; Latin 91, 115n
  unsuitable for Cahuilla 137–38; Egyptian Radio Arabic 132; Hindi 167; Latin 180; Maithili 160–61; Palestinian Arabic 130
Unstressable Word Syndrome **110–13**, 403. *See also* Incorporation
  and metrical coherence 114
  and top-down stressing 117–18
  in Asheninca 291; Hixkaryana 208
unsyllabified segments
  and extrametricality 106, 401
  in Cyrenaican Bedouin Arabic 230, 233; Dakota 267n; English 108n; Latin 108n; Negev Bedouin Arabic 227; Palestinian Arabic 125–26, 129

verbs, resistance to phrasal stress 375, 376
verse 28, 80–81, 163
vowel length
  in closed and open syllables 220
  three-way distinctions
    avoid neutralization 269
    in Choctaw and Chickasaw 211; Estonian 318; Malecite-Passamaquoddy 216; St. Lawrence Island Yupik 241
  usually present when quantity distinction present 205
vowel reduction. *See* reduction
vowels, English, possible stress levels 15

weak local parsing 308–9, 315–16
  alternatives **312–14**, 316
  and extrametricality 314
  in Old English 118
  list of cases 315
  proximity to strong local equivalent 315
    in Auca 365; Estonian 322, 328; Finnish 330; Hungarian dialects 330; Karelian 330; Koniag Pacific Yupik 333, 344; Mantjiltjara 365; Winnebago 350; Yupik languages 346
weak vs. strong prohibitions on degenerate feet 87
Weight, of syllables. *See* prominence; quantity
Weight by Position **52**, 270
  in Cahuilla 132; Eastern Ojibwa 217; Norton Sound Yupik 242; Pacific Yupik 334
Weight-bearing segments. *See also* moras
  as evidence for location of onset in moraic theory 54
  CRV in Winnebago 362
  in Pacific Yupik 343; Yupik languages 247

X theory (of segment length) 51, **52–53**